Oxford Handbook of
Clinical Dentistry

Oxford Handbook of Clinical Dentistry

DAVID A. MITCHELL
and
LAURA MITCHELL

Second edition

Oxford New York Tokyo
OXFORD UNIVERSITY PRESS
1995

Oxford University Press, Walton Street, Oxford OX2 6DP

Oxford New York
Athens Auckland Bangkok Bombay
Calcutta Cape Town Dar es Salaam Delhi
Florence Hong Kong Istanbul Karachi
Kuala Lumpur Madras Madrid Melbourne
Mexico City Nairobi Paris Singapore
Taipei Tokyo Toronto
and associated companies in
Berlin Ibadan

Oxford is a trade mark of Oxford University Press

Published in the United States
by Oxford University Press Inc., New York

A catalogue record for this book is available from the British Library

Library of Congress Cataloging in Publication Data
Mitchell, David A.
Oxford handbook of clinical dentistry/David A. Mitchell and Laura
Mitchell. – 2nd ed.
Includes bibliographical references and index.
1. Dentistry—Handbooks, manuals, etc.   I. Mitchell, Laura
[DNLM 1. Tooth Diseases—therapy—handbooks.   2. Dentistry—
handbooks.   WU 39 M6810 1995]
RK56.M58 1995   617.6—dc20   94–37577
ISBN 0 19 262602 7

Typset by
Joshua Associates Ltd, Oxford
Printed in Great Britain by
Harper Collins Manufacturing, Glasgow

Preface to the second edition

It would appear that our 'baby' is now a toddler and rapidly out-growing his previous milieu. Caring for such a precocious child is hard work and therefore we have again relied on the help of understanding friends and colleagues who have contributed their knowledge and expertise.

The pace of change in dentistry, both scientifically and politically, is so fast that although the first edition was only published in 1991, this second edition has involved extensive revision of all chapters. Advances in dental materials and restorative techniques have necessitated major revision of these sections and we are indebted to Mr Andrew Hall, who has helped update the chapter on restorative dentistry.

Since the first edition was published, political changes in the UK have resulted in a shift towards private dentistry. This changing emphasis is reflected in the practice management chapter, which now includes a new page on independent and private practice. In addition recent developments in cross-infection and UK health and safety law control have been included.

That old favourite, temporomandibular pain dysfunction syndrome, has also been given the treatment and is now situated on a newly devised page in the chapter on oral medicine. Non-accidental injury, guided tissue regeneration, AIDS, ATLS, and numerous other topical issues have been expanded in this edition.

One aspect of this developing infant remains, however, unchanged. The sole purpose of this book is to enable you, the reader, to gain easy access to the sometimes confusing conglomerate of facts, ideas, opinions, dogma, anecdote, and truth that constitutes clinical dentistry. To this framework you should add, on the blank pages provided, the additional information which will help you treat the next patient or pass the next exam, or more importantly the practical hints and tips which you will glean with experience. It is the potential for that interaction which makes this book distinctive in clinical dentistry. It is participating in that interaction which makes your book unique.

Preface to the first edition

Dental students are introduced to real live patients at an early stage of their undergraduate course in order to fulfil the requirements for clinical training, with the result that they are expected to absorb a large quantity of information in a relatively short time. This is often compounded by clinical allocations to different specialties on different days, or even the same day. Given the obvious success of the Oxford handbooks of clinical medicine and clinical specialties, evidenced by their position in the white coat pockets of the nation's medical students, the extension of the same format to dentistry seems logical. However, it is hoped that the usefulness of this idea will not cease on graduation, particularly with the introduction of Vocational Training. While providing a handy reference for the recently qualified graduate, it is envisaged that trainers will also welcome an *aide mémoire* to help cope with the enthusiastic young trainee who may be more familiar with recent innovations and obscure facts. We also hope that there will be much of value for the hospital trainee struggling towards FDS.

The Handbook of Clinical Dentistry contains those useful facts and practical tips that were stored in our white coat pockets as students and then postgraduates; initially on scraps of paper, but as the collection grew, transferred into notebooks to give a readily available reference source.

The dental literature already contains a great number of erudite books which, for the most part, deal exclusively, in some depth, with a particular branch or aspect of dentistry. The aim of this handbook is not to replace these specialist dental texts, but rather to complement them by distilling together theory and practical information into a more accessible format. In fact, reference is made to sources of further reading where necessary.

Although the authors of this handbook are not the specialized authorities usually associated with dental textbooks, we are still near enough to the coal-face to provide, we hope, some useful practical tips based on sound theory. We were fortunate whilst compiling this handbook in being able to draw on the expertise of many colleagues; the contents, however, remain our sole responsibility.

The format of a blank page opposite each page of text has been plagiarized from the other Oxford handbooks. This gives space for the reader to add his own comments and updates. Please let us know of any that should be made available to a wider audience.

We hope that the reader will find this book to be a useful addition to their white coat pocket or a companion to the BNF in the surgery.

Acknowledgements

In addition to those readers whose comments and suggestions have been implemented into this second edition, we would like to thank the following for giving their time and expertise: Ms F. Carmichael, Ms S. Dowsett, Dr I. D. Grime, Mr A. Hall, Mr H. Harvie, Dr J. Hunton, Dr C. H. Lloyd, and Professor R. Yemm.

Also the first edition would not have been possible without the help of Mr B. S. Avery, Mr N. E. Carter, Mrs J. J. Davison, Mr D. Jacobs, Dr I. D. Grime, Mr A. Hall, Mr P. J. Knibbs, Professor J. F. McCabe, Mr R. A. Ord, Professor A. Rugg-Gunn, Professor R. A. Seymour, Professor J. V. Soames, Professor R. Yemm.

We are grateful to the editor of the *BMJ*, the *BDJ*, and Professor M. Harris, and the Royal National Institute of the Deaf, for granting permission to use their diagrams and to VUMAN for allowing us to include the Index of Orthodontic Treatment Need.

Again the staff of OUP deserve thanks for their help and encouragement.

Note Although this is an equal opportunity publication, the constraints of space have meant that in some places we have had to use 'he' to indicate 'he/she'.

Contents

Symbols and Abbreviations   x

1   History and examination   1

2   Preventive and community dentistry   26

3   Paedodontics   58

4   Orthodontics   130

5   Periodontology   200

6   Restorative dentistry   250

7   Prosthetics and gerodontology   332

8   Oral surgery   378

9   Oral medicine   432

10   Maxillofacial surgery   482

11   Medicine relevant to dentistry   516

12   Therapeutics   598

13   Analgesia, anaesthesia, and sedation   626

14   Dental materials   652

15   Law and ethics   690

16   Practice management   710

17   Syndromes of the head and neck   740

18   Useful information and addresses   750

Index   765

Symbols and abbreviations

Some of these are included because they are in common usage, others because they are big words and we were trying to save space.

▶	this is important
S	supernumerary
−ve	negative
+ve	positive
↑	increased
↓	decreased
<	lesser than
>	greater than
#	fracture
&/or	and/or
∴	therefore
1°	primary
2°	secondary
$\overline{5}$, $\overline{\text{inc}}$	lower second premolar, lower incisor
$\underline{2}$, inc	upper lateral incisor, upper incisor
ACTH	adrenocorticotrophic hormone
ADH	antidiuretic hormone
ADJ	amelo-dentinal junction
Ag	antigen
AIDS	acquired immune deficiency syndrome
AP	anteroposterior
ARF	acute renal failure
ASAP	as soon as possible
AUG	acute ulcerative gingivitis
b/w	bitewing radiograph
bd	twice daily
BDA	British Dental Association
BDJ	British Dental Journal
BIPP	bismuth iodoform paraffin paste
BNF	British National Formulary
BP	blood pressure
BSS	black silk suture
Ca^{2+}	calcium
CDS	Community Dental Service
Class I	Class I relationship
Class II/1	Class II division 1 relationship
Class II/2	Class II division 2 relationship
Class III	Class III relationship
C/I	contraindications
CLP	cleft lip and palate
CNS	central nervous system
C/O	complaining of
CSF	cerebrospinal fluid
CT	computed tomography
CVS	cardiovascular system
DNA	did not attend

DSA	dental surgery assistant
DPF	Dental Practioner's Formulary
DPB	Dental Practice Board
DPT	dental panoramic tomogram (politically correct: OPT/OPG)
ECG	electrocardiograph
EDTA	ethylene diamine tetraacetic acid
e.g.	for example
ENT	ear, nose, and throat
EO	extra-oral
EUA	examination under anaesthesia
F	female
F/-	full upper denture (and -/F for lower)
FA	fixed appliance
FABP	flat anterior bite plane
f/s	fissure sealant
FBC	full blood count
Fe	iron
FESS	functional endoscopic sinus surgery
FHSA	Family Health Service Authority
FWS	freeway space
g	gram
GA	general anaesthesia
GDC	General Dental Council
GDP	general dental practitioner
GDS	General Dental Services
GI	glass ionomer
GMP	general medical practitioner
GP	gutta-percha
h	hour
Hb	haemoglobin
HBeAg	hepatitis B e antigen (high-risk marker)
HBsAg	hepatitis B surface antigen
Hep B/C	hepatitis B/C
Hg	mercury
HIV	human immunodeficiency virus
HPV	human papilloma virus
i.e.	that is
ICP	intercuspal position
ID	inferior dental
IDB	inferior dental block
Ig	immunoglobulin (e.g. IgA, IgG, etc.)
IM	intramuscular
IMF	intermaxillary fixation
inc	incisor
IO	intra-oral
IV	intravenous
JP	juvenile periodontitis
K^+	potassium
KCT	kaolin–cephalin clotting time
l	litre
LA	local anaesthesia
LFH	lower face height

LFT	liver function tests
LLS	lower labial segment
M	male
mand	mandible/mandibular
max	maxilla/maxillary
mg	milligrams
MHz	megahertz
MI	myocardial infarction
min	minute
ml	millilitre
mmol	millimoles
MMPA	maxillary mandibular planes angle
MRI	magnetic resonance imaging
NAD	nothing abnormal detected
NAI	non-accidental injury
NHS	National Health Service
nm	nanometre
nocte	at night
NSAID	non-steroidal anti-inflammatory drug
o/b	overbite
O/E	on examination
od	once daily
OH	oral hygiene
OHCM	*Oxford Handbook of Clinical Medicine*
OHI	oral hygiene instruction
o/j	overjet
OPG/	
OPT	orthopantomograph radiograph
OVD	occlusal vertical dimension
P/-	partial upper denture
PA	posteroanterior
PDH	past dental history
PDL	periodontal ligament
PJC	porcelain jacket crown
PM	premolar
PMH	past medical history
PO	per orum (by mouth)
ppm	parts per million
PR	per rectum
prn	as required
PRR	preventive resin restoration
PU	pass urine
qds	four times daily
RA	relative analgesia
RCP	retruded contact position
RCT	root canal treatment/therapy
Rx	treatment
SBE	subacute bacterial endocarditis
sc	subcutaneous
sec	second
SS	stainless steel
STD	sexually transmitted diseases
TANI	target average net income

TC	tungsten carbide
tds	thrice daily
TMJ	temporomandibular joint
TTP	tender to percussion
TMPDS	temporomandibular pain dysfunction syndrome
ULS	upper labial segment
μm	micrometre
URA	upper removable appliance
URTI	upper respiratory tract infection
US	ultrasound
UTI	urinary tract infection
U&Es	urea and electrolytes
Xbite	crossbite
X-rays	either X-rays or radiographs
yrs	years
ZOE	zinc oxide eugenol

1 History and examination

Listen, look, and learn   2
Presenting complaint   4
The dental history   6
The medical history   8
Medical examination   10
Examination of the head and neck   12
Examination of the mouth   14
Investigations—general   16
Investigations—specific   18
Radiology and radiography   20
Advanced imaging techniques   22
Differential diagnosis and treatment plan   24

Relevant pages in other chapters: It could, of course, be said that all pages are relevant to this section, because history and examination are the first steps in the care of any patient. However, as that is hardly helpful, the reader is referred specifically to the following: dental charting, p. 752; medical conditions, Chapter 11; the child with toothache, p. 66; pre-operative management of the dental patient, p. 576; the cranial nerves, p. 542; orthodontic assessment, p. 138; pulpal pain, p. 254.

Principal sources: Experience.

Listen, look, and learn

Much of what you need to know about any individual patient can be obtained by watching them enter the surgery and sit in the chair, their body language during the interview, and a few well-chosen questions. One of the great secrets of health care is to develop the ability to actually listen to what your patients tell you and to use that information. Doctors and dentists are often concerned that if they allow patients to speak rather than answer questions, history-taking will prove inefficient and prolonged. In fact, most patients will give the information necessary to make a provisional diagnosis, and further useful personal information, if allowed to speak uninterrupted. Most will lapse into silence after two to three minutes of monologue. History-taking should be conducted with the patient sitting comfortably; this rarely equates with supine! In order to produce an all-round history it is, however, customary and frequently necessary to resort to directed questioning, here are a few hints:

- Always introduce yourself to the patient and any accompanying person, and explain, if it is not immediately obvious, what your role is in helping them.
- Remember that patients are (usually) neither medically nor dentally trained, so use plain speech without speaking down to them.
- Questions are a key part of history-taking and the manner in which they are asked can lead to a quick diagnosis and a trusting patient, or abject confusion with a potential litigant. Leading questions should, by and large, be avoided as they impose a preconceived idea upon the patient. This is also a problem when the question suggests the answer, e.g. 'is the pain worse when you drink hot drinks?' To avoid this, phrase questions so that a descriptive reply rather than a straight yes or no is required. However, with the more reticent patient it may be necessary to ask leading questions to elicit relevant information.
- Notwithstanding earlier paragraphs, you will sometimes find it necessary to interrupt patients in full flight during a detailed monologue on their grandmother's sick parrot. Try to do this tactfully, e.g. 'but to come more up to date' or 'this is rather difficult—please slow down and let me understand how this affects the problem you have come about today'.

Specifics of a medical or dental history are described on p. 8 and p. 6. The object is to elicit sufficient information to make a provisional diagnosis for the patient whilst establishing a mutual rapport, thus facilitating further investigations and/or treatment.

Presenting complaint

The aim of this part of the history is to have a provisional differential diagnosis even before examining the patient. The following is a suggested outline, which would require modifying according to the circumstances:

C/O (complaining of) In the patient's own words. Use a general introductory question, e.g. 'Why did you come to see us today? What is the problem?'

If symptoms are present:

Onset and pattern When did the problem start? Is it getting better, worse or staying the same?

Frequency How often, how long does it last? Does it occur at any particular time of day or night?

Exacerbating and relieving factors What makes it better, what makes it worse? What started it?

If pain is the main symptom:

Origin and radiation Where is the pain and does it spread?

Character and intensity How would you describe the pain: sharp, shooting, dull, aching, etc.

Associations Is there anything, in your own mind, which you associate with the problem?

The majority of dental problems can quickly be narrowed down using a simple series of questions such as these to create a provisional diagnosis and judge the urgency of the problem.

The dental history

It is important to assess the patient's dental awareness and the likelihood of raising it. A dental history may also provide invaluable clues as to the nature of the presenting complaint and should not be ignored. This can be achieved by some simple general questions:

How often do you go to the dentist?
(this gives information on motivation, likely attendance patterns, and may indicate patients who change their GDP frequently)

When did you last see a dentist and what did he do?
(this may give clues as to the diagnosis of the presenting complaint, e.g. a recent RCT)

How often do you brush your teeth and how long for?
(motivation and likely gingival condition)

Have you ever had any pain or clicking from your jaw joints?
(TMJ pathology)

Do you grind your teeth or bite your nails?
(TMPDS, personality)

How do you feel about dental treatment?
(dental anxiety)

What do you think about the appearance of your teeth?
(motivation, need for orthodontic treatment)

What is your job?
(socio-economic status, education)

Where do you live?
(fluoride intake, travelling time to surgery)

What types of dental treatment have you had previously?
(previous extractions, problems with LA or GA, orthodontics, periodontal treatment)

What are your favourite drinks/foods?
(caries rate, erosion)

The medical history

There is much to be said for asking patients to complete a medical history questionnaire, as this encourages more accurate responses to sensitive questions. However, it is important to use this as a starting point, and clarify the answers with the patient.

Example of a medical questionnaire

QUESTION   YES/NO

Are you fit and well?
Have you ever been admitted to hospital?
  If yes, please give brief details:

Have you ever had an operation?
  If so, were there any problems?

Have you ever had any heart trouble or high blood pressure?
Have you ever had any chest trouble?
Have you ever had any problems with bleeding?
Have you ever had asthma, eczema, hayfever?
Are you allergic to penicillin?
Are you allergic to any other drug or substance?
Have you ever had:
  rheumatic fever?
  diabetes?
  epilepsy?
  tuberculosis?
  jaundice?
  hepatitis?
  other infectious disease?

Are you pregnant?
Are you taking any drugs, medications, or pills?
  If yes, please give details:

Who is your doctor?

▶ Check the medical history at each recall.
▶ If in any doubt contact the patient's GMP, or the specialist they are attending, before proceeding.

NB A complete medical history (as required when clerking in-patients) would include details of the patient's family history (for familial disease) and social history (for factors associated with disease, e.g. smoking, drinking, and for home support on discharge). It would be completed by a systematic enquiry:

Cardiovascular chest pain, palpitations, breathlessness.

Respiratory breathlessness, wheeze, cough—productive or not.

Gastrointestinal appetite and eating, pain, distension, and bowel habit.

Genitourinary pain, frequency (day and night), incontinence, straining, or dribbling.

Central nervous system fits, faints, and headaches.

Medical examination

For the vast majority of dental patients attending as out-patients to a practice, community centre, or hospital, simply recording a medical history should suffice to screen for any potential problems. The exceptions are patients who are to undergo general anaesthesia (as in-patients or out-patients) and anyone with a positive medical history undergoing extensive treatment under LA or sedation. The aim in these cases is to detect any gross abnormality so that it can be dealt with (by investigation, by getting a more experienced or specialist opinion, or by simple treatment if you are completely familiar with the problem).

General Jaundice: look at sclera in good light, anaemia ditto. Cyanosis, peripheral: blue extremities, central: blue tongue. Dehydration, lift skin between thumb and forefinger.

Cardiovascular system Feel and time the pulse. Measure blood pressure. Listen to the heart sounds along the left sternal edge and the apex (normally 5th intercostal space midclavicular line on the left), murmurs are whooshing sounds between the 'lup dub' of the normal heart sounds. Palpate peripheral pulses and look at the neck for a prominent jugular venous pulse (this is difficult and takes much practice).

Respiratory system Look at the respiratory rate (12–18/min) is expansion equal on both sides? Listen to the chest, is air entry equal on both sides, are there any crackles or wheezes indicating infection, fluid, or asthma? Percuss the back, comparing resonance.

Gastrointestinal system With the patient lying supine and relaxed with hands by their sides, palpate with the edge of the hand for liver (upper right quadrant) and spleen (upper left quadrant). These should be just palpable on inspiration. Also palpate bimanually for both kidneys in the right and left flanks (healthy kidneys are not palpable) and note any masses, scars, or hernia. Listen for bowel sounds and palpate for a full bladder.

Genitourinary system Mostly covered by abdominal examination above. Patients with genitourinary symptoms are more likely to go into post-operative urinary retention. Pelvic and rectal examinations are neither appropriate nor indicated and should not be conducted by the non-medically-qualified.

Central nervous system Is the patient alert and orientated in time, place, and person? Examination of the cranial nerves, p. 542. Ask the patient to move their limbs through a range of movements, then repeat passively and against resistance to assess tone, power, and mobility. Reflexes: brachioradialis, biceps, triceps, knee, ankle, and plantar are commonly elicited (stimulation of the sole normally causes plantar flexion of the great toe).

Musculoskeletal system Note limitations in movement and arthritis, especially affecting the cervical spine, which may need to be hyperextended in order to intubate for anaesthesia.

Examination of the head and neck

This is an aspect of examination that is both undertaught and overlooked in both medical and dental training. In the former, the tendency is to approach the area in a rather cursory manner, partly because it is not well understood. In the latter it is often forgotten, despite otherwise extensive knowledge of the head and neck, to look beyond the mouth. For this reason the examination below is given in some detail, but so thorough an inspection is only necessary in selected cases, e.g. suspected oral cancer, facial pain of unknown origin, trauma, etc.

Head and facial appearance Look for specific deformities (p. 198), facial disharmony (p. 196), syndromes (p. 740), traumatic defects (p. 486), and facial palsy (p. 470).

Assessment of the cranial nerves is covered on p. 542.

Skin lesions of the face should be examined for colour, scaling, bleeding, crusting, palpated for texture and consistency and whether or not they are fixed to, or arising from, surrounding tissues.

Eyes Note obvious abnormalities such as proptosis and lid retraction, (e.g. hyperthyroidism) and ptosis (drooping eyelid). Examine conjunctiva for chemosis (swelling), pallor, e.g. anaemia or jaundice. Look at the iris and pupil. Ophthalmoscopy is the examination of the disc and retina via the pupil. It is a specialized skill requiring an adequate ophthalmoscope and is acquired by watching and practising with a skilled supervisor. However, direct and consensual (contralateral eye) light responses of the pupils are straightforward and should always be assessed in suspect head injury (p. 488).

Ears Gross abnormalities of the external ear are usually obvious. Further examination requires an auroscope. The secret is to have a good auroscope and straighten the external auditory meatus by pulling upwards, backwards, and outwards using the largest applicable speculum. Look for the pearly grey tympanic membrane; a plug of wax often intervenes.

The mouth, p. 14

Oropharynx and tonsils These can easily be seen by depressing the tongue with a spatula, **the hypopharynx and larynx** are seen by indirect laryngoscopy, using a head-light and mirror, and the **post-nasal space** is similarly viewed.

The neck Inspect from in front and palpate from behind. Look for skin changes, scars, swellings, and arterial and venous pulsations. Palpate the neck systematically, starting at a fixed standard point, e.g. beneath the chin, working back to the angle of the mandible and then down the cervical chain, remembering the scalene and supraclavicular nodes. Swellings of the thyroid move with swallowing. Auscultation may reveal bruits over the carotids (usually due to atheroma).

TMJ Palpate both joints simultaneously. Have the patient open and close and move laterally whilst feeling for clicking, locking, and crepitus. Palpate the muscles of mastication for spasm and tenderness. Auscultation is not usually used.

Examination of the mouth

Most dental textbooks, quite rightly, include a very detailed and comprehensive description of how to examine the mouth. These are based on the premise that the examining dentist has never before seen the patient, who has presented with some exotic disease. Given the constraints imposed by routine clinical practice, this approach needs to be modified to give a somewhat briefer format that is as equally applicable to the routine dental attender who is symptomless as to the new patient attending with pain of unknown origin.

The key to this is to develop a systematic approach, which becomes almost automatic, so that when you are under pressure there is less likelihood of missing any pathology. As any abnormal findings indicate that further investigation is required, the reader is referred to the page numbers in parenthesis, as necessary.

EO examination (p. 12). For routine clinical practice this can usually be limited to a visual appraisal, e.g. swellings, asymmetry, patient's colour, etc. More detailed examination can be carried out if indicated by the patient's symptoms.

IO examination
- Oral hygiene.
- Soft tissues. The entire oral mucosa should be carefully inspected. Any ulcer of >3 weeks' duration requires further investigation (p. 450).
- Periodontal condition. This can be assessed rapidly, using a periodontal probe. Pockets >5 mm indicate the need for a more thorough assessment (p. 210).
- Chart the teeth present (p. 752).
- Examine each tooth in turn for caries (p. 28) and examine the integrity of any restorations present.
- Occlusion. This should involve not only getting the patient to close together and examining the relationship between the arches (p. 138), but also looking at the path of closure for any obvious prematurities and displacements (p. 176). Check for evidence of tooth wear (p. 304).

For those patients complaining of pain, a more thorough examination of the area related to their symptoms should then be carried out, followed by any special investigations (p. 16).

Investigations—general

▶ Do not perform or request an investigation you cannot interpret.

▶ Similarly, always look at, interpret, and act on any investigations you have performed.

Temperature, pulse, blood pressure, and respiratory rate
These are the nurses' stock in trade. You need to be able to interpret the results.

Temperature (35.5–37.5 °C). ↑ physiologically post-operatively for 24 h, otherwise may indicate infection or a transfusion reaction. ↓ in hypothermia or shock.

Pulse Adult 60–80 beats/min; child is higher (up to 140 beats/min in infants). Should be regular.

Blood pressure (120–140/60–90 mmHg), ↑ with age. Falling BP may indicate a faint, hypovolaemia, or other form of shock. High BP may place the patient at risk from a GA. An ↑ BP + ↓ pulse suggests ↑ intracranial pressure (p. 488).

Respiratory rate (12–18 breaths/min). ↑ in chest infections, pulmonary oedema, and shock.

Urinalysis is routinely performed on all patients admitted to hospital. A positive result for:

Glucose or *ketones* may indicate diabetes.

Protein suggests renal disease especially infection.

Blood suggests infection or tumour.

Bilirubin indicates hepatocellular &/or obstructive jaundice.

Urobilinogen indicates jaundice of any type.

Blood tests (sampling techniques, p. 578). Reference ranges vary.

Full blood count (EDTA, pink tube) measures:

Haemoglobin (M 13–18 g/dl, F 11.5–16.5 g/dl). ↓ in anaemia, ↑ in polycythaemia and myeloproliferative disorders.

Haematocrit (packed cell volume) (M 40–54%, F 37–47%), ↓ in anaemia, ↑ in polycythaemia and dehydration.

Mean cell volume (76–96 fl), ↑ in size (macrocytosis) in B12 and folate deficiency, ↓ (microcytosis) iron deficiency.

White cell count ($4-11 \times 10^9/1$) ↑ in infection, leukaemia, and trauma, ↓ in certain infections, early leukaemia and after cytotoxics.

Platelets ($150-400 \times 10^9/1$), see also p. 522.

Biochemistry Urea and electrolytes are the most important:

Sodium (135–145 mmol/l) large fall causes fits.

Potassium (3.5–5 mmol/l) Must be kept within this narrow range to avoid serious cardiac disturbance. Watch carefully in diabetics, those on IV therapy, and the shocked or dehydrated patient. Suxamethonium (muscle relaxant) ↑ potassium.

Urea (2.5–7 mmol/l) Rising urea suggests dehydration, renal failure, or blood in the gut.

Creatinine (70–150 micromol/l) Rises in renal failure.

Various other biochemical tests are available to aid specific diagnoses, e.g. bone, liver function, thyroid function, cardiac enzymes, folic acid, Vitamin B12, etc.

Glucose (fasting 4–6 mmol/l) ↑ suspect diabetes, ↓ hypo-glycaemic drugs, exercise. Competently interpreted proprietary tests e.g. 'BMs' ≡ well to blood glucose (p. 592).

Virology Viral serology is costly and rarely necessary. If you must, use 10 ml clotted blood in a plain tube.

Immunology Similar to above but more frequently indicated in complex oral medicine patients; 10 ml in a plain tube.

Bacteriology

Sputum and pus swabs are often helpful in dealing with hospital infections. Ensure they are taken with sterile swabs and transported immediately or put in an incubator.

Blood cultures are also useful if the patient has septicaemia. Taken when there is a sudden pyrexia and incubated with results available 24–48 h later. Take two samples from separate sites and put in paired bottles for aerobic and anaerobic culture (i.e. 4 bottles, unless your lab indicates otherwise).

Biopsy See p. 410.

Cytology With the exception of smears for candida, cytology is little used and not widely applicable in the dental specialties.

Investigations—specific

Vitality testing It must be borne in mind when vitality testing that it is the integrity of the nerve supply that is being investigated. However, it is the blood supply which is of more relevance to the continued vitality of a pulp. Test the suspect tooth and its neighbours.

Application of cold This is most practically carried out using ethyl chloride on a pledget of cotton wool.

Application of heat Vaseline should be applied first to the tooth under test to prevent the heated GP sticking. No response suggests that the tooth is non-vital, but an ↑ response indicates that the pulp is hyperaemic.

Electric pulp tester The tooth to be tested should be dry, and prophy paste used as a conductive medium. Most machines ascribe numbers to the patient's reaction, but these should be interpreted with caution as the response can also vary with battery strength or the position of the electrode on the tooth.

For the above methods misleading results may occur:

False-positive	*False-negative*
Multi-rooted tooth with vital + non-vital pulp	Nerve supply damaged, blood supply intact
Canal full of pus	Secondary dentine
Apprehensive patient	Large insulating restoration

Test cavity Drilling into dentine without LA is an accurate diagnostic test, but as tooth tissue is destroyed it should only be used as a last resort. Can be helpful for crowned teeth.

Percussion is carried out by gently tapping adjacent and suspect teeth with the end of a mirror handle. A positive response indicates that a tooth is extruded due to exudate in apical or lateral periodontal tissues.

Mobility of teeth is ↑ by ↓ in the bony support (e.g. due to periodontal disease or an apical abscess) and also by # of root or supporting bone.

Palpation of the buccal sulcus next to a painful tooth can help to determine if there is an associated apical abscess.

Biting on to gauze or rubber can be used to try and elicit pain due to a cracked tooth.

Radiographs (pp. 20 and 738.)

Area under investigation	*Radiographic view*
General scan of teeth and jaws (retained roots, unerupted teeth)	OPG, Lateral obliques
Localization of unerupted teeth	Parallax periapicals
Crown of tooth and interdental bone (caries, restorations)	Bitewing
Root and periapical area	Periapical
Submandibular gland	Lower occlusal view
Sinuses	Occipito-mental, OPG
TMJ	OPG, Trans-pharyngeal

Skull and facial bones

Occipito-mental
PA and lateral skull
Submento-vertex

Local anaesthesia can help localize organic pain.

Radiology and Radiography

Radiography is the taking of radiographs, *radiology* is their interpretation. Referring to a radiologist as a radiographer ensures upset.

Radiographic images are produced by the differential attenuation of X-rays by tissues. Radiographic quality depends on the density of the tissues, the intensity of the beam, sensitivity of the emulsion, processing techniques, and viewing conditions.

Intra-oral views

Uses a stationary anode (tungsten), direct current ↓ dose of self-rectifying machines. Direct action film (↑ detail) using D or E speed. E speed is double the speed of D hence ↓ dose to patient. Rectangular collimation ↓ unnecessary irradiation of tissues.

Periapical shows all of tooth, root, and surrounding periapical tissues. Performed by:

1 Paralleling technique Film is held in a film holder parallel to the tooth and the beam is directed (using a beam-aligning device) at right angles to the tooth and film. Focus to film distance is increased to minimize magnification, the optimium distance is 30 cm. Most accurate and reproducible technique.

2 Bisecting angle technique Older technique which can be carried out without film holders. Film is placed close to the tooth and the beam is directed at right angles to the plane bisecting the angle between the tooth and film. Normally held in place by patient's finger. Not as geometrically accurate a technique as more coning off occurs and needlessly irradiates the patient's finger.

Bitewings shows crowns and crestal bone levels, used to diagnose caries, overhangs, calculus, and bone loss <4 mm. Patient bites on wing holding film against the upper and lower teeth and beam is directed between contact points perpendicular to the film in the horizontal plane. A 5° tilt to vertical accommodates the curve of Monson.

Occlusals demonstrate larger areas. May be oblique, true or special. Used for localization of impacted teeth, salivary calculi. Film is held parallel to the occlusal plane. Vertex occlusal requires an intensifying screen and is now obsolete. Oblique occlusal is similar to a large bisecting angle periapical. True occlusal of the mandible gives a good cross-sectional view.

Key points
- Use paralleling technique.
- Use film holders.
- Rectangular collimation.
- E speed film.

Extra-oral views

Skull and general facial views use a rotating anode and grid which ↓ scattered radiation reaching the film but ↑ dose to patient. Screen film is used for all extra-orals (intensifying screens are now rare earth, e.g. gadolinium and lanthanum). X-rays act on screen which fluoresces and light interacts with emulsion. There is loss of detail but ↓ the dose to patient. Darkroom techniques and film storage are affected due to the properties of the film.

Lateral oblique Largely superseded by panoramic but can use dental X-ray set.

PA mandible Patient has nose/forehead touching film. Beam perpendicular to film. Used for diagnosing/assessing # mandible.

Reverse Townes position, as above, but beam 30° up to horizontal. Used for condyles.

Occipitomental Nose/chin touching the film beam parallel to horizontal unless OM prefixed by, e.g. 10°, 30°, which indicates angle of beam to horizontal.

Submentovertex Patient flexes neck vertex touching film, beam projected menton to vertex. ↓ use due to ↑ radiation and risk to cervical spine.

Cephalometry (pp. 140, 142) uses cephalostat for reproducible position. Use Frankfort plane or natural head position. Wedge (aluminium or copper and rare earth) to show soft tissues. Lead collimation to reduce unnecessary dose to patient and scatter leading to ↓ contrast.

Panoramic Generically referred to as DPT (dental panoramic tomograph), sometimes by make, e.g. OPT/OPG. The technique is based on tomography (i.e. objects in focal trough are in focus, the rest is blurred). The state of the art machine is a moving centre of rotation (previously 2 or 3 centres) which accommodates the horseshoe shape of the jaws. Correct patient positioning is vital. Blurring and ghost shadows can be a problem (ghost shadows appear opposite to and above the real image due to 5–8° tilt of beam). Relatively low-dose technique and sectional images can be obtained.

Lead aprons (0.25 mm lead equivalent)
The 10-day rule is now defunct. In well-maintained, well-collimated equipment where the beam does not point to the gonads the risk of damage is minimal. Apply all normal principles to pregnant women (use lead apron if primary beam is directed at fetus), otherwise do not treat any differently.

There is no risk in dentistry of deterministic/certainty effects (e.g. radiation burns). Stochastic/change effects are more important (e.g. tumour induction). The thyroid is the principal organ at risk. Follow principles of ALARP, p. 736.

Advanced imaging techniques

Computed tomography (CT)

Images are formed by scanning a thin cross-section of the body with a narrow X-ray beam (120 kV), measuring the transmitted radiation with detectors and obtaining multiple projections which a computer then processes to reconstruct a cross-sectional image ('slice'). Three-dimensional reconstruction is also possible on some machines. Modern scanners consist of either a fan beam with multiple detectors aligned in a circle, both rotating around the patient or a stationary ring of detectors with the X-ray beam rotating within it. The image is divided into pixels which represent the average attenuation of blocks of tissue (voxels). The CT number (measured in Hounsfield units) compares the attenuation of the tissue with that of water. Typical values range from air at -1000 to bone at $+400$ to $+1000$ HU. As the eye can only perceive a limited gray scale the settings can be adjusted depending on the main tissue of interest (i.e. bone or soft tissues). These 'window levels' are set at the average CT number of the tissue being imaged and the 'window width' is the angle selected. The images obtained are very useful for assessing extensive trauma or pathology and planning surgery. The dose is, however, higher compared with conventional films and NRPB recommend that all radiologists be made aware of the high dose implications.

Magnetic resonance imaging (MRI)

The patient is placed in a machine which is basically a large magnet. Protons then act like small bar magnets and point 'up' or 'down' with a slightly greater number pointing 'up'. When a radio-frequency pulse is directed across the main magnetic field the protons 'flip' and align themselves along it. When the pulse ceases the protons 'relax' and as they re-align with the main field they emit a signal. The hydrogen atom is used because of its high natural abundance in the body. The time taken for the protons to 'relax' is measured by values known as T1 and T2. A variety of pulse sequences can be used to give different information. T1 is longer than T2 and times may vary depending on the fluidity of the tissues (e.g. if inflamed). MRI is not good for imaging bone as the protons are held firmly within the bony structure although bone margins are visible. It is useful, however, for the TMJ and facial soft tissues.

Problems are: patient movement, expense, the claustrophobic nature of the machine, noise, magnetizing, and movement of instruments or metal implants and foreign bodies. Credit cards near the machine are also affected.

Digital imaging

This technique has been used extensively in general radiology where it has great advantages over conventional methods in that there is a marked dose reduction and less concentrated contrast media may be used. The normal X-ray source is used but the receptor is a charged coupled device linked to a computer. The image is practically instantaneous and eliminates the problems of processing. However, the sensor is difficult to position and the smaller than normal film means the dose reduction is not always obtained. Gives ↓ resolution. Popular in some European countries.

Ultrasound (US)

Ultra-high frequency sound waves (1–20 MHz) are transmitted through the body using a piezoelectric material (i.e. the material distorts if an electric field is placed across it and vice versa). Good probe/skin contact is required (gel) as waves can be absorbed, reflected or refracted. High frequency (short wavelength) waves are absorbed more quickly whereas low frequency waves penetrate further. US has been used to image the major salivary glands and the TMJ.

Doppler US is used to assess blood flow as the difference between the transmitted and returning frequency reflects the speed of travel of red cells. Doppler US has also been used to assess the vascularity of lesions and the patency of vessels prior to reconstruction.

Sialography

This is the imaging of the major salivary glands after infusion of contrast media under controlled rate and pressure using either conventional radiographic films, or CT scanning. The use of contrast media will reveal the internal architecture of the salivary glands and show up radiolucent obstructions, e.g. calculi within the ducts of the imaged glands. Particularly useful for inflammatory or obstructive conditions of the salivary glands. Patients allergic to iodine are at risk of anaphylactic reaction if an iodine-based contrast medium is used.

Arthrography

Just as the spaces within salivary glands can be outlined using contrast media so can the upper and lower joint spaces of the TMJ. Although technically difficult, both joint compartments can be injected with contrast media under fluoroscopic control and the movement of the meniscus can be visualized on video. Stills of the real-time images can be made although interpretation is often unsatisfactory.

Differential diagnosis and treatment plan

Arriving at this stage is the whole point of taking a history and performing an examination, because by narrowing down your patient's symptoms into possible diagnoses you can, in most instances, formulate a series of investigations and/or treatment that will benefit them.

Suggested approach

1 History and examination (as above).
2 Preliminary investigations.
3 Differential diagnosis.
4 Specific investigations which will confirm or refute the differential diagnoses.
5 Ideally, arrive at the definitive diagnosis(es).
6 List in a logical progression the steps which can be undertaken to take the patient to oral health.
7 Then carry them out.

Simple really!

This is the ideal, but life, as you are no doubt well aware, is far from ideal, and it is not always possible to follow this approach from beginning to end. The principles, however, remain valid and this general approach, even if much abbreviated, will help you deal with every new patient safely and sensibly.

An example

Mr Ivor Pain, 25, an otherwise healthy young man has 'toothache'.

C/O Pain, left side of mouth.

HPC Lost MOD amalgam $\overline{5}$ three weeks ago. Had twinges since then which seemed to go away, then two days ago tooth began to throb. Now whole jaw aches and can't eat on that side. The pain radiates to his ear and is worse if he drinks tea. He has a foul taste in his mouth. Little relief from analgesics.

PMH Well. Medical history NAD, i.e. no 'alarm bells' on questionnaire.

PDH Means well, but is an irregular attender, 'had some bad experiences', 'don't like needles'.

O/E

EO Medical examination inappropriate in view of PMH. Some swelling left side of face due to left submandibular lymphadenopathy. Looks distressed and anxious.

IO Moderate OH, generalized chronic gingivitis, no mucosal lesions, caries

$$\frac{7\ 6\quad\ \ |\ 4\ 6}{7\ 6\ 5\ |\ 5\ 7},$$

partially erupted $\overline{8}$ with pus exuding,

$\overline{5}$ large cavity, but seems periodontally sound, no fluctuant soft tissue swelling. Otherwise complete dentition with Class I occlusion.

General investigations Temperature 38 °C.

Differential diagnoses
1 Acute apical abscess $\overline{5}$
2 Acute pericoronitis $\overline{8}$
3 Chronic gingivitis? periodontitis.
4 Caries as charted.

Specific investigations
1 Vitality test $\overline{5}$. (Non-vital)
2 Periapical X-ray $\overline{5}$. (Patent canal, apical area).

Treatment plan
1 Drain $\overline{5}$ via root canal (does not require LA as pulp is necrotic, hence won't unduly distress anxious patient, but will relieve pain and infection).
2 Irrigate operculum of $\overline{8}$.
3 Antibiotics (as patient is pyrexic with two sources of infection: usually due to mixed anaerobic/aerobic organisms ∴ use amoxycillin and metronidazole) and analgesics (NSAID for 24–48 h).
4 Explain the problems and arrange a review appointment for OHI, periodontal charting, and a DPT.

Future plan
5 OHI, scaling
6 RCT $\overline{5}$
7 Plastic restorations as indicated
8 Post/core crown $\overline{5}$
9 Remove third molars as indicated (clinically and from DPT).

Treatment at the first visit is kept at a minimum to relieve patient's pain and thereby gain his trust and future attendance.

2 Preventive and community dentistry

Dental caries 28
Caries diagnosis 32
Fluoride 34
Planning fluoride therapy 36
Bacterial plaque and dental decay 38
Fissure sealants 40
Sugar 42
Dietary analysis and advice 46
Dental health education 48
Provision and receipt of dental care 50
Dentistry for the handicapped 52
Hygienists and other ancillaries 54
Lies, damn lies, and statistics 56

Relevant pages in other chapters: Plaque control, p. 224; prevention of secondary caries, p. 272; prevention of trauma to anterior teeth, p. 107.

Principal sources: J. J. Murray 1989 *The Prevention of Dental Disease*, OUP. E. A. M. Kidd 1987 *Essentials of Dental Caries: The Disease and its Management*, Wright. R. J. Elderton 1987 *Positive Dental Prevention*, Heinemann. R. J. Elderton 1990 *Clinical Dentistry in Health and Disease*, Vol. 3, *The Dentition and Dental Care*, Heinemann. A. Rugg-Gunn 1993 *Nutrition and Dental Health*, OUP.

Dental caries

Dental caries is a sugar-dependent infectious disease.[1] Acid is produced as a by-product of the metabolism of dietary carbohydrate by plaque bacteria, which results in a drop in pH at the tooth surface. In response, calcium and phosphate ions diffuse out of enamel, resulting in demineralization. This process is reversed when the pH rises again. Caries is ∴ a dynamic process characterized by episodic demineralization and remineralization occurring over time. If destruction predominates, disintegration of the mineral component will occur, leading to cavitation.

Enamel caries The initial lesion is visible as a white spot. This appearance is due to demineralization of the prisms in the subsurface layer, with the surface enamel remaining more mineralized. With continued acid attack the surface changes from being smooth to rough, and may become stained. As the lesion progresses, pitting and eventually cavitation occur. The carious process favours repair, as remineralized enamel concentrates fluoride and has larger crystals, with a ↓ surface area. Fissure caries often starts as two white spot lesions on opposing walls, which coalesce.

Dentine caries comprises demineralization followed by bacterial invasion, but differs from enamel caries in the production of secondary dentine and the proximity of the pulp. Once bacteria reach the ADJ, lateral spread occurs, undermining the overlying enamel.

Rate of progression of caries Although it has been suggested that the mean time that lesions remain confined radiographically to the enamel is 3–4 years,[2] there is great individual variation and lesions may even regress.[3] The rate of progression through dentine is unknown; however, it is likely to be faster than through enamel. Progression of fissure caries is usually rapid due to the morphology of the area.

Arrested caries Under favourable conditions a lesion may become inactive and even regress. Clinically, arrested dentine caries has a hard or leathery consistency and is darker in colour than soft, yellow active decay. Arrested enamel caries can be stained dark-brown.

Susceptible sites The sites on a tooth which are particularly prone to decay are those where plaque accumulation can occur unhindered, e.g. approximal enamel surfaces, cervical margins, and pits and fissures. Host factors, e.g. the volume and composition of the saliva, can also affect susceptibility.

Saliva and caries Saliva acts as an intra-oral antacid, due to its alkali pH at high flow-rates and buffering capacity. In addition:
- ↓ plaque accumulation and aids clearance of foodstuffs.

1 E. A. M. Kidd 1987 *Essentials of Dental Caries: The Disease and its Management*, Wright.   **2** N. B. Pitts 1983 *Comm Dent Oral Epidemiol* **11** 228.   **3** N. B. Pitts 1991 *BDJ* **171** 313.

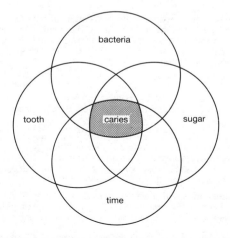

Diagram to show the factors involved in the development of caries.

Dental caries (cont.)

- Acts as a reservoir of calcium, phosphate, and fluoride ions, thereby favouring remineralization.
- Has an antibacterial action because of its IgA, lysozyme, lactoferritin, and lactoperoxidase content.

An appreciation of the importance of saliva can be gained by examining a patient with a dry mouth.

Some manufacturers are now promoting the remineralizing potential of chewing gum, effected by an increase in salivary production. Chewing sugar-free gum regularly after meals does appear to ↓ caries, but the reduction is small.[1]

Root caries With gingival recession root dentine is exposed to carious attack. Rx requires, first, control of the aetiological factors and for most patients this involves dietary advice and OHI. Topical fluoride may aid remineralization and prevent new lesions developing. However, active lesions will require restoration with GI cement (p. 269).

Caries prevention

Classically three main approaches are possible:
1 Tooth strengthening or protection.
2 Reduction in the availability of microbial substrate.
3 Removal of plaque by physical or chemical means.

In practice this means dietary advice, fluoride, fissure sealing, and regular toothbrushing (which is also important in the prevention of periodontal disease). The relative value of these varies with the age of the individual.

Of equal importance with the prevention of new lesions is a preventive philosophy on the part of the dentist, so that early carious lesions are given the chance to arrest and a minimalistic approach is taken to the excision of caries where primary prevention has failed.

1 *Drug and Therapeutics Bulletin* 1992 **30** 23.

Caries diagnosis

As caries can be arrested or even reversed, early diagnosis is important.

Aids to diagnosis

- Good eyesight (and a clean, dry tooth). Detection of inter-proximal caries is aided by using an elastic orthodontic separator to open the contact.
- A blunt probe (as a sharp probe may actually damage an incipient lesion).
- Bitewing radiographs are useful in the detection of occlusal and interproximal caries. The clinical situation is more advanced than the radiographic appearance. However, it is thought that the probability of cavitation is low when a lesion is confined to enamel on X-ray.
- Transillumination probes with a 0.5 mm tip have been advocated for diagnosing interproximal caries. The value of these over conventional methods has not been proven, and problems with eyestrain have been reported.[1]
- Electronic caries detectors (e.g. Vanguard) work on the principle that conductivity increases with demineralization, but have not been fully evaluated.

Diagnosis and its relevance to management

▶ Remember that caries diagnosis is inaccurate. Even trained epidemiologists are <80% reliable ∴ if in doubt be conserva-tive.

Smooth surface caries is relatively straightforward to diagnose. The chances of remineralization are ↑ as it is obvious, and accessible for cleaning. Restoration is indicated if prevention has failed and the lesion is active, or if the tooth is sensitive or aesthetics poor.

Pit and fissure caries is difficult to diagnose reliably, especially in the early stages. A sharp probe is of limited value as stickiness could be due to the morphology of the fissure. The anatomy of the area also tends to favour spread of the lesion, which often occurs rapidly. As fissure caries is less amenable to fluoride and ↑ OH, fissure sealing is preferable to watching and waiting. Occlusal caries evident on b/w radiographs should be excised. If in doubt about a fissure it is justifiable to investigate the area with a small round bur. The 'cavity' can be aborted if no caries found and the surface sealed.

Approximal caries Currently accepted practice:
- If lesion confined to enamel on b/w, institute preventive measures and keep under review.
- If lesion has penetrated dentine radiographically, a restora-tion is indicated unless serial radiographs show that it is static.

---

1 A. D. Sidi 1988 *BDJ* **164** 15.

Recall intervals[1]

This subject has evoked considerable controversy, some arguing that regular attendance puts a patient more at risk of receiving replacement fillings, while others contend that regular and frequent check-ups are necessary to monitor prevention. In fact, it would appear that only a minority of the British public attend for 6-monthly check-ups. The available evidence suggests that there is no clear benefit for recall intervals of less than 1 yr for *healthy* patients, although the at risk patient often needs to be seen more frequently.[2] In addition, as changing dentist ↑ the likelihood of replacement restorations the profession has to re-examine its criteria for replacement.

1 BASCD 1988 *The desirable frequency of attendance for dental clinical examination—a policy statement.* **2** N. B. Pitts 1992 *BDJ* **172** 225.

Fluoride

The history of fluoride is covered well in other texts.[1]

Mechanisms of the action of fluoride in reducing dental decay

Enamel deposition → Enamel maturation → Eruption into oral
and calcification   environment

↑   ↑   ↑

Fluoride in blood   Fluoride in tissue   Fluoride in saliva
                         fluid   and crevicular fluid

The concentration of fluoride in enamel ↑ with ↑ fluoride content of water supply and ↑ towards the surface of enamel.

Pre-eruptive effects Enamel formed in the presence of fluoride has:

- Improved crystallinity and ↑ crystal size, and therefore ↓ acid solubility.
- More rounded cusps and fissure pattern, but effect small.

Discontinuation of systemic fluoride results in an ↑ in caries, therefore pre-eruptive effects must be limited.

Post-eruptive effects NB Newly erupted teeth derive the most benefit.

- Inhibits demineralization and promotes remineralization of early caries. Fluoride enhances the degree and speed of remineralization and renders the remineralized enamel more resistant to subsequent attack.
- Decreases acid production in plaque. A drop in pH results in an ↑ in ionic fluoride, which augments this action.
- An ↑ concentration of fluoride in plaque inhibits the synthesis of extracellular polysaccharide.
- It has been suggested that fluoride affects pellicle and plaque formation, but this is unsubstantiated.

NB Fluoride is more effective in ↓ smooth surface than pit and fissure caries.

Safety and toxicity of fluoride

Fluoride is present in all natural waters to some extent. Many simple chemicals are toxic when consumed in excess, and the same is true of fluoride.

Fluoride is absorbed rapidly mainly from the stomach. Peak blood levels occur 1 h later. It is excreted via the kidneys, but traces are found in milk and saliva. The placenta only allows a small amount of fluoride to cross, ∴ pre-natal fluoride is relatively ineffective.

Fluorosis (or mottling) occurs due to a long-term excess of fluoride. It is endemic in areas with a high level of fluoride occurring naturally in the water. Clinically, it can vary from faint white opacities to severe pitting and discoloration. Histologically, it is caused by ↑ porosity in the outer third of the enamel.

1 J. J. Murray 1991 *Fluorides in Caries Prevention*, Wright.

Concentration of fluoride (ppm) in water supply	Degree of mottling
<0.9	++
0.9	+
2	++
>2	+++

Toxicity Certainly lethal dose = 32–64 mg F/kg body weight. To reach 5 mg F/kg threshold (requiring hospitalization) a 5 yr-old would have to ingest 95 (1 mgF) tablets, 95 ml of (0.8% F) toothpaste or 7.6 ml of 1.23% APF gel.

Antidotes: <5 mg F/kg body weight—large volume of milk. >5 mg F/kg body weight—refer to hospital quickly for gastric lavage. If any delay give IV calcium gluconate and an emetic.

Cancer there is no evidence to support the contention that fluoridated communities experience a higher incidence of cancer.

Planning fluoride therapy

Many consider that the most important action of fluoride is to favour remineralization of the early carious lesion. Although fluoride incorporated within developing enamel results in a high local concentration following acid attack, the maximum benefit appears to be derived from frequent low-concentration topical administration.[1] Fluoridated water is the most effective method, as it provides both a systemic and topical effect.

Systemic fluoride

▶ To prevent mottling only one systemic measure should be used at a time.

Water fluoridation in a concentration of 1 ppm (1 mg F per litre) gives a caries reduction of 50%. The two main advantages of this measure are that no effort is required on the part of the individual, and the low cost. Yet despite the proven benefits only 10% of the UK population receive fluoridated water. In some countries school water has been fluoridated, but a concentration of 5 ppm is required to offset the less frequent intake.

Fluoride drops and tablets Regimen (mg F per day) depends upon drinking water content (see table opposite).

Milk with 2.5 to 7 ppm F has been tried successfully.

Salt is cheap and effective for rural communities in developing countries where water fluoridation is not feasible.

Topical fluoride

Professionally applied fluorides A wide variety of solutions, gels, and application protocols are available. Overall, caries reductions of 20–40% are reported. If these are applied in trays without adequate suction the systemic dosage can be high ∴ it is better to apply to a few, well isolated teeth at a time. Fluoride varnish (e.g. Duraphat) is useful for applying directly to individual lesions to aid arrest, but care is required in young children, as it contains 23 000 ppm fluoride.

Rinsing solutions Mouth-rinses are C/I in children <7 yrs. The concentration prescribed depends upon the frequency of use: 0.2% fortnightly/weekly or 0.05% daily. Daily use is the most beneficial. Caries reductions of the order of 16–50% have been reported with rinsing alone. The most widely used solution is sodium fluoride, but interest in stannous fluoride has rekindled following suggestions that it has an ↑ antiplaque effect.

Toothpastes aid tooth cleaning and polishing, but, most importantly, act as a vehicle for fluoride delivery. In the UK they contain abrasives (to a specified abrasivity standard), detergents, humectants, flavouring, binding agents, preservatives, and active agents, including:

1 E. A. M. Kidd 1987 *Essentials of Dental Caries: The Disease and Its Management*, Wright.

1 Fluoride. Most toothpastes contain sodium monofluoro-phosphate and/or sodium fluoride, in concentrations of 1000 to 1450 ppm (i.e. 1 mg to 1.45 mg per $\frac{1}{2}$ inch of paste). Caries reductions of 15% (in fluoridated areas) to 30% (in non-fluoridated areas) are reported. Recently, junior formulations, e.g. Macleans' Milk Teeth Formula, have been introduced with a lower fluoride content (525 ppm), to ↓ risk of mottling.

2 Anticalculus agents, e.g. sodium pyrophosphate, can ↓ calculus formation by 50%.

3 De-sensitizing agents, e.g. 10% strontium or potassium chloride, or 1.4% formaldehyde.

4 Antibacterial agents, e.g. Triclosan. Not yet proved in clinical trials.

Recommended daily fluoride supplementation (mg F)
Concentration of fluoride in water (ppm)

Age	<0.3	0.3–0.7	
6 months to 2 yrs	0.25	0	drops
2 to 4 yrs	0.5	0.25	drops
>4 yrs	1.0	0.5	tablets

For maximum benefit should be continued from birth to 13 yrs, and the tablets sucked slowly.

Suggested guidelines

1 Where the fluoride level in the drinking water is low, systemic supplements are valuable only until the child starts to use fluoride toothpaste. Then the child should be encouraged to spit out the remaining toothpaste, but not rinse after brushing. However, for children particularly at risk of caries or whose general health would be compromised, the advantages of additional fluoride outweigh the small risk of mottling (and this can be reduced by limiting the amount of toothpaste used and rinsing after brushing).

2 Where the fluoride level is >0.7 ppm, no additional systemic method should be employed. For pre-school children the amount of toothpaste used should be limited to 0.3 g (i.e. the size of a pea).

3 Probably only those with a high caries rate will benefit from topical fluoride rinsing, in addition to brushing with a fluoridated toothpaste.

4 Fluoridation of water still remains the most cost-effective method.

Bacterial plaque and dental decay

Evidence for role of bacteria in dental caries

1 *In vitro*. Incubating teeth with plaque and sugar in saliva results in caries.
2 Animal experiments, e.g. germ-free rodents fed a cariogenic diet do not develop caries, but following the introduction of *Streptococcus mutans*, caries occurs.
3 Epidemiological evidence showing that a supply of bacterial substrate results in caries.
4 Clinical experiments, e.g. stringent removal of plaque ↓ decay.

A correlation has been found between the presence of *Strep. mutans* and caries. This is not surprising, because this organism is acidophilic, can synthesize acid rapidly from sugar and produces a sticky extracellular polysaccharide which helps bind it to the tooth. However, caries can develop in the absence of *Strep. mutans*, and its presence does not inevitably lead to decay, e.g. root caries has been associated with *Strep. salivarius* and *Actinomyces* species. *Lactobacilli* are also acidophilic and have been implicated in fissure caries. In addition, plaque prevents acid diffusing away from the enamel and hinders the neutralizing effect of salivary buffers.

Methods of preventing caries by bacterial control

Physical removal of plaque

1 By a professional. If sufficiently frequent can ↓ caries,[1] but rarely practical.
2 By the individual. Unfortunately, at the standard employed by the majority of the general public, toothbrushing *per se*, is not an effective method of caries control. However, brushing with a fluoridated toothpaste provides regular topical fluoride. Also ↓ gingivitis.

Chemical removal of plaque To achieve more than a transitory effect, an antiseptic needs to be retained in the mouth. The only chemical capable of this at present is chlorhexidine, a positively charged bacteriocidal and fungicidal antiseptic, which is attracted to the negatively charged proteins on the surface of teeth and oral mucosa, and in saliva from where it gradually leaches out. It is available as a 0.2% mouthwash and a 1% gel (Corsodyl) which are cheaper over the counter than by prescription. Although the main application of chlorhexidine is in the management of gingivitis, it has been shown to be effective at ↓ caries when used regularly.[2] While its widespread use for this purpose is not practical, it can be helpful in the management of handicapped patients or those with ↓ salivary flow. Unwanted effects include staining, disturbance of taste, and parotid swelling (which is reversible). It is less effective in the presence of a large build-up of plaque and is inactivated by commercial toothpastes.

---

1 J. Lindhe 1975 *Comm Dent Oral Epidemiol* **3** 150. **2** H. Loe 1972 *Scand J Dent Res* **80** 1.

A variety of pre-brushing rinses are now available. Research suggests that these do have a small beneficial effect if used in conjunction with toothbrushing.[1]

Immunization against caries As no vaccine is completely safe, the ethics of vaccinating against caries, an avoidable non-lethal disease, have been hotly debated.[2] Yet despite considerable research, efforts to produce a viable vaccine have been unsuccessful due to a number of problems:

- Which species of *Strep. mutans* to target, and whether pathogenicity would then shift to another species.
- Differing modes of action in monkeys and rodents, $\therefore$? relevance of experiments to humans.
- Cross-reactivity with heart muscle in animal experiments.
- Duration of effect and acceptance by public. Some patients may prefer caries to repeated injections of a vaccine.

1 H. V. Worthington 1993 *BDJ* **175** 322.   **2** W. Sims 1985 *Comm Dent Health* **2** 129.

Fissure sealants

Occlusal fissures provide a sheltered niche for bacterial proliferation. By providing an impervious barrier to the fissure system, fissure sealants can help prevent occlusal caries.

Historical Several approaches to ↓ fissure caries have been tried:
- Chemical treatment of the enamel, e.g. with silver nitrate.
- Prophylactic odontotomy. This involved restoring the fissure with amalgam (hardly a preventive approach!).
- Sealing of the fissures. Several materials have been tried, including black copper cement (not retained), cyanoacrylate (toxic), polyurethane, and glass ionomer (viscosity often necessitates preparation of the fissure).

The modern f/s is a composite resin used with an acid-etch technique.

Is there a need for sealants? Although developed countries have enjoyed a reduction in dental decay in recent years this has not been uniform for all tooth surfaces. Given that part of this reduction is thought to be due to an increased availability of fluoride, it is not surprising that there has been a greater reduction in approximal, rather than in pit and fissure caries. If decay is to be eliminated, then the need for a method of ↓ occlusal caries is even more pressing.

Are sealants effective? To be effective, sealants need to be carefully applied to susceptible teeth. Unfortunately, those situations where they are most valuable (recently erupted first molars), are often where moisture control is the most difficult, therefore sealants should be monitored and replaced, if lost. For maximum benefit, teeth should be sealed as soon as practicable after eruption and certainly within 2 years. Guide-lines for placement of f/s have been described.[1] Briefly, sealing of molars is indicated in patients with extensive caries of the 1° dentition &/or caries of the 6s. Sealing of premolars is only indicated in patients with special needs or high caries risk. It is better to investigate any suspect areas and place a PRR rather than seal over caries knowingly.

The accepted figures for sealant retention are >85% after 1 yr and >50% after 5 yrs.[2]

Discussion of the cost-effectiveness of sealants compared to restoration has been well aired over the years, which is surprising given that the end results are not comparable. However, recent studies which have indicated that amalgam restorations have a rather more finite life than was once assumed (p. 272), has deflated this debate.

Types of fissure sealant Sealants can be classified by polymerization method (light or self-cure), resin system (Bis—GMA or urethane diacrylate), colour (clear or tinted), and whether they are filled or unfilled. The choice is one of personal preference;

1 BPS 1987 *BDJ* **163** 42.   **2** National Institutes of Health 1984 *J Am Dent Assoc* **108** 233.

however, it has been pointed out that coloured sealants are more readily obvious to the patient (and the Dental Practice Board!). The retention rates of the different types are similar: success depends upon maintaining an absolutely dry field during application.

Fissure sealant technique In most cases the manufacturer's instructions have not kept up with recent research. This regimen should be quicker:

1 Prophylaxis can be omitted, unless the tooth is covered in plaque.
2 Isolate and dry the tooth.
3 Etch for 20 sec with 30–50% phosphoric acid.
4 Wash thoroughly, re-isolate and dry very, very well. If salivary contamination occurs for >0.5 sec, re-etch.
5 Apply f/s (method depends upon delivery system).
6 After polymerization try to remove the sealant. If satisfactory, occlusal adjustment is usually not required unless a large volume has inadvertently been applied or a filled resin is used.

Sugar

The term sugar is commonly used to refer to the mono- and di-saccharide members of the carbohydrate family. Monosaccharides include glucose (dextrose or corn sugar), fructose (fruit sugar), galactose, and mannose. Disaccharides include lactose (in milk), maltose, and sucrose, (cane or beet sugar). Polysaccharides (starch) are composed of chains of glucose molecules and are not readily broken down by the oral flora. Dietary sugars have been classified as intrinsic when they are part of the cells in a food (vegetables and fruit) or extrinsic (milk sugar or, the real baddy, non-milk extrinsic sugar, e.g. table sugar).

Evidence for the role of sugar in dental caries[1]
1 Epidemiological evidence:
  • World-wide comparison of sugar consumption and caries levels.
  • Low caries experience of people on low-sugar diet, e.g. wartime diet; patients with hereditary fructose intolerance.
  • ↑ caries experience following ↑ availability of sugar, e.g. Eskimos.
  • Cross-sectional studies relating caries experience to sugar intake.
2 Clinical studies, e.g. Vipeholm study, Turku sugar study (xylitol).
3 Plaque pH studies, *in vivo* and *in vitro*. See Stephan curve opposite.
4 Animal experiments, e.g. rats fed by stomach tube do not develop caries.

Sucrose is considered the major culprit, due in part to being the most commonly available sugar, but also because of its ability to facilitate production of extracellular polysaccharide in plaque. However, other sugars can also cause caries. In ↓ cariogenicity:
1 Sucrose, glucose, fructose, maltose (honey).
2 Galactose, lactose.
3 Complex carbohydrate (e.g. starch in rice, bread, potatoes).

The frequency of sugary intakes and the interval between them, the total amount of sugar eaten in the diet and the concentration of sugar and stickiness of a food have been shown to be important. The acidogenicity of a sugar-containing food can be modified by other items in the food or meal. Foods that stimulate salivary flow can speed the return of plaque pH to normal, e.g. cheese, sugar-free gum, salted peanuts.

Sugar and health In 1989 the COMA panel on Dietary Sugars and Human Disease reported that dental decay is positively associated with the frequency and amount of non-milk extrinsic sugar consumption. However, while sugar may contribute to the excess calorific intake which causes obesity and predisposes towards diabetes or coronary heart disease, there is no direct evidence between sugar intake and these medical conditions.[2]

1 A. Rugg-Gunn 1993 In *Nutrition and Dental Health*, OUP.
2 COMA 1989 *Dietary Sugars and Human Disease*, HMSO.

Diagram of a Stephan curve showing the pH drop that occurs after a sugary drink is consumed (shown by arrow). The dashed line indicates the critical pH; below this pH demineralization will occur. The shape of the curve is affected by a number of factors, including the type of sugary food, buffering potential of the saliva, and foods or drinks ingested after the sugary challenge.

Sugar (*cont.*)

Prevention of caries by ↓ the availability of microbial substrate

The following aims take into account the modern habit of 'snacking' (also known as 'grazing'):

- Remove sugar from selected foods.
- Substitute non-cariogenic sweeteners.
- Modify sugar-containing foods so that they are less cariogenic.

Modification of only a restricted number of snack foods would probably be necessary to have a significant effect.

Alternative sweeteners

(In this table sweetness of sucrose ≡ 1)

Sweetener	Type	Sweetness	Cariogenicity	Comments
Sorbitol	bulk sweetener	0.5	low	isocalorific to sugar
Mannitol	bulk sweetener	0.7	low	
Xylitol	bulk sweetener	1	none	diarrhoea
Isomalt	bulk sweetener	0.5	low	
Lycasin*	bulk sweetener	0.75	low	
Acesulfame	intense	130	none	
Aspartame	intense	200	none	C/I phenylketonuria
Saccharin	intense	500	none	bitter aftertaste
Thaumatin	intense	4000	none	

* Lycasin is the trade name for hydrogenated glucose syrup (which didn't fit in the table!).

The bulk sweeteners (largely polyols) can cause osmotic diarrhoea if consumed in large amounts and are therefore C/I in small children. However, it is probably wise to avoid all artificial sweeteners in pre-school children. The bulk sweeteners are isocalorific with sucrose, whereas the intense sweeteners are low calorie.

Dietary analysis and advice

Diet can affect teeth:

Pre-eruptively Fluoride is the most important. The effect of calcium, phosphate, vitamins, and sugar is unclear, but is unlikely to be great.

Post-eruptively Again, fluoride is important, as is sugar. Acidic foods or drinks can cause erosion (p. 304).

Dietary analysis

Aim To reduce the time for which the teeth are at risk of demineralization and increase the potential remineralization period.

Indications **1** high caries activity **2** unusual caries pattern **3** suspected dietary erosion.

Dietary advice should be tailored to the individual. This is most easily done after analysing the patient's present eating pattern.

Method A consecutive 3-day analysis (1 weekend and 2 weekdays) is the most widely used, with the patient recording the time, content, and quantity of food/drink consumed. In addition, toothbrushing and bedtime should be indicated. When the form is returned the entries should be checked with the patient.

Analysis
1 Ring the main meals. If in any doubt, identify those snacks that contain complex carbohydrate. Assess nutritional value of meals.
2 Underline all sugar intakes in red.
3 Identify between meal snacks and note any associations, e.g. following insubstantial meals or at school.
4 Decide on a maximum of three recommendations.

Dietary advice should include an explanation of the effect of between meals eating and sugary drinks. It must also be personal, practical, and positive! The suggestion that a child should select crisps when friends are buying sweets is more likely to be followed than total abstinence.

Some helpful hints:
- Suggest saving sweets to be eaten on 1 day, e.g. Saturday dinner-time.
- All-in-one chocolate bars are preferable to packets of individual sweets.
- Foods which stimulate salivary flow (e.g. cheese, sugar-free chewing gum) can help to reverse the pH drop due to sugar, if eaten afterwards.
- Treacle and honey are cariogenic.
- Artificial sweeteners should be avoided in pre-school children.
- Fibrous foods, e.g. apples, are preferable to a sucrose snack, but there is no evidence that they can clean teeth.

Where the nutritional content of meals is inadequate considerable tact is necessary. It may be possible to suggest that larger meals would reduce the temptation to eat snacks.

BUT Remember that while cheese, peanuts, and crisps may constitute a safe snack in dental terms, they are all high in fat, and peanuts can be inhaled by small children. Also 'diet' cola is sugar-free, but can still cause erosion if large quantities are drunk.

Therefore, dental dietary advice should be given in the wider context of the general health of the individual, i.e. ↓ consumption of sugars and fats, and ↑ consumption of fibre-rich starchy foods, fresh fruit, and vegetables. Meals provide a better nutritional balance than snacks.

Dental health education

What is it? The objective of dental health education is to influence the attitude and behaviour of the individual to maintain oral health for life and prevent oral disease.

Primary prevention: seeks to prevent the initial occurrence of a disease or disorder and is aimed at healthy individuals.

Secondary prevention: aims to arrest disease through early detection and treatment.

Tertiary prevention: helps individuals to deal with the effects of the disease and to prevent further recurrence.

Who should give it? All health professionals. In practice, many patients relate better to advice from a hygienist or nurse.

What information should be given? It is important that the information given is factual and that different sources do not give conflicting advice. In order to unify the profession's approach, the Health Education Authority have published a policy document[1] laying out 4 simple messages:
- Restrict sugar-containing foods to meal times.
- Clean teeth and gums thoroughly every day with a fluoride toothpaste.
- Attend the dentist regularly.
- Water fluoridation is beneficial.

How? The way in which the advice is imparted is as important as its content. There are three main routes for dental health education:
- The mass media. This is an expensive alternative and, whilst commercial advertisers tempt the consumer, the success of a dental health education which is exhorting the public to stop doing something they find pleasurable is not guaranteed.
- Community programmes. These need to be carefully planned, targeted, and monitored.[2]
- One-to-one in the clinical environment. This is usually the most successful approach, because the message can be tailored to the individual and reinforcement is facilitated. However, it is expensive in terms of manpower.

Individual dental health education Because many patients find the dental surgery threatening, it may be better to choose a more neutral environment, e.g. a dental health or preventive unit. It is important that the information is given by someone the patient trusts and can relate to—this is not always the dentist! It is important also to have adequate time, as a hurried approach is of dubious value, and to choose words that the patient will understand.

1 *The Scientific Basis of Dental Health Education* 3rd edn 1989 Health Education Authority, Hamilton House, Marbledon Place, London WC1H 9TX. **2** *Notes on Dental Health Education*, Scottish Health Education Group, Woodburn House, Canaan Lane, Edinburgh.

The following approach has been used successfully:
1 Define the problem and its aetiology. For example, poor OH which has resulted in periodontal disease—is it because the patient lacks motivation or the appropriate skills? This stage includes questioning the patient to discover how often and for how long he brushes.
2 Set realistic objectives. It is better to start with trying to motivate the patient to brush well once a day, than teaching them how to floss.
3 Demonstrate on the patient, as this makes the advice more relevant, and more likely to be remembered.
4 Monitor by comparing plaque scores before and after.
5 Remember that everyone responds well to praise, so if a patient is doing well, tell him.

Keys to successful dental health education
• Relevant to the individual, their life-style and problems.
• Keep message simple. Too much information may be counter-productive.
• Repetition of message.
• Positive reinforcement.

Where to go for help or information Advice on preparing a talk on dental health education, setting up a preventive unit, or even a health programme can be obtained from:
1 Local Health Education (or Promotion) Unit. These centres will be happy to provide leaflets, educational packs, slides, videos or just advice.
2 Consultant in Dental Public Health, or District Dental Officer, whose telephone number should be in the telephone directory.

Provision and receipt of dental care

Delivery of care

General Dental Service This is the main source of dental care for the majority of the population (whether NHS or private).

Community Service The Community Service was formed from the School Dental Service in 1974, but is now designed to act as a safety net for children and adults who are unable to attend the GDS because of handicap or geographic problems. The epidemiological and managerial skills of the service are to be expanded to define priority groups and areas so as to better target resources and health education.[1]

Hospital Service The role of the consultant service is to provide specialist advice and treatment, in addition to postgraduate training.

Receipt of care

Two factors are important:

- Availability and accessibility of dental services. Research shows that a greater proportion of the public visit the dentist regularly where the dentist to population ratio is high. This ratio tends to follow a geographical pattern with the greatest number of dentists in the south-east.
- Social class affects both the incidence of dental disease and the uptake of dental care. Interestingly, the differences in caries experience between the social classes are much lower in fluoridated regions.

Because dentists prefer to practise in leafy suburbs rather than poor inner city areas, the effects of these are often compounded.

Barriers to the uptake of dental care[2]

A recent survey found that the two main barriers to regular uptake of dental care by the general public were anxiety and cost.

Anxiety This was manifest as fear of pain or a particular procedure, or a feeling of vulnerability brought about by relinquishing control to the dentist in the sensitive area of the mouth. The importance of first impressions was highlighted, because the reception received from staff and the environment in which the patient had to wait to be seen could either allay or reinforce their anxieties. The attitude of the dentist was also a significant factor: a 'good' dentist had a friendly, personal touch and explained what he was doing.

Cost Respondents to the survey reported that they thought dental treatment was expensive and the way in which the charges are calculated, confusing, but welcomed an estimate of the costs prior to treatment.

1 National Health Service 1989 *General Dental Services—The Future Development of the Community Dental Services*, FPN 468. **2** Barriers to the receipt of Dental Care *BDJ* 1988 **164** 195.

The results suggested that the pattern of attendance varies throughout life, with children now enjoying a visit to the dentist, but adolescents breaking the habit of regular attendance due to apathy and/or other pressures on their time. A return to the dentist may be triggered by pregnancy and a desire to provide a good example to the children, or a need for urgent treatment and a fear of becoming edentulous.

Dentistry for the handicapped

A handicap is a disability that retards, distorts, or otherwise adversely affects growth, development, or adjustment to life.

Mental handicap Prevalence 3% Classified into mild (IQ 50–70), severe (IQ <50), and Down syndrome.

Physical handicap Most common is cerebral palsy, which is the motor manifestation of cerebral damage. Many patients with cerebral palsy have normal IQs, but ↑ muscle tone and hyperactive reflexes can make treatment difficult. Many can be treated by GDS provided there is wheelchair access.

Medical handicap 1% of children have either heart disease, bleeding disorders, diabetes, or kidney disease.

Sensory handicap i.e. blindness, deafness.

Many have more than one type of handicap.

The above groups are general handicaps. We also need to consider those that are orally handicapped, i.e. have a gross oral problem or deficit which necessitates special dental treatment (e.g. cleft lip &/or palate).

Problems

It is difficult to generalize, but usually mental handicap provides the biggest challenge. Difficulties ↑ in patients with >1 handicap.
- Delivery of care. This has three aspects: **1** ↓ demand, due to low priority placed on dental health; **2** lack of provision made to provide the necessary care; **3** practical difficulties in carrying out dental work.
- In general, handicapped patients have ↓ plaque control and ∴ ↑ periodontal problems.
- Although caries incidence is not sigficantly ↑ compared to the normal population, the amount of untreated caries is.
- Long-term sugared medications.
- ↑ prevalence of hepatitis in institutionalized and Down syndrome patients.
- Dentures may be impractical ∴ extractions not a realistic solution to the problems of providing dental treatment.

Management

Again, it is difficult to generalize. Patients with less severe handicaps can be treated in GDS along with other members of the family. Those with severe medical and/or mental handicaps are probably best managed by a specialist who will have greater access to facilities. However, with the changing emphasis in the CDS, better care should become available to institutionalized patients.

Treatment planning An initial plan should be formulated ignoring the handicap. This can then be discussed with the patient, parent or carer and modified for the individual. It is advisable to start with OHI and prevention, then re-assess treatment requirements in the light of the response. For those patients for whom a satisfactory standard of OH is not possible, restorative treatment should aim to ↓ plaque accumulation (e.g. P/P are C/I).

OHI Those patients who can brush their own teeth should be encouraged to do so. Modification of toothbrush handles (p. 376) or purchase of an electric one may be helpful. Where patients are unable to brush their teeth, instruction should be given to their carer. The best method is to stand behind the patient and cradle the head with one arm, leaving the other free to brush. However, if possible this should be supplemented with regular professional cleaning. Chemical control of plaque with chlorhexidine may be helpful.

Restorative care Greatest problems are posed by mentally handicapped. Kind but firm restraint may be necessary—ideally, get the patient's carer to help. A prop (e.g. McKesson rubber) may be needed. It is often easier to use intraligamentary LA technique. Sedation may help reduce the spontaneous movements of cerebral palsy. In some cases there is no alternative but to carry out examination and treatment under GA. In addition, for those patients who can tolerate out-patient treatment, but only a little at a time, it may be kinder to clear a back-log under GA, thus allowing concentration on prevention subsequently. However, this approach requires special facilities and no medical C/I.

Down syndrome, p. 743.

The Royal National Institute for the Deaf

STANDARD MANUAL ALPHABET

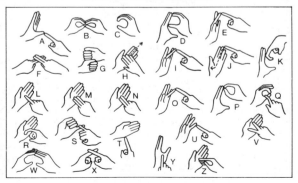

Reprinted by kind permission of the RNID (105 Gower St., London).

Hygienists and other ancillaries

Dental auxiliary personnel are defined as those persons who are involved to a greater or lesser extent in the practice of dentistry in its widest forms, but who are not qualified with a degree or diploma in dentistry. At present ancillaries comprise 1 in 7 dental staff.[1]

The WHO classify dental auxiliary personnel as follows:
1 Non-operating auxiliaries: Type I Dental technician, Type II DSA, Type III Dental preventive worker.
2 Operating auxiliaries: Type IV Hygienist, Type V Dental therapist.

The advantage of employing operating ancillary staff in dentistry is that they are able to undertake some of the simpler tasks, thereby freeing the dentist for more complex procedures. In addition to improved job satisfaction, it is more cost-effective to employ ancillary staff for those duties that do not require 5 yrs of University training.

Training costs of dental personnel (£K):[1] Dentist—82, Hygienist—7.

Hygienists At present training is 9 to 12 months and hygienists can carry out the following unsupervised procedures under the written direction of a dentist:
• Administer infiltration LA.
• Scaling and polishing.
• Application of topical fluoride and sealants.
• OHI and preventive advice, e.g. diet.

Therapists This type of auxiliary is no longer trained. In addition to the duties of a hygienist, therapists are permitted to carry out deciduous tooth extraction and simple restorations.

Orthodontic auxiliaries This grade of auxiliary is widely employed in many countries, including the USA and Scandinavia. Their work includes placement of fixed appliances, changing archwires, and taking impressions.

The future — a dental team

With increasing demand for dental (and orthodontic) treatment and restraints on health care costs, the advantages of delegating more routine tasks to dental auxiliaries is obvious. It would appear inevitable that the duties of present auxiliary grades will expand, as will the types of ancillary worker. The Nuffield Report published in 1993 advocated the introduction of two new grades of clinical auxiliaries. Oral health therapists would include existing therapists and hygienists, but further recruitment would be from DSAs. Orthodontic auxiliaries would be included within this grade. Clinical dental technicians would be trained from qualified technicians and be allowed to undertake limited clinical work, e.g. taking impressions.

1 K. D. O'Brien 1988 *Br J Orthod* **15** 286.

With the introduction of more auxiliaries the role of the dentist will inevitably change, to become more centred on diagnosis and treatment planning, and more complex treatment, with team leadership skills becoming increasingly important.

Lies, damn lies, and statistics

Sugar

- The UK per capita consumption of sugar is 2 lb/week.
- UK children receive about $\frac{1}{5}$ to $\frac{1}{4}$ of their energy intake from sugar. Of these $\frac{2}{3}$rds are added sugars, and of which >$\frac{2}{3}$rds come from sweets, table sugar, and soft drinks.[1]
- 65% of all soft drink sales are to <15 yr-olds.
- Low-income families consume more sugar/person/day than higher income families.

Fluoride

- Water fluoridation ↓ caries experience by about 50%.
- A cup of tea contains about 1 ppm. (1 in 3 people in UK take tea-bags abroad with them on holiday!)
- At equivalent concentrations there is no difference in the efficacy of sodium fluoride or sodium monofluorophosphate-containing toothpastes.[2]

Caries

- A reduction of 10–60% in the caries experience of developed countries has been widely reported. This is thought to be due to a variety of factors including: fluoride toothpaste, increased public awareness, changes in infant feeding practises, ↓ sugar consumption, and antibiotics in the food chain.
- In addition, there has been a change in the pattern of carious attack with a greater ↓ in smooth surface than fissure caries (perhaps reflecting the influence of fluoride).
- Small occlusal lesions appear to be becoming the predominant type of lesion.[3]
- BUT, there is some evidence to suggest that the ↓ in caries may have halted in Britain.[4]

Adult dental health 1988[5]

	1978	1988
Proportion of adults edentulous	30%	21%
Average condition of teeth:		
• missing	9 teeth	7.8 teeth
• decayed	1.9 teeth	1 tooth
• filled	8.1 teeth	8.4 teeth
• sound	13 teeth	14.8 teeth

A marked regional variation was noted with Scotland and N Ireland having the smallest number of sound untreated teeth (13 and 12.6 respectively). The average number of filled teeth was highest in Southern England (9.1).

1 A. J. Rugg-Gunn 1986 *Human Nutrition: Applied Nutrition* **40A** 115. **2** *American Journal of Dentistry*, Special Issue, 1993. **3** A. Sheiham 1989 *BDJ* **165** 240. **4** A. Rugg-Gunn 1990 *Dental Update* **17** 24. **5** OPCS Monitor 1990 *Adult Dental Health United Kingdom*, HMSO.

Periodontal disease

- Prevalence in UK:

	17–24 yrs	25+ yrs
Gingivitis	77%	87%
Periodontitis	3%	64%

- Recent estimates indicate that approximately 10–15% of adults may be at high risk of developing progressive periodontal disease.

Child dental health 1983

- The proportion of caries-free 5 yr-olds in England and Wales rose from 29% in 1973 to 52% in 1983, thus achieving one of the goals set by the WHO for the year 2000.[1]
- Since the early 1970s caries experience in 5 yr-olds has declined by about 50%, by 40% in 12 yr-olds, and by a third in 15 yr-olds.[2]
- Overall 56% of 9 yr-old children are in need of orthodontic treatment.
- 64% of 12 yr-olds attend for regular check-ups, 12% occasionally, and 24% only when in trouble.

Indices

DMFT decayed, missing, and filled permanent teeth.
dmft decayed, missing, and filled deciduous teeth.
deft decayed, exfoliated, and filled deciduous teeth.
dft decayed and filled deciduous teeth.
DMFS decayed, missing, and filled surfaces in permanent teeth.

1 J. E. Todd 1985 *Children's Dental Health*, HMSO. **2** M. C. Downer 1984 In *Cariology Today*, Kargar.

3 Paedodontics

The child patient   60
Treatment planning for children   62
The anxious child   64
The child with toothache   66
Abnormalities of tooth eruption and
    exfoliation   68
Calcification and eruption times   69
Abnormalities of tooth number   72
Abnormalities of tooth structure   74
Abnormalities of tooth form   76
Abnormalities of tooth colour   78
Anatomy of primary teeth (and relevance to
    cavity design)   80
Extraction versus restoration of deciduous
    teeth   82
Local anaesthesia for children   84
Restoration of carious deciduous teeth   86
Class I in deciduous molars   88
Class II in deciduous molars—amalgam   90
Class II in deciduous molars—alternative
    techniques   92
Stainless steel crowns   94
Class III, IV, and V in primary teeth   96
Rampant caries   98
Deciduous molar pulp therapy   100
Pulpotomy techniques for vital pulps   102
Pulpotomy techniques for non-vital pulps   104
Pulp therapy for primary anterior teeth   104
Dental trauma   106
Non-accidental injury   108
Injuries to primary teeth   110
Injuries to permanent teeth—
    crown fractures   112
Root fractures   114
Luxation, subluxation, intrusion, and
    extrusion   116
Splinting   118
Management of the avulsed tooth   120
Pulpal sequelae following trauma   122

Management of missing incisors 124
Common childhood ailments affecting the
  mouth 126
Sugar-free medications 128

59

Principal sources: R. Andlaw 1992 *A Manual of Paedodontics*, Churchill Livingston. J. O. Andreasen 1981 *Traumatic Injuries of the Teeth*, Munksgaard. J. A. Hargreaves 1981 *The Management of Traumatised Anterior Teeth in Children*, Churchill Livingston. P. J. Holloway 1982 *Child Dental Health*, Wright. J. O. Andreasen 1992 *Atlas of Replantation and Transplantation of Teeth*, Mediglobe.

The child patient

▶ Treat the patient not the tooth.

Principal aims of treatment

- Freedom from pain and infection.
- A happy and co-operative patient.
- Prevention.
- Development of a healthy and attractive permanent dentition.

Points to remember

- Praise good behaviour, ignore bad.
- Involve parents (they determine whether child will return).
- Do not offer choice where there is none (shall we send your tooth to sleep?).
- Children have short attention spans (↑ with age).
- Children have ↓ sensory acuity (may confuse pressure with pain, vitality tests less reliable).
- Children have ↓ manual dexterity, ∴ need help with tooth-brushing <7 yrs.
- Start with easy procedures (e.g. OHI) and progress, at child's pace, to more complicated treatment.
- Set attainable targets for each visit and attain them.

The first visit

- The younger the better, although watching other members of the family receive treatment prior to first appointment preferable.
- Keep brief.
- Let parent accompany child: check medical history and reason for attendance.
- Talk to child: communication is the key to success!
- Show patient chair, mirror, light, and explain purpose.
- Count the patient's teeth.
- If good progress, polish a few teeth, but don't tire child by attempting too much.
- Show parent child's teeth and what has been done that visit.
- If child in pain the source of this needs to be determined and dealt with as quickly as possible.
- Younger children can be more successfully examined if mum sits with child facing her and then lowers child back on to dentist's lap.

3 Paedodontics
The child patient

Treatment planning for children

The underlying principles are the same as for adults. Importance should be placed on prevention, this can be emphasized by the priority given in the treatment plan. However, the first step should always be relief of pain. Any non-urgent extractions are best carried out as the final item.

For example:

1 Oral hygiene instruction with patient using their own tooth-brush.
2 Fissure sealing of $\frac{6}{6}$
3 Fissure sealing of $\frac{|6}{|6}$
4 Restoration of $\overline{|E}$

▶ Remember to consider the developing occlusion:
- Long-term prognosis for first permanent molars (p. 152).
- DPT at age 8 to check position/presence of permanent teeth.
- Palpate for $\underline{3|3}$ at age 9–10 yrs (p. 160).
- Beware disturbances in eruption sequence (p. 68) and asymmetry.
- Early referral to specialist for skeletal discrepancies, and for any abnormal findings.

▶ Look out for any signs of underlying medical or social problems.
- small stature
- failure to thrive
- bleeding tendency
- systemic disease
- child abuse (p. 108)

Should parent accompany child into surgery? Essential on first visit, thereafter depends upon child's age. If in doubt ask child's preference! However, if parent is a dental phobic, their anxiety on being in a dental environment may adversely affect child, so in these cases it is probably wiser to leave Mum or Dad in the waiting room. Some children will play up to an over-protective parent in order to gain sympathy or rewards and may prove more co-operative by themselves.

The anxious child

Techniques for behaviour management

Most of these are fancy terms to describe techniques that come with experience of treating children over a period of time. However, for the student they may prove useful for answering essay questions as well as for handling their first few child patients.

General principles

- Show interest in child as a person.
- Touch > facial expression > tone of voice > what is said.
- Don't deny patient's fear.
- Explain—why, how, when.
- Reward good behaviour, ignore bad.
- Get child involved in treatment, e.g. holding saliva ejector.
- Giving the child some control over the situation will also help them to relax, e.g. raising their hand if they want you to stop for any reason.

Tell, show, do Self-explanatory, but use language the child will understand.

Densitization Used for child with pre-existing fears. Involves helping patient to relax in dental environment, then constructing a hierarchy of fearful stimuli for that patient. These are introduced to the child gradually, with progression on to the next stimulus only when the child is able to cope with previous situation.

Modelling Useful for children with little previous dental experience, who are apprehensive. Encourage child to watch other children of a similar age or siblings receiving dental treatment happily.

Behaviour shaping This is planning treatment so that it is carried out in small steps from simple to more complicated procedures.

Re-inforcement Rewarding good behaviour by approval and praise. If a child protests and is uncooperative during treatment, do not immediately abandon session and return them to the consolation of their parent, as this will negatively reinforce the undesirable behaviour. It is better to try and ensure that some phase of the treatment is completed, e.g. placing a dressing.

Sedation

Indicated for the genuinely anxious child who wishes to co-operate with treatment.

Oral: unpredictable response in children ∴ C/I.
Intramuscular: rarely used in children.
Intravenous: rarely used in children.
Per rectum: popular in some Scandinavian countries.
Inhalation: uses nitrous oxide/oxygen mixture to produce relative analgesia (RA) and is the most popular technique for use with children. Its use is C/I with the very young, mentally handi-

capped, and in the presence of upper airways obstruction, e.g. a cold. Technique, see p. 636. It is a good idea not to carry out any treatment during the visit when the child is introduced to 'happy air'. Let child position nose-piece themselves.

Hypnosis

Hypnosis produces a state of altered consciousness and relaxation, though it cannot be used to make subjects do anything that they do not wish. Although many good books[1] and articles are available, attendance at a course is necessary to gain experience with susceptible subjects, so the operator has confidence in his abilities. It can be described as either a way of helping the child to relax, or as a special kind of sleep.

General anaesthesia

Allows dental rehabilitation to be achieved at one visit for handicapped children and, as a last resort, for the uncooperative child. In the latter case as subsequent treatment is preventive rather than restorative, progress can then be made on gaining the child's confidence.

Other behaviour problems and their management

- The questioner: attempts to delay treatment by a barrage of questions. Firm, but gentle handling is needed. Tell the patient that you understand their anxieties and that you will explain as you go along.
- The temper tantrum: ignore the tantrum and try to complete treatment regardless. If this is not possible wait until child tires himself out and then try to complete a phase of treatment.

1 J. Hartland 1982 *Medical and Dental Hypnosis*, Ballière Tindall.

The child with toothache

When faced with a child with toothache the dentist has to use his clinical acumen to try and determine the pulpal state of the affected tooth/teeth as this will decide the treatment required. To that end the following investigations may be employed:

History Unfortunately children are rather unreliable when it comes to giving an accurate history. If this appears to be the case, and with small children, question mum or dad regarding disturbed sleep, difficulty with eating, duration of symptoms, etc.

Examination Look for caries, abscess formation, mobility (? due to exfoliation or apical infection), and erupting teeth.

Percussion Unreliable in children.

Vitality testing Again, this is unreliable in deciduous teeth, but for permanent teeth, a cotton-wool roll, ethyl chloride, and considerable ingenuity may provide some useful information.

Radiographs Bitewing radiographs are most useful, because not only are they less uncomfortable for small mouths than periapicals, but they show up the bifurcation area where most deciduous molar abscesses start.

Remember the only 100% accurate method is histological!

With a fractious child keep examination and operative intervention to a minimum, doing only what is necessary to alleviate pain and win child's trust.

If extractions under a GA are required consider carefully the long-term prognosis of remaining teeth to try and avoid a repeat of the anaesthetic in the near future.

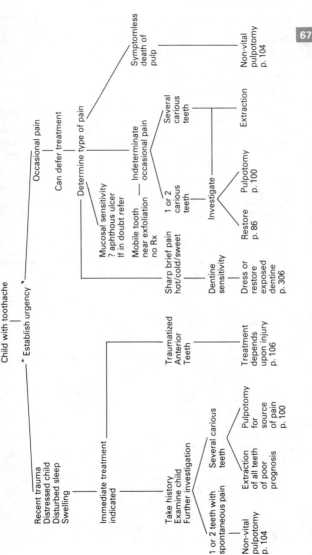

Child with toothache — Establish urgency

Occasional pain — Can defer treatment — Determine type of pain

Determine type of pain:
- Mucosal sensitivity ? aphthous ulcer If in doubt refer
- Mobile tooth near exfoliation no Rx
- Sharp brief pain hot/cold/sweet — Dentine sensitivity — Dress or restore exposed dentine p. 306
- Indeterminate occasional pain
  - 1 or 2 carious teeth — Investigate — Restore p. 86 / Pulpotomy p. 100
  - Several carious teeth — Extraction
- Symptomless death of pulp — Non-vital pulpotomy p. 104

Immediate treatment indicated
- Recent trauma
- Distressed child
- Disturbed sleep
- Swelling

Traumatized Anterior Teeth — Treatment depends upon injury p. 106

Take history Examine child Further investigation
- 1 or 2 teeth with spontaneous pain — Non-vital pulpotomy p. 104
- Several carious teeth
  - Extraction of all teeth of poor prognosis
  - Pulpotomy for source of pain p. 100

Abnormalities of tooth eruption and exfoliation

Natal teeth are usually members of the deciduous dentition, not supernumerary teeth and so should be retained if possible. Most frequently affects mandibular incisor region and, because of limited root development at that age, are mobile. If in danger of being inhaled or causing problems with breast-feeding they can be removed under topical anaesthesia.

Teething As eruption of the 1° dentition coincides with a diminution in circulating maternal antibodies, teething is often blamed for systemic symptoms. However, local discomfort, and so disturbed sleep, may accompany the actual process of eruption. A number of proprietary 'teething' preparations are available which usually contain a combination of an analgesic, an antiseptic, and anti-inflammatory agents for topical use. Having something hard to chew may help, e.g. teething ring.

Eruption cyst is caused by an accumulation of fluid or blood in the follicular space overlying an erupting tooth. The presence of blood gives a bluish hue. Most rupture spontaneously, allowing eruption to proceed. Rarely it may be necessary to marsupialize the cyst.

Failure of/delayed eruption It must be remembered that there is a wide range of individual variation in eruption times. Developmental age is of more importance in assessing delayed eruption than chronological age.

▶ Disruption of normal eruption sequence and asymmetry in eruption times of contralateral teeth >6 months warrants further investigation.

General causes Hereditary gingival fibromatosis, Down syndrome, cleido-cranial dysostosis, ricketts.

Local causes
- Congenital absence. Is the most likely cause for failure of appearance of $\underline{2}$ (p. 72).
- Crowding. Rx: extractions.
- Retention of primary tooth. Rx: extraction of 1° tooth.
- Supernumerary tooth. Is the most likely reason for failure of eruption of $\underline{1}$ (p. 72).
- Dilaceration (p. 76).
- Abnormal position of crypt. Rx: extraction or orthodontic alignment. See options for palatally displaced $\underline{3}$ (p. 160).
- Primary failure of eruption usually affects molar teeth. The aetiology is not understood. Although bone resorption proceeds above the unerupted tooth, they appear to lack any eruptive potential. Rx: keep under observation, but ultimately extraction may be necessary.

Submerged primary molars Caused by preponderance of repair in normal resorptive/repair cycle of exfoliation. This is usually self-correcting and the affected tooth is exfoliated at the normal

Calcification and eruption times

	E	D	C	B	A	1	2	3	4	5	6	7	
Eruption (months)	24 to 36	12 to 15	18 to 20	7 to 8	6 to 7	7 to 8	8 to 9	11 to 12	10 to 11	10 to 12	5 to 6	12 to 13	Eruption (years)
Calcifctn begins (weeks *in utero*)	16 to 23	14 to 17	15 to 18	13 to 16	12 to 16	3 to 4	10 to 12	4 to 5	18 to 21	24 to 27	0	30 to 36	Calcifctn begins (months)

E	D	C	B	A	1	2	3	4	5	6	7
E	D	C	B	A	1	2	3	4	5	6	7

	E	D	C	B	A	1	2	3	4	5	6	7	
Calcifctn begins (weeks *in utero*)	16 to 23	14 to 17	15 to 18	13 to 16	12 to 16	3 to 4	3 to 4	4 to 5	21 to 24	27 to 30	0	30 to 36	Calcifctn begins (months)
Eruption (months)	24 to 36	12 to 15	18 to 20	7 to 8	6 to 7	6 to 7	7 to 8	9 to 10	10 to 12	11 to 12	5 to 6	12 to 13	Eruption (years)

Root calcification complete in 1° dentition 1 to 1.5 yrs after eruption.

Root calcification complete in 2° dentition 2 to 3 yrs after eruption.

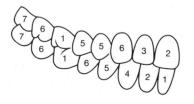

Normal sequence of eruption (permanent dentition)

Abnormalities of tooth eruption and exfoliation (*cont.*)

time.[1] However, where the premolar is missing or where the submerging molar appears in danger of disappearing below the gingival level, extraction may be indicated.

Ectopic eruption of the upper first permanent molars resulting in impaction of the tooth against the E occurs in 2–5% of children. It is an indication of crowding. In younger patients (<8 yrs) it may prove self-correcting. If still present after 4–6 months or in older children, insertion of an orthodontic separating spring (or a piece of brass wire tightened around the contact point) may allow the 6 to jump free. More severe impactions should be kept under observation. If the E becomes abcessed or the 6 is in danger of becoming carious then the deciduous tooth should be extracted. The resulting space loss can be dealt with as part of the overall orthodontic treatment plan later.

Premature exfoliation Most common reason for early tooth loss is extraction for caries. Traumatic avulsion is less common. More rarely systemic disease may result in an abnormal periodontal attachment and thus premature tooth loss (p. 222).

Abnormalities of tooth number

Anodontia

Means complete absence of all teeth. Rare. Partial anodontia is a misnomer.

Oligodontia

American term indicating absence of one or more teeth.

Hypodontia

British equivalent of oligodontia.

Prevalence 1° dentition 0.1–0.9%, 2° dentition 3.5–6.5%.[1] In Caucasians most commonly affected teeth are 8 (25–35%), 2̲ (2%), 5̅ (3%). Affects F>M and is often associated with smaller than average tooth-size in remainder of dentition. Peg-shaped 2̲ often occurs in conjunction with absence of contralateral 2̲.

Aetiology Often familial—polygenic inheritance. Also associated with ectodermal dysplasia and Down syndrome.

Rx: 1° dentition—none. 2° dentition depends on crowding and malocclusion.
8—none.
2̲—see p. 124.
5̅—(be warned, late development of 5̅ not unknown). If patient crowded, extraction of E̅, either around 8 yrs for spontaneous space closure or later if space is to be closed as a part of orthodontic treatment. If lower arch well-aligned or spaced, consider preservation of E̅ and bridgework later.

Hyperdontia

Better known as supernumerary teeth.

Prevalence 1° dentition 0.8%, 2° dentition 2%.[1] Occurs most frequently in premaxillary region. Affects M>F. Associated with cleido-cranial dysostosis and CLP. In about 50% cases $ in 1° dentition followed by $ in 2° dentition, so warn Mum!

Aetiology Theories include: offshoot of dental lamina, third dentition.

Classification either by

Shape	or	*Position*
Conical (peg-shaped)		Mesiodens
Tuberculate (barrel-shaped)		Distomolar
Supplemental		Paramolar
Odontome		

Effects on dentition and treatment
- No effect. If unerupted keep watch, if erupts—extract.
- Crowding. Rx: extract, if supplemental extract tooth with most displaced apex.
- Displacement. Can cause rotation &/or displacement. Rx: extraction of $ and fixed appliance, but tendency to relapse.

1 A. H. Brooks 1974 *J Int Assoc Dent Child* **5** 32–53.

- Failure of eruption. Most likely cause of 1̱ to fail to erupt. Rx: extract S̶ and ensure sufficient space for unerupted tooth to erupt. May require extraction of deciduous teeth &/or permanent teeth and appliances. Then *wait*. Average time to eruption in these cases is 18 months.[1] If after 2 yrs unerupted tooth fails to erupt despite sufficient space may require conservative exposure and orthodontic traction.

1 D. DiBiase 1971 *Dental Practitioner* **22** 95.

Abnormalities of tooth structure

Disturbances in structure of enamel

Enamel usually develops in two phases, first an organic matrix and second mineralization. Disruption of enamel formation can therefore manifest as:

Hypoplasia Caused by disturbance in matrix formation and is characterized by pitted &/or grooved enamel.

Hypomineralization (hypocalcification) is a disturbance of calcification. Affected enamel appears white and opaque, but post-eruptively may become discoloured. Most disturbances of enamel formation will produce both hypoplasia and hypomineralization, but clinically one type usually predominates.

Aetiological factors (not an exhaustive list)
Localized causes: Infection, trauma, irradiation, idiopathic (see enamel opacities, p. 78).
Generalized causes:
1 Environmental (chronological hypoplasia)
 (a) Pre-natal, e.g. rubella, syphilis
 (b) Neo-natal, e.g. prolonged labour, premature birth
 (c) Post-natal, e.g. measles, congenital heart disease, fluoride
2 Hereditary
 (a) Affecting teeth only—Amelogenesis imperfecta
 (b) Accompanied by systemic disorder, e.g. Down syndrome

Chronological hypoplasia So called because the hypoplastic enamel occurs in a distribution related to the extent of tooth formation at the time of the insult. Characteristically, due to their later formation, $\underline{2}$ are affected nearer to their incisal edge than $\underline{1}$ or $\underline{3}$.

Fluorosis , p. 34.

Rx of hypomineralization/hypoplasia depends on extent and severity:
Posterior teeth: small areas of hypoplasia can be fissure sealed or restored conventionally, but more severely affected teeth will require crowning. SS crowns (p. 94) can be used in children as a temporary measure.
Anterior teeth: Small areas of hypoplasia can be restored using composites, but larger areas may require veneers (p. 288) or crowns. For treatment of fluorosis, see p. 78.

Amelogenesis imperfecta Many classifications exist, but generally these are divided into hypomineralized and hypoplastic types and classified as to their mode of inheritance. Usually both 1° and 2° dentitions and all the teeth are affected. The different subgroups give rise to a wide variation in clinical presentation, ranging from discoloration to soft and/or deficient enamel. It is therefore difficult to make general recommendations, but it is wise to seek specialist advice for all but the mildest forms. Rx: in more severe cases, SS crowns and composite resin can be used to maintain molars and 2° incisors, prior to more permanent restorations when child is older.

Disturbances in the structure of dentine

Disturbances in dentinogenesis can follow disruption of amelogenesis. Examples include shell teeth, dentinal dysplasia, regional odontodysplasia (ghost teeth), and vitamin-D-resistant ricketts—all of which are rare. A commoner defect is hereditary opalescent dentine which is often referred to as dentinogenesis imperfecta, though strictly speaking this term should only be used when associated with osteogenesis imperfecta.

Dentinogenesis imperfecta affects 1 in 8000 people. Both 1° and 2° dentitions are involved, although later formed teeth less so. Affected teeth have an opalescent brown or blue hue, bulbous crowns, short roots, and narrow flame-shaped pulps. The ADJ is abnormal, which results in the enamel flaking off, leading to rapid wear of the soft dentine. Rx: along similar lines as for severe amelogenesis.

▶ Early recognition and treatment of amelogenesis and dentinogenesis imperfecta important to prevent rapid tooth wear.

Disturbances in the structure of cementum

Hypoplasia and aplasia of cementum are uncommon. The latter occurs in hypophosphatasia and results in premature exfoliation. Hypercementosis is relatively common and may occur in response to inflammation, mechanical stimulation, Paget's disease, or be idiopathic. Concrescence is the uniting of the roots of two teeth by cementum.

Abnormalities of tooth form[1]

Normal width $\underline{1}$ = 8.5 mm, $\underline{2}$ = 6.5 mm.

Double teeth

Gemination occurs by partial splitting of a tooth germ. *Fusion* occurs as a result of the fusion of two tooth germs. As fusion can take place between either two teeth of the normal series or, less commonly, with a $ tooth, then counting the number of teeth will not always give the correct aetiology.

As the distinction is really only of academic interest the term 'double teeth' is to be preferred. Both 1° and 2° teeth may be affected and a wide variation in presentation is seen. The prevalence in the 2° dentition is 0.1–0.2%.

Treatment for aesthetics should be delayed to allow pulpal recession. If the tooth has separate pulp chambers and root canals, separation can be considered. If due to fusion with a $ tooth, the $ portion can be extracted. Where a single pulp chamber exists either the tooth can be contoured to resemble two separate teeth or the bulk of the crown reduced.

Macrodontia/megadontia Generalized macrodontia rare, but is unilaterally associated with hemifacial hypertrophy. Isolated megadont teeth are seen in 1% of 2° dentitions.

Microdontia

Prevalence 1° dentition <0.5%. In 2° dentition overall prevalence is 2.5%. Of this figure 1–2% is accounted for by diminutive $\underline{2}$. Peg-shaped $\underline{2}$ often have short roots and are thought to be a possible factor in the palatal displacement of $\underline{3}$ (p. 160). $\underline{8}$ also commonly affected.

Dens in dente This is really a marked palatal invagination, which gives the appearance of a tooth within a tooth. Usually affects $\underline{2}$, but can also affect premolars. Where the invagination is in close proximity to the pulp, early pulp death may ensue. Fissure sealing of the invagination as soon as possible after eruption may prevent this, but is often too late. Conventional RCT is difficult and extraction is usually required.

Dilaceration describes a tooth with a distorted crown or root. Usually affects $\underline{1}$. Two types seen dependent upon aetiology.

Developmental[2]	Traumatic
Crown turned upward and labially	Crown turned palatally
Regular enamel and dentine	Disturbed enamel and dentine formation seen
Usually no other affected teeth	
Affects F>M	No sex predilection

The traumatically induced type is caused by intrusion of the 1° incisor resulting in displacement of the developing 2° incisor tooth germ. The effects depend upon the developmental stage at the time of injury.

1 A. H. Brooks 1974 *J International Association Dentistry for Children*. **2** D. J. Stewart 1978 *BDJ* **145** 229.

Rx: depends upon severity and patient co-operation. If mild it may be possible to expose crown and align orthodontically provided the apex will not be positioned against the labial plate of bone at the end of treatment, otherwise extraction indicated.

Turner tooth Term used to describe the effect of a disturbance of enamel and dentine formation by infection from an overlying 1° tooth ∴ usually affects premolar teeth. Rx: as for hypoplasia, p. 74.

Taurodontism Of academic interest only, but seems to crop up on X-rays in exams much more frequently than in clinical practice. Means bull-like, and radiographically an elongation of the pulp chamber area of the root is seen. Rx: none required.

Abnormalities of tooth colour

Extrinsic staining By definition this is caused by extrinsic agents and can be removed by prophylaxis. Green, black, orange, or brown stains are seen and may be formed by chromogenic bacteria or be dietary in origin. Chlorhexidine mouthwash causes a brown stain by combining with dietary tannin. Where the staining is associated with poor oral hygiene, demineralization and roughening of the underlying enamel may make removal difficult. Rx: A mixture of pumice powder and toothpaste or an abrasive prophylaxis paste together with a bristle brush should remove the stain. Give OHI to prevent recurrence.

Intrinsic staining This can be caused by
- Changes in the structure or thickness of the dental hard tissues, e.g. enamel opacities.
- Incorporation of pigments during tooth formation, e.g. tetracycline staining (blue/brown), prophyria (red).
- Diffusion of pigment into hard tissues after formation, e.g. pulp necrosis products (grey), root canal medicaments (grey).

Enamel opacities are localized areas of hypomineralized (or hypoplastic) enamel. Fluoride (p. 34) is only one of a considerable number of possible aetiological agents.

Rx: four possible approaches:
1 Acid pumice abrasion technique is used only for surface enamel defects.
▶ Rubber dam, protective eyewear, bicarbonate of soda placed around teeth to be treated and care, essential.

A mixture of 18% hydrochloric acid and pumice is applied to the affected area using a wooden stick. The mixture is rubbed into the surface for 5 secs and then rinsed away. These two steps are repeated (max 10 times—removing <0.1 mm enamel) until the desired colour change is achieved. The enamel is then polished and a fluoride solution applied.[1]
2 Bleaching, p. 310.
3 Veneers, p. 288.
4 Crowns, p. 280.

1 T. P. Croll and R. R. Cavanagh 1986 *Quintessence Int* **17** 81.

Anatomy of primary teeth (and relevance to cavity design)

Deciduous teeth differ in several respects from permanent teeth, affecting both the sequelae of dental disease and its management.

Thinner enamel(1) Enamel in 1° teeth is approximately 1 mm thick, which is $\frac{1}{2}$ that of 2° teeth.

Larger pulp horns(2) The pulp chamber in 1° teeth is proportionately larger, with more accentuated pulp horns. $\overline{DE}$—3 pulp horns MB, DB and palatal. $\overline{DE}$—4 pulp horns MB, ML, DB, and DL. These features mean that caries will affect the pulp sooner and there is a greater likelihood of pulp exposure during cavity preparation. Aim for 0.5–1.0 mm penetration of dentine only.

Pulpal outline(3) follows the amelo-dentinal junction more closely in 1° teeth, therefore cavity floor should follow external contour of tooth sinuously to avoid exposure.

Narrower occlusal table Greater convergence of the buccal and lingual walls results in a proportionately narrower occlusal table. This is more pronounced in D than E. Therefore, over-extension of an occlusal cavity or lock can lead to weakening of the cusps.

Broad contact points(4) This makes detection of interproximal caries more difficult, and means that in 1° molars divergence of the buccal and lingual walls towards the approximal surface is necessary to ensure cavity margins are self-cleansing. Isthmus should not extend $>\frac{1}{3}$ intercuspal distance.

Bulbous crown(5) 1° molars have a more bulbous crown form than 2° molars making matrix placement more difficult.

Inclination of the enamel prisms(6) In the cervical $\frac{1}{3}$rd of 1° molars the enamel prisms are inclined in an occlusal direction so there is no need to bevel the gingival floor of a proximal box.

Cervical constriction(7) is more marked in 1° molars, therefore if the base of the proximal box is extended too far gingivally it will be difficult to cut an adequate floor without encroaching on the pulp.

Alveolar bone permeability This is increased in younger children, thus it is usually possible to achieve local anaesthesia of 1° mandibular molars by infiltration alone, up to 6 yrs of age.

Thin pulpal floor and accessory canals(8) may explain the greater incidence of inter-radicular involvement following pulp death.

Root form(9) 1° molars have proportionately longer roots than their permanent counterparts. They are also more flared to straddle the developing premolar tooth. The roots are flattened MD, as are canals within.

Radicular pulp(10) follows a tortuous and branching path, making cleansing and preparation of the root canal system almost impossible. In addition, as the roots resorb, a different approach to RCT is needed for the 1° dentition.

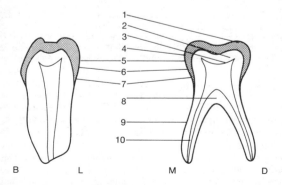

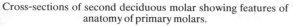

Cross-sections of second deciduous molar showing features of anatomy of primary molars.

Extraction versus restoration of deciduous teeth

Despite a welcome reduction in the prevalence of dental decay, the dilemma of whether to restore or extract a deciduous tooth is still all too familiar. In making a decision a number of factors should be considered:

Age This will influence the likely co-operation for restorative procedures, the expected remaining length of service of the affected tooth, and also the severity of sequelae following early tooth loss.

Medical history Possible sources of blood-borne infection should be avoided in patients with a history of cardiac disease and in the immunocompromised (extractions should be carried out under antibiotic cover). In haemophiliacs, extractions should be avoided and 1° teeth preserved, if possible, until their exfoliation. Prevention is particularly important in these patients.

Motivation and co-operation of parents As it is the parents that bring the child to the surgery, we must explain to them the benefits of maintaining the deciduous dentition. Unfortunately, a small proportion of the population still regard a dentist that fills deciduous teeth as a cowboy—after all, everyone knows that baby teeth fall out!

Caries rate In a child with an otherwise caries-free mouth every attempt should be made to preserve an intact dentition. However, heroic attempts to try and conserve several carious 1° molars will exhaust both the child's and dentist's patience.

Pain If a child is suffering pain from one or more teeth, this needs to be alleviated as soon as possible. If symptom-free, then the dentist will have more time to explore the extent of the lesion(s) and the child's co-operation.

Extent of lesion(s) In 1° molars it is said that destruction of the marginal ridge indicates a high probability of pulpal involvement.[1] If required, pulp therapy will make restoration more difficult and ↓ the prognosis. If more than two 1° molars require pulp therapy, serious thought should be given to extraction rather than restoration.

Position of tooth Although early loss of 1° incisors will have little effect, extraction of C, D, or E will, in a crowded patient, lead to localization of the crowding. Extraction of Es, particularly in the upper arch, should be deferred, if possible, until the first permanent molar has erupted.

Presence/absence of permanent successor Bear in mind the amount of crowding present and the likelihood of spontaneous space closure.

---

1 P. Hobson 1970 *BDJ* **128** 275.

Malocclusion If still undecided, it is worth considering the occlusion. In a particularly crowded case, restoration of a decayed tooth may be indicated if further space loss would mean that extraction of more than one premolar per quadrant, would be required. Much has been written about compensating (same tooth in opposing arch) and balancing (contralateral tooth) extractions, but little consensus emerges.[1] The rationale is that a symmetrical problem is easier to deal with later, but if taken to its logical conclusion, gross caries of D͟| and |E͞ will result in a clearance! Therefore, should be avoided with one exception, loss of C͟, C͞, or D͞ in a crowded patient should be balanced to prevent a centre-line shift.

So much for the theory; in practice, it should be remembered that a happy and co-operative patient is more important long-term. For some children this may mean that the extraction of several carious teeth at the one visit, is preferable to prolonged open combat in the dental chair. For most, restoration is better than running the risk of a frightening GA. Inevitably, the wrong decision will sometimes be made, but we are all human.

1 I. A. Ball *Int J Paed Dent* **3** 179.

Local anaesthesia for children

Although there is no scientific evidence to suggest that 1° teeth are less sensitive than 2° teeth, clinically it is often possible to complete cavity preparation without LA, provided extensive dentine removal is not required. However, Walls *et al.* found that restorations placed without LA did not survive as long as those where LA was used.[1]

General principles
- Explain to patient in terms they will understand what you are trying to do and why.
- Use topical anaesthesia.
- Use fine-gauge disposable needle.
- Always have DSA to assist.
- Hold mucosa taut.
- Slow rate of injection.
- Warn re post-op numbness.

Choice of anaesthetic agent
1st choice: lignocaine 2% with 1:80 000 adrenalin.
2nd choice: prilocaine 3% with felypressin (0.31 iu/ml)—gives less profound anaesthesia.

Dosage depends upon body weight. Under 12 yrs, 1 ml is usually sufficient for an infiltration for 1° tooth. For >12 yrs and for IDB give 2 ml.

Infiltration injection Used for maxillary teeth, mandibular incisors, and for lower deciduous molars before 6̲ has erupted. After 6 yrs of age bone permeability is reduced and an IDB is required. Technique as for adults (p. 633). In children, the malar buttress overlies 6̲, so it is often advisable to deposit some solution over the more permeable bone mesial and distal to this tooth.

Block injection

Inferior dental block Using thumb and forefinger, find the shortest width of ramus. Penetrate about 1 cm into lingual tissues from internal oblique ridge, on a line between thumb and finger. An aspirating syringe is essential.

Posterior superior alveolar block is rarely required in children. If necessary due to failure of infiltration for 6̲, the technique should be modified by depositing solution in the space above buccinator muscle (achieved by advancing needle upwards 1.5 cm in line with the estimated position of 7̲) and massaging it backwards towards the posterior superior alveolar foramen.

Alternative techniques

Intraligamentary injection The purpose-designed syringe has an ultra-short needle and a 'gun' appearance. This makes it helpful for children with a needle phobia, or as a more acceptable alternative to an IDB. In addition, as the lips and tongue are

1 A. W. G. Walls *et al.* 1985 *BDJ* **158** 133.

not anaesthetized it is useful for young or handicapped children, in whom there is a greater risk of post-operative soft tissue trauma.

Jet injection In this technique a jet syringe (e.g. Syrijet) is used to inject LA solution under pressure through mucosa and bone to a depth of about 1 cm. Useful for producing soft tissue analgesia prior to conventional LA injection or for infiltration analgesia.

Restoration of carious deciduous teeth

▶ Primary aim is a relaxed, happy patient rather than the ideal restoration.

Isolation Some authoritative texts advise the routine use of rubber dam (p. 258) for children for all restorative procedures. While rubber dam is well accepted by the good patient where moisture control is less of a problem, it is usually less acceptable to the uncooperative patient in whom it would be most useful. Rubber dam, however, is essential for RCT of permanent teeth which is usually carried out in an older age-group who will accept it. Plastic disposable saliva ejectors are better tolerated than the flanged metal type which are often too big for a child's mouth, restricting access for the dentist.

Local anaesthesia, p. 84.

Instruments

Burs High-speed: pear-shaped bur numbers 330, 525, and short fissure bur number 541. Slow-speed: Eastman pattern numbers 1, 2, are useful for first and second deciduous molars respectively. This design has a collar which limits penetration allowing preparation of a cavity floor that follows the external contour of the tooth. For access use a small bur, and for caries removal use the largest round bur which fits into the cavity.

Handpiece A miniature head handpiece is invaluable. Some children are apprehensive of the aspirator tip, making use of a high-speed water-cooled handpiece difficult, others find the vibration of the slow-speed handpiece distressing, and may confuse it with pain. In these cases a vivid imagination and considerable ingenuity help. It is possible, but time-consuming, to complete cavity preparation with hand instruments.

Materials

As restorations in 1° molars only have to last for the lifetime of the tooth not the patient, GI cements are replacing amalgam as the most popular material. This is due to the adhesive nature of GI which means that cavity preparation can be more conservative and in addition its fluoride releasing ability ↓ recurrent caries.

Principles of cavity design

Outline form should include any undermined enamel. Extension for prevention is now outmoded, but any suspect adjacent fissures should be included. Do not cross transverse marginal ridges unless they are undermined.

Caries removal Caries should be excavated from the ADJ first. If necessary, may need to re-establish outline form to improve access to ensure ADJ is caries free. Then progress to carefully removing caries from floor.

Resistance form/Retention form The completed restoration must be able to adequately resist dislodgement. Usually a 90° cavosurface angle and caries removal suffice.

87

Reasons for failure of restorations in primary teeth

- Recurrent caries, often due to failure to adequately complete caries removal because of flagging patient co-operation. Sometimes it is necessary to be cruel to be kind, but if unable to finish cavity it is better to place a temporary dressing and try again at another visit.
- Cavity preparation does not satisfy the mechanical requirements of the filling material.
- Inadequate moisture control, especially true of GI cements.
- Presence of occlusal high spot.

Many others, but these are the most common.

Useful tips

- Let the child participate by 'looking after' the saliva ejector or cotton wool.
- If the child is nervous give them some control by asking them to signal, e.g. by raising their hand, if they want you to stop.
- If the child's co-operation runs out before the cavity is completed, try and ensure all caries is removed from the amelo-dentinal junction and place a dressing of either ZnO or GI cement. This can then be left for several visits until you are ready to try again.
- Vibration is less of a problem with lower teeth, therefore if possible, start with a lower tooth.
- Don't try to do too much at one visit, quadrant conservation is not really feasible in an 8 yr-old!

Communication

▶ It is important to explain to the child what you are trying to do, and why, in terms they can understand.

It may be helpful to describe some of the instruments we use in ways that can make them seem less threatening to a child, e.g.

handpiece and bur	bumble bee
	tooth tickler
handpiece and prophylaxis cup	electric toothbrush
aspirator tip	vacuum cleaner/hoover
rubber dam	tooth raincoat
saliva ejector	straw
	curly-wurly (coiled type only)
air from 3-in-1	wind
fissure sealant	plastic coating
etchant solution	tooth shampoo/cleaner
	lemon juice
cotton wool roll	snowman
dental light	the sun

Class I in deciduous molars

See p. 80 for anatomy of 1° molars and effect upon cavity design. Have all necessary instruments and filling materials ready so that appointment is kept as short as possible.

- Explain and show child (and parent) what you are trying to do.
- LA if required (p. 84).
- In small cavity will need to gain access and this is most easily done with a high speed handpiece and pear-shaped bur. The outline can then be established and caries removed.
- In larger cavities an excavator or large round bur can be used to start caries removal from the walls. Any undermined enamel should be cut back.
- If caries deep, stop and re-assess whether pulpotomy (p. 100) required.
- Check retention and walls are caries-free.
- Wash and dry cavity.
- Line with hard-setting calcium hydroxide, unless cavity is shallow.
- Place amalgam incrementally, and carve. If using glass ionomer see p. 92.
- Check occlusion.
- This is usually a good opportunity to reinforce any preventive advice, but keep it brief.
- ▶ Praise child, and if bribery has been used don't forget the sticker/badge/toothbrush.

Polishing of amalgam restorations in 1° molars is unnecessary.

Cross-sectional view Class I restoration (bucco-lingually).

Class II in deciduous molar—amalgam

See p. 80 for anatomy of 1° molars and effect upon cavity preparation. Class II cavities are designed for the treatment of interproximal caries and consist of three parts.

Occlusal key is designed to retain the restoration and eliminate any occlusal caries. Should be prepared first and is identical to Class I cavity preparation, p. 88.

Isthmus joins the occlusal key with the interproximal box. It is the part of the filling most prone to fracture. Dimensions of the isthmus are a balance between:

adequate depth without risking pulpal exposure (1.5–2 mm)	adequate width without weakening cusps ($\frac{1}{3}$–$\frac{1}{2}$ distance between cusps)

Eastman pattern burs (1 for D, 2 for E) are useful for getting balance right.

Interproximal box To allow access for caries removal. Ideally should just extend into embrasures and walls should converge occlusally.

Minimal caries with marginal ridge intact.
- Follow steps for small occlusal cavity.
- When occlusal cavity (key) complete extend it towards approximal surface. Most texts advise retaining some enamel interproximally to protect adjacent tooth, but this is often easier said than done.
- Establish floor of box, taking care not to extend beyond maximum bulbosity of tooth.
- Fracture away remaining approximal enamel with hand instruments.
- Complete preparation of box following external contours of tooth and 90° cavosurface angle.
- Remove caries.
- Check retention.
- Place hard-setting calcium hydroxide lining, unless shallow.
- Position narrow matrix band and wedge.
- Condense amalgam and carve with matrix in place.
- Check occlusion.

More advanced caries marginal ridge broken down (↑ likelihood of pulpal involvement)
- Enquire about symptoms.
- Take X-ray to assess pulpal position and root morphology.
- If pulp healthy proceed (if suspect pulpal involvement, see p. 100).
- LA preferable, but less essential than for minimal preparation.
- Establish outline, including occlusal key, and define walls and floor of interproximal box. If cusps severely weakened consider stainless steel crown, p. 94.
- Excavate caries carefully, moving from walls to floor.

- Check retention (if insufficient available consider adhesive restorative material).
- Place lining, matrix, and amalgam as for minimal cavity.

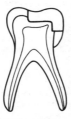

Cross-section of D̄mo restoration (mesio-distally).

Class II in deciduous molars—alternative techniques

Glass ionomer and cermet restorations

Advantages: adhesion to enamel and dentine, leaches fluoride.

Disadvantages: more wear seen compared to amalgam, but for definitive life of 1° molar this is relatively unimportant. Technique is more critical than most dentists realize.

Class I Follow directions for amalgam restoration, except line only the smallest area necessary for pulp protection to maximize adhesion. Place varnish when filling complete. Cermet restorations (Ketac Silver) may be finished immediately, but delay a further 5 min for Chemfil and 10 min for Ketac. Alternatively, can protect with light-cure resin which provides lubrication for finishing and is then cured when complete. Or place occlusal detection wax over tooth to protect from moisture (this can be lightly burnished to help establish occlusal contour). Varnish should be placed once cement is set and wax removed.

Class II involving approximal surface only
- Gain access through marginal ridge and remove caries and unsupported enamel.
- Undercut slightly the buccal and lingual walls and place a retention groove in the dentine of the floor.
- Wash and dry cavity.
- Wipe a little Vaseline around a narrow matrix band and position with wedge.
- Insert GI (using powder as a separator on instruments) and remove any gross excess.
- Apply varnish.
- Wait before finishing restoration (see above) and check occlusion. GI may also be used to restore a conventional Class II cavity as described in the section for amalgam provided occlusal loading on restoration will not be excessive.

Composite restorations

Modifications of conventional cavity design have been found to be less successful than hoped.[1] However, when used in classical cavities encouraging results over the short time-frame of the deciduous dentition have been achieved.[2] Unfortunately, very technique sensitive which restricts their application in children and they lack the fluoride releasing ability of GI.

1 T. R. Odenburg *et al.* 1985 *Pediatric Dentistry* 7 96.   **2** M. W. Roberts *et al.* 1985 *Pediatric Dentistry* 7 14.

Stainless steel crowns

This term is a misnomer, as only nickel chromium crowns are now manufactured, but the name has stuck.

Indications
- Badly broken down 1° molar.
- After pulp therapy in 1° molars.
- As interim measure for 2° molars where crowns are required but the patient is too young.
- Temporary coverage during preparation of cast crown for premolar or 2° molar.

SS crowns were used as a temporary measure following trauma to permanent incisors, but since the introduction of composites, not surprisingly, their use has declined.

Instruments High-speed tapered diamond bur (e.g. 582) and diamond occlusal wheel. Straight handpiece and a stone. Slow-speed handpiece and burs as required. Johnstone contouring pliers (number 114) and Abel pliers (number 112). Crown scissors, dividers, and selection of suitable crowns.

Technique
Stainless steel crowns rely for retention only on a tight adaptation at the gingival margin of the preparation, ∴ taper of walls is not critical.
- LA, not required post pulpotomy, but still wise to use a little topical LA on gingivae.
- Measure M–D length with dividers to aid crown selection.
- Remove caries.
- Occlusal reduction (approx 1 mm) with occlusal wheel, following cuspal planes.
- Approximal reduction (approx 20° from vertical) using tapered diamond, without producing a ledge at gingival margin.
- Remove buccal and lingual bulbosites only sufficient to seat crown.
- Select and trim crown so that it just extends into gingival margin.
- Check height and occlusion.
- Use 112 plier to adapt contact points and 114 plier to crimp margins and smooth margins with stone.
- Cement with zinc polycarboxylate cement.

Technique for 2° molars is similar.

Success rates Dawson *et al.* after 3 yrs follow-up found 75% of 2-surface amalgams were replaced, compared with 13% of SS crowns on deciduous molars.[1]

---

1 C. R. Dawson *et al.* 1981 *Journal of Dentistry for Children* **48** 420.

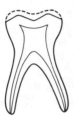

Occlusal reduction

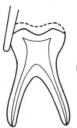

Mesial (and distal) reduction

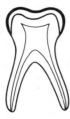

Completed crown

Preparation for stainless steel crown.

Class III, IV, and V in primary teeth

Carious 1° incisors and canines are seen less frequently than molars and are therefore indicative of a high caries rate (see rampant caries).

Management Objectives are relief of pain and prevention. Aesthetics are less important.

Treatment options include
- Extraction
- Topical fluoride (2% sodium fluoride) and observation. Intervene if caries progresses.
- Discing (safer to use flat fissure No. 1 bur, than disc) plus topical fluoride.
- Restoration, usually there is insufficient hard tissue for adequate retention, ∴ adhesive materials are preferable.

Class III restoration Similar technique as for permanent incisors, but omit incisal retention groove and use glass ionomer.

Class IV restoration If restoration is essential the greater strength of composite is required. Placement of polycarboxylate crowns is difficult to justify in the caries-prone child.

Class V restoration Remove caries with inverted cone and restore with glass ionomer.

Rampant caries

Definition Rapid carious attack involving several teeth, including those surfaces that are usually caries-free.

Can affect both 1° and 2° dentitions.

Aetiology Frequent ingestion of sugar +/− reduced salivary flow.

Nursing bottle caries Associated with prolonged bottle feeding of a sugar-containing drink. Classically, questioning of the parent reveals a history of the child being given a bottle at night containing a well-known blackcurrant drink, but this is less common now. Characteristically, starts with the maxillary 1° incisors, but in more severe cases the molars are also involved. The mandibular incisors are relatively protected by the tongue and saliva.

Nursing caries Attributed to prolonged on-demand breast-feeding, especially at night, due to the lactose in breast milk.[1]

Radiation caries Radiation for head and neck cancer may result in fibrosis of salivary glands and ↓ salivary flow. Patients often resort to sucking sweets to alleviate their dry mouth, which exacerbates the problem.

Rampant caries may also be caused by the prolonged and frequent intake of sugar-based medications; however, both pharmaceutical companies and doctors are becoming more aware of the problem and the number of alternative sugar-free preparations is increasing. See p. 128 for list.

Management
- Removal of aetiological factors (education, artificial saliva).
- Fluoride rinses for older age groups (daily 0.05%).
- 1° dentition—may need to extract teeth of poor prognosis and concentrate on prevention for permanent dentition.
- 2° dentition—need assessment of long-term prognosis for teeth. Final treatment plan should be drawn up in consultation with orthodontist.

1 G. J. Roberts 1982 *Journal of Dentistry* **10** 346.

Deciduous molar pulp therapy

NB Deciduous molar roots resorb.

Where the carious process has jeopardized pulpal vitality there are two alternatives (1) extraction (2) pulp therapy.

Indications and contraindications See Extraction versus restoration, p. 82.
▶ Any medical condition where a focus of infection is potentially dangerous (e.g. congenital heart disease, rheumatic fever) is an absolute contraindication to pulp therapy. Extraction under antibiotic cover as advised by the child's physician is necessary.

Pulp therapy is preferable to extraction in children with bleeding disorders. Tooth must be restorable following pulp therapy.

Diagnosis of pulpal state is difficult, as not only is a child's perception of pain less precise than an adults, but the clinical picture may be complicated by death of one root-canal whilst the other(s) remain vital.
Consider: history, clinical examination (? abscess present), X-rays, condition of pulp on exposure (no bleeding—non-vital, profuse haemorrhage—irreversible pulpitis), size of carious lesion.

Definitions

Pulpotomy: removal of coronal pulp and treatment of radicular pulp.

Pulpectomy: removal of entire coronal and radicular pulp.

Principles of treatment Attempting to retain the vitality of the pulp in 1° molars is not recommended because (1) pulpal involvement is more likely, (2) it is difficult to accurately determine the likely condition of the pulp, and (3) calcium hydroxide frequently leads to internal resorption. Therefore, direct pulp capping is only advisable for small traumatic exposures. Pulpectomy is C/I because of the ribbon-shaped roots, numerous accessory canals, physiological resorption of roots, and the risk of damage to the underlying tooth germ. Pulpotomy remains the treatment of choice for 1° molars:

```
                              ⟋ one-visit pulpotomy (with LA)
VITAL PULP—devitalization
                              ⟍ two-visit pulpotomy (without LA)
```

NON-VITAL PULP—non-vital (mortal) pulpotomy

Materials The most commonly used medicaments are Formocreosol (for one-visit pulpotomy), can be mixed by a pharmacist, (a 1 to 5 dilution has been used successfully)
    formalin 19 ml creosol 35 ml glycerin 25 ml water 21 ml
Easlick's devitalizing paste (for two-visit pulpotomy)
    paraformaldehyde 1 g lignocaine 0.06 g
    carmine(colour) 0.01 g carbowax 1500 1.3 g propylene glycol 0.5 ml

Beechwood creosote (for non-vital pulpotomy) Standard
   British Pharmacopoeia preparation.

▶ These materials are caustic, ∴ take care.

101

Success rates vary from 50% for non-vital teeth to over 90% for
vital pulps.[1]

1 M. E. J. Curzon and B. K. Drummond 1986 *Dental Advertiser* 51 14.

Pulpotomy techniques for vital pulps

In 1° molars the relatively larger pulp results in earlier pulpal involvement, therefore devitalization and fixation of the pulpal tissues gives more consistent results than techniques that attempt to retain vitality, e.g. indirect pulp capping. There are two alternative approaches:

- one-visit formocreosol pulpotomy which requires LA.
- two-visit pulpotomy for which LA is not required.

Choice of technique depends upon whether the child will accept LA, and less importantly, upon the time available.

One-visit formocreosol pulpotomy

This method fixes most of the radicular pulp, but the apical part may be unaffected by the medicament.

- Give LA.
- Complete cavity preparation and excavate caries.
- Remove roof of pulp chamber.
- Amputate coronal pulp with a large excavator or sterile large, round bur.
- Wash chamber and arrest bleeding with cotton wool dampened with sterile saline.
- Place cotton-wool pledget moistened with formocreosol on exposed pulp stumps for 5 min and then remove.
- Apply antiseptic dressing (eugenol + formocreosol 50:50 mixed with zinc oxide powder).
- Place zinc phosphate lining (to provide firm base for amalgam).
- Restore tooth.

Problems *Inadequate LA*: either repeat LA or change to two-visit technique.
Necrotic pulp: proceed as for non-vital tooth.
Profuse haemorrhage of pulp: keep applying pressure until bleeding stops and then continue with procedure except that pledget should be sealed in place for 1 week. Alternatively, can change to non-vital technique.

Alternative medicaments Beechwood creosote, N2, and gluteraldehyde have also been successfully used.

Two-visit devitalization pulpotomy

First visit

- Complete cavity preparation if possible and remove caries.
- Apply devitalizing paste to exposure on a small pledget of cotton wool and seal with a temporary dressing (either ZOE or glass ionomer).
- Re-appoint 1–2 weeks later (warn parent that some minor discomfort that day is to be expected, but if child experiences any other symptoms or dressing lost, return immediately).

Second visit

- Remove temporary dressing and cotton wool.
- Remove roof of pulp chamber.
- Proceed as for one-visit formocreosol technique above.

Problems Pulp bleeding and/or sensitive at second visit: will need to repeat first stage, but to ↑ likelihood of success, enlarge exposure site.

Alternative medicaments Ledermix, contains steroid, but not in sufficient concentration to have a systemic effect.

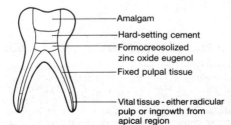

—Amalgam

—Hard-setting cement

—Formocreosolized zinc oxide eugenol

—Fixed pulpal tissue

—Vital tissue - either radicular pulp or ingrowth from apical region

Vital pulpotomy—one-visit technique.

Pulpotomy techniques for non-vital pulps

This method removes infected coronal pulp and disinfects radicular pulp, so allowing normal root resorption to proceed.

First visit
- LA may be required as part of pulp could still be vital.
- Complete cavity preparation and removal of caries.
- Remove roof of pulp chamber and excavate pulpal debris.
- Place pledget of cotton wool moistened with beechwood creosote in pulp chamber.
- Seal with temporary dressing (glass ionomer or ZOE).
- Arrange next appointment 1–2 weeks later.

Second visit
- Enquire re symptoms; if none proceed.
- Remove temporary dressing and cotton wool.
- Place antiseptic dressing (50:50 formocreosol and eugenol mixed with zinc oxide powder) and press down into root canals.
- Restore tooth.

Problems *Vital &/or sensitive tissue encountered*: place devitalizing paste and seal for 1–2 weeks, before proceeding with non-vital pulpotomy.

Abscess formation during treatment: either repeat (consider whether need to incise abscess) or extract tooth.

Alternative medicaments Camphorated monochloral phenol, formocreosol, and 'Kri' liquid have all been suggested.

Technique for abscessed teeth Acute abscesses require drainage to relieve symptoms. This can be achieved either by leaving tooth on open drainage for 1 week before proceeding as above (this is more applicable to upper teeth), or by incising abscess under topical LA. If chronic abscess drainage may be occurring through a sinus; if so, proceed straight to first-visit technique. If drainage occurring through occlusal cavity, placement of a seal may lead to an exacerbation in symptoms, therefore always warn parents to return if any problems.

Pulp therapy for primary anterior teeth

Usual treatment is extraction, as A and B are exfoliated before patient is able to co-operate satisfactorily with more complicated treatment. However, C is exfoliated later and unilateral loss may result in a centre-line shift, ∴ pulp treatment is indicated for some patients. The root canal morphology is amenable to pulpectomy and the canal should be cleaned using files, with care (remember underlying successor). A resorbable filling material, e.g. calcium hydroxide or ZOE should be used.

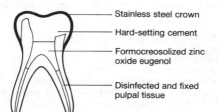

Non-vital pulpotomy with beechwood creosote.

Dental trauma

▶ If evidence of head injury transfer patient to hospital immediately.

Note
- By 15 yrs of age 33% of children have experienced at least one episode of dental trauma.[1]
- Prognosis ↑ with good immediate treatment, ∴see patient as soon as possible.
- Avulsed permanent teeth should be re-planted immediately.
- Child and parent may be upset, ∴ handle accordingly and defer any non-urgent treatment.
- Take good notes for future reference and medico-legal purposes.
- If crown # this will have dissipated most of the energy of impact, ∴ root # less likely.

History Need important and relevant details only, can leave fuller history to second visit.
- Was there any loss of conciousness or amnesia? Any neurological signs (p. 488)?
- How long ago?
- Where? Does patient need a tetanus booster, if so refer to GMP or hospital.
- How? Be alert to the possibility of other injuries.
- Whereabouts of any tooth fragments? They may have been inhaled or be embedded in the lip.
- Medical and dental history may modify treatment plan.

Examination Gently wash away any blood first.
- Check for facial # &/or lacerations.
- Intra-orally look for soft-tissue lacerations, dento-alveolar #, and damage to teeth. Examine traumatized teeth for mobility.
- Check occlusion especially if any teeth displaced.
- Vitality testing is of little value immediately post-trauma as may take 3 months for vital pulp to recover and respond.
- Take radiographs of affected teeth to check for root #. Soft-tissue films and facial views if indicated.

Aims of treatment

$1°$ dentition—preserve integrity of permanent successor.
$2°$ dentition—(1) preserve intact dentition, (2) maintain pulp vitality.

Principles of treatment

Emergency Rx
- Elimination of pain.
- Protection of pulp.
- Reduction and immobilization of mobile teeth.
- Suturing of soft tissue lacerations (IO—3/0 resorbable suture (Dexon, Vicryl) EO—refer to hospital).

---

1 J. E. Todd and T. Dodd 1985 *Children's Dental Health in UK*, HMSO.

- ?antibiotics, ?tetanus, ?analgesics, ?chlorhexidine mouthwash.

Intermediate Rx
- Pulp therapy.
- Consider orthodontic requirements and long-term prognosis of damaged teeth.
- Semi-permanent restorations.
- Keep under review, usually 1 month, 3 months, and then 6-monthly for 2 yrs.

Permanent Rx
- Usually deferred until >16 yrs (to allow pulpal and gingival recession and ↓ likelihood of further trauma), e.g. PJC, post and core crown.

Classification of tooth injuries

Several exist, but none is comprehensive and all classify using Roman numerals, which results in confusion. Better to describe in words the actual injuries sustained.

Prevention

- Prevalence ↑ as the o/j ↑ (>9 mm prevalence doubles), ∴ ? early orthodontics.
- Mouthguard for sports (vacuum formed thermoplastic vinyl best, triple thickness).

▶ Be alert for evidence of non-accidental injury (p.108).

Non-accidental injury (NAI)

All professionals involved with children need to be alert to the possibility of NAI (a term which is now favoured rather than child abuse).

The following signs are associated with NAI:

- Usually younger children are involved.
- The presenting injuries may not match the parent's account of how they were sustained.
- Attendance at a surgery or clinic for treatment of the injury often delayed.
- Bruises of different vintages found on examination.
- Ear pinches, and frenal tears in children <1 year of age are pathognomic.

Management

In most areas local guidelines will have been drawn up. These can usually be obtained from Social Services or the Paediatric Department at the local hospital. A copy of these should be kept in every practice. The BDA also has included advice about handling NAI in their advice sheet *Ethical and Legal Obligations of Dental Practitioners.*[1]

If the possibility of child abuse is suspected, the practitioner should refer to the local protocol, or contact either the Consultant in Public Dental Health or the duty Social Worker at the local Social Services Department for advice. If the referral is made to the Social Services Department this will need to be confirmed in writing. The child's medical practitioner should also be informed. If a child presents with serious injuries, which are suspicious, they should be referred to the nearest Accident and Emergency department, and the Department informed in advance of the situation before the child's arrival.

Tact is required in dealing with the patient's family. It is better to concentrate on treating the patient's injuries, referring them on to the experts who will fully evaluate the case before making a diagnosis of NAI.

---

1 BDA Advice Sheet *Ethical and Legal Obligations of Dental Practitioners.*

Injuries to primary teeth

Eight per cent of 5 yr-olds have experienced dental trauma,[1] mainly at toddler stage. As alveolar bone is more elastic the younger the child, the most common injuries are loosening +/− displacement. Crown and root # rare.

Management

If radiographs are required may have to get mother to hold child and film. Alternatively try placing a periapical film between the teeth (like an occlusal view) and angle the beam at 45°.

Need to consider the effect of any proposed treatment upon permanent successor. Splinting of 1° incisors is exceedingly difficult. When in doubt—extract 1° tooth! (For definitions, see p. 116.)

Concussion of tooth Rx: reassurance and soft diet.

Subluxation If tooth near to exfoliation extract. Otherwise, soft diet (for about 1 week). May become non-vital, ∴ keep under observation.

Luxation Extraction indicated unless crown displaced palatally (away from permanent tooth), tooth is not in danger of being inhaled, and does not interfere with occlusion. If crown displaced labially ↑ risk of damage to underlying 2° incisor.

Intrusion Most common injury (>60%[2]). If X-ray suggests that tooth has been forced into follicle of 2° tooth, extract 1° incisor. If not, can afford to leave tooth and wait to see if spontaneous eruption will occur (as it usually does). Unfortunately, pulpal necrosis often follows, necessitating either pulp treatment (p. 104) or extraction. Should the tooth fail to erupt, then extraction is indicated. It is prudent to warn parents about possible damage to underlying permanent tooth.

Extrusion If >1–2 mm, extract as difficult to splint and will probably become non-vital.

Avulsion Do not replant.

Crown # Rare. Minimal # can be smoothed and left under observation. Larger # either restore with composite +/− RCT if pulp involved, or extract.

Root # Provided not displaced and little mobility, advise soft diet and keep under review. If coronal fragment displaced or mobile, extract, but leave apical portion as it will usually resorb.

Sequelae of trauma—*Deciduous dentition*

Discoloration Grey—if in early post-trauma period pulp may be vital and discoloration reversible. Greying later indicates pulp necrosis. Yellowing of tooth is suggestive of calcification of pulp—no treatment required.

1 J. E. Todd and T. Dodd 1985 *Children's Dental Health in UK*, HMSO. **2** J. O. Andreasen 1981 *Traumatic Injuries of the Teeth*, Munksgaard.

Ankylosis Rx: extraction to prevent displacement of 2° incisor.

Pulp death Rx: RCT or extraction.

Permanent dentition

In about 40% of cases trauma to 1° tooth affects underlying developing successor.[1] Effect depends upon stage of development, type of injury and severity, treatment and pulpal sequlae. Can cause hypomineralization, hypoplasia (↑ likelihood <4 yrs and more severe injury), dilaceration, severe malformation, and arrest of development.

1 J. A. Hargreaves 1981 *The Management of Traumatized Anterior of Children*, Churchill Livingstone.

Injuries to permanent teeth—
crown fractures

Prevalence: 26–76% injuries.

Enamel only

For small enamel # smooth with white stone.

Enamel and dentine

Need to protect exposed dentine, preferably with a hard-setting calcium hydroxide cement and an acid-etch retained composite. If time permits, this can be done with a crown former to restore tooth contour. Keep under review. Veneer or PJC can be considered later. If # near to pulp, treat as for pulp involvement.

Acid-etch composite tip technique
- Place rubber dam, if possible.
- Place hard-setting calcium hydroxide on exposed dentine. No need to bevel enamel.
- Using contralateral tooth as guide select a cellulose acetate crown former.
- Trim crown former to within 1–2 mm of # line.
- Etch enamel for 20 sec, wash, and dry.
- Place bonding resin and cure.
- Put sufficient composite and a little extra into crown former and position.
- Allow to cure and remove crown former.
- Trim, using soflex discs. Shofu points are useful for palatal aspect.
- Check occlusion.

Enamel, dentine, and pulp

Rx depends upon size of exposure, state of root development (1 root radiographically complete 10–11 yrs, histologically 14–15 yrs), time since injury and other injuries (e.g. root #). If apex open ↑ blood supply to pulp, ∴ likelihood of pulp death ↓. This is advantageous as treatment should be directed towards retaining vitality of apical ⅓rd to allow root closure to continue. If pulp non-vital, see p. 122. Otherwise, treatment alternatives are:

| | | SIZE OF EXPOSURE | |
		<1 mm	>1 mm
ROOT	*Apex open*	pulp cap	pulpotomy
DEVELOPMENT	*Apex closed*	pulp cap if seen immediately, otherwise pulpectomy	pulpectomy

Pulp cap with non-setting calcium hydroxide (mixed with water) followed by hard-setting calcium hydroxide (e.g. Dycal), and place composite tip. Review vitality.

Pulpotomy requires rubber dam and LA.
- Open up pulp chamber and amputate coronal pulp to cervical constriction with sterile bur/sharp excavator.

- Wash with sterile water.
- Place non-setting calcium hydroxide and restore tooth with polycarboxylate cement and composite.
- Leave 6–8 weeks, then review symptoms and vitality. Investigation of the presence or not of a calcific barrier is not necessary. However, if desired, a hard-setting calcium hydroxide can be substituted.
- If tooth becomes non-vital, see p. 122.

All pulpotomized teeth should be kept under long-term review as pulp necrosis and calcification are common sequelae. Even if the tooth maintains its vitality RCT may be necessary once apexification is complete, if a post and core crown is required. Success rates of 79% for pulpotomies have been reported.[1]

1 M. J. Gelbier and G. B. Winter 1988 *BDJ* **164** 319.

Root fractures

Prevalence: <10% injuries permanent dentition.

NB

One periapical X-ray—75% chance of detecting root #
Two periapical X-rays—90% chance of detecting root #

The prognosis for this type of injury depends upon whether the # line communicates with the gingival crevice. Actual treatment depends on position of #.

Apical $\frac{1}{3}$rd Usually no treatment required. However, tooth should be kept under observation as death of coronal $\frac{2}{3}$rds of pulp may occur. Only need to prepare canal to # line as apical $\frac{1}{3}$rd usually retains vitality. Prognosis good. If extraction required, apical $\frac{1}{3}$rd can be left *in situ* to preserve bone.

Middle $\frac{1}{3}$rd In majority of cases tooth is loosened, ∴ in order to achieve repair of # line with calcific tissue, the tooth should be splinted for 8–12 weeks. If coronal part not displaced loss of vitality unlikely. Where coronal fragment displaced, re-position, splint, and RCT to # line. Delay in treatment ↓ prognosis. If extraction required, consider leaving apical portion *in situ*.

Coronal $\frac{1}{3}$rd By definition, # in this group communicate with the gingival crevice allowing ingress of bacteria into pulp. Emergency treatment consists of a choice between either extraction of both parts of tooth or, preferably, removal of the coronal fragment, RCT of remainder and then placement of a dressing which will prevent gingival tissues overgrowing root surface. This can be achieved by placing a temporary post-retained crown, but replacement of the coronal fragment using a dentine-bonding agent has been described. For permanent treatment can place a post and core crown. However, if # extends below alveolar crest need improved access for crown fabrication, there are two alternatives:

Ostectomy/gingivectomy	*Orthodontic extrusion*
Gives quicker result	Cervical circumference of
Need post and diaphragm	crown ↓ compared to
Tend to get perio pocket	contralateral tooth
Leads to ↓ gingival width	Better crown:root ratio
	May ↑ attached gingiva

Orthodontic extrusion can be accomplished, using a URA with a buccal arm which engages either an attachment bonded on to labial surface of a temporary post and core crown or any available enamel. Forces of 50–100 g should be used. When sufficient extrusion has been achieved, retain for at least 3–6 months before fabricating a permanent restoration.

If tooth extracted a P/- will need to be fabricated (p. 366).

Oblique Provided # extends <4 mm below alveolar crest can treat as coronal #. Otherwise, if possible, extract coronal portion only, leaving apical portion *in situ* to preserve bone.

Vertical Extract.

Luxation, subluxation, intrusion, and extrusion

Prevalence: 15–40% injuries.

Definitions

Concussion Injury to supporting tissues of tooth, without displacement.

Luxation Displacement of tooth (laterally, labially, or palatally).

Subluxation Actually means partial displacement, but is commonly used to describe loosening of a tooth without displacement.

Intrusion Is displacement of tooth into its socket. Often accompanied by # of alveolar bone.

Extrusion Is partial displacement of tooth out of its socket.

Treatment

Concussion Reassurance and soft diet.

Luxation Need to re-position tooth as soon as possible. Give LA and use fingers to push back into place. Then tooth should be splinted for 2–3 weeks. If there has been a delay of more than 24 hrs since the injury, manual reduction is unlikely to be successful. In these cases the tooth can be re-positioned orthodontically. If the displaced tooth is interfering with the occlusion an URA, with buccal capping, should be fitted as soon as possible. Loss of vitality is a common sequelae following luxation, leading to inflammatory resorption (p. 121), therefore keep under review.

Subluxation If minor, no treatment other than advising a soft diet is necessary. If mobile, splint for 1–2 weeks and watch vitality.

Intrusion Teeth with immature roots are likely to erupt and therefore no immediate treatment is required. However, teeth with closed apices will need orthodontic extrusion and this should be started as soon as possible to facilitate access for RCT. Pulp death +/− root resorption can ensue rapidly after injury and early pulp extirpation and placement of a calcium hydroxide dressing is advisable. In immature teeth the ↑ blood supply means that loss of vitality is less likely, but not impossible.

Extrusion The affected tooth should be re-positioned under LA with digital pressure and splinted for 1–2 weeks. Again, loss of vitality is a common sequelae, so the tooth should be observed for any signs of resorption or pulp death.

If any of the above occur in conjunction with # of the alveolar bone, the splinting period should be ↑ to 3–4 weeks to aid bony healing. If, however, the socket is comminuted, splinting needs to be extended to 6–8 weeks.

Splinting

Indications
- To stabilize a loosened tooth to allow periodontal fibres to repair. To encourage fibrous rather than bony healing (ankylosis), a short splinting time is recommended (<3 weeks).
- To stabilize a root # and encourage healing with calcified tissue. Splinting for 8–12 weeks is indicated.

Methods
Direct Constructed on patient. An almost infinite variety have been described, but the following are the most popular:
- Acid-etch splint with composite/acrylic/epimine resin +/− wire/orthodontic attachments.
- Interdental wiring. Of historical interest only.
- Lead foil +/− cement. Useful in Australian Outback, but better alternatives available in most dental surgeries!

Indirect Requires an impression of traumatized mouth and involves some delay (few hrs/days) before splint can be fitted. However, this type of splint is removable, allowing an assessment of mobility or firmness, which is valuable in cases of re-implantation. The more common types are:
- URA with cribs 6|6 and occlusal coverage.
- Vacuum formed thermoplastic polyvinyl acetatepolyethylene (!) 'Drufomat' type.

Factors affecting choice of splint
- Type of injury and therefore length of time splint required. For example, root # will need 8–12 weeks of splinting, ∴ composite and wire splint advisable. For re-planted tooth need short splinting time, ∴ thermoplastic mouthguard type may be indicated.
- Dental status of patient, e.g. if 6EDC 1|1 CDE6 present and both 1|1 traumatized need full coverage acrylic splint.
- Facilities and time available.

Management of the avulsed tooth

Exarticulation = avulsion. Prevalence: 0–16% injuries.

Factors affecting prognosis Success depends upon re-establishment of a normal periodontium.

- *Time from loss to re-implantation*. As PDL cells rarely survive >60 min extra-orally, immediate replacement (by whoever is available at the scene) is the treatment of choice.
- *Storage medium*. Prognosis saliva > milk > water > air.
- *Splinting time*. Prolonged splinting will promote ankylosis.
- *Viability of pulp*. Seepage of pulp breakdown products into PDL will contribute towards the development of inflammatory resorption. Although re-vascularization is possible in a tooth with an open apex which is replaced within 30 min, those teeth with closed apices and longer extra-alveolar times should be considered non-vital.

Immediate treatment (if avulsed tooth not already replaced).
- Avoid handling root surface. If tooth contaminated, hold crown and agitate gently in saline.
- Place tooth in socket. If does not readily seat, get patient to bite on gauze for 15–20 min.
- Compress buccal and lingual alveolar plates.
- Take impression for thermoplastic splint (will need to support tooth with a piece of wire curved over incisal edge and held by DSA). If facilities for indirect splint not readily available, splint a curved piece of 0.7 mm wire to acid-etched enamel of affected and adjacent teeth using temporary crown material as this is less traumatic to remove than composite.
- Prescribe antibiotics, chlorhexidine mouthwash, and arrange tetanus booster.

Intermediate treatment (1 week later)
- Review splinting. Stop if tooth appears firm, continue for further week if still mobile. (If still mobile after 2 weeks, check nothing has been overlooked, e.g. root # or loss of vitality—in these cases prognosis is poor).
- If apex closed (or tooth with open apex, but extra-alveolar period >30 min) extirpate pulp, clean canal, and place calcium hydroxide to try and prevent inflammatory resorption.
- Keep teeth with open apices under close observation, so that at the first sign of pulp death RCT can be instituted. Waiting for radiographic evidence of inflammatory resorption is too late.
- Keep tooth under review. If calcium hydroxide placed in canal should be renewed every three months until apical barrier formed and then GP filling placed.

Prognosis If above procedure followed:[1]
- Incomplete root formation 66% survive 5 years.
- Complete root formation 90% survive 5 years.

1 J. O. Andreasen 1992 *Atlas of Replantation and Transplantation of Teeth*, Mediglobe.

Sequelae

Surface resorption occurs as a result of minor trauma to PDL cells. Usually is self-limiting and affected areas are repaired by cementum. No treatment.

Replacement resorption (ankylosis) caused by damage to PDL cells during extra-alveolar period and promoted by prolonged splinting. It appears that the absence of vital periodontal ligament allows resorption of the root and replacement by bone. In growing child results in infra-occlusion of affected tooth. Once started is usually progressive, resulting in the eventual loss of the tooth.

Inflammatory resorption Development of inflammatory resorption is dependent upon the presence of both damage to the periodontal ligament and breakdown products from pulp necrosis diffusing through the dentinal tubules to the PDL. Occurs rapidly, as soon as 1–2 weeks after injury. Once evident radiographically prognosis is poor, as it is progressive and treatment is not always successful. Inflammatory resorption can be prevented by extirpation of the pulp as soon as is practicable after injury and placement of non-setting calcium hydroxide. If resorption is halted a GP root filling can be placed.

Delayed presentation Where viability of PDL cells doubtful, Andreasen has suggested chemical treatment of the root surface with F to limit resorption.[1] Following RCT with GP, the tooth is immersed in 2.4% sodium fluoride solution for 20 mins. Then tooth is replanted and splinted for 6 weeks. As some replacement resorption is inevitable, perhaps best limited to adults. If the extra-alveolar period is greater than 24 h, leave and consider instead whether the resulting space should be maintained with a P/- (p. 336).

Pulpal sequelae following trauma

Damage to the pulp can occur either as a result of disruption of the apical vessels, exposure of the pulp by a crown or root # or by haemorrhage and inflammation of coronal pulp resulting in strangulation.

Pulp death Remember that no response to vitality testing indicates damage to the nerve supply of a tooth, but not necessarily to the blood supply. Therefore following trauma, should assess vitality in the light of symptoms, tooth colour, mobility, presence of buccal swelling and radiographic evidence. Except where a tooth has been re-planted it is best to adopt a wait and see approach, if in doubt about vitality. When pulp death has occurred, subsequent treatment depends upon whether the apex is closed (p. 112) or open.

RCT of teeth with immature apices As achievement of an apical seal is difficult in a tooth with an open apex, treatment should aim to allow apexification to continue. Under rubber dam, the necrotic pulp should be extirpated. The working length is set 1–2 mm short of the radiographic apex (unless vital pulp tissue is encountered earlier) and narrow files used in order to negotiate any undercuts. The canal should then be filled with a radiopaque non-setting calcium hydroxide, e.g. Reogan (alternatively can use the catalyst from Dycal), to the apex and sealed. The calcium hydroxide should be replaced every 3 months, until a calcific apical barrier is detectable by gentle probing with a paper point. Then the canal can be filled. Usually, because of the width of the canal, a large GP point (can use a conventional point upside down) is required. This should be warmed in a flame before pressing into place and then lots of laterally condensed points used to obtain a good seal. The average time for a calcific barrier to be formed is 9 months.[1] A 5-year survival rate of 86% has been reported.[2] Clinical experience would suggest that RCT of incisors in children is often complicated by intractable infection of the canal. This may be due to the ↑ patentcy of the dentinal tubules. Polyantibiotic pastes can be tried, but a cheaper alternative is to crush metronidazole tablets with saline and place in the canal for 1 week.

Resorption (commonly seen after avulsion, luxation, intrusion, or extrusion).

Internal resorption is associated with chronic pulpal inflammation, which results in resorption of dentine from the pulpal surface. Is progressive, therefore the pulp needs to be carefully extirpated. Dressing the tooth with calcium hydroxide appears to help arrest the resorption and, once controlled, a GP filling may be placed. If perforation has occurred the prognosis is ↓ considerably. Raising a flap, removal of granulation tissue, and placement of an amalgam seal is indicated.

1 I. C. Mackie and V. N. Warren 1988 *Dental Update* **15** 155.
2 I. C. Mackie 1993 *BDJ* **175** 99.

External resorption Three types are seen surface, replacement, and inflammatory (p. 121).

Calcification occurs in 6–35% of luxation-type injuries. Prophylactic endodontic treatment is not necessary as pulp necrosis occurs in only 13–16% of cases. A high rate of success (80%) has been reported for subsequent RCT despite a hairline or no root-canal detectable on X-ray.

Management of missing incisors

1 Rarely congenitally absent, usually lost following trauma or because of dilaceration.

2 Congenitally absent in approx. 2% population (with ↑ likelihood of displacement 3), but may also be lost following trauma. Both can occur unilaterally, bilaterally or together.

Missing upper anterior teeth are noticed by the general public before other types of malocclusion, e.g. ↑ o/j. Therefore the aim of treatment is to provide 321|123 smile. Although Cary Grant did well enough with a missing upper central incisor, symmetry is usually preferable. The management of missing incisors involves either recovery or maintenance of space for a prosthetic replacement, or orthodontic space closure. Nordquist[1] found that space closure was better aesthetically and periodontally than prosthetic replacement; however, with the introduction of newer materials and techniques this finding may be outdated. For each patient a number of factors need to be considered:

Skeletal relationship In a Class III case, space closure in the upper arch could compromise the incisor relationship, whereas, in a class II/1, it would facilitate o/j ↓. Consider also the vertical relationship as space closure is easier in patients with ↑ LFH and vice versa in ↓ LFH.

Crowding/spacing In a patient with no crowding, space closure is difficult and requires prolonged retention. Before opening space it is important to ensure that sufficient will be available at the end of treatment for an aesthetic replacement (minimum width of 2 is 5 mm), for which a Keslings set-up is useful.

Colour and form of adjacent teeth Although much can be done with composite additions and grinding, if 3 is significantly darker &/or caniform in shape, it will be difficult to turn it into a convincing 2 if space closure planned. 2 can only be used to mimic 1 if root length and circumference at gingival margin are not significantly shorter.

Inclination of adjacent teeth This will influence the type of appliance required to open or close space. The final axial inclination of the teeth with determine the aesthetics of the finished result.

Buccal occlusion If a good buccal interdigitation exists this may C/I bringing the posterior teeth forward to close space.

Unilateral loss A symmetrical result is more pleasing, ∴ maintenance or opening of space is preferable. If a 2 is missing and the contralateral tooth is peg-shaped, thought should be given to extracting this tooth to achieve symmetry.

Gingival level Can alter with periodontal surgery.

1 G. G. Nordquist *et al*. 1975 *Journal of Periodontology* **46** 139.

Patient's wishes and co-operation Only after assessing the above factors, can the patient be given an informed choice. If the patient refuses fixed appliances this may alter the treatment plan.

Kesling's set-up Require duplicate models of both arches for this procedure, including at least two of the upper arch. Using a small hack-saw, the teeth which will require orthodontic movement are removed from the model and re-positioned using wax. As many alternatives as desired can be tried to find the best result.

Space closure This can be facilitated by early extraction of the deciduous teeth on the affected side, therefore the earlier the decision is made to close space, the better. Almost invariably involves the use of fixed appliances, as even though spontaneous space closure may occur in a crowded mouth, over-correction of the axial inclination is advisable. It is better to carry out any masking procedures before orthodontic treatment e.g. contouring $\underline{3}$ to resemble $\underline{2}$ (by removal of enamel incisally, interproximally and from the palatal aspect ± composite addition) as this will facilitate final positioning and occlusion. Retain with a bonded retainer.

NB the average difference in width between $\underline{3}$ and $\underline{2}$ is 1.2 mm, which can easily be removed mesially and distally from $\underline{3}$. If the lower arch is crowded extraction of a lower premolar will allow a Class I buccal segment relationship to be established.

Space-maintenance/opening If an incisor is electively extracted and space maintenance is desired, a P/- or acid-etch bridge should be fitted immediately. Where a $\underline{2}$ is congenitally missing, this will not be possible and space may need to be opened orthodontically. The inclination of the teeth to be moved will determine whether fixed or removable appliances will be required. Following tooth movement, retention with a P/- for 3–6 months is advisable to allow the teeth to settle. If an acid-etch retained prosthesis is planned, ensure that there is sufficient room occlusally for the wings.

Acid-etch retained prostheses, p. 302.

Transplantation of a lower premolar into the socket of an extracted incisor, popular in Scandinavia.

Implant when growth complete, becoming more widely available (p. 428).

Common childhood ailments affecting the mouth

▶ Refer any patient with an ulcer that doesn't heal within 3 weeks or with any soft tissue lesion of unknown aetiology.

See chapter on oral medicine.

Most common disease is gingivitis (p. 212).

Viral

Primary herpetic gingivostomatitis Occurs >6 months of age. *Symptoms*: febrile, cervical lymphadenitis, vesicles → ulcers on gingiva and oral mucosa. Rx: soft diet with plenty of fluids. Self-limiting, lasts for approximately 10 days.

Secondary herpes labialis Vesicles form around the lips, and crust. Self-limiting, but 5% acyclovir cream will speed healing.

Hand, foot, and mouth disease Rash on hands and feet plus ulcers on oral mucosa and gingiva. Little systemic upset. Self-limiting.

Herpangina Febrile illness with sore throat due to ulcers on soft palate and throat. Usually lasts about 3–5 days. Soft diet.

Warts Check hands, usually self-limiting.

Also chickenpox (vesicles → ulcers), mumps (inflamed parotid duct), glandular fever (ulcers), measles.

Bacterial

Impetigo Very infectious staphylococcal (± streptococci) rash. Starts around mouth and may be mistaken for 2° herpes.

Streptococcal sore throat Can get associated streptococcal gingivitis.

Acute ulcerative gingivitis Rare <16 yrs (p. 220).

Fungal

Candida Commensal of the mouth which may become pathogenic when oral environment favours its proliferation. Two types of manifestation are seen in children. **Acute pseudo-membranous candidiasis** (thrush) Seen in newborn, under-nourished infants after prolonged use of antibiotics or steroids. Presents as white patches that rub off. Rx: Miconazole (25 mg/ml) and correct underlying problem. **Chronic atrophic candidiasis** Most commonly URA + poor OH ± high sugar intake. Rx: OHI for appliance and teeth. Chlorhexidine mouth-wash or miconazole gel.

Miscellaneous

Apthous ulceration,[1] p. 440.

---

1 E. A. Field 1992 *Int J Paed Dent* **2** 1.

Common causes of oral ulceration in children (in order of frequency) apthous; traumatic; acute herpetic gingivostomatitis; herpangina; hand, foot, and mouth disease; glandular fever. If in doubt refer.

Common causes of soft tissue swellings in children
Abscess; mucocele; eruption cyst; epulides; papilloma.

▶ Oral cancer does occur in children, ∴ if in doubt refer for biopsy.

Sugar-free medications

The cariogenic effect of long-term medication sweetened with sugar is now well-recognized. As the Medicines Commission has instructed that liquid formulations for use in chronic conditions in children should be free of cariogenic sugars, the future looks bright, but little progress has been made on the re-formulation of commonly used drugs. Unfortunately, there is no evidence that rinsing out, or brushing the teeth after use of a sugar-based medicine will significantly reduce the incidence of caries. Current medical advice is for liquid medicines to be given to children by disposable syringes. This approach has the advantage that an accurate dose can be directed at the back of the mouth.

Below is a list of some sugar-free medicines. It is not exhaustive and where required reference should be made to the *British National Formulary* or *Pharmaceutical Journal* 1988 **241** 16–17.

Analgesics

aspirin (>12 years)	Dispersable aspirin tablets BP
paracetamol	Disprol paediatric
	Junior Disprol suspension/tablets
	Junior Q-Panol
	Panadol baby and infant elixir
	Panadol Junior sachets
	Panadol soluble
	Junior Panaleve/Panaleve elixir
paracetamol and codeine	Panadeine soluble
	Paracodol dispersable tablets
	Solpadeine

Antacids

aluminium and magnesium	Mucogel
	Maalox suspension
cimetidine	Dyspamet chewable tablets
ranitidine	Zantac dispersable tablets
	Zantax suspension

Anticonvulsants

carbamazepine	Tegretol liquid
phenobarbitone	Phenobarbitone elixir BP 30 mg/10 ml
sodium valproate	Epilim crushable tablets/elixir

Antihistamines

terfenadine	Triludan suspension

Anti-infectives

acyclovir	Zovirax suspension
amoxycillin	Amoxil dispersable tablets/ syrup SF
amphotericin	Fungilin oral suspension/lozenges
ampicillin and cloxacillin	Ampiclox neonatal suspension

co-trimoxazole	Bactrim dispersable tablets/ SF syrup
	dispersible co-trimoxazole tablets BP
	Septrin dispersable tablets/ suspension
erythromycin	Erythroped sugar-free sachets
miconazole	Daktarin oral gel

Respiratory agents

salbutamol	Cobutolin syrup
	Salbutamol syrup (Evans)
	Ventolin syrup

Miscellaneous

choline salicylate	Teejel
	Bonjela
folic acid	Lexpec syrup
iron edetate	Sytron
vitamins A, B, C, D, and E	Boots' Plurivite syrup for children

Rules for calculating drug dosages for children:
Young's rule

$$\text{Dosage} = \frac{\text{age} \times \text{adult dose}}{\text{age} + 12}$$

Clark's rule

$$\text{Dosage} = \frac{\text{weight in lb} \times \text{adult dose}}{150}$$

4 Orthodontics

What is orthodontics?	132
The Index of Orthodontic Treatment Need	134
Definitions	136
Orthodontic assessment	138
Cephalometrics	140
More cephalometrics	142
Treatment planning	144
Management of the developing dentition—	146
Management of the developing dentition—2	148
Extractions	150
Extraction of the first permanent molars	152
Distal movement of the upper buccal segments	154
Spacing	156
Buccally displaced maxillary canines	158
Palatally displaced maxillary canines	160
Increased overjet	162
Management of increased overjet	164
Increased overbite	166
Management of increased overbite	168
Anterior open bite	170
Reverse overjet	172
Management of reverse overjet	174
Crossbites	176
Anchorage	180
Removable appliances—design	182
Removable appliances—active components	184
Removable appliances—fitting and monitoring progress	186
Fixed appliances	188
Functional appliances—rationale and mode of action	192
Types of functional appliance and practical tips	194
Orthodontics and orthognathic surgery	196
Cleft lip and palate	198

Relevant pages in other chapters: Surgical management of CLP, p. 502; abnormalities of eruption, p. 68; supernumerary teeth, p. 72; orthognathic surgery, p. 504.

Principal sources: W. J. B. Houston and W. J. Tulley 1986 *A Textbook of Orthodontics*, Wright. J. R. E. Mills 1987 *Principles and Practice of Orthodontics*, Churchill Livingstone.

It has been said that orthodontists forget to ask the patient to open wide, thus missing any dental pathology, whilst generalists forget to ask the patient to close together, thus missing any malocclusion! The aim of this chapter is to help ensure that this is not true. A problem-orientated approach has been used (rather than by classification) for simplicity.

What is orthodontics?

Orthodontics has been defined as that branch of dentistry concerned with growth of the face, development of the dentition, and the prevention and correction of occlusal anomalies.[1]

Prevalence of malocclusion[2] At age 12: crowding 69%, o/j >5 mm 22%, at least one instanding incisor 6%.

Why do orthodontics? Research shows that individual motivation has more effect upon the presence of plaque than alignment of the teeth. Therefore, the main indications for orthodontic treatment are aesthetics and function. Functional reasons for treatment include crossbites (as associated occlusal interferences may tend to predispose towards TMPDS); deep traumatic overbite; increased overjet (↑ risk of trauma) and labial crowding of a lower incisor (as this reduces periodontal support labially). While it is accepted that severe malocclusion may have a psychologically debilitating effect, the impact of more minor anomalies and indeed, perceived need, are influenced by social and cultural factors. The Index of Orthodontic Treatment Need (IOTN) has been developed to try and standardize and quantify this difficult issue (p. 134).

It is important that patients realize that orthodontic treatment is not without its disadvantages. Even with good OH, a small loss of periodontal attachment and root resorption is common and in susceptible patients this may be significant. With poor OH greater loss of periodontal support and decalcification may result. Therefore, the potential benefit to the patient must be sufficient to counter-balance these drawbacks.

Who should do orthodontics? All dentists should be concerned with growth and development. Unless anomalies are detected early and any necessary steps taken at the appropriate time, then provision of the best possible treatment for that patient is unlikely. It is more cost-effective if the patient's GDP or a specialist practioner provide orthodontic treatment, with the hospital consultant service acting as a source of advice and a referral point for more complex and multidisciplinary problems.

When should we do orthodontics? This depends upon the particular anomaly. In the early mixed dentition, treatment is only indicated to correct incisor Xbites and posterior Xbites with displacement. Functional appliance treatment is often started in the mixed dentition to coincide with the pubertal growth spurt; however, the majority of orthodontic treatment is not started until the 2° dentition has erupted. Although reaching adulthood does not preclude orthodontics, treatment during the early teens is preferable because the response to orthodontic forces is more rapid, appliances are better tolerated and, most importantly, growth can be utilized to help effect sagittal or vertical change. This means that in adults, tooth movement will

1 W. J. B. Houston 1986 *A Textbook of Orthodontics*, Wright. **2** J. E. Todd 1983 *Children's Dental Health in the UK*, HMSO.

be slower and lack of growth will limit the type of malocclusion
that can be tackled by orthodontics alone.

▶ If in doubt, refer a patient earlier rather than later, especially
in cases with marked skeletal discrepancies.

The Index of Orthodontic Treatment Need

The Index of Orthodontic Treatment Need (IOTN) was developed to quantify and standardize an individual patient's need for orthodontic treatment, so that the potential benefits can be weighed against the possible disadvantages.[1] The Index consists of two components:

The dental health component was developed from an index used by the Swedish Dental Health Board (which was used to determine the amount of financial help that would be given by the state towards treatment costs). The Dental Health Component of IOTN (is reproduced by kind permission of VUMAN Ltd.) has 5 categories of treatment need ranging from little need, to very great need, for treatment. A patient's grade is determined by recording the single worst feature of their malocclusion.

The aesthetic component is based on a series of 10 photographs of the labial aspect of different Class I or Class II malocclusions which are ranked according to their attractiveness. A patient's score is determined by the photograph which is deemed to have an equivalent degree of aesthetic impairment.

The Index of Orthodontic Treatment Need

Grade 1 (None)

1 Extremely minor malocclusions including displacements less than 1 mm.

Grade 2 (Little)

2a Increased overjet 3.6–6 mm with competent lips.
2b Reverse overjet 0.1–1 mm.
2c Anterior or posterior crossbite with up to 1 mm discrepancy between retruded contact position and intercuspal position.
2d Displacement of teeth 1.1–2 mm.
2e Anterior or posterior openbite 1.1–2 mm.
2f Increased overbite 3.5 mm or more, without gingival contact.
2g Pre-normal or post-normal occlusions with no other anomalies. Includes up to half a unit discrepancy.

Grade 3 (Moderate)

3a Increased overjet 3.6–6 mm with incompetent lips.
3b Reverse overjet 1.1–3.5 mm.
3c Anterior or posterior crossbites with 1.1–2 mm discrepancy.
3d Displacement of teeth 2.1–4 mm.
3e Lateral or anterior openbite 2.1–4 mm.
3f Increased and complete overbite without gingival trauma.

1 W. C. Shaw 1991 *BDJ* **170** 107.

Grade 4 (Great)

4a Increased overjet 6.1–9 mm.

4b Reversed overjet greater than 3.5 mm with no masticatory or speech difficulties.

4c Anterior or posterior crossbites with greater than 2 mm discrepancy between retruded contact position and inter-cuspal position.

4d Severe displacement of teeth, greater than 4 mm.

4e Extreme lateral or anterior openbites, greater than 4 mm.

4f Increased and complete overbite with gingival or palatal trauma.

4h Less extensive hypodontia requiring pre-restorative ortho-dontic space closure to obviate the need for a prosthesis.

4l Posterior lingual crossbite with no functional occlusal contact in one or both buccal segments.

4m Reverse overjet 1.1–3.5 mm with recorded masticatory and speech difficulties.

4t Partially erupted teeth, tipped and impacted against adjacent teeth.

4x Supplemental teeth.

Grade 5 (Very great)

5a Increased overjet greater than 9 mm.

5h Extensive hypodontia with restorative implications (more than 1 tooth missing in any quadrant) requiring pre-restorative orthodontics.

5i Impeded eruption of teeth (with the exception of third molars) due to crowding, displacement, the presence of supernumerary teeth, retained deciduous teeth, and any pathological cause.

5m Reverse overjet greater than 3.5 mm with reported masticatory and speech difficulties.

5p Defects of cleft lip and palate.

5s Submerged deciduous teeth.

Definitions

Ideal occlusion Anatomically perfect arrangement of the teeth. Rare.

Normal occlusion Acceptable variation from ideal occlusion.

Competent lips Lips meet together at rest.

Incompetent lips Lips do not meet together at rest.

Frankfort plane Line joining porion (superior aspect of external auditory meatus) with orbitale (lowermost point of bony orbit).

Lower facial height (LFH) Clinically it is the distance from the base of nose to point of the chin and in a normally proportioned face is equal to the middle facial third (eyebrow line to base of nose). Cephalometrically, it is the distance from anterior nasal spine to menton as a percentage of the total face height (from nasion to menton).

Class I The lower incisor edges occlude with or lie immediately below the cingulum of upper incisors.

Class II The lower incisor edges lie posterior to the cingulum of the upper incisors. *Division 1* the upper central incisors are upright or proclined and the overjet is ↑ *Division 2* the upper central incisors are retroclined and the overjet is usually ↓ but may be ↑.

Class III The lower incisor edges lie anterior to the cingulum of the upper incisors and the overjet is ↓ or reversed.

Bimaxillary proclination Both upper and lower incisors are proclined.

Overjet Distance between the upper and lower incisors in the horizontal plane.

Overbite Overlap of the incisors in the vertical plane.

Complete overbite The lower incisors contact the upper incisors or the palatal mucosa.

Incomplete overbite The lower incisors do not contact the upper incisors or the palatal mucosa.

Anterior open bite When the patient is viewed from the front and the teeth are in occlusion, a space can be seen between the upper and lower incisor edges.

Crossbite A deviation from the normal bucco-lingual relationship. May be anterior/posterior &/or unilateral/bilateral.

Buccal crossbite Buccal cusps of lower premolars or molars occlude buccally to the buccal cusps of the upper premolars or molars.

Lingual crossbite Buccal cusps of lower molars occlude lingually to the lingual cusps of the upper molars.

Dento-alveolar compensation The position of the teeth has compensated for the underlying skeletal pattern, so that the occlusal relationship between the arches is less severe.

Leeway space The difference in diameter between C, D, E, and 3, 4, 5. Greater in lower than upper arch.

Mandibular deviation Path of closure starts from a postured position of the mandible.

Mandibular displacement When closing from the rest position, the mandible displaces (either laterally or anteriorly) to avoid a premature contact.

Balancing extraction Extraction of the same (or adjacent) tooth on the opposite side of the arch to preserve symmetry.

Compensating extraction Extraction of the same tooth in the opposing arch.

137

Orthodontic assessment

Equipment A mirror, probe, and an engineer's SS rule (these are cheaper from ironmongers).

Brief screening procedure

The purpose of this is to ensure early detection and treatment of any abnormality, prepare the patient for any later treatment and influence the management of any teeth of poor prognosis.

At every visit Once the 2° incisors have erupted and until 2° dentition established (if in doubt refer).
1 Keep the eruption sequence in mind (p. 69). Any deviations from this should be observed for a few months only and then investigated.
2 Failure of a tooth to appear >6 months after the contralateral tooth has erupted should ring alarm bells.
3 Ask child to close together and look for Xbites, reverse or ↑ o/j.

At age 8 take X-rays to check for the position and presence of unerupted teeth. Consider the long-term prognosis of the first permanent molars (p. 152).

From age 9 and until they erupt, palpate for $\underline{3}$ in the buccal sulcus. A definite hollow ± asymmetry warrants further investigation.

Detailed orthodontic examination

Should be carried out in a logical sequence so that nothing is missed.
- Who wants treatment (patient or parent) and what for?
- What complexity of treatment is patient prepared to accept. Have their peers worn braces and if not, will the child make a good pioneer?
- Enquire about any previous extractions and orthodontic treatment.

EO examination (with Frankfort plane horizontal)
- Assess skeletal pattern:
    1 antero-posteriorly (max = mand Class I, max>mand Class II, max<mand Class III);
    2 vertically (Frankfort-mandibular planes angle approx. 28°, lower ⅓rd face usually 50% of total face height);
    3 transversely (? asymmetry).
- Soft tissues: Lips are only competent if they meet at rest. Check the position of the lower lip relative to the $\underline{inc}$ and how the patient achieves an oral seal (? lip to lip, lip to tongue, or by the lower lip being drawn up behind the incisors). Note also the length of the upper lip, the amount of $\underline{inc}$ seen and lip tonicity.
- Check rest position of mandible and for any displacement on closure.
- Habits? Does patient suck a thumb/finger, bite finger-nails or brux?

IO examination

- Record OH, gingival condition and teeth present. Any of poor prognosis?
- LLS: Inclination to mandibular base, crowding/spacing, displaced teeth, angulation of $\overline{3|3}$.
- ULS: Inclination to maxillary base, crowding/spacing, rotations, displaced teeth, presence and angulation of $3|3$.
- Measure o/j (mm), o/b (↑ or ↓, complete or incomplete). Check centre-lines coincident and correct within face.
- Buccal segments: crowding/spacing, displaced teeth.
- Check molar and canine relationship. Any Xbites?

X-rays Usually require a DPT (or 2 lateral obliques) and an intra-oral of the inc. A lateral skull view is indicated if the patient has a skeletal discrepancy or A–P movement of the incisors is anticipated).

- Look for unerupted, missing or \$ teeth, root resorption, or other pathology.
- Cephalometric analysis, p. 140.

Summary should include a description of the patient's incisor relationship and the main points of the malocclusion, e.g. skeletal pattern, crowding, Xbites. This gives a 'problem list' from which the aims of treatment can be derived (p. 144).

Study models are not obligatory for orthodontic diagnosis, but they certainly help. Taking models allows you to mull over the possibilities, rather than have to make a snap decision. If no treatment is planned unless the malocclusion deteriorates, give the models to the patient, who will usually guard them well.

▶ It is important to take models before and after orthodontic treatment so that progress can be monitored and for medico-legal purposes.

Cephalometrics

Cephalometric analysis is the interpretation of lateral skull radiographs. It is not obligatory for orthodontic diagnosis, and where no A–P change in incisor position is planned, the X-ray exposure is not justified for the information gained. However, where A–P movement is required, a lateral skull radiograph will back-up the clinical assessment of skeletal pattern, help to determine the degree of difficulty and whether fixed or removable appliances are indicated. Serial lateral skulls allow assessment of growth and/or treatment.

Tracing

Although this has been done using grease-proof paper and a TV as background illumination, the quality of diagnostic information so obtained is limited. Tracing paper (secured to the film with masking tape), a sharp pencil, and good background illumination are essential. Spotting orthodontic landmarks is infinitely easier if carried out in a darkened room. If a point is hard to see, block off the rest of the film so that only that area is illuminated. If this fails, try holding up to a bright spotlight, but if still not clear, make a guesstimate! Due to the slight magnification (7–8%), two images of the mandibular border are usually seen. Both should be traced and an average taken for gonion. The most prominent image should be traced, i.e. the most anterior in the face so that the difficulty of treatment is not underestimated.

Pitfalls

- Consider the cephalometric values for a particular patient in conjunction with the clinical assessment, as variation from the normal in a measurement may be compensated for elsewhere in the face or cranial base.
- Angle ANB varies with the relative prominence of nasion and the lower face. If SNA significantly ↑ or ↓ this could be due to the position of nasion, in which case an additional analysis should be used, e.g. Ballard's conversion.
- For landmarks which are bilateral (unless superimposed exactly) the midpoint between the two should be taken to correspond with those reference points which are in the midline.
- Tracing errors, with careful technique these should be of the order of ± 0.5° and 0.5 mm. Errors are compounded when comparing tracings, ∴ changes of 1 or 2 degrees should be interpreted with caution.

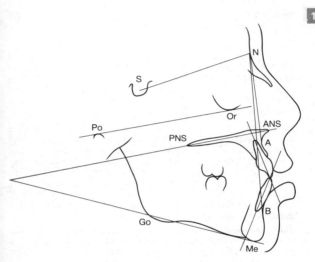

Most commonly used cephalometric points:

S = Sella: mid-point of sella turcica.
N = Nasion: most anterior point on fronto-nasal suture.
Or = Orbitale: most inferior anterior point on margin of orbit (take average of two images).
Po = Porion: uppermost outermost point on bony external auditory meatus.
ANS = Anterior nasal spine.
PNS = Posterior nasal spine.
Go = Gonion: most posterior inferior point on angle of mandible.
Me = Menton: lowermost point on the mandibular symphysis.
A = A point: position of deepest concavity on anterior profile of maxilla.
B = B point: position of deepest concavity on anterior profile of mandibular symphysis.

Frankfort plane = Po–Or.
Maxillary plane = PNS–ANS.
Mandibular plane = Go–Me.

More cephalometrics

Analysis and interpretation

The analysis of lateral skull tracings is carried out by comparing a number of angular measurements and proportions with average values for the population as a whole. Normal values for Caucasians (UK)[1] standard deviations in parentheses:

SNA = 81° (±3)

SNB = 79° (±3)

ANB = 3° (±2)

$\underline{1}$-Max = 109° (±6)

$\overline{1}$-Mand = 93° (±6) or 120 minus MMPA

MMPA = 27° (±4)

Facial proportion = 55% (±2)

Inter-incisal angle = 133° (±10)

If SNA or SNB are more than 2 standard deviations from the normal (and the patient looks human), check your tracing! However, the ANB difference is not an infallible assessment of skeletal pattern as it assumes (incorrectly in some cases) that there is no discrepancy in the cranial base and that A and B are indicative of basal bone position. When a cephalometric tracing seems at odds with your clinical impression it is worth doing another analysis which avoids reliance on the cranial base, such as a Ballard's conversion.

Before deciding on a treatment plan it is helpful to consider what factors have contributed to a particular malocclusion. E.g., in a patient with a Class II/1 incisor relationship on a Class I skeletal pattern, the prognosis for simple treatment is much better if the ↑ o/j is due to proclination of the upper rather than retroclination of the lower incisors. The relative contribution of the maxilla and mandible to the skeletal pattern may indicate possible lines of treatment. E.g., if an increased o/j is due to a retrusive mandible, a better aesthetic result may be achieved by use of a functional appliance.

A quick method of deciding whether simple tilting movements will suffice to correct an increased o/j or whether bodily movement is required, is to assume that there is 2.5° of retroclination of the upper incisors for every millimetre of o/j reduction. If the resultant angle is <95° then some bodily movement is required (as further retroclination of the upper incisors is unaesthetic).

1 J. R. E. Mills 1987 *Principles and Practice of Orthodontics*, Churchill Livingstone.

Ballard's conversion. Used to assess skeletal pattern. The incisors are tilted to their correct inclination about a point $\frac{1}{3}$ away down the root from the apex. In this example:

upper incisors corrected to 109°
lower incisors corrected to 120—MMPA = 120–26.5 = 93.5°
resultant overjet = 2.5 mm; that is, Class I.
(dashed outline = corrected position)

Treatment planning

First draw up a list of the aims of treatment, e.g.
1 Relieve upper and lower arch crowding.
2 Expand upper arch to correct posterior crossbite.
3 Reduce overbite.
4 Reduce overjet.

This usually gives an outline of the ideal treatment for that patient. It may be necessary to modify these aims after consideration of patient co-operation, who is going to do the treatment and where. If this is the case, it is important to note in the patients records the ideal treatment plan and why it had to be compromised. With the exception of mixed dentition problems, the majority of malocclusions are amenable to the method outlined below.

Plan LLS The LLS is in a zone of balance between the lips and the tongue, ∴ it is safer to consider its position as immutable. This gives a starting point around which to plan treatment. The first step is to decide if the LLS is crowded sufficiently to warrant extractions. If the crowding is likely to ↑ (patient in early teens, 8|8 present), then extractions are probably indicated (p. 142). In Class II/2 cases it may be advisable to accept a little crowding (p. 162). If the LLS is well aligned go straight to the next step.

Mentally correct 3̄ The inclination of 3̄, the patient's age, and the amount of movement will determine whether appliances are required or driftodontics will suffice.

Reposition 3 in mind's eye into a Class I relationship with 3̄ (in corrected position). This will give an indication of the space required and the amount and type of movement necessary. In the upper arch space for retraction of 3 can be gained by (1) extractions, (2) expansion (only indicated if a Xbite exists), (3) distal movement of the upper buccal segments (p. 154), (4) a combination of these. Should extractions be indicated in both arches, a good reason is required (e.g. poor long-term prognosis) not to extract the same tooth in the upper as in the lower.

NB if the LLS is spaced, or the upper incisors are small (e.g. peg-shaped 2) or the ULS is retroclined, it is not necessary to retract 3 as far as class I.

Plan ULS What tooth movements are required. URA or FA?

Consider molar relationship Decide what it should be at the end of treatment (extractions upper arch only—Class II, extractions lower arch only—Class III, otherwise—Class I) and how it will be achieved (spontaneously or appliances required?).

Prognosis Will the proposed treatment be stable? Beware of o/j ↓ in a patient with grossly incompetent lips, or proclination of upper incisors in a Class III where there is no o/b.

Practical treatment By now an outline of the proposed treatment should be emerging, next consider the mechanics of carrying it out. If the patient refuses the FA indicated, will using URA only leave the patient worse off or result in an acceptable compromise? The practicalities of providing treatment also warrant consideration. Often more than one treatment plan can be offered to the patient, with a hierarchy of complexity and finesse of finished result.

145

▶ **Beware** If the malocclusion under consideration contains one of the following features, refer them to a specialist for treatment.

- Marked skeletal discrepancy, Class II/III or vertically.
- If the o/j is ↑ and the upper incisors are upright.
- If the o/j is reversed and there is no o/b to retain correction of the incisor relationship.
- Severe Class II/2 malocclusions.
- Class II/1 incisor relationship, with molars a full unit Class II and a crowded lower arch.

Treatment planning is the most important, and most difficult, part of orthodontics. If at all doubtful, seek advice.

Management of the developing dentition—1

See also delayed eruption, p. 68.

The way in which mixed dentition problems are approached will often affect the ease or difficulty of subsequent treatment.

Normal development of dentition The 1° incisors are usually upright and spaced. If there is no spacing warn the parents that the 2° incisors will probably be crowded. Overbite reduces throughout the 1° dentition until the incisors are edge to edge. All 2° incisors develop lingual to their predecessors, erupt into a wider arc and are more proclined. It is normal for $\underline{1|1}$ to erupt with a median diastema which reduces as $\underline{2|2}$ erupt. Later, pressure from the developing canines on the roots of $\underline{2|2}$, results in their being tilted distally and spaced. This has been called the 'ugly duckling stage', but it is better to describe it as normal development to parents. As the $\underline{3}$ erupts the $\underline{2}$ upright and the spaces usually close.

The majority of Es erupt so that their distal edges are flush. The transition to the normal stepped (Class I) molar relationship usually occurs during the 1° dentition as a result of greater mandibular growth &/or the leeway space.

Development of dental arches In the average (!) child, the size of the dental arch is more or less established once the 1° dentition has erupted, except for an increase in inter-canine width (2–3 mm up to age 9) which results in a modification of arch shape.

Retained deciduous teeth if deflecting eruption of 2° tooth, extract.

Submerging deciduous molars Prevalence 8–14%. Provided there is a successor, a submerged 1° molar will probably be exfoliated at the same time as the contralateral tooth.[1] Extraction is only indicated if there is no successor or the submerged tooth is likely to disappear below the gingival margin.

Impacted upper first permanent molars Prevalence 2–6% Indicative of crowding. Spontaneous disimpaction rare after 8 yrs. Can try dislodging $\underline{6}$ by tightening a piece of brass wire round the contact point with $\underline{E}$ over several visits. Otherwise just observe, extracting $\underline{E}$ if unavoidable and dealing with resultant space loss in 2° dentition.

Habits Effects produced depend upon duration of habit and intensity. It is best not to make a great fuss of a finger-sucking habit. If parents concerned reassure them (in presence of child) not to worry, as only little girls/boys suck their fingers. Appliances to break the habit don't help; children will stop when they are ready. However, this is no reason to delay the start of treatment for other aspects of the malocclusion.

1 J. Kurol 1985 *Am J Orthod* **87** 46.

Management of the developing dentition—2

Effects of premature loss of deciduous teeth

Unfortunately often the first time a child attends is with toothache. In the rush to relieve pain, it is all too easy to extract the offending tooth without consideration of the consequences. The major effect of early 1° tooth loss is localization of crowding, in crowded mouths. The extent to which this occurs depends upon patient's age, degree of crowding and the site. In a crowded mouth the adjacent teeth will move round into the extraction space, ∴ unilateral loss of a C (and to a lesser degree a D) will result in a centre-line shift. This is also seen when a C is prematurely exfoliated by an erupting 2. As correction of a centre-line discrepancy often involves fixed appliances, prevention is better than cure, so loss of Cs should always be balanced. If Es are lost the 6 will migrate forward. This is particularly marked if it occurs before eruption of the permanent tooth, so if extraction of an E is unavoidable try to defer until after the 6s are in occlusion and do not balance or compensate. The effect of early loss of 1° teeth on the eruption of the permanent successor is variable.

Serial extraction Originally proposed by Kjellgren in 1948[1] because of a shortage of dentists which is less applicable today! The aim is to shift crowding from the labial to the buccal segments where it can be dealt with by extraction of 4s. Classically involves:

1 extraction of the Cs as the 2s are erupting to allow incisors to align,
2 extraction of Ds to hasten the eruption of 4s,
3 extraction of 4s.

In selected cases it works well, but the technique has its pitfalls. Following extraction of Cs and then Ds, forward drift of the buccal segments occurs, which in a severely crowded mouth could mean that extraction of four first premolars will not suffice. Serial extraction is really indicated for moderate crowding; however, it is often difficult to assess how crowded the patient will be.

The lower incisors tip lingually slightly after loss of $\overline{C|C}$ resulting in an increase in o/b, thus C/I this approach in Class II/2. In Class III, however, extraction of $\overline{C|C}$ only could be advantageous.

For damage limitation the following guidelines are suggested:

1 Choose patients with Class I malocclusions and mild to moderate crowding. All teeth should be present on X-ray and in a good position.
2 Measure inter-canine width and review the patient in 4 months. If the inter-canine distance is unchanged, proceed.
3 Extract Cs when two-thirds of the roots of 3s calcified.

1 B. Kjellgren 1948 *Acta Odontol Scand* **8** 17.

4 Extract 4s on eruption.

5 Review ? space maintenance. Yes if space 2–6 <16 mm.

Better still avoid serial extraction and deal with the crowding in the 2° dentition. However, timely loss of Cs is still indicated for

- 2 erupting palatally due to crowding. Extraction of C|C as the $\overline{2}$ erupting may allow the tooth to escape labially and prevent a Xbite.

- Extraction of $\overline{C|C}$ when a lower incisor is being crowded labially will help to ↓ loss of periodontal support.

149

Extractions

In orthodontics, teeth are extracted either to relieve crowding or to compensate for a skeletal discrepancy.

▶ Before planning the extraction of any permanent teeth a thorough orthodontic and radiographic examination should be carried out.

▶ In a Class I or II should extract at least as far forward in the upper arch as the lower; vice versa in a Class III.

Lower incisors Following the extraction of a $\overline{inc}$, the LLS tends to tilt lingually followed by the ULS. In addition, it is difficult to arrange 6 ULS teeth around 5 LLS teeth, ∴ try to avoid. However, if necessary, delay if possible until growth complete and use -/FA.

Upper incisors are never the teeth of choice for extraction, but if traumatized or dilacerated, there may be no alternative (p. 124).

Lower canines should only be extracted if severely displaced, as the resulting contact between $\overline{2}\ \overline{4}$ is unsatisfactory.

Upper canines, p. 160.

First premolars are the most popular choice, due to their position in the arch and because a good contact point between the canine and second premolar is more likely. For maximum spontaneous improvement 4s should be extracted just as the 3s are appearing, but if appliance therapy is planned, defer until the canines have erupted.

Second premolars Preferred in cases with mild crowding, as their extraction alters the anchorage balance, favouring space closure by forward movement of the molars. FA are required, especially in the lower arch. If 5s hypoplastic or missing there may be no choice. Early loss of an E will often lead to forward movement of the 6 and lack of space for 5s. In the upper arch this results in $\underline{5}$ being displaced palatally, and provided 4 is in a satisfactory position, extraction of $\underline{5}$ on eruption may ↓ the need for appliance therapy. In the lower arch $\overline{5}$ are usually crowded lingually. Extraction of $\overline{4}$ is easier and will give $\overline{5}$ space to upright spontaneously, hopefully avoiding a lower FA.

First permanent molars, p. 152.

Second permanent molars Extraction indicated (1) to relieve mild lower premolar crowding; (2) to try and prevent further LLS crowding; (3) to avoid difficult extraction of impacted $\overline{8}$; (4) to facilitate distal movement of the upper buccal segments.

To increase likelihood of $\overline{8}$ erupting successfully to replace $\overline{7}$ need: posterior crowding, $\overline{8}$ formed to bifurcation and at an angle of between 15° and 30° to long axis $\overline{6}$. Even so, may still require appliance therapy to align $\overline{8}$.

Third permanent molars Early extraction of these teeth is sometimes advocated to prevent LLS crowding, but, as this can occur even in their absence, wisdom teeth are only a part of the aetiology.

Space can also be provided in selected cases by:
1 expansion (only in upper arch with a Xbite, otherwise not stable);
2 distal movement of the upper buccal segments (p. 154);
3 reducing the width of the teeth approximally (LLS only, in adults with <2 mm crowding and a Class I occlusion).

Extraction of the first permanent molars

First permanent molars are never the first choice for extraction, as even if removed at the optimal time, a good spontaneous alignment of the remaining teeth is unlikely. However, when a two-surface (or more) restoration is required in a molar tooth for a child, the long-term prognosis should be considered. A well-timed extraction ± orthodontic treatment, may be better for the child (and your BP) than heroic attempts to restore hopeless molars. Points to note:

- Check the remaining teeth are present and in a good position. If not avoid extraction of $\overline{6}$ in affected quadrant.
- In the lower arch good spontaneous alignment is more likely following extraction $\overline{6}$ if (1) $\overline{7|7}$ development has reached bifurcation, (2) Angulation between crypt $\overline{7}$ and $\overline{6}$ is <30°, (3) $\overline{7|7}$ crypts overlap $\overline{6|6}$ roots.
- There is a greater tendency for mesial drift in the maxilla, therefore the timing of loss of 6 is less critical.
- Assess the prognosis for remaining 6s. If they are all restored then extraction of all four is probably indicated. If only one poor $\underline{6}$, do not extract corresponding lower tooth. If $\overline{6}$ of poor prognosis it is advisable to extract opposing $\underline{6}$ as otherwise this tooth will overerupt and prevent $\overline{7}$ moving forward. Balancing with extraction of a corresponding sound $\overline{6}$ is inadvisable, better to deal with other side of arch on its merit.
- In Class I with anterior crowding and Class II $\underline{6|6}$ need to be preserved until $\underline{7|7}$ have erupted and can be held back by an appliance and the extraction space utilized.
- In Class III if $\underline{6|6}$ of poor prognosis try to preserve until incisor relationship corrected (to provide retention for URA). In cases with poor quality $\underline{6|6}$, extract at optimal time to aid space closure.
- If the dentition is uncrowded, avoid extraction of 6s as space closure will be difficult.
- Extraction of 6s will relieve buccal segment crowding, but will have little effect on labial segment crowding. Impaction of 8s less likely but not impossible.
- ▶ In a child with poor quality 6s, remember that the premolars may well be in a similar condition 6 yrs on unless the caries rate drops.

153

Distal movement of the upper buccal segments

This is usually thought of as an alternative to extraction, but in practice often results in the crowding being shifted distally, requiring the loss of $\underline{7}$ or $\underline{8}$. It is only applicable to the upper arch, in the following situations:

- Either Class I with mild upper arch crowding or Class II/1, with well-aligned lower arch and molars <1 unit Class II.
- Where extraction of $\underline{4|4}$ does not provide sufficient space to align upper arch (if this is the case unilaterally, then less head-gear wear is required if an URA, with a screw to move $\underline{6}$ distally and extra-oral anchorage, is used).

Can be achieved either by an En Masse appliance, or by EOT directly to molar bands on $\underline{6|6}$. As $\underline{6|6}$ move distally will need some expansion. Greater chance of success with growing child. Can expect $\frac{1}{2}$ unit change in 3–4 months.

En masse appliance
cribs 6/6 0.7 mm SS wire
4/4 0.6 mm SS wire

EOT tubes soldered to cribs 6/6
coffin spring 1.25 mm SS wire
plus facebow and headgear

Spacing

Uncommon in UK, crowding is the norm.

Generalized spacing is due either to hypodontia or small teeth ± large jaws. Note that hypodontia is associated with small teeth p. 72. Treatment of this problem is difficult, a purely orthodontic approach is liable to relapse and requires prolonged retention. In milder cases try and encourage the patient to accept the situation. In more severe cases a combined restorative/orthodontic approach will be required. This may involve composite additions to increase the width of the teeth and/or orthodontics to localize the space for provision of a prosthesis.

Median diastema

Prevalence 6 yr-olds = 98%, 11 yr-olds = 49%, 12–18 yr-olds = 7%.

Aetiology Small teeth in large jaws; absent or peg-shaped 2|2; midline \$; proclination of ULS; physiological (caused by pressure of developing teeth on upper incisor roots which resolves as 3 erupt) or due to a frenum.

The upper incisive frenum is attached to the incisive papilla at birth. As 1|1 erupt the frenum recedes, but this is less likely if the arch is spaced. A frenum contributes to a diastema in a small number of cases and is associated with the following features:

- Blanching of incisive papilla when frenum put under tension.
- Radiographically there is a V-shaped notch in the interdental bone between 1|1 indicating the attachment of the frenum.
- Anterior teeth may be crowded.

Management Always take a periapical X-ray to exclude presence of a \$.

1 Before 3 erupted: If diastema <3 mm—review after eruption of canines as will probably close unaided. If >3 mm—may need to approximate incisors to provide space for canines to erupt, but care is required not to resorb roots of 2|2 against crowns of 3|3. If the incisors are tilted distally use an URA, otherwise FA. Retain.

2 After 3 erupted: Orthodontic closure will require prolonged retention as has high tendency to relapse. If frenum undoubtedly a major aetiological factor perform a frenectomy during closure, but retention still wise. Safest approach is to measure width of 1 2, and if they are narrower than average (1 = 8.5 mm, 2 = 6.5 mm) consider composite additions to close space. If teeth of normal width and no other orthodontic treatment required, try and talk patient into accepting their diastema.

Buccally displaced maxillary canines

▶ Width $\underline{3}$ > width $\underline{4}$ > width $\underline{C}$.

$\underline{3}$ is usually the last tooth to erupt anterior to $\underline{6}$. If the upper arch is crowded, $\underline{3}$ may be squeezed buccal to its normal position in which case space needs to be created for its alignment. Usually $\underline{4}$ is the tooth of choice for extraction and, if so, this should be carried out just as $\underline{3}$ is about to erupt. If there is plenty of space it is sufficient to keep the patient under review, otherwise fit an appliance.

Don't be tempted to start retraction of a partially erupted $\underline{3}$, as a spring active against the inclined plane of the mesial cusp will tend to intrude the tooth. Where possible, palatal finger springs should be used for retraction as they are easier to adjust, but if palatal movement is required then a buccally approaching spring will be necessary. The most efficient design of buccal spring is shown diagrammatically opposite. This should be fabricated in 0.5 mm wire for ease of adjustment and be tubed up to the coil for strength. The spring is activated (half a tooth width) by winding up the coil in the direction of desired movement.

Where $\underline{2}$ and $\underline{4}$ are in contact, extraction of $\underline{4}$ alone will not provide sufficient space to accommodate the canine and thought should be given to extracting $\underline{3}$ and accepting the status quo.

Less commonly, a canine may develop well-forward over the root of $\underline{2}$. In the latter case orthodontic treatment to align $\underline{3}$ will be prolonged and often results in a poor gingival margin due to lack of attached gingivae. If the arch is crowded it may be simpler to extract $\underline{3}$ and align remaining teeth.

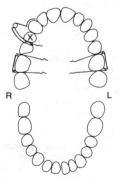

Upper removable appliance to retract and move 3| palatally
buccal canine retractor 3|0.5 mm tubed SS, cribs 6|6 0.7 mm SS

Palatally displaced maxillary canines

▶ Early detection is essential. ▶ Width $\underline{3}$ > width $\underline{4}$ > width $\underline{C}$.

Prevalence Up to 2%. Occurs bilaterally in 17–25% of cases. F>M.

Aetiology The maxillary canine develops palatal to $\underline{C}$ and then migrates labially to erupt down the distal aspect of $\overline{2}$ root. The aetiology is not fully understood, but research would suggest a lack of guidance is the reason behind the association with missing or short-rooted $\underline{2}$.[1]

Prevention Early detection may allow corrective interceptive Rx, ∴ when examining any child >9 yrs, palpate for unerupted $\underline{3}$. If there is a definite hollow and/or asymmetry between sides, further investigation is warranted. Extraction of $\underline{C}$ may result in improvement of a displaced $\underline{3}$,[2] but this eliminates the possibility of maintaining $\underline{C}$ should sufficient improvement in the position of $\underline{3}$ not materialize, ∴ confine to those cases where there is some evidence of resorption of the deciduous tooth and the $\underline{3}$ is not too far displaced. More markedly displaced $\underline{3}$ should be referred to a specialist.

Assessment Clinically by palpation and from inclination of $\underline{2}$ and by X-rays. An OPG and two intra-oral views taken in parallax (remember your PAL goes with you!) provide a good assessment. Consider also position and prognosis of adjacent teeth (including $\underline{C}$), the malocclusion, and available space.

Management If the canine is only very slightly palatally displaced or impacted between $\underline{2}$ and $\underline{4}$, provision of space should result in eruption. The majority of palatally displaced canines, however, do not erupt spontaneously, so hopeful watching and waiting may only result in an older patient who is less willing to undergo the prolonged treatment required to align the displaced tooth. Rx alternatives available:

1 Interceptive extraction of $\underline{C}$ in mixed dentition (see above).
2 Maintain $\underline{C}$ and keep unerupted canine under radiographic review. Provided no evidence of cystic change or resorption, removal of $\underline{3}$ can be left until GA required, e.g. for extraction of 8s. Patient must understand that $\underline{C}$ will eventually be lost, necessitating a prosthesis.
3 No Rx, if $\underline{2}$ and $\underline{4}$ are in contact and appearance is satisfactory, or if patient refuses other options. Again, $\underline{3}$ will require removal in due course.
4 Exposure and orthodontic alignment only feasible if (a) canine crown not displaced further towards midline than half-way across $\underline{1}$ on same side (this is the limit of orthodontic alignment), (b) sufficient space available for $\underline{3}$, or can be created, (c) patient willing to undergo surgery and prolonged orthodontic Rx (usually 2+ yrs). Sequence is to provide space, arrange exposure and allow tooth to erupt for 3

1 A. Becker 1981 *Angle Orthodontist* 51 24.   2 S. Ericson 1988 *Eur J Orthod* 10 283.

months and then commence orthodontic traction to move tooth towards arch. FA are usually required for a good result.
5 Transplantation is not an instant solution, as space is needed to accommodate $\underline{3}$ which may involve appliances and/or extractions. Poor long-term results have been reported, e.g. only $\frac{1}{3}$ still functional after 10 yrs.[1] However, shorter splinting times (1–2 weeks) and RCT for teeth with closed apices within 3 weeks of transplantation may ↑ prognosis.

Resorption Unerupted and impacted canines can cause resorption of incisor roots. For this to occur a 'head-on' collision between the two is required. If detected on X-ray, a specialist opinion should be sought, quickly. Extraction of the canine may be necessary to limit resorption, but if extensive, removal of the affected incisor may be preferable, thus allowing the canine to erupt.

Transposition always involves a canine tooth. In the maxilla 3 is usually transposed with $\underline{4}$ and in the mandible with $\overline{2}$.

1 J. P. Moss 1975 *Br J Oral Surg* **12** 268.

Increased overjet

When is an o/j increased? This is really a matter of opinion, but provided the arches are well-aligned, an o/j <6 mm is acceptable. If treatment is required for other reasons consider reduction of o/j >4 mm.

Aetiology

Skeletal pattern Increased o/j can occur in association with Class I, II, or even III skeletal patterns. If Class II, is often due to a normally sized mandible being positioned posteriorly on the cranial base. Be wary of patients with vertical proportions at either extreme of the range, as they are difficult to treat.

Soft tissues The effects of the soft tissues are usually determined by the skeletal pattern, as the greater the discrepancy the less likely it is that the patient will have competent lips. Where the lips are incompetent, the way an anterior oral seal is achieved will influence incisor position. E.g. if the lower lip is drawn up behind the upper incisors this may have contributed to the increased o/j, but if the incisors can be retracted within control of lower lip at the end of treatment the prognosis for stable o/j reduction is good. This is less likely if the LFH is increased and the lower lip lies beneath the upper incisors, as it will be unable to control their position following o/j reduction. The soft tissues can also help to compensate for the skeletal pattern by proclining the lower and/or retroclining the upper incisors.

Dental Crowding may contribute to an increased o/j ∴ relief of crowding may aid stability. Digit sucking can cause proclination of the upper and retroclination of the lower incisors, but this will resolve once habit is stopped unless maintained by adverse soft-tissue activity to achieve an oral seal.

In the majority of cases skeletal pattern will determine ease of treatment, but the soft tissues will influence the stability of the end result.

Stability of overjet reduction

Provided the inc have been retracted to a position of balance within the lower lip, this should not be a problem. Nevertheless, a period of retention is usually necessary to allow for periodontal fibre and soft-tissue adaption. However, prolonged retention will not make stable an inherently instable position. A common mistake is to stop treatment before o/j reduction is complete and the lips competent. If the patient returns to retracting the lower lip behind the ULS to achieve an anterior oral seal, the o/j is likely to increase.

Management of increased overjet

(See also Functional appliances, p. 197).

Principles

1 Provision of space for o/j reduction ± relief of crowding.
2 Reduce o/b before o/j reduction (p. 168).
3 ? reduce o/j by tilting (URA) or bodily movement (FA).

Class I or mild Class II skeletal pattern

Stability of o/j reduction ↑ with age as the lips mature and are held together by the patient, ∴ early treatment is likely to require prolonged retention. Unless a functional appliance is indicated, it is advisable to await the 2° dentition before embarking on treatment to reduce an o/j. Provided ↑ o/j due to proclination of the upper and not retroclination of the inc, the majority of patients in this group can be treated with an URA. For a Class II/1 with moderate crowding a likely sequence of treatment is

1 Fit an URA with palatal finger springs to retract 3|3, cribs 6|6, a long labial bow 6–6 and a flat anterior bite plane to ↓ o/b.
2 Once the patient can wear appliance satisfactorily, extract all 4s.
3 Retract 3|3 into class 1 relationship with 3|3 (allowing for any expected improvement in 3|3 position).
4 Fit new appliance to reduce o/j (or add strap spring to labial bow).
5 Retain, usually 3–6 months full-time, then nights only for 3–6 months.

Following extraction of 4|4 will get spontaneous improvement in alignment of lower inc provided patient still growing and 3|3 mesially inclined (if not FA indicated). If lower arch not crowded consider either distal movement of the upper buccal segments (p. 154), or functional (p. 192), or extractions in upper arch only.

Moderate to severe Class II skeletal pattern

Approaches available:
1 Modification of growth—either by restraint of maxillary growth with headgear, or by encouraging mandibular growth with a functional appliance.
2 Orthodontic camouflage—by extractions in upper arch and bodily movement of inc. Requires FA for good result.
3 Surgical correction.

Because mandibular growth predominates during teens, a greater proportion of Class II skeletal problems are amenable to orthodontic correction than Class III. Research would suggest that the amount of growth modification that can be achieved is limited, but every little helps and in practice the majority of growing children in this category are treated by a combination of approaches 1 and 2. This usually takes the form of an initial phase of functional appliance therapy, followed by FA in the permanent dentition ± extractions. Adults whose skeletal pattern is not too severe may be treated by orthodontic camouflage, but in cases with a more severe skeletal problem &/or an ↑ o/b a surgical correction may be the only option.

Upper removable appliance to retract 3|3
palatal finger springs 3|3 0.5 mm SS
cribs 6|6 0.7 mm SS
long labial bow 6–6 0.7 mm SS
flat anterior bite plane–½ height 1|1 o/j = 6 mm

Increased overbite

Normal o/b is between $\frac{1}{3}$ to $\frac{1}{2}$ overlap of the lower incisors. It is more practical to record o/b in terms of whether it is increased, decreased or normal, than to try to measure it with a ruler. Increased o/b is associated with Class II/2 incisor relationship, where typically 1|1 are retroclined and 2|2 are proclined, reflecting their relationship to the lower lip. But the o/b can also be ↑ in Class III and II/1 malocclusions. Increased o/b *per se* is not an indication for treatment, unless it is traumatic and this is relatively rare, but ↓ of o/b may be necessary before correction of other anomalies. In Class III cases an ↑ o/b is advantageous as this will help to retain the corrected incisor position.

Aetiology ↑ o/b occurs because the incisors are able to erupt past each other due to a combination of some or all of the following interrelated factors: ↓ LFH, high lower lip line, retroclined incisors, ↑ inter-incisal angle. Normal inter-incisal angle is 135°. Highest acceptable angle is 145°. Above this value the tendency for the lower incisors to erupt may be inadequately resisted.

Approaches to reducing overbite

1 *Extrusion/eruption of molars*. Passive eruption of lower molars occurs when an URA incorporating a biteplane is worn. Active extrusion of molars in either arch is possible using a FA. However, unless the patient grows vertically to accommodate this increased dimension, the molars will re-intrude under the forces of occlusion once appliances are withdrawn. This approach is of limited value in adults.

2 *Intrusion of incisors*. This is difficult, requires FA and in most cases the major effect is extrusion of the buccal segments. Can use high-pull headgear attached to an upper archwire by J-hooks to try and hold upper incisors whilst the rest of the face grows vertically.

3 *Proclination of lower incisors*. This will only be stable if the LLS has been trapped behind the ULS, in which case, provision of a removable biteplane may allow the lower incisors to spontaneously procline. Active proclination should only be attempted by the experienced orthodontist who will be better able to judge those cases where this is indicated.

4 *Surgery*. This approach should only be considered as a last resort, but lack of vertical growth in adults may result in this being the only alternative in severe cases.

Stability of overbite reduction depends upon removing the aetiological factors, but ↓ LFH and high lower lip line can only be altered if growth is favourable. Reduction of the inter-incisal angle is necessary to provide a 'stop' to the incisors re-erupting, but requires FA to move incisor apices lingually.

Management of increased overbite

Class II/2 In milder cases it may be wiser to accept ↑ o/b and confine any treatment towards relief of crowding and alignment. It is often prudent to accept a mild degree of lower arch crowding, as extractions may be followed by lingual tipping of the lower incisors resulting in a further ↑ of the o/b. In these cases treatment can be directed towards alignment of the upper arch only. Where 2|2 are mesiolabially rotated a FA will be required to de-rotate them, followed by prolonged retention, so unless FA are indicated for another aspect of the malocclusion, acceptance of a mild degree of 2|2 rotation may be easier. Cases with sufficient crowding to warrant premolar extractions in the lower arch and moderately to severely increased overbite are best treated with FA to close space by forward movement of the buccal segments.

Where o/b reduction is required, the inter-incisal angle will need to be reduced in order to achieve a stable result. Usually this necessitates FA, but in growing patients with a skeletal II pattern and no or mild crowding an alternative approach is to procline inc with an URA and then use a functional appliance.

Class II/1 O/b ↓ is required before o/j ↓. In cases amenable to using an URA, o/b ↓ should be commenced with a flat anterior biteplane as the canines are retracted. The biteplane should be checked at every visit and when the lower molars have erupted into contact with the opposing arch, re-activated by the addition of some cold-cure acrylic at the chairside. If o/b ↓ is slow, check that the patient is wearing the appliance for meals. In some patients (particularly those with a reduced LFH) intrusion of the incisors with a fixed appliance will be necessary; however, nothing is lost by trying a biteplane first.

Class III (p. 172) Avoid reducing o/b as it will aid retention of the corrected incisor position.

Retention Where the incisal relationship has been changed, retention ideally should be continued until growth is complete. In most patients this is not practicable.

Anterior open bite (AOB)

Vertical overlap of incisors (o/b)

↑o/b	normal o/b	incomplete o/b	AOB

AOB can occur in Class I, II, and III malocclusions.

Aetiology Either *skeletal*—vertical > horizontal growth (↑LFH &/or ↑MMPA), or *environmental*—habits, tongue thrust, iatrogenic, or a combination. If the distance between maxilla and mandible is sufficiently increased such that even if incisors develop to their full potential they do not meet, an AOB will result. This is often associated with incompetent lips and a lip to tongue anterior oral seal, which may exacerbate the AOB. Tongue thrusts are almost always adaptive and can maintain an AOB due to a habit even after the habit has stopped. Localized failure of maxillary dento-alveolar development resulting in an open bite is seen in CLP.

Treatment is generally difficult except where due mainly to a habit, ∴ it is wise to refer patient to a specialist for advice.

Skeletal In milder cases can align arches and accept, or try to restrain vertical development of the maxilla &/or upper molars with headgear &/or a functional appliance with posterior bite blocks. Extrusion of the incisors is unstable. For more severe cases the only alternative is surgery, but even this is not easy and liable to relapse.

Habits Better to await natural cessation of habit, but not if that means defering treatment for other aspects of malocclusion. Once habit stops o/b should re-establish within 3 yrs, unless perpetuated by soft tissues or skeletal in origin.

Tongue thrust None.

Hints for cases with ↑ vertical dimension and ↓o/b or AOB.
- Avoid extruding molars, e.g. cervical pull headgear to 6|6, URA with a biteplane.
- Avoid upper arch expansion as this will tip down the palatal cusps of buccal segment teeth, reducing o/b.
- Extraction of molars will not 'close down bite'.
- Space closure occurs more readily in patients with ↑ LFH and ↑ MMPA.

Reverse overjet

This will include only those cases with >2 teeth in linguo-occlusion, i.e. Class III cases. For management of 1 or 2 teeth in Xbite, see p. 176.

Aetiology

Skeletal Reverse overjets are usually associated with an underlying Class III skeletal pattern. This is most commonly due to either a large mandible &/or a retrusive maxilla. Class III malocclusions occur in association with the whole range of vertical patterns. Crossbites are a common feature due either to a large mandible or the anterior position of the mandible relative to the maxilla.

Soft tissues A patient's efforts to achieve an anterior oral seal often result in dento-alveolar compensation, i.e. retroclination of the lower and proclination of the upper incisors. Therefore the incisor relationship is often less severe than underlying skeletal pattern.

Dental crowding This is usually greater in the upper than the lower arch.

Assessment (p. 138) Consider also the following:
- Patient's opinion about their facial appearance (be tactful!).
- Severity of skeletal discrepancy.
- Amount of dento-alveolar compensation. If upper incisors already markedly proclined, further proclination is undesirable.
- Amount of overbite. Remember that proclination of ULS will reduce o/b and retroclination of LLS will increase o/b.
- Can patient achieve an edge-to-edge contact of the incisors. If not, simple treatment C/I.

Treatment planning, see p. 144.

▶ Class III malocclusions tend to become worse with growth.
▶ In severe cases seek a specialist opinion before embarking on treatment or extractions, as if surgery is necessary, de-compensation will probably involve the reverse of orthodontic camouflage.

Major factors determining treatment approach are skeletal discrepancy and overbite.

Skeletal pattern	Normal or ↑ o/b	↓ o/b
mild to normal	procline ULS	accept
moderate	FA to procline ULS and retrocline LLS	accept or FA to retrocline LLS ± procline ULS
severe	surgery	align and accept if possible or surgery but ? relapse

If child on borderline between groups, assume more severe as growth will probably prove you right.

Management of reverse overjet

Relief of crowding Extractions in the upper arch alone may run the risk of worsening the incisor relationship, ∴ it is advisable to extract at least as far forward in lower arch as upper.

Moderate crowding responds best to extraction of premolars. Extraction of 5|5 will maintain 4|4 to support the ULS, but often crowding in the upper arch necessitates extraction of 4|4. If using URA only to procline ULS, extraction of lower premolars may result in some residual spacing, but if adequate o/j and o/b exist this will provide space for further dento-alveolar compensation by retroclination of LLS if mandibular growth continues. If using FA to procline ULS and retrocline LLS, space will be required in the lower arch to accomplish this. Distal movement of the upper buccal segments is C/I in Class III as restraint of maxillary growth is undesirable.

Practical treatment

Accept This may be the wisest option for those patients with increased LFH and reduced o/b, with treatment directed towards achieving alignment within the arches only.

Proclination of ULS only This is only suitable for milder cases where ULS is not already proclined and where sufficient o/b will be present to retain the corrected incisor position. Best carried out in the mixed dentition, provided 3|3 are not sitting labial to the roots of 2|2. If extraction of C|C necessary for space, it is advisable to match this with loss of C̄|C̄ to avoid compromising the incisor relationship. In the mixed dentition retention of URA is often a problem, ∴ use a screw appliance so that the teeth to be moved can also be cribbed. If there is sufficient o/b, stability is not usually a problem, but if the incisors are not fully erupted the appliance can be worn as a retainer until an o/b develops. Often there is adequate o/b for the 1|1, but not 2|2; if so, a bonded retainer is advisable, but remember that no further eruption will occur once it is in place.

Retroclination of LLS ± proclination of ULS requires FA. By interchanging the position of the incisors within the neutral zone, stability is not compromised. Class III elastic traction from the back of the upper arch to the 3̄ region helps to retrocline LLS. However, in addition to extruding the incisors which is desirable, also results in extrusion of the upper molars, which reduces o/b. In order to try and limit this tendency can run elastics from the middle of the upper arch to the lower canine area.

Orthodontics and orthognathic surgery, p. 196.

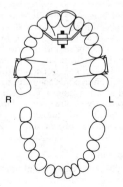

Screw appliance to procline $\underline{21|12}$
cribs $\underline{6|6}$ 0.7 mm SS
southend clasp $\underline{1|1}$ 0.7 mm
buccal capping

Crossbites

By convention should describe the lower teeth relative to the upper (p. 136). Crossbites can be anterior or posterior (unilateral/bilateral), with displacement, or no displacement.

Aetiology Xbites can be skeletal &/or dental in origin. For posterior Xbites, the skeletal component is usually the major factor. Antero-posterior discrepancies obviously play a part in anterior Xbites, but can also result in posterior Xbites in Class II (lingual Xbite) and Class III (buccal Xbite) skeletal patterns.

Displacement may occur when a premature or deflecting cuspal contact is encountered on closure and the mandible is postured either anteriorly or laterally to achieve better interdigitation. This new path of closure becomes learned and the patient closes straight into maximum interdigitation. To help detect displacement on closure, try to get the patient to close on a hinge axis by asking them to curl their tongue back to touch the back of the palate and then close together slowly, whilst guiding the mandible back via the chin. In addition, look for other clues like a centre-line shift (of lower in direction of displacement) in association with a posterior unilateral Xbite. Displacements are one of the few functional indications for orthodontic treatment, as it has been shown that displacing contacts may predispose to TMPDS.[1]

Anterior crossbites

Class III malocclusions are considered on p. 172. Should be treated early, especially if associated with a displacement, provided sufficient o/b exists to retain the result. If not, probably best to defer until the 2° dentition and use FA. Correction of 1 or 2 teeth with reverse overjet and a +ve overbite can usually be accomplished using an URA; however, need to have space available to accommodate the tooth in the arch (or create with extractions). Application of a force to procline the ULS results in an URA being unseated anteriorly, ∴ good anterior retention is usually required. In the mixed dentition, the morphology of the deciduous teeth makes this difficult. A screw appliance has the advantage that the teeth to be moved can also be clasped. Buccal capping should be added to free the tooth to be moved from contact with lower arch. Upper lateral incisors that are displaced bodily and palatally due to lack of space are not amenable to simple proclination. Refer patient to a specialist.

Posterior crossbites

Unilateral Generally the greater the number of teeth involved the greater the skeletal contribution to the aetiology. If only 1 or 2 teeth, movement of opposing teeth in opposite directions for correction may be required. This can be achieved by cross-elastics attached to bands on the affected teeth. 5 are often

---

1 R. W. Wassell 1989 *J Dent* **17** 101.

crowded palatally, but are easily aligned using a T-spring on an URA. Unilateral Xbite from the canine region distally is usually associated with a displacement, as true skeletal asymmetry is rare. If the arches are of a similar width, displacement to the right or left will give better interdigitation. In these cases treatment should be directed towards expanding the upper arch so that it fits around the lower. Provided the upper teeth are not already buccally tilted, this can be accomplished with an URA incorporating a midline screw to which buccal capping or a FABP should be added (depending on amount of overbite) to free the occlusion with the lower arch and prevent reciprocal expansion.

Bilateral buccal crossbite This suggests a greater underlying transverse skeletal discrepancy. Rarely, associated with displacement. Correction of a bilateral Xbite should be approached with caution, because partial relapse may result in the teeth occluding cusp to cusp and development of a unilateral Xbite with displacement.

Bilateral lingual crossbite (or scissor bite) occurs due to either a narrow mandible or a wide maxilla. In milder cases 4|4 may only be involved, and if these teeth are extracted to relieve crowding or for retraction of 3|3, so much the better. Where all of the buccal segments are involved, treatment will probably involve expansion of the lower and/or contraction of the upper, therefore refer to a specialist.

Rapid maxillary expansion

Involves a screw appliance comprising bands attached to 64|46 and connected to a midline screw. The object is to expand the maxilla by opening the midline suture and is ∴ more successful in younger patients. Large forces are required to accomplish this, ∴ the screw is turned 0.2 mm twice a day for about two weeks. Over-expansion is necessary as the teeth relapse about 50% under soft tissue pressure. Not to be attempted by the inexperienced!

Quad helix appliance

This is a very efficient fixed, slow expansion appliance. Suitable for mixed or 2° dentition. Attaches to upper teeth by bands on 6 and is W-shaped.

Upper removable appliance to expand upper arch with midline screw.

cribs <u>6|6</u> 0.7 mm SS

<u>4|4</u> 0.6 mm SS

buccal capping or FABP

Anchorage

Anchorage is defined as the source of resistance to the reaction from the active component(s) in an appliance. In practice, it is the balance between the applied force and the available space. E.g., in a case where 3|3 are being retracted following extraction of 4|4, an equal, but opposite force will also be acting on 65|56. The amount of forward movement of these anchor teeth will depend upon their root surface area and the force used.

Anchorage loss can be minimized by limiting the number of teeth being moved at any one time, applying the correct force for the movement required and increasing the resistance of the anchor teeth (e.g. by permitting only bodily movement). In some situations, movement of the anchor teeth is desirable, e.g. in a Class III where space is being opened up for an unerupted 5. However, it is important to assess the anchorage requirements of a particular malocclusion before embarking on treatment, and if indicated reinforce from the start.

Re-inforcing anchorage

Although it is tempting in some patients to re-inforce anchorage by means of a nail through their URA, the more accepted methods include:

Intra-maxillary (teeth in same arch) By including the maximum number of teeth in the anchorage unit. Applicable to both fixed and removable appliances.

Inter-maxillary (teeth in opposing arch) This is achieved by running elastics from one arch to the other. It is mainly applicable to FA, as an URA will be dislodged. The direction of elastic pull is described according to the type of malocclusion to which it is applicable.

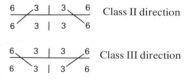

Increased mucosal coverage By virtue of its palatal coverage an URA has more potential anchorage than a FA.

Extra-oral Anchorage is transmitted to the appliance by elastic or spring force from a head or neck strap. This involves either (1) a face-bow which engages tubes soldered on to molar cribs (URA) or bands (FA), or (2) J-hooks which engage hooks soldered on to a short labial bow (URA) or the archwire (FA). The direction of pull can be selected according to the malocclusion for FA, but for URA a force above the occlusal plane will aid retention. For EO anchorage, 250 g for 10 h/day should suffice. For EO traction (i.e. using the head-gear force to achieve movement rather than just resist it) forces in the region of 500 g for 14–16 h/day are necessary.

Commercially made headgear components, particularly the types with built-in safety features are worth the extra outlay.

Anchorage loss

May occur because of (1) insufficient wear of an URA allowing anchor teeth to come forward into extraction space, (2) active force exceeding available anchorage (often due to over-activation or too many teeth being moved at a time).

Removable appliances—design

URA can be used as the active appliance for some, or all, of the tooth movements required for a malocclusion; or as a passive space-maintainer or retainer.

Advantages of URA	Disadvantages of URA
Easier to clean than FA	Patient can leave appliance out
Easy to adjust, therefore $\downarrow$ chairside time	Only tipping movement possible
Contact with mucosa $\uparrow$ anchorage	Affects speech
Can be used for o/b $\downarrow$	Good technician required

Lower removable appliances are difficult to wear, -/FA preferable.

Design Four components need to be considered for every appliance.

Active component Exerts force required for desired movement (p. 184).

Anchorage Resists force generated by active component (p. 180).

Retention is required to retain appliance in mouth. The best retention posteriorly is provided by the Adams crib, which is made in SS in either 0.7 mm (for 6, 7, 8) or 0.6 mm (for 3, 4, 5, D, E) wire. These clasps are very technician sensitive and no amount of adjustment will compensate for a badly made crib. Should engage about 1 mm of undercut, which on a child's molar may be at, or just under, the gingival margin and in an adult may only be part-way down the clinical crown. The versatility of the crib can be increased by soldering tubes for EOT, labial bows, or buccal springs on to the bridge of the clasp. Anterior retention can be gained by a labial bow or either an Adams or Southend Clasp. The latter are made of 0.7 mm wire and follow the gingival contour round 1|1. They are adjusted by bending the wire between 1|1 towards the teeth.

Baseplate not only holds other elements together, but may itself be active. Heat-cure acrylic is more robust than self-cure.
- A flat anterior bite plane should only be prescribed if o/b $\downarrow$ is required. In case your technician isn't telepathic, it is wise to specify the height (e.g. $\frac{1}{2}$ height of 1|1) and how far back the biteplane should extend (e.g. o/j = 8 mm). In order to $\uparrow$ the likelihood of the URA being worn the molars should only be separated by 1–2 mm by the biteplane, $\therefore$ self-cure acrylic should be added during treatment in order to continue o/b $\downarrow$.
- Buccal capping frees the occlusion on the tooth being moved and allows further relative eruption of the incisors ($\therefore$ is C/I if o/b is already $\uparrow$).
- ▶ The easier it is for the patient to insert and wear their URA, the more likely it is that treatment will be successful, $\therefore$ keep design simple.

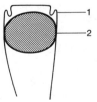

Adjustment of Adams crib
1 arrowhead moves horizontally in towards tooth.
2 arrowhead moves in towards tooth and also vertically towards gingival crevice.

Removable appliances—active components

Springs

The relationship between the distance (d) that an orthodontic spring is deflected on activation, the length of the spring (l), the diameter or radius (r) of the wire it is composed of, and the force generated is:

$$F \propto \frac{d\,r^4}{l^3}.$$

A commonly made mistake is to activate a spring in 0.7 mm wire the same distance as one of 0.5 mm wire. In fact, for the same design of buccal spring, the 0.5 mm spring should be activated about 3 mm, whereas the 0.7 mm spring requires only 1 mm activation to deliver 25–50 g force, leaving little margin for error. An almost infinite variety of designs exist, some of the more useful types are described below.

Palatal finger springs are the most commonly used active components for mesial or distal movement along the arch. They are fabricated in 0.5 mm wire and are easy to adjust and activate. If boxed out from the acrylic and made with a guard wire they are more stable in the vertical plane than a buccal spring and ∴ should be used in preference. Where palatal movement is required, a **buccally approaching spring** is necessary (p. 158).

For buccal movement of incisors use a **Z-spring** (for 1 tooth in 0.5 mm, for >1 tooth in 0.6 mm) and **T-springs** for premolars (0.5 mm) and molars (0.6 mm). These are all activated only 1–2 mm by pulling them away at 45° from baseplate in the desired direction of movement.

Springs are also used for o/j↓. **The Roberts' retractor** comprises an apron spring of 0.5 mm wire supported in 0.5 mm tubing, which requires fabrication of a new appliance for o/j↓ and then a separate retainer. **Strap springs** (0.5 mm wire) can be wound on to the labial bow of the appliance used for canine retraction. They are activated by tightening the supporting labial bow and can be removed once the o/j is reduced so the same appliance is used as a retainer. Strap springs are said to flatten the labial segment, but in practice this does not seem to be a problem.

Elastics

Can also be used to reduce an o/j. Either attached to a purpose-designed appliance or by converting a labial bow by cutting it and bending the ends into hooks adjacent to 3|3. Whilst popular with patients, they do have a tendency to slide up proclined incisors and retract the gingivae instead. Elastics are also useful for aligning displaced teeth, e.g. 3, from an attachment bonded on to the tooth surface and a hook soldered on to a labial bow or crib (alternatively, place under the arrowhead of an Adams crib).

Screws

Basically two types, (1) spring-loaded piston (e.g. Landin)—activated by advancing the whole assembly forward, (2) the screw type (e.g. Glenross) which is opened (or closed) by means of a key, a quarter turn separating the two halves by 0.2 mm. The screw type is useful when the teeth to be moved need to be clasped for retention (e.g. proclining incisors in the mixed dentition), and for expanding the upper arch. It is advisable to start patients turning the screw only once a week, progressing on to a maximum of two turns per week. If worn intermittently, screw appliances will become progressively ill-fitting, in which case the screw should be turned back until the URA fits and ↑ wear encouraged. Remember that these screws have about 18 activations and if necessary make a second appliance.

Removable appliances—fitting and monitoring progress

Instruments For each visit: Universal pliers (No. 64), Maun's wire cutters, spring forming pliers (No. 65), dividers, steel rule, straight handpiece and acrylic bur. Also the patient's study models and records. Hollow-chop pliers are ideal for adjusting springs with rubber gloves on, as you just insert the wire and squeeze.

Fitting an URA Explain again what treatment will involve. A good idea is to have this in the form of a letter which the patient's parent signs.

▶ It is advisable to fit the appliance and check on wear before carrying out any extractions.

- Check it is the correct appliance for the correct patient, and the prescription has been followed.
- Examine fitting surface and smooth any sharpness.
- Show patient appliance and how it works.
- Try in URA, adjusting retention until it clicks positively into place.
- Trim biteplane or buccal capping if required.
- Activate appliance slightly (if not relying on extractions to provide space).
- Show patient how to insert and remove appliance, using cribs.
- Go through instructions (see opposite) with patient and parent. It is helpful to explain that URA will feel uncomfortable at first, like a new pair of shoes, and will need to be 'worn in'.
- Arrange next appointment (ideally, review every 3–4 weeks).
- If appliance made of self-cure acrylic, keep working model for repairs.

Monitoring progress See instruments above.
- Refer to treatment plan to jog memory.
- Check patient's OH and if necessary reinforce. Poor OH can contribute to candidal infection of the mucosa underlying the appliance, Rx: OHI of teeth and URA, and antifungal.
- Look for anchorage loss by recording o/j and buccal segment relationship. If necessary reinforce anchorage with headgear.
- Measure tooth movement since last appointment, using dividers, and indent records for reference.
- Adjust retention. Some children click their URA in and out which flexes the cribs and makes the URA loose, ∴ should be discouraged.
- Add self-cure acrylic to bite-plane or buccal capping if required. For trimming of biteplane during o/j ↓ see diagram opposite.
- Re-activate appliance.
- Record what you have done and mention what action you think will be necessary next visit, e.g. 'add to b/plane NV'.

Causes of slow rate of tooth movement Normally expect about 1 mm/month for child, less for adult.

- Patient not wearing appliance full-time. Clues include URA still looks new, no marks in mouth, lisping, frequent breakages. Asking patient 'how long are you managing to wear your brace' is more likely to elicit a truthful response than 'are you wearing your brace all the time?' Try and find out why patient is not complying; remotivation or even a change of design may help.
- Spring inactive or distorted. Check how patient is removing and cleaning appliance, before carefully re-adjusting spring.
- Spring inserted incorrectly. Explain purpose to patient.
- Tooth movement obstructed by wires or acrylic of appliance. Remove or adjust.
- Tooth movement may also be slowed by occlusion with opposing arch. May need lower appliance if building up bite-plane or capping doesn't suffice.

During o/j reduction the bite plane needs to be trimmed to allow space for the retraction of upper incisors (and palatal gingivae) whilst maintaining o/b reduction.

Instructions to patient

A sheet or sticky label with instructions for the patient about their URA, are invaluable. It is important to note in the patient's records that written and verbal instructions have been issued, in the eventuality of something going wrong. A specimen set of instructions are outlined below.

1 Wear your appliance all the time, including in bed at night and at meal times.
2 Your appliance should be removed after meals for tooth-brushing, and replaced immediately.
3 It is usual to experience some discomfort initially; this will pass as you get used to appliance.
4 Do not eat hard or sticky foods.
5 Let us know immediately (on phone number 1234) if you cannot wear your appliance.

Fixed appliances

▶ FA should only be used in co-operative patients with good OH, to minimize damage.

As the name implies, FA are attached to the teeth. They vary in complexity from a single bracket used in conjunction with an URA, to attachments on all teeth. URA are limited to tilting movements, but FA can tilt, rotate, intrude, extrude, and move teeth bodily. Not surprisingly, FA have a greater propensity for things to go wrong.

Principles

- Treatment planning (p. 144), but with ↑ attention to anchorage requirements, especially if apical movement is planned.
- As FA are able to achieve bodily movement it is possible (within limits) to move teeth to compensate for a skeletal discrepancy.
- FA can be used in conjunction with other appliances &/or headgear.
- For initial alignment flexible archwires are used, but to minimize unwanted movements, progressively more rigid archwires are necessary.
- Archwires should be based on the pre-treatment lower arch-form for stability.
- Mesio-distal movement is achieved either by (1) sliding the teeth along the archwire with elastic force (sliding mechanics), or (2) moving the teeth with the archwire.
- Intermaxillary traction is often used to aid anteroposterior correction and increase anchorage.

Components of fixed appliances

Bands Usually used on molar teeth so that the end of the archwire is retained even if band becomes loose. Indicated for other teeth if bonds fail or lingual attachment is required for de-rotation. If tooth contacts are tight these will need to be separated prior to band placement using an elastic doughnut stretched around the contact point for 1–7 days. Use of GI cements helps to ↓ decalcification.

Bonds are attached to enamel with (acid-etch) composite. There are three types: (1) metal (poor aesthetics), (2) plastic (become stained), (3) ceramic (prone to # and can cause enamel wear).

Archwires SS remains the most widely used material, but nickel titanium, tungsten molybdenum, and cobalt chromium alloys are also popular. The greater flexibility of these newer wires means that rectangular archwires can be used earlier in treatment.

Auxiliaries Elastic rings or wire ligatures are used to tie the archwire to the brackets. Forces can be applied to the teeth by auxiliary springs or elastics.

Fixed appliances (*cont.*)

Types of fixed appliance

Almost infinite variety but most are based on:

Begg uses round wires which fit loosely into a vertical slot in the bracket, thus allowing the teeth to tip freely. Auxillaries are required to achieve apical and rotational movements.

Edgewise uses rectangular brackets that are wide mesio-distally for rotational control. Round wires are used initially for alignment, but rectangular wires are necessary for apical control.

Pre-adjusted systems These 'pre-programmed' brackets allow ↑ use of pre-formed archwires. As each tooth has its own individual bracket these systems are more expensive, but that is offset by savings in operator time. Andrews' Straight Wire appliance is to pre-adjusted systems what the Hoover is to vacuum cleaners.

Lingual appliances Popular with patients, but not with orthodontists as they are difficult to adjust!

Whip spring appliance[1] is useful for derotating 1 or 2 teeth. An edgewise bracket is bonded to the tooth to be moved and a whip spring is fabricated, as shown opposite, and ligated into the bracket. This is engaged on to the labial bow of an URA by means of a hook. A bonded retainer is advisable.

1 T. I. McCartney 1981 *BJO* **8** 37.

Diagram to show whip spring and URA for derotation of 2.
Whip spring is fabricated in 0.016-inch SS wire and incorporates a hook to engage on to the labial bow of URA.

Functional appliances—rationale and mode of action

Definition Functional appliances utilize, eliminate, or guide the forces of muscle function, tooth eruption and growth to correct a malocclusion.

Philosophy The term functional appliance dates back to a belief that by eliminating abnormal muscle function, normal growth and development would follow. Nowadays, the importance of both genetic and environmental factors in the aetiology of malocclusion is acknowledged, but functional appliances are still successfully used to correct Class II malocclusions by a combination of skeletal and dental effects. Functionals can also be used in the treatment of AOB and Class III, but generally alternative approaches are more successful, ∴ we shall only consider Class II malocclusions.

Mode of action In the average child the face grows forward relative to the cranial base and mandibular growth predominates. Functional appliances help to harness this change to correct Class II malocclusions by a combination of force application and force elimination. The relative contribution of which depends upon the design of appliance. Force application usually takes the form of inter-maxillary traction i.e. a restraining effect on the maxilla and maxillary teeth and a forward pressure on the mandible and mandibular teeth. A similar effect is produced with Class II elastics (p. 180). Functional appliances are ineffective for individual tooth movement.

Application

1 The main treatment of a malocclusion. Ideally, for Class II/1 with mandibular retrusion, average or reduced LFH, upright or retroclined LLS, and well-aligned arches. A useful test is to examine the profile with patient postured forward to a Class I incisor relationship and if not improved consider another appliance.

2 To achieve some antero-posterior correction for a severe Class II malocclusion prior to FA ± extractions.

Changes produced by functional appliances

Skeletal
- Restraint or redirection of forward maxillary growth.
- Optimizing of mandibular growth. Some proponents of functional appliances have claimed increased mandibular growth. Others have replied that the changes seen are small and unsubstantiated long-term. The debate continues! It is therefore better to consider functional appliances as providing an environment for achieving an individual's maximum mandibular growth potential.
- Forward movement of the glenoid fossa.[1]
- Increase in LFH.

---

1 G. Woodside 1987 *Am J Orthod Dentofac Orthop* **92** 181.

Dental
- Lingual tipping of the upper incisors.
- Labial tipping of the lower incisors (not a consistent finding).
- Inhibition of forward movement of the maxillary molars.
- Mesial and vertical eruption of the mandibular molars.

Keys to success with functional appliances

- Co-operative and keen patient. Remember that co-operation is finite.
- Favourable growth, therefore coincide treatment with pubertal growth spurt (girls 11–13, boys 13–15).
- Confident operator, so that child believes appliance will work and will persevere with wearing it.

Types of functional appliance and practical tips

Choice of appliance

Except for those designed for cases with ↑ LFH, the effects produced by the different types of appliance are similar. It is wiser to become familiar with one particular design. For each appliance well-extended upper and lower impressions are required and a construction wax bite. Some of the more popular types are:

Harvold Originally designed with a construction bite opened to 11 mm (although as little as 5 mm has also been successfully used) and postured forward as far as possible. A second appliance may be required to complete sagittal correction. Worn 14–16 h per day. It is useful to incorporate cribs on 6|6 to aid retention initially, but if used some preliminary expansion of upper arch may be needed. Bulky appliance.

Bionator Less bulky with an edge to edge construction bite and no lower incisor capping. Worn full-time, except meals.

Frankel advocated for cases with abnormal soft-tissue pattern, e.g. lower lip trap. Several sub-types of which FRII is the most popular. Buccal shields allow expansion of the arches, but the long-term stability of this is unsubstantiated. Construction bite is forward 6 mm and open 3–4 mm in premolar region. Worn full-time. Difficult to repair or adjust, but can be re-activated by dividing buccal shields and advancing.

Medium Opening Activator A preliminary phase of upper arch expansion is required in most patients as appliance incorporates cribs to aid retention. Construction bite as for Frankel. Worn full-time. Must be made in heat-cure acrylic as lower arch extensions prone to fracture.

Functionals in Class II/1 with ↑ LFH

Need appliance with molar capping to try and prevent molar eruption, encourage auto-rotation of the mandible and thus ↓ of LFH. Increasing the bite opening of an appliance, is thought to result in a more forward direction of mandibular growth.[1] Some designs include high-pull headgear to restrain vertical maxillary growth.

Practical tips

- Advise patient to wear appliance for first week evenings only, then at weekend progress to wearing in bed as well. Subsequently build up to 14 h per day or full-time (except meals) depending upon the design. See every 2 months.
- If some preliminary expansion necessary, give patient expansion plate to re-insert should they be unable to wear functional for any reason.

1 S. E. Bishara 1989 *Am J Orthod Dentofac Orthop* **95** 250.

- Appliances that fall out at night are often cured by increasing wear during the day. If a problem despite good wear, add cribs or change design.
- Expect at least 1 mm o/j reduction per month.
- Wise to continue until o/j almost edge-to-edge.
- Retain by wearing nights only, ideally until growth rate has slowed and permanent dentition established.

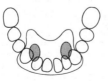

Medium Opening Activator

labial and palatal wires 0.9 mm SS

cribs 6|6 0.8 mm SS

occlusal stops 6|6 0.8 mm

deep lower incisor capping

heat-cure acrylic

(shaded area indicates acrylic posts joining two halves of appliance)

Orthodontics and orthognathic surgery

Orthognathic surgery is the correction of skeletal discrepancies outwith the limits of orthodontic treatment, either because of their severity or a lack of growth. Usually deferred until growth complete.

Diagnosis and treatment planning

This is best undertaken jointly by orthodontist and oral surgeon. Require the following information:

Patient's perception of problem c/o Appearance of jaws or teeth, speech, or mastication? Are patient's expectations realistic?

Clinical examination Assessment of the balance and proportions of full face and profile.[1]

Study models Duplicate for model orthodontics/surgery. For bimaxillary procedure mount on semi-adjustable articulator.

Radiographs require OPG and lateral skull plus PA skull for asymmetries. It is helpful to compare patient's cephalometric tracing with the ideal (Bolton Standard) to visually assess areas of discrepancy. A number of computer programs are available to aid in diagnosis and planning. The most useful include a facility to look at the result of a particular surgical option on the screen before printing. However, these should not supersede clinical assessment in planning.

Photographs Required as pre-treatment record and can also be enlarged to match 1:1 with lateral skull for visual predictions.

It is important to correlate desired facial changes with patient's occlusion. Pre-surgical orthodontics may be required to decompensate teeth so that a full surgical correction is possible.

Include the patient in treatment planning so they understand what is involved. It is also helpful if the prospective patient can meet a previous (successful!) candidate.

Sequence of treatment

Pre-surgical orthodontics This is not always necessary, but FA does provide a means of fixation at surgery. Aim of orthodontic treatment is to align and co-ordinate the arches so that the teeth will not interfere when the jaws are placed in their correct position. This usually involves decompensation, i.e. removal of any dento-alveolar compensation for the skeletal discrepancy so that the teeth are at their correct axial inclinations and a full surgical correction can be achieved. If a segmental procedure is planned space will be needed interdentally for surgical cuts. It is inefficient to carry out movements that can be accomplished more readily at surgery (e.g. expansion of upper arch if Le Fort 1

---

1 N. P. Hunt 1984 *Br J Orthod* **11** 126.

planned), or following surgery (e.g. levelling of lower arch in Class II/2).

Surgery, p. 504.

Post-surgical orthodontics Lighter round wires and inter-maxillary traction are used to detail occlusion. Then retention, usually with removable retainers.

Relapse This can be surgical or orthodontic or both. Relapse more likely in treatment of deficiencies as soft tissues are under greater tension post-operatively.

Cleft lip and palate CL(P)

Prevalence CL(P) Varies with racial group and geographically. Occurs 1:750 Caucasian births, but prevalence ↑. M>F. If unilateral L>R. Family history in 40% of cases.

Isolated cleft palate (CP) occurs 1:2000 births. F>M. Family history in 20%.

Aetiology Polygenic inheritance with a threshold. Environmental factors may precipitate susceptible individual towards threshold.

Classification Many exist, but best approach is to describe cleft: primary &/or secondary palate; complete or incomplete; unilateral or bilateral. Submucous cleft is often missed until poor speech noticed, as overlying mucosa is intact.

Problems

Embryological anomalies Tissue deficit, displacement of segments, abnormal muscle attachments.

Post-surgical distortions Unrepaired clefts show normal growth. In repaired clefts maxillary growth is ↓ anteroposteriorly, transversely and vertically. Mandibular growth also ↓.

Hearing and speech are impaired.

Other congenital anomalies occur in up to 20% of cases with CL(P) and are more likely in association with isolated clefts of palate than lip.

Dental anomalies In CL(P) ↑ prevalence of hypodontia and $ teeth (especially in region of cleft). Also ↑ incidence of hypoplasia and delayed eruption.

Management (of unilateral complete CLP)

Need central co-ordinator, usually cleft surgeon or orthodontist, plus established CLP team (at least: speech therapist, plastic surgeon, ENT surgeon, oral and maxillofacial surgeon, social worker). Centralization of care gives better results.

Birth Parents need reassurance and help with feeding (Rosti bottle and Gummi teat with additional holes). Pre-surgical orthopaedics now out of vogue as benefits not proven. Take visual records ± study models.

Lip closure The majority of centres do lip repair at about 3 months, but neonatal repair is ↑ popular. Delaire or Millard ± modifications are the most popular. Some surgeons do Vomer flap at same time. Bilateral lips are closed either in 1 or 2 operations.

Palatal closure Usually between 9–18 months. Delaire or Von Langenbeck ± modifications are the most popular. Deferring repair until patient older reduces growth disturbance, but resultant poor speech has a greater psychological impact.

Deciduous dentition Lip revision may be carried out before patient starts school. Speech and hearing assessments are required.

Mixed dentition May be necessary to procline upper incisors if they erupt into linguo-occlusion, otherwise orthodontic treatment is better deferred until just prior to secondary bone grafting between 8–11 yrs.

Secondary bone grafting Involves grafting cancellous bone from the iliac crest into the cleft. Advantages:
1 Provides bone for $\underline{3}$ to erupt through (therefore, ideally before eruption of $\underline{3}$, but can be done after).
2 Allows tooth movement into cleft site ∴ intact arch possible.
3 ↑ bony support for alar base.
4 Aids closure of oro-nasal fistulae.

Orthodontic expansion of collapsed arches and alignment of the upper incisors may be required prior to grafting to improve access.[1] Closure is with local keratinized flaps; therefore, if any 1° extractions are planned these should be carried out in advance.

Permanent dentition Once permanent teeth have erupted, FA are usually required for alignment and space closure. Ideally, if $\underline{2}$ missing, Rx should aim to bring $\underline{3}$ forward to replace it, thus avoiding a prosthesis.

Growth complete A final nose revision is often performed at this stage. Orthognathic surgery to improve facial aesthetics (p. 504) may also be considered, in which case it is preferable to postpone the nose revision until bony surgery complete.

1 O. Bergland 1986 *Cleft Palate J* **23** 175.

5 Periodontology

Oral microbiology   202
Plaque   204
Calculus   206
Aetiology of periodontal disease   208
Epidemiology of periodontal disease   210
Chronic gingivitis   212
Chronic periodontitis   214
Pocketing   216
Diagnostic tests and monitoring   218
Acute periodontal disease   220
Periodontitis in children   222
Prevention of periodontal disease   224
Principles of treatment   226
Non-surgical therapy—1   228
Non-surgical therapy—2   230
Minimally invasive therapy—1   232
Minimally invasive therapy—2   234
Periodontal surgery   236
Guided tissue regeneration   238
Peri-implantitis   238
Mucogingival surgery   240
Reattachment/New attachment   242
Occlusion and splinting   244
Perio-endo lesions   246
Furcation involvement   248

Principal sources: J. Lindhe 1989 *Textbook of Clinical Periodontology*, Munksgaard. *Journal of Clinical Periodontology*, Munksgaard. Current Opinion in Dentistry 1991 (I) and 1992 (II).

Oral microbiology

The mouth is colonized by microorganisms a few hours after birth, mainly by aerobic and facultative anaerobic organisms. The eruption of teeth allows the development of a complex ecosystem of microorganisms (>300 species have been identified)[1] and the healthy mouth depends on maintaining an environment in which these organisms coexist without damaging oral structures.

Microorganisms worth noting

Streptococcus mutans: Aerobic. Synthesizes dextran. Colony density on tooth surface plaque rises to >50% in presence of high dietary sucrose. Able to produce acid from most sugars. The most important organism in the aetiology of dental caries.

Streptococcus sanguis: Accounts for up to half the streptococci in plaque and is strongly implicated in around half the cases of subacute bacterial endocarditis. Two recognized species.

Streptococcus salivarius: Accounts for half the streptococci isolated from saliva. Inconsistent producer of dextran.

Streptococcus milleri group: Common isolates from dental abscesses, also implicated in abscess formation at other sites in the body. Three recognized species.

Lactobacillus: Secondary colonizer in caries (mainly dentine).

Porphyromonas gingivalis: Obligate anaerobe. A member of the 'black pigmented bacteroides' group which is associated with rapidly progressive periodontitis. Others include **Prevotella intermedia** and **P. denticola**.

Fusobacterium: Obligate anaerobes. Originally thought to be principal pathogens in AUG. Remain a major periodontal pathogen as a collective group. The terminology for specific organisms is confusing and unimportant.

Borrellia vincenti: (refringens) The largest oral spirochaete and once thought to be the major co-pathogen in AUG. This disease is best now thought of as simply an anaerobic, fusospirochaetal complex infection.

Actinobacillus actinomycetemcomitans: Microaerophilic, capnophilic, gram −ve rod. Found particularly in juvenile periodontitis and rapidly progressive periodontitis.

Actinomyces israelii: Filamentous organism, major cause of actinomycosis, a persistent rare infection which occurs predominantly in the mouth and jaws and the female reproductive tract. Implicated in root caries.

Candida albicans: Yeast-like fungus famous as an opportunistic oral pathogen, probably carried as a commensal by most people.

Spirochaetes: Obligate anaerobes much implicated in periodontal disease, present in virtually all adult mouths, *Borrelia*, *Treponema*, and *Leptospira* all belong to this family.

1 J. Hardie 1992 *BDJ* **172** 271.

Plaque

Dental plaque is a firmly adherent mass of bacteria in a muco-polysaccharide matrix. It cannot be rinsed off but can be removed by brushing. It is the root of most dental evils.

Attachment Although it is possible for plaque to collect on irregular surfaces in the mouth, to colonize smooth tooth surfaces it needs the presence of *acquired pellicle*. This is a thin layer of salivary glycoproteins, formed on the tooth surface within minutes of polishing. The pellicle has an ion-regulating function between tooth and saliva and contains immuno-globulins, complement, and lysozyme.

Development Up to 10^6 viable bacteria per mm^2 of tooth surface can be recovered 1 h after cleaning,[1] these are selectively adsorbed streptococci. Bacteria recolonize the tooth surface in a predictable sequence. *Streptococcus mutans* synthesizes extracellular polysaccharides (glucan and fructan) specifically from sucrose and promotes its early colonization in this way. Cocci predominate in plaque for the first two days following which rods and filamentous organisms become involved. This is associated with increasing numbers of leucocytes in the gingival margin. Between 6–10 days, if no cleaning has taken place, vibrios and spirochaetes appear in plaque and this is associated with clinical gingivitis. It is generally felt that the move towards a more gram −ve anaerobe-dense plaque is associated with the progression of gingivitis and periodontal disease.

Plaque in caries (p. 38) As several oral streptococci, most notably *S. mutans*, secrete acids and the matrix component of plaque, there is a clear relationship between the two. However, various other factors complicate the picture, including saliva, other microorganisms, and the structure of the tooth surface.

Plaque in periodontal disease There is a direct correlation between the amount of plaque at the cervical margin of teeth and the severity of gingivitis[2] and experimental gingivitis can be produced and abolished by suspending and reintroducing oral hygiene.[3] It is commonly accepted that plaque accummulation causes gingivitis, the major variable being host susceptibility. While there are numerous interacting components which determine the progression of chronic gingivitis to periodontitis, particularly host susceptibility, the presence of plaque, particularly 'old' plaque with its high anaerobe content, is widely held to be crucial, and most treatment is based on the meticulous, regular removal of plaque.

1 R. Gibbons 1973 *J Periodont* **44** 347. **2** M. Ash 1964 *J. Periodont* **35** 424. **3** H. Loe 1965 *J Periodont* **36** 177.

Calculus

Calculus (tartar) is a calcified deposit found on teeth (and other solid oral structures) and is formed by mineralization of plaque deposits. It can be subdivided into:

Supragingival calculus, most often found opposite the openings of the salivary ducts, i.e. $\underline{76|67}$ opposite the parotid (Stensens) duct and the lingual surface of $\overline{76|67}$ the lower anterior teeth opposite the submandibular/sublingual (Whartons) duct. It is usually yellow, but can become stained a variety of colours.

Subgingival calculus is found, not surprisingly, underneath the gingival margin and is firmly attached to tooth roots. It tends to be brown or black, is extremely tenacious and is most often found on interproximal and lingual surfaces. It may be identified visually, by touch using a calculus probe, or on radiographs. With gingival recession it can become supragingival.

Composition Consists of up to 80% inorganic salts, mostly crystalline, the major components being calcium and phosphorus. The microscopic structure is basically that of randomly orientated crystal formation.

Formation is always preceeded by plaque deposition, the plaque serving as an organic matrix for subsequent mineralization. Initially, the matrix between organisms becomes calcified with, eventually, the organisms themselves becoming mineralized. Subgingival calculus usually takes many months to form, whereas friable supragingival calculus may form within 2 weeks.

Pathological effects Calculus (particularly, subgingival calculus), is invariably associated with periodontal disease. This may be because it is invariably covered by a layer of plaque. Its principal detrimental effect is probably that it acts as a retention site for plaque and bacterial toxins. The presence of calculus makes it difficult to implement adequate oral hygiene.

Aetiology of periodontal disease

Plaque is the principal aetiological factor in virtually all forms of periodontal disease. Periodontal damage is almost certainly the direct consequence of colonization of the gingival sulcus by organisms within dental plaque. However, the progression from gingivitis to periodontitis is far more complex than this statement suggests, as it involves host defence, the oral environment, the pathogenicity of organisms, and plaque maturity. It is probably easiest to regard periodontal disease as a complex multifactorial infection complicated by the inflammatory response of the host. Various elements of this process are worthy of special note:

Microbiology

The changing microbiology of dental plaque has already been referred to (p. 204). The inflammatory response of gingiva to the presence of initial young plaque creates a minute gingival pocket which serves as an ideal environment for further bacterial colonization, providing all the nutrients required for the growth of numerous fastidious organisms. In addition, there is an extremely low oxygen level within gingival pockets which favours the development of obligate anaerobes, several of which are closely associated with the progression of periodontal disease. High levels of carbon dioxide favour the establishment of the capnophilic organisms, some of which are associated with juvenile periodontitis (JP).

Briefly, clinically healthy gingivae are associated with a high proportion of gram +ve rods and cocci which are facultatively anaerobic or aerobic. Gingivitis is associated with an ↑ number of facultative anaerobes, strict anaerobes and an increasing number of gram negative rods. Established periodontitis is associated with a majority presence of anaerobic gram negative rods. Specific organisms involved in periodontal disease worthy of note include: **Porphyromonas gingivalis**, **Prevotella intermedia**, **P. denticola** (previously Bacteroides), and 'Spirochaetes' have many properties which ↑ pathogenicity, particularly activity against neutrophils. *Actinobacillus actinomycetemcomitans* a capnophilic organism thought to be involved in the aetiology of juvenile periodontitis, is also active against neutrophils. However, to date, it has not been possible to identify one particular organism or group of microorganisms which is solely responsible for the initiation and progression of periodontal disease, although the general concepts outlined reflect current working data.

Immunopathology

The inflammatory response to the presence of dental plaque is detectable both clinically and histologically and is certainly responsible for at least some of the periodontal destruction which occurs. Both inflammatory and immunologically mediated pathways can contribute to periodontal damage. Antigenic substances released by plaque organisms elicit both cell-mediated and humoral responses which, while designed to be protective, also cause local tissue damage, usually by complement activation. Non-immune mediated damage is caused by one or all of the major endogenous mediators of inflammation: vasoactive amines (histamine), plasma proteases (complement), prostaglandins and leukotrienes, lysosomal acid hydrolases, proteases, free radicles, and cytokines.

Host

Local and systemic modifying factors influence progress of disease.

Systemic factors include: immune status, stress, endocrine function, smoking, drugs, age, and nutrition.

Local factors are: tooth position and morphology, calculus, overhangs and appliances, occlusal trauma, and mucogingival state.

Epidemiology of periodontal disease

Epidemiology is the study of the presence and effect of disease upon a population. In order for this to be of value it is essential to be able to quantify the prevalence and degree of severity of any given disease, in a reproducible manner. The search for suitable **indices** in periodontology has left the literature replete with often confusing and redundant scoring systems.

The original purpose of periodontal indices was to study the extent of disease within population groups; however, the value of indices in the screening and management of individual patients soon became apparent. The following two indices fulfil both these criteria and are simple and easy to perform:

Debris or Oral Hygiene Index can be modified for personal use by using disclosing agents.[1]
  0 No debris or stain.
  1 Soft debris covering not more than $\frac{1}{3}$rd of the tooth surface.
  2 Soft debris covering more than $\frac{1}{3}$rd but less than $\frac{2}{3}$rds.
  3 Soft debris covering over $\frac{2}{3}$rds of tooth surface.

Basic Periodontal Examination (BPE) also known as **Community Periodontal Index of Treatment Needs (CPITN)**[2] This technique is used to screen for those patients requiring more detailed periodontal examination. It examines every tooth in the mouth, thus taking into account the site specific nature of periodontal disease. A World Health Organization (WHO) periodontal probe (ball-ended with a coloured band 3.5–5.5 mm from the tip) should be used. The mouth is divided into sextants, i.e. 2 buccal and 1 labial segment per arch. 6 sites on each tooth are explored and the highest score per sextant recorded, usually in a simple 6-box chart.

1 = Gingival bleeding but no pockets, no calculus, no overhanging restoration. Rx: OHI.

2 = Deepest pocket <3 mm, subgingival calculus present or subgingival retention site, e.g. overhang. Rx: OHI, scaling, and correction of any iatrogenic problems.

3 = Deepest pocket 4 or 5 mm. Rx: OHI and deep scaling.

4 = One or more tooth in sextant has a pocket >6 mm. Rx: curettage and root planing +/− flap.

Individuals identified as having areas of advanced periodontal disease will require a full probing depth chart, together with recordings of recession and furcation involvement and X-ray examination. BPE cannot be used for monitoring the progress of treatment.

Also useful, particularly for patient motivation are:

Bleeding index Score 1 or 0 depending on whether or not bleeding occurs after a probe is gently run around the gingival sulcus. A percentage score is obtained by dividing by the number of teeth and multiplying the result by 100.

1 J. Greene 1960 *J Am Dent Assoc* **61** 172. **2** J. Ainamo 1988 *Int Dent J* **32** 281. **3** C. Douglas 1991 *Current Op Dent* **1** 12.

Plaque index This is based on the presence or absence of plaque on the mesial, distal, lingual, and buccal surfaces revealed by disclosing.

$$\text{percentage score} = \frac{\text{number of surfaces with plaque}}{\text{total number of teeth} \times 4} \times 100$$

Recent research using relatively reliable indices of peridontal disease suggests that it is one of, if not, the commonest disease on the planet. European data suggests up to 50% of dentate people in the 50–65 yr-old category have advanced periodontal disease (pockets >6 mm); however, of these, more than half have only a limited number of sites affected. There is some evidence to suggest ↓ prevalence in the USA[3] in the *overall* population but this may miss ↑ in significant minority groups within the population.

The direct association between the presence of tooth surface plaque and periodontal damage has been confirmed, but the rate of destruction has been shown to vary not only between individuals, but between different sites in the same mouth and at different times in the same individual.

WHO probe for use in BPE/CPITN.

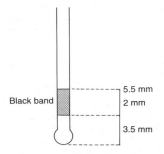

Black band

5.5 mm
2 mm
3.5 mm

If black band disappears on probing pocket perform *full* periodontal examination in that sextant.

Chronic gingivitis

Chronic gingivitis is, as the name suggests, inflammation of the gingival tissues. It is **not** associated with alveolar bone resorption or apical migration of the junctional epithelium. Pockets >2 mm can occur in chronic gingivitis due to an increase in gingival size because of oedema or hyperplasia (false pockets). Three different types of gingivitis are described; however, they are not mutually exclusive.

Chronic (oedematous) gingivitis This is present in virtually all mouths to some extent. The classic triad of redness, swelling and bleeding on gentle probing are diagnostic and are usually associated with a complaint by the patient that their 'gums bleed on brushing'. False pocketing may also be present. Gingivitis occurs as a result of low-grade infection caused by the presence of undisturbed dental plaque which is associated with a change in the flora from gram +ve aerobes to gram −ve anaerobes (p. 204). This gives rise to inflammatory changes in the associated gingivae, which are detectable histologically prior to the appearance of overt clinical gingivitis, which is observed after about 7 days of undisturbed plaque accummulation. The inflammatory response seen comprises an alteration in the integrity of the gingival microcirculation, an ↑ in the numbers of inflammatory cells in the gingival connective tissue (i.e. plasma cells, lymphocytes, macrophages, and neutrophils), a ↓ in the number of fibroblasts and a ↓ in collagen density. These inflammatory changes are easily reversible after institution of effective plaque control. While gingivitis is reversible, it should be remembered that calculus and other factors which promote plaque retention (e.g. overhanging restorations) will make adequate oral hygiene difficult. These factors should ∴ be corrected by scaling and appropriate restorative treatment in addition to OHI. Gingivitis may be a precursor to, or marker of, chronic periodontitis, and this must be excluded by measuring probing attachment levels (looking for 'true' pockets) and excluding the presence of alveolar bone loss by X-ray, if indicated.

Chronic hyperplastic gingivitis Hyperplasia of the gingivae is caused by an ↑ in collagen and fibroblasts produced in response to the inflammatory reaction to plaque. In most cases of gingivitis there is both an oedematous and hyperplastic response; however, the relative proportion of these is rather variable. Where fibrous tissue proliferation predominates the term 'hyperplastic gingivitis' is used. Again, the lesions are gingivally contained and any pocketing is 'false'. This condition is also found as an unwanted response to certain drugs, notably phenytoin, cyclosporin A, and nifedipine. A familial condition known as hereditary gingivofibromatosis exists and hyperplasia is a feature of Crohn's disease, sarcoidosis, and Wegener's granulomatosis. Chronic hyperplastic gingivitis which does not fully resolve after meticulous plaque control is one of the few indications for conventional gingivectomy.

Chronic atrophic (desquamative) gingivitis is gingival inflammation in the presence of gingival atrophy and is usually a manifestation of mucous membrane pemphigoid or lichen planus.

Chronic periodontitis

Chronic periodontitis (CP) can be regarded as a progression of the combination of infection and inflammation of gingivitis into the deep tissues of the periodontal membrane. It is characterized by breakdown of periodontal fibre bundles at the cervical margin, resorption of alveolar bone and apical proliferation of junctional epithelium beyond the amelocemental junction. The progression of chronic gingivitis to periodontitis is by no means straightforward, as it varies in rate and progression not only between individuals, but between sites within the same mouth, and with time. It is now thought that periodontal destruction occurs in acute bursts of disease activity, followed by a quiescent phase. The active phase is characterized by a rapid loss of attachment. This lasts for a variable period of time and is probably induced by a change in the quantity and/or quality of the subgingival microflora in undisturbed plaque, although other parameters may be involved. These include host response due to systemic disease, and local factors such as occlusal trauma, iatrogenic damage, or an as yet undetermined variable in the inflammatory response. The quiescent phase is associated with no advance in disease on either clinical or radiographic grounds and may last for extended periods of time. However, complete healing of the lesion does **not** occur because plaque remains on the root surface, and inflammation therefore persists in the connective tissues.

Microbiology In the majority of cases the microbial pathogens remain within the plaque in the periodontal pocket and do not invade the periodontal tissues. It has already been mentioned (p. 208) that sampling of pockets in cases of CP consistently recovers a greatly ↑ proportion of gram −ve anaerobic rods and spirochaetes. Debate still continues between adherents of the specific (CP caused by specific organisms), and the non-specific plaque (CP caused by quantity rather than quality of plaque) theories. Nevertheless, certain gram −ve anaerobes appear particularly prominently in active disease sites; namely, *Porphyromonas gingivalis*, *Prevotella intermedius*, *Treponema* sp, *Wolinella recta* and *Eikenella corrodens* (apologies to the reader, microbiologists appear to have a pathological need to rename organisms). Some of these organisms synthesize enzymes such as proteases which would exert a deleterious effect on the periodontal membrane, and all gram −ve organisms release endotoxin (their cell wall) on death.

Diagnosis is based on:
- Probing: to elicit bleeding (which is the single most useful indicator of disease activity), measure pocket depth, and detect subgingival calculus
- Testing teeth for mobility and vitality
- X-ray examination (DPT or full mouth periapicals).

Treatment, pp. 226 to 248.

Rapidly progressive periodontitis is a severe form of generalized periodontitis affecting young adults (20–35 yrs). Affects 1–2% of the Western population with an ↑ in Afro-Caribbeans. Microbiologically—round-up the usual suspects plus *A actinomycetomcomitans*. Immunologically there may be an associated neutrophil defect. Rx as for JP.

Pocketing

Periodontal pockets can be divided into:

False pockets Which are due to gingival enlargement with the pocket epithelium at or above the amelo-cemental junction.

True pockets Which imply apical migration of the junctional epithelium beyond the amelo-cemental junction and can be divided into suprabony and infrabony pockets. Infrabony are described according to the number of bony walls: three-walled defect is the most favourable, as it is surrounded on three sides by cancellous bone and on one side by the cementum of the root surface. Two-walled defect may be either a crater between teeth having bone on two walls and cementum on the other two, or have two bony walls, the root cementum, and an open aspect to the overlying soft tissues. One-walled defects may be hemiseptal through and through defects, or one bony wall, two root cementum, and one soft tissue.

Periodontal probes are the key instruments in detecting pockets. Numerous designs exist, and while individual preference will influence choice, it is sensible to reduce variability by selecting a single type of probe and using that type of probe throughout any one individual's treatment. The use of the WHO probe for screening using the BPE index is described on p. 210. Patients who are identified as having advanced CP should then be investigated further, including probing around each tooth. The main other indicator of periodontal disease, bleeding, is also detected using a probe (gently), and again consistency with a single type of probe is necessary.

Probing variables The depth of penetration depends upon:
- Type of probe and its position.
- Amount of pressure used.
- Degree of inflammation.[1]

It is now apparent that the measurement obtained with a probe does not correspond to sulcus or pocket depth. In the presence of inflammation a probe tip can pass through the inflamed tissues until it reaches the most coronal dento-gingival fibres, about 0.5 mm apical to the apical extent of the junctional epithelium, i.e. an over-estimation of the problem. The amount of penetration into the tissues varies directly with the degree of inflammation, so that, following resolution of inflammation, an underestimate of attachment levels may be given. Formation of a tight, long junctional epithelium following treatment may also give a false sense of security if probing measurements are not interpreted with a degree of caution. For this reason the term 'probing attachment level' is preferred to pocket depth.

---

1 M. A. Listgarten 1980 *J Clin Perio* **7** 165.

Diagnostic tests and monitoring

It is widely accepted that disease active and inactive pockets exist. Progression is episodic and more likely in susceptible patients. Bleeding on probing has traditionally been the most useful indicator of disease activity.

With ↑ emphasis on specific periodontopathic bacteria p. 208 and availability of assays for components of immunological response have some chairside diagnostic tests using gingival crevicular fluid.

Two types exist:
1. Based on detection of antibodies to specific periodontopaths, e.g. Kodak Evalusite.
2. Based on components of immune response, e.g. elastase in Dentsply Prognostik.

Both aim to predict sites of future and actual disease progression and the former may indicate need for specific antibiotic therapy p. 230.

There is a huge amount of ongoing research into improving and refining these tests but evidence is still required to demonstrate predictive ability and a higher level of accuracy than bleeding on probing.

Other peripheral techniques worthy of note are pocket temperature probes and computerized subtraction radiovisiography.

Radiographs are useful in comparing degree of bone loss and root surface deposits with pocket depth. Standardized sequential radiographs allow monitoring of disease.
- Bitewings provide a good view of interproximal bone, useful for relatively minor degrees of bone loss and to detect calculus deposits.
- OPG is an excellent screening X-ray but only gives an indication of periodontal destruction.
- Full mouth periapicals (long cone technique) is the X-ray assessment of choice for patients with significant periodontal disease, as it can clearly demonstrate root surface deposits, furcation involvement, extensive bone loss, infrabony pocketing, and perio-endo lesions.

The results of X-ray examination, clinical assessment, and pocket depth can all be marked on an updatable periodontal chart to monitor progress with treatment.

Acute periodontal disease

Acute ulcerative gingivitis (AUG) is also known as acute necrotizing ulcerative gingivitis, Vincent's gingivitis, Vincent's gingivostomatitis, ulceromembraneous gingivitis, or trench mouth. It should not be confused with Vincent's angina which is also a fusospirochaetal infection, but is typically localized over the tonsils.

AUG is characterized by painful papillary yellowish-white ulcers which bleed readily. Patients also often complain of a metallic taste and the sensation of their teeth being wedged apart. Regional lymphadenitis, fever, and malaise may occur in some cases. AUG is associated with poor oral hygiene, but stress and smoking act as co-factors. Inadequately treated AUG will lapse into a less symptomatic form which has become known as chronic ulcerative gingivitis, in which a slower rate of destruction occurs. AUG is usually a limited gingival condition, but a rare and more serious form known as cancrum oris or noma is found in patients who are malnourished, and in this form can lead to extensive destruction of the jaws and face.

Microbiology Originally *Borellia vincenti* (*refringens*) and *Fusobacterium fusiformis* were held to be the major culprits in a mixed anaerobic infection. Further research has, however, served to confuse the picture, with *Bacteroides melaninogenicus* being implicated, as well as *Treponema* sp, *Selenomonas* sp, and *Borrelia intermedius*. The crucial aspect of the microbiology of AUG is that it is a gram −ve anaerobic infection which has been shown to actually invade the tissues, but usually responds to local debridement. In recent years, the tendency for patients who are HIV +ve to develop this condition has renewed interest in it. From the practical point of view it should be remembered that AUG presenting in an otherwise apparently healthy young adult may be a presenting sign of HIV infection. Practically, consider control of cross-infection, examining the mouth for other signs of infection (p. 472), and advising the patient about HIV testing.

Rx In most cases local measures, i.e. thorough debridement, and adequate oral hygiene will suffice; however, if there is evidence of systemic upset (lymphadenopathy) metronidazole 200 mg tds for 3 days is indicated. It is wise to get the patient to rinse with chlorhexidine gluconate 0.2% prior to ultrasonic scaling to reduce aerosol spread. Chlorhexidine rinses may also be prescribed as an adjunct to brushing, which is painful initially. Later treatment such as gingivectomy for persistent craters is only rarely required.

Periodontal abscess is a localized collection of pus within a periodontal pocket. It occurs either due to the introduction of virulent organisms into an existing pocket or ↓ drainage potential. The latter classically occurs during treatment as reduction of inflammation in the coronal gingival tissues occludes drainage by a tighter adaptation to the tooth. May also occur due to impaction of a foreign body such as a fishbone in a pre-existing pocket or even in an otherwise healthy periodontal membrane.

Diagnosis Need to distinguish from apical abscess.

Apical abscess	*Periodontal abscess*
Non-vital	Usually vital
TTP vertically	Pain on lateral movements
May be mobile	Usually mobile
Loss of lamina dura on X-ray	Loss of alveolar crest on X-ray

Insertion of a GP point into an associated sinus and an X-ray may be helpful.

221

Rx *Emergency*: Incision and drainage under LA; debridement of the pocket (e.g. ultrasonic scaler); systemic antibiotic, e.g. metronidazole 400 mg tds &/or amoxycillin 250–500 mg tds for 5 days. *Follow-up*: conventional treatment for periodontal pockets (p. 236), combined periodontal—endodontic lesion (p. 246).

Acute herpetic gingivostomatitis, p. 436.

Acute streptococcal gingivitis Rare. Beefy, red, painful gingivae usually caused by a Lancefield A streptococcus. Rx: penicillin V 500 mg qds for 7 days and OHI.

Periodontitis in children

Two main forms exist: juvenile periodontitis (JP) and pre-pubertal periodontitis.

Juvenile periodontitis

Has been recognized since about 1942 when it was known as periodontosis. It is an aggressive periodontitis occurring in an otherwise healthy adolescent, characterized by rapid loss of connective tissue attachment and alveolar bone of >1 permanent tooth. It can occur in both localized (affecting

$$\frac{6 \quad 1 \;|\; 1 \quad 6}{6 \quad 1 \;|\; 1 \quad 6}$$

only) and generalized forms. It is not known if these are separate disease entities or part of a spectrum. It is, however, accepted that JP is a well-defined clinical entity which is quite different from the adult disease. The gingivae around affected teeth may appear entirely normal, but deep periodontal pockets can be detected by careful probing. The degree of periodontal destruction seems out of proportion to the deposits of plaque and calculus.

Prevalence is difficult to define, due to varying diagnostic criteria in published studies, but is probably <7%. F>M and often familial. There is a racial variation: Afro-Caribbean and Asian > Caucasians. Onset usually 11–13 yrs.

Microbiology There have been a number of rare and unusual organisms consistently cultivated from pockets in juvenile periodontitis, these include: *Actinobacillus actinomycetem-comitans*, *Capnocytophaga species*, *Eikenella corrodens* and *Bacteroides intermedius*. Some studies have implicated *A. actinomycetemcomitans* as the main pathogen, on the basis of a 90% occurrence in sampled lesions, and significantly ↑ serum antibodies to the organism in affected individuals. In addition, *A. actinomycetemcomitans* possesses a number of virulent mechanisms which allow periodontal destruction to occur, e.g. leukotoxin, chemotactic inhibition factor, endotoxic cell wall, and fibroblast growth inhibitor. These fit well with the findings that both the local and general host-response is inhibited in juvenile periodontitis.

Rx Meticulous oral hygiene, conventional scaling, and root planing, often in conjunction with access flap surgery. Consider tetracycline either locally (e.g. by slow release mechanisms) or systemically (e.g. minocycline 100 mg bd for 4 weeks). 3-monthly monitoring of subgingival microflora for *A. actino-mycetemcomitans* is advised, with combination drug/mechanical Rx targetted at sites of recurrence.

Prepubertal periodontitis

Is a rare form of periodontitis which affects the deciduous dentition and appears to have a familial tendency. It occurs in both localized or generalized forms, which probably reflect either ends of the spectrum of a single disease rather than separate entities.

Prevalence <1% of the child population.

Microbiology Usual periodontal pathogens, e.g. *B. intermedius*, *A. actinomycetemcomitans*, and other black, pigmented bacteroides.

Pathology is as yet poorly defined, and some confusion exists between this condition and the aggressive root resorption seen in hypophosphatasia (a condition termed cementopathia). An abnormal local response of white cells, as occurs in most severe periodontal disease, has sometimes been demonstrated, but it is not clear whether this is an intrinsic flaw or induced by the infecting organisms.

223

Rx: as for juvenile periodontitis, with the important exception of avoiding tetracycline, due to its unwanted effects on developing teeth, use amoxycillin 125–250 mg tds instead.

Chediak–Higashi, p. 743.

Papillon–Lefevre, p. 746.

Prevention of periodontal disease

Following the comments on the pages on the aetiology and epidemiology of periodontal disease, it is quite clear that dental plaque is the cause of the problem and its elimination will prevent gum disease. This is easier said than done—remember most of the world's population has gingivitis &/or periodontitis. The key to prevention is regular and thorough plaque removal, ∴ 'Oral Hygiene Instruction' (OHI) is probably the most useful advice you can give to your patients.

OHI should include an explanation of the reasons for good oral hygiene. Identify and demonstrate to the patient the disease (swollen gingivae, bleeding on probing) using a hand mirror, then demonstrate the cause (plaque) either directly by scraping off a deposit, or by disclosing. Explain how plaque starts to grow immediately after toothbrushing, so that regular removal is necessary, and that it cannot be rinsed away. Then demonstrate how to remove it, avoiding overt criticism of the patient's present efforts, as this is often counter-productive.

Toothbrushing requires a brush, ideally with a small head and even nylon bristles (3–4 tufts across by 10–12 lengthways), which should be renewed at least monthly. Toothpaste makes the process more pleasant and is a useful medium for topical fluoride and other agents. Anticalculus pastes ↓ its formation by about 50%. Chlorhexidine gel is effective against plaque organisms but has an astringent taste and causes staining. Numerous methods are described, based on the movement of the brush stroke: roll, vibratory, circular, vertical, horizontal. The best is the one which works for that patient (as demonstrated by absence of plaque on disclosing after brushing) and does no harm to tooth or gingivae. The horizontal scrub is notorious for creating abrasion cavities and gingival recession. Modifications to toothbrushes and brushing technique are often required in children, the elderly, and those with disabling diseases.

Interdental cleaning Brushing alone is unlikely to clean the interdental spaces adequately; however, it must be mastered first by the patient, before interdental cleaning is taught. Flossing, mini interdental brushes (particularly good for concave root surfaces), and interspace brushes are available for cleaning interproximally. The use of dental floss is something of an art form and must be learned by demonstration.

Professional preventive techniques

Regular periodic examination for periodontal disease, which is largely asymptomatic, is essential. This requires a full oral examination including probing for pockets as part of routine examination (p. 14). In patients who have been demonstrated to have periodontal disease, 3-monthly follow-ups are advisable. Routine scaling and polishing is of little value unless accompanied by intensive education and motivation of the patient (however tedious this may be), as it is the patients' efforts which will prevent the re-formation of plaque and their own periodontal disease. Employment of a hygienist has obvious benefits in this area. Elimination, avoidance, and treatment of iatrogenic problems such as overhanging margins on restorations, ill-fitting crown margins, poorly designed appliances, etc., are obviously mandatory.

Principles of treatment

- Establish diagnosis.
- Periodontal disease is an infection due to the presence of plaque, ∴ control of plaque is the key to success. More complex treatments will always fail in the absence of effective plaque control.
- The aims of corrective techniques such as scaling, root planing, periodontal surgery, restorative work, endodontics, and occlusal adjustment etc. are:
  1 to eliminate pathological periodontal pockets, or to create a tight epithelial attachment where the pocket once existed;
  2 to arrest loss of, and in some cases improve, the alveolar bone support;
  3 to create an oral environment which is relatively simple for the patient to keep plaque-free.

The overall aim could be summarized as the creation of a healthy mouth which the patient is both capable of, and willing to, maintain.

It is often convenient to divide the principles of periodontal therapy into three phases:

1 The *initial* (cause related) phase, where the aim is to control or eliminate gingivitis and arrest any further progression of periodontal disease by the removal of plaque and other contributory factors.
2 The *corrective* phase is designed principally to restore function and where relevant aesthetics.
3 The *maintenance* phase which aims to reinforce patient motivation so that their OH is adequate to prevent recurrence of disease. This phase is receiving ↑ attention due to the relative ease with which disease activity can be monitored by probing and chairside diagnostic assays (p. 218).

Non-surgical therapy—1

Non-surgical therapy in periodontal disease consists of debridement, restorative treatment (to correct coexisting or exacerbating factors, e.g. periapical infection, overhanging margins), and the use of antiseptics and antibiotics.

Scaling is the removal of plaque and calculus from the tooth surface either with hand instruments (e.g. currettes, hoes, chisels, and jaquettes) or mechanically (e.g. 'Cavitron'). Scaling can be sub- or supragingivally depending upon the site of the deposits. LA is usually not necessary. It is customary to make use of an ultrasonic scaler for the bulk of the work and finish off, particularly subgingivally, with hand instruments. The precise use of hand instruments is largely a matter of personal preference; however, it is essential to use controlled force and a secure finger-rest. Ultrasonic instruments are quicker, but they can be uncomfortable and leave an uneven root surface (though the significance of the latter is controversial). Ultrasonic scaling employs a frequency of 25–40 000 cycles/sec. Another instrument known as a sonic scaler vibrates at 2.3–6.3000 cycles/sec has been shown to be equally effective at removing calculus and may result in smoother root surfaces. With both it is essential to use a copious coolant spray.

The teeth are usually polished after scaling, preferably using a rubber cup and a fluoride-containing paste (e.g. toothpaste). Patients can then appreciate the feeling of a clean mouth, which they must then maintain.

Local delivery of medicaments

As there is controversy over efficacy and unwanted effects from systemic antibiotics, methods of direct delivery into the pocket are ↑ This can be done using injected paste or gel or by impregnated fibre. This gives a high local dose, low systemic uptake and prolonged exposure of the pathogens to the drug. Examples: Lederle 'Dentomycin' (minocycline), Dumex 'Elyzol' (metronidazole).

Antimetabolites

Non-steroidal anti-inflammatory drugs (NSAIDs) are the main group being investigated. Experimental gingivitis and alveolar bone loss in animals are ↓ by flurbiprofen. This is *not* only due to cyclo-oxygenase pathway inhibition. Evidence of significant long-term benefit in humans is awaited. A future possibility is inclusion of NSAIDs in topical form in a toothpaste.

Non-surgical therapy—2

Antiseptics and antibiotics

As the mechanical control of plaque is a tedious and time-consuming procedure the use of chemicals either as antiseptics or antibiotics would be a significant aid. Problems with their use include the unwanted side-effects of such chemicals, development of resistance, mode of delivery and most importantly the fact that no single organism has been identified as the main pathogen in periodontal disease. Nevertheless, elimination of gram −ve anaerobic bacteria has been demonstrated to result in improved periodontal health.

Antiseptics The antiseptic of greatest proven value is chlorhexidine gluconate. This is commonly used in a 0.2% mouthwash or gel, although some studies suggest that this could be reduced to as low as 0.12%. A standard regimen is 10 ml of solution rinsed for 1 min bd. The gel can be used instead of toothpaste. NB An interaction between conventional toothpaste and chlorhexidine reduces the antiseptic's efficacy. Other proprietary antiseptic rinses have yet to meet this 'gold standard'.

Antibiotics A recent review of the use of antibiotics in periodontology recommended their use for patients with recent, or a high risk of, periodontal breakdown.[1] It was suggested that antibiotics should be used as an adjunct to conventional pocket treatment, in conjunction with pocket sampling to identify the major pathogens and their sensitivities. Short term use in relatively high dosage, or high locally delivered doses are used. The two most useful drugs are:

- *Tetracycline*. Can be administered either systemically or directly into the pocket via a slow release mechanism (e.g. incorporation of the drug into cellulose acetate or methacrylate strips). It is active against spirochaetes, *A. actinomycetemcomitans*, and many other periodontal pathogens. Most useful in the treatment of JP. Minocycline is the most popular of this group.

- *Metronidazole*. Effective against protozoa and strict anaerobes and is capable of eliminating all the strict anaerobes found in periodontal pockets. As it does not interfere with aerobes or facultative anaerobes it is highly unlikely to allow the development of opportunist pathogens. There is frequent mention of the mutagenicity and teratogenicity of this drug in animal studies but, despite widespread use in medicine, no clinical evidence has emerged to support this. It is, in fact, an extremely safe and useful drug. When combined with mechanical pocket therapy, significant improvements in terms of probing depth have been demonstrated.[2] There is an emetic interaction with alcohol.

1 J. Slots 1990 *J Clin Periodontol* **17**, 479. **2** W. J. Loesche 1984 *J Clin Periodontol* **55** 325.

There is a clear place for antibiotics/antiseptics as adjuncts in selected types of periodontal disease, e.g. JP, refactory/severe CP, rapidly progressive adult periodontitis.

Minimally invasive therapy—1

There has been a progressive move away from pocket elimination surgery, towards the creation of a healthy periodontium which gives access for both professional and home cleaning. Recognition that periodontal disease is a localized infection due to the presence of dental plaque and that its arrest and prevention is dependent on the removal of plaque and plaque-retaining factors, has allowed a far more rational and conservative approach to therapy to develop. The prime aims of periodontal surgery are now:

- Remove deposits from root surfaces.
- Create subgingival root surfaces which are accessible for cleaning, either by professionals or by the patient.
- Maximize the potential for healing of damaged periodontal tissues.

Minimally invasive techniques include deep subgingival scaling/root planing and the modified Widman flap (open flap curettage). These are suitable for those patients who despite good supragingival plaque control still have true pockets, but do not need recontouring of the gingival margins.

Scaling and root planing Deep subgingival scaling and root planing are best carried out under LA. Deep scaling is a procedure which removes plaque and calculus from the root surface, it differs from supragingival scaling simply in its thoroughness, and the discomfort it causes. Root planing is a technique whereby damaged or softened cementum is removed from the root surface by scraping the root. The same instruments are used for both procedures (p. 228), and in practice it is probably impossible to differentiate between subgingival scaling and root planing, as in both techniques plaque, calculus, root cementum, and small amounts of dentine will be removed. These are the core procedures to **all** periodontal techniques, the only real difference with formal surgery is that it allows treatment under direct vision, osseous recontouring and sculpting if desired, and re-positioning of the gingiva.

It is often most effective to treat one quadrant at a time under LA, partly because it is painful and partly because it is tedious when performed meticulously, which is the only worthwhile way to do it. The choice of instruments is less important than the end result, as effective debridement of the root surface is essential. It is **not** necessary to currette the epithelial lining of the pocket. Success will allow tight adaptation of the pocket epithelium to the root creating a *long junctional epithelium*.

Practical tips for periodontal surgery

Local anaesthesia The infiltration, block and/or lingual/palatal injections required will be determined by the site of surgery. Both LA and haemostasis are improved by injecting directly into the gingival margin and interdental papillae until blanching is seen.

Suturing techniques Interrupted interproximal sutures are used when buccal and lingual flaps are being reapposed at the same level. When flaps are repositioned at different levels, a suspensory suture is used, where the suture only passes through the buccal flap and is suspended around the cervical margins of the teeth.

Periodontal packs These are essential after gingivectomy to ↓ post-operative discomfort. Many favour them after all periodontal surgery to help reappose the flap to bone.

Classified:
- Eugenol dressings, e.g. ZnO, have the advantage of being mildly analgesic but can cause sensitivity reactions.
- Eugenol-free dressings, e.g. Coe-pack are more popular.

Minimally invasive therapy—2

The modified Widman flap

This is a technique which enables open debridement of the root surface, with a minimal amount of trauma. There is no attempt to excise the pocket although a superficial collar of tissue is removed. This has the advantage of allowing close adaptation of the soft tissues to the root surface with minimal trauma to, and exposure of, underlying bone and connective tissue, thus causing fewer problems with post-operative sensitivity and aesthetics. It is useful in conjunction with GTR.

Technique (after Ramfjord and Nissle 1974).[1] A scalloped incision is made parallel to the long axis of the teeth involved 1 mm from the crevicular margin, except when pockets are <2 mm deep when an intracrevicular incision is made. This incision is extended interproximally as far as possible, separating the pocket epithelium from the flap to be raised, and then extended mesially and distally, allowing the flap to be raised as an envelope without relieving incisions. The flap should be as conservative as possible and only a few millimetres of alveolar bone exposed by a second incision intercrevicularly to release the collar of pocket epithelium and granulation tissue. A third incision at 90° to the tooth separates the pocket epithelium, and this is removed along with accompanying granulation tissue with curettes and hoes. The root surface is then thoroughly scaled and root planed. Although bony defects can be curetted **no** osseous surgery is carried out. The flaps are then repositioned to cover all exposed alveolar bone and sutured into position. Post-operatively chlorhexidine 0.2% 10 ml rinsed bd is given, and most periodontologists prefer to use a periodontal pack to improve patient comfort post-operatively.

Notes There continues to be considerable confusion about nomenclature with this type of flap. The original Widman flap described in 1918 by Leonard Widman used relieving incisions, osseous surgery, and attempted to remove the pocket. The Neumann flap used an intracrevicular incision, again attempting to excise the pocket, but the 'modified flap operation' described by Kirkland in 1931 did not require sacrifice of inflamed tissues or apical displacement of the gingival margin. The latter was, of course the precursor of the 'modified Widman flap'.

1 S. Ramfjord 1974 *J Periodont* **45** 601.

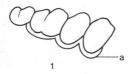

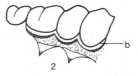

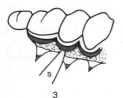

Modified Widman flap
1 Design of flap
 a Incision
2 Flap elevated
 b Gingival cuff to be discarded
3 Excision of supra-alveolar pocket
 s Scalpel blade
4 Flap repositioned and sutured in place

Periodontal surgery

Gingivectomy

Has ↓ in use over the last decade, its remaining indications are in cases with persistence of deep supra-alveolar pockets (e.g. gingival hypertrophy), to reshape severely damaged gingivae into an easily manageable contour and crown lengthening prior to restorative procedures. It is **not** suitable for the management of deep 'true' pocketing as excision of the pocket will remove the entire thickness of keratinized gingivae. It is of no value in the treatment of infrabony lesions.

Technique Pockets are delineated by use of pocket marking forceps, e.g. Crane–Kaplan forceps. This marks out a line of incision, which may be either smooth or scalloped, made with the blade angled at 100°–110° to the long axis of the tooth. This bevelled incision excises supragingival pockets and allows for gingival re-contouring. Once the incision has been made the strip of gingiva remaining is released by an intercrevicular incision. The root surfaces are then curretted and an open area of freshly cut granulation tissue left to heal under a periodontal pack. Prescribe chlorhexidine mouthwash 10 ml bd. The pack is left in place for about 1 week.

Disadvantages Loss of attached gingiva, raw wound, exposed root surface (which ↑ likelihood of sensitivity and caries). Some remodelling of alveolar bone occurs, despite there being no operative interference.

Apically repositioned flap

This procedure is used to expose alveolar bone and includes the option for osseous surgery to correct infrabony defects. It allows excellent access to the root surface for debridement. The principal difference between this procedure and the modified Widman flap is the deliberate exposure of alveolar bone, and the apical repositioning of the flap with post-operative exposure of the root surfaces. This is primarily a buccal procedure, and although it can be performed on lingual pockets, it is obviously impossible on the palate where a conventional or reverse bevel gingivectomy approach has to be used.

Technique A reverse bevel incision is made in the attached gingiva angled to excise the periodontal pocket in a scalloped outline with vertical relieving incisions at either end. A split thickness flap is made down to bone and then converted to full thickness leaving a residual collar of tissue around the root surfaces. This combination of pocket epithelium and granulation tissue is removed with a curette. If indicated the alveolar crest can be remodelled.

Advantages Include exposure of alveolar bone with controlled bone loss, exposure of furcation area, minimal postoperative pocket depth, ability to reposition the flap and primary closure of the wound. In addition, keratinized gingiva is preserved.

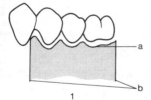

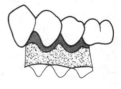

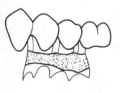

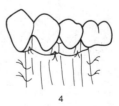

Apically repositioned flap

1 Design of flap
   a Reverse bevel incision
   b Relieving incisions
2 Elevating the flap. Tissue enclosing pocketing which is to be
   discarded is hatched
3 Flap elevated, pockets excised. Osseous surgery can be
   performed at this stage
4 Flap apically repositioned and sutured in position

Periodontal surgery (*cont.*)

Disadvantages Exposure of root surface (leading to ↑ susceptibility to caries and sensitivity) and ↑ loss of alveolar bone height which accompanies full exposure of the bone at operation.

Osseous surgery

Bone re-contouring has become less popular as it is always accompanied by some degree of alveolar resorption and therefore ↓ support for the tooth. **Osteoplasty** is conservative recontouring of the bone margin (i.e. non-supporting bone). **Ostectomy** is excision of bone aimed at eliminating infra-alveolar pocketing, but unfortunately it also ↓ alveolar support. The aim of osseous surgery should be to establish a more anatomically correct relationship between bone and tooth while maintaining as much alveolar support as possible.

Other flap procedures

These include simple replaced flaps which give ↑ bony access compared to the modified Widman flap. Also crown-lengthening procedures which can range from a simple gingivectomy to an apically repositioned flap ± bone removal. In addition, many periodontologists have their own modification of the aforementioned techniques.

Guided tissue regeneration

The recognition that epithelium migrated along the root surface before any other cell type, after periodontal surgery, and created the *long junctional epithelium* which prevented new attachment created the possibility that prevention of migration of epithelium would allow new connective tissue attachment. Guided tissue regeneration (GTR) is essentially interposing a barrier to epithelial migration prior to completion of surgical or non-surgical therapy. Original barriers were Millipore and PTFE. Goretex membranes in a variety of shapes are now widely used and infrabony and furcation defects can be infilled using a combination of bone/hydroxyapatite/biocoral covered by the membrane. Goretex has to be removed at 4–6 weeks. Biocompatible resorbable materials, such as Vicryl and lyophilized collagen, are being developed for a similar role. Tissue growth factor β-2 and bone morphogeneic protein may have a role in stimulating bone and connective tissue deposition. Main disadvantage is cost.

Peri-implantitis

Osseo-integrated implants are now well established p. 428. Implants which have achieved full osseo-integration may fail by overloading, peri-implantitis (similar to periodontitis) or a combination of both. Implant salvage in the failing stage consists of ↓ overloading and the entire arsenal of periodontal therapies. Local antibiotics and bone-supplemented GTR may be particularly useful.

Mucogingival surgery

Mucogingival surgery encompasses those techniques aimed at the correction of local gingival defects. The rationale for this type of surgery has been hotly debated over many years. Initially, it was felt that a margin of attached gingiva of around 3 mm was required to protect the periodontium during mastication and to dissipate the pull to the gingival margin from fraenal attachments. In fact, data from properly conducted experimental work has demonstrated that the width of attached gingiva and the presence or absence of an attached portion are **not** of decisive importance for the maintenance of gingival health.[1] As a result of this, the indications for mucogingival surgery have been rationalized:

1 where change in the morphology of the gingival margin would improve plaque control, e.g. presence of high fraenal attachments or deep areas of recession;

2 areas where recession creates root sensitivity or aesthetic problems;

3 very thin layer of attached gingiva overlying a tooth which is to be moved orthodontically: the evidence for this is somewhat anecdotal.

Gingival recession

Gingival recession is one of the commonest reasons for carrying out mucogingival surgery. The two commonest causes are plaque-induced gingival inflammation and toothbrush trauma. Therefore, basic periodontal care and correction of faulty toothbrushing technique are the first lines of treatment. While anatomical features may contribute, these and trauma from occlusion, high fraenal attachments, and impingement from restorations, etc., are a secondary consideration.

Mucogingival techniques

Can be divided into two main groups:

Vestibular extension procedures essentially aim to increase the area of attached gingivae and deepen the buccal or labial sulcus. As they are always accompanied by a degree of bone resorption these methods are not recommended as a therapeutic periodontal technique.

Grafting is subdivided into

1 *Free grafts*, which are completely removed from their donor area. Free gingival grafts, commonly of palatal mucosa and connective tissue are taken and grafted to donor sites prepared by incising between attached and alveolar mucosa. While this technique may successfully cover exposed root surfaces of around 2 mm square and will certainly increase the width of keratinized gingiva, long-term cross-over studies suggest that in the presence of meticulous OH there is no significant difference

1 J. Wenstrom 1987 *J Clin Periodont* **14** 181.

between attachment levels in grafted and non grafted sites with similar degrees of recession.[1]

2 *Pedicle grafts* are not separated from their blood supply. Commonly used pedicle grafts are the laterally re-positioned flap, coronally re-positioned flap and the double papilla flap. These techniques may be of some value in very narrow areas of isolated gingival recession.

241

1 H. S. Dorfmann 1980 *J Clin Periodont* **7** 316.

Reattachment/New attachment

Definitions[1]

Reattachment is defined as the reunion of connective tissue and root separated by incision or injury. *New attachment* is defined as the reunion of connective tissue with a root surface which has been pathologically exposed (i.e. due to periodontal disease). It is the ideal aim of periodontal therapy.

There are two ways in which new attachment may occur, these can be subdivided anatomically as those occurring within bony pockets and those occurring between the soft tissue of the previously existing periodontal pocket and the root surface. There have been numerous claims and counterclaims as to which of these may occur, following virtually all forms of periodontal intervention. However, there is very little satisfactory evidence to support the contention that new attachment takes place above the level of the alveolar crest. More recently, re-entry procedures in animals have suggested that new attachment formation is inhibited by the apical migration of dento-gingival epithelium forming a long, but apparently quite healthy, *long junctional epithelium*. Some newer procedures now describe methods to inhibit apical migration of dento-gingival junction epithelium cells in an attempt to obtain new attachment. However, animal experiments[2] suggest that granulation tissue originating from bone or gingival connective tissue is unable to establish new connective tissue attachment, even when the intervening dento-gingival epithelium was prevented from migrating into the treated area. In fact, when close apposition of connective tissue to root surface occurred, resorption and ankylosis often resulted, suggesting that the migration of dento-gingival epithelium into the treatment area may be a protective mechanism. Experiments to assess whether this behaviour also applied to the periodontal ligament (PDL) have been carried out.[3] Interestingly, it would appear that PDL cells may develop new attachment if dento-gingival junction epithelial cells are prevented from migrating into the area, without the occurrence of resorption and ankylosis. This has been confirmed in clinical studies, using a variety of materials to form a barrier against ingress of epithelial cells by a process known as guided tissue regeneration, or GTR (p. 238). This involves placement of a porous Teflon membrane (e.g. Goretex) underneath the flap, extending from the outer surface of the alveolar process to the crown of the tooth above the gingival margin. This allows preferential colonization by PDL cells. A second operation is then required to remove the membrane.

Bony infill in osseous defects There are a number of studies which suggest that complete regeneration can occur in up to 70% of 3-walled infra-bony defects, this success rate, however, is not consistent, especially in combined and 2-walled defects.

1 K. I. Kalkwarf 1974 *Perio Abstracts* **2** 53. **2** S. Nyman 1980 *J Clin Periodont* **7** 394. **3** S. Nyman 1982 *J Clin Periodont* **9** 157.

Occlusion and splinting

All matters relating to occlusion seem to have developed a high degree of mysticism about them in the dental world, this is also true of the relationship between occlusion and periodontal therapy.

It used to be claimed that angular bony defects and increased mobility were directly attributable to trauma from the occlusion. This belief is less commonly held nowadays as angular defects can be found around both occlusally traumatized and non-traumatized teeth.[1] It is, however, self-evident that an already periodontally diseased tooth can change its relationship in the arch to become traumatized, or a tooth which is already in traumatic occlusion can develop periodontal disease, and that the two factors can exacerbate one another.

Tooth mobility Increased tooth mobility may simply be a result of loss of periodontal attachment and bony support. It may also result purely as a localized effect due to a heavy occlusal loading, causing a widening of the periodontal membrane space, though this is usually iatrogenic in origin. It is now felt that the diagnosis of occlusal trauma should only be made where progressive *increasing* tooth mobility is observed, but in order to do this it is necessary to have an objective method of measuring tooth mobility. The most widely used is Miller's classification:

Grade I   = Slight mobility
Grade II = Mobility <1 mm in a horizontal direction
Grade III = Mobility >1 mm in any direction.

Rx First priority should be to diagnose and treat any existing periodontal disease and correct any pre-existing iatrogenic causes e.g. poor crowns or bridges, high restorations. If tooth mobility persists as a direct result of diagnosable occlusal trauma, occlusal adjustment is a sensible treatment modality. If the tooth is mobile as a result of lack of alveolar bone support this is not automatically an indication for splinting (see below).

Splinting Is indicated in the following situations:
- Tooth with healthy but reduced periodontium where mobility is increasing.
- Tooth with increased mobility which patient finds uncomfortable during function.

It is very easy to design splints which are impossible for the patients to keep clean as all additions to the natural tooth surface will ↑ plaque retention. A wide range of different techniques and materials have been described, including orthodontic wire fastened to teeth by composite, composite alone, fixed bridges, partial prostheses, or acid-etch retained splints.

1 J. Waerhaug 1970 *J Periodont* **50** 355.

Perio-endo lesions

▶ It is essential to vitality test any heavily restored tooth with periodontal involvement.

Given the relative frequency of both periodontal disease and periapical pathology, it is not surprising that both may occur together, which can result in diagnostic confusion. In fact, there is little evidence to support the popular notion that periodontitis leads to pulp necrosis. However, there is no doubt that pulp pathology can exacerbate periodontal problems.

Pulpal problems

Acute pulpitis, p. 254.

Non-vital pulp (p. 254) may cause asymptomatic periapical lesion or periapical abscess.

Lateral canal ± non-vital pulp may mimic periodontal abscess, as can a root perforation following endodontic therapy.

Vertical root fracture ± non-vital pulp can lead to periodontal inflammation and may mimic periodontal abscess.

Horizontal root fracture may mimic periodontal abscess.

Periodontal pathology and its effect on the pulp

Deep pocketing may encroach on lateral canals in the apical $\frac{1}{3}$rd of the root, but is otherwise unlikely to cause direct pulpal pathology.

Gingival recession is directly associated with hypersensitivity of root dentine.

Root planing and furcation procedures actively involve dentine and can clearly lead to hypersensitivity and sometimes acute pulpal changes.

Differential diagnosis

	Primarily periodontal	*Primarily pulpal*
History	No preceeding toothache	Often toothache
Percussion	TTP especially lateral	TTP especially vertical
Probing	Pocketing always	May be no pockets
Probing sinus	May lead to pocket	May lead to apex
Vitality test	Usually +ve	−ve
X-rays	Vertical bone loss	Apical area

Combined perio-endo lesions

may be either
- Co-existing, but separate from each other, in which case standard endodontic and periodontal therapy is used as indicated.
- Interconnected, in which case probing both pocket and sinus will reach the apex. This can be confirmed by taking a periapical film with a gutta-percha point inserted into the pocket.

Rx of combined interconnected lesion First, resolve the acute infection and inflammation by drainage (± antibiotics), then treat with orthograde RCT (the greater the pulpal component the better the prognosis). The apparent periodontal lesion will often be seen to resolve to a substantial degree, over a period of months, ∴ the decision to carry out surgery should be deferred. Combined apicectomy and periodontal surgery is quite feasible but carries a poorer long-term prognosis. The worst prognosis applies to those teeth where the periapical/pulpal pathology has been due entirely to apical extension of the periodontal pocket. These are often diagnosed after the fact, when endodontics completely fails to resolve the lesion.

Furcation involvement

The extension of periodontal disease into the bi- or trifurcation of multirooted teeth is known as furcation involvement.

Diagnosis is established by probing into the furcation and by X-ray. The possibility of pulpal pathology is ↑ in teeth with furcation involvement and vitality testing is essential. X-rays give a guide to the degree of alveolar bone loss both mesially and distally and in the furcation area.

Classification

248

1st degree—Horizontal loss of support not exceeding ⅓rd tooth width. Requires scaling and root planing, possibly with furcation plasty.

2nd degree—Horizontal loss of support exceeding ⅓rd but not encompassing the total width of the furcation area. Requires furcation plasty, &/− tunnel preparation, &/− root resection, &/− extraction.

3rd degree—Horizontal through and through destruction in the furcation area. Requires tunnel preparation, &/− root resection, &/− extraction.

Treatment techniques

Scaling and root planing (p. 228) Unless the post-treatment morphology can be kept clean by the patient it will not be successful.

Furcation plasty An open procedure involving a muco-periosteal flap to allow root planing and scaling, followed by the removal of tooth structure in the furcation area in order to achieve a widened entrance to give access for cleaning. Osseous recontouring may be used if indicated. The flap is re-positioned and sutured to ↑ access post-operatively. There is an obvious risk of pulpal damage and post-operative dentine sensitivity.

Tunnel preparation is a similar procedure to furcation plasty using buccal and lingual flaps, the main difference being that the entire furcation area is exposed and the flaps are sutured together intra-radicularly to leave a large exposed furcation. There is a high risk of post-operative caries, dentine sensitivity, and pulpal exposure, making this a method to be used with caution. It is of most value for mandibular molars in patients with optimal OH. In many cases considered for furcation plasty or tunnelling, it may be more sensible to proceed to a more radical approach such as root resection.

Root resection Involves amputation of one (or even two) of the roots of a multi-rooted tooth leaving the crown and the root stump. It is important to ensure that the root to be retained can be treated endodontically, is in sound peridontal state with good bony support, is restorable and will be a viable tooth in the long term. At operation it is wise to raise a flap to enable direct visualization of the root surface. Resection of the root with a high speed bur is followed by smoothing, re-contouring, and restoration of any residual pulp cavity. It is sometimes not possible to proceed with root resection, despite apparently favourable X-rays, especially in maxillary molars, so warn patient pre-operatively.

Hemisection Involves dividing a two-rooted tooth in half to give two smaller units each with a single root. Again, RCT is necessary pre-operatively and restoration of the divided crown is required post-operatively.

Extraction Ensures removal of peridontal disease but carries its own problems.

249

Guided tissue regeneration,[1] p. 238.

It is important to note that the techniques described above are of less significance to long-term outcome than the degree of plaque control that can be achieved and maintained by the patient. Mini interdental brushes are a valuable aid in cleaning furcation defects and are available in a variety of sizes and shapes.

1 R. Ponteriero 1989 *J Clin Periodont* **16** 170.

6 Restorative dentistry

Treatment planning	252
Dental pain	254
Isolation and moisture control	258
Principles of cavity preparation	260
Class I	262
Class II—amalgam and glass ionomer	264
Class II—composite and inlays	266
Class III, Class IV, Class V, and root surface caries	268
Management of the deep carious lesion	270
Survival and failure of restorations	272
Occlusion—1	274
Occlusion—2	276
Anterior crowns for vital teeth—1	280
Anterior crowns for vital teeth—2	282
Anterior post and core crowns	284
Anterior post and core crowns—practical tips	286
Veneers	288
Posterior crowns	290
Bridges	292
Bridges—treatment planning and design	296
Bridges—practical stages	298
Bridge failures	300
Resin-bonded bridges	302
Attrition, abrasion, and erosion	304
Temporary restorations	306
Pinned restorations	308
Bleaching	310
Root canal therapy	312
Root canal therapy—instruments	314
Root canal therapy—materials	316
Canal preparation—1	318
Canal preparation—2	322
Canal obturation	326
Some endodontic problems and their management	328
Four-handed dentistry	330

Relevant pages in other chapters: Caries diagnosis, p. 32; amalgam, p. 656; composite, p. 660; the acid-etch technique, p. 662; dentine adhesives, p. 664; glass ionomers and cermets, p. 668; cements, p. 670; impression materials, p. 672; casting alloys, p. 676; acid-etch tip, p. 112.

Principal sources: *Dental Update. The British Dental Journal.* T. R. Pitt Ford 1985 *The Restoration of Teeth*, Blackwell. D. N. Allan 1986 *Crown and Bridge Prosthodontics*, Wright. B. G. N. Smith 1990 *Planning and Making Crowns and Bridges*, Dunitz. C. J. R. Stock 1985 *Endodontics in Practice*, BDA. J. L. Gutmann 1988 *Problem Solving in Endodontics*, Year Book Medical Publishers.

To attempt to resolve the problem of caries by preparing and filling cavities is comparable to trying to resolve the problem of poliomyelitis by manufacturing more attractive and better quality crutches, more quickly and more cheaply.

Treatment planning

A proper treatment plan can only result from a thorough patient assessment which must include a history, examination, relevant special tests, and ultimately, a diagnosis.

Under ideal circumstances a comprehensive treatment plan is formulated for each patient at the start of a course of treatment. Very often, however, the treatment plan will need to be revised in the light of clinical findings as the treatment progresses, e.g. patient co-operation, response to periodontal therapy, investigation of teeth of doubtful prognosis, etc. When dealing with patients with a range of problems it is therefore wise for the inexperienced operator to formulate a treatment plan which has a small number of readily achievable goals, and then on completion of this to reassess the patient to decide on what further work is necessary.

Sequence of treatment

This list is obviously an over-simplification but should serve as a general guide to the order in which treatment should be carried out.

1. Relief of pain.
2. Control and prevention of disease—e.g. OHI, dietary advice, topical fluoride, and initial periodontal therapy (scale and polish).
3. Design partial dentures and bridges.
4. Treatment of large and active carious lesions.
5. Remaining routine restorative procedures.
6. Root canal therapy.
7. Further periodontal therapy (if required).
8. Crown and bridgework.
9. Denture construction.
10. Review and recall.

Practical points

- The priority of items of treatment must be taken into account when formulating a treatment plan which may lead to deviations from the scheme described above.
- Deviations from the scheme may also be necessary to take into account patient attitudes, e.g. in an apprehensive patient it would be more appropriate to complete small restorations before dealing with the large ones.
- Explain what the treatment plan involves to the patient and the role they will have to play in ↓ disease.
- When considering complex items ask the patient what they want done—and tell them how much it will cost!
- If need to check a medical history with patient's GMP or refer patient to a specialist, allow sufficient time to elapse before arranging to carry out any treatment that is dependent upon the outcome.
- It is important to bear in mind subsequent items on a treatment plan, e.g. the design of a P/− may influence the choice of material and contour of restorations.

- For complex cases, several short treatment plans, each ending with a reassessment are more logical and efficient than one long one that keeps changing.
- When formulating a treatment plan group items together into appointments and decide how long you will need for each visit.
- Although it is usually advantageous to complete as much work as possible at each visit, in a proportion of patients (or if carried to extreme in any patient) this can be counter-productive. If in doubt about how much treatment to do at a visit, ask the patient.
- Regularly reinforce the OH throughout the treatment (e.g. whilst waiting for LA to take).
- At the end of each visit carefully note what has been done and the materials used (including sizes and shades). Cross that item off the treatment plan and adjust patient's chart. Note what is to be done next visit, this will save time by orientating you.

Stabilization or caries control In patients with multiple carious lesions it may take several weeks/months to complete the permanent restorations necessary. In these cases it may be advisable to prevent an ↑ in size of any symptomless large lesions by placing temporary dressings. The cavities should be rendered caries-free at the margins, and temporarily restored with a strong cement, e.g. glass ionomer or polycarboxylate.

Dental pain

When a patient attends the surgery and complains of toothache the pain may be arising from a variety of different structures and may be classified as follows:

Pulpal pain.
Periapical/periradicular pain.
Non-dental pain.

Dental pain can be very difficult to diagnose and the clinician must first gather as much information as possible from the history, clinical and radiographic examinations, and other special tests (see Chapter 1).

Pulpal pain

The pulp may be subject to a wide variety of insults, e.g. bacterial, thermal, chemical, traumatic, the effects of which are cumulative and can ultimately lead to inflammation in the pulp (pulpitis) and pain. The dental pulp does not contain any pro-proceptive nerve endings ∴ a characteristic of pulpal pain is that the patient is unable to localize the affected tooth. The ability of the pulp to recover from injury depends upon its blood supply, not the nerve supply, which must be borne in mind when vitality testing (p. 18).[1] It is impossible to reliably achieve an accurate diagnosis of the state of the pulp on clinical grounds alone, the only 100% accurate method is histological section.

Although numerous classifications of pulpal disease exist, only a limited number of clinical diagnostic situations require identification before effective treatment can be given.

Reversible pulpitis (hyperaemia)

Symptoms: Fleeting sensitivity/pain to hot, cold or sweet with immediate onset. Pain is usually sharp and may be difficult to locate. Quickly subsides after removal of the stimulus.

Signs: Exaggerated response to pulp testing. Carious cavity/leaking restoration.

Rx: Remove any caries present and place a sedative dressing (e.g. ZOE) or permanent restoration with an adequate lining.

Irreversible pulpitis

Symptoms: Spontaneous pain which may last several hours, be worse at night and is often pulsatile in nature. Pain is elicited by hot and cold at first, but in later stages heat is more significant and cold may actually ease symptoms. A characteristic feature is that the pain remains after the removal of the stimulus. Localization of pain may be difficult initially, but as the inflammation spreads to the periapical tissues the tooth will become more sensitive to pressure.

Signs: Application of heat (e.g. warm GP) elicits pain. Affected tooth may give no or a reduced response to electric pulp tester. In later stages may become TTP.

1 A. H. R. Rowe 1990 *Int. Endo. J* **23** 77.

Rx: Extirpation of the pulp and RCT is the treatment of choice (assuming the tooth is to be saved). If time is short or if anaesthesia proves elusive then removal of the coronal pulp and a Ledermix dressing can often control the symptoms until the remaining pulp can be extirpated under LA at the next appointment.

Dentine hypersensitivity

This is pain arising from exposed dentine in response to a thermal, tactile or osmotic stimulus (but not all exposed dentine gives rise to symptoms). It is thought to be due to dentinal fluid movement stimulating pulpal pain receptors. Prevalence is approximately 1 in 7 adults with a peak in young adults, then ↓ with age.[1] Diagnosis is by elimination of other possible causes and by evoking symptoms.

Rx: Involves ↓ aetiological factors (i.e. OHI including toothbrushing technique) and by ↓ permeability of dentinal tubules (e.g. by toothpaste containing strontium, formalin &/or fluoride; placement of varnishes or restorations).

Cracked tooth syndrome

Symptons: Sharp pain on biting—short duration.
Signs: Often relatively few ∴ diagnosis difficult. Tooth usually has a large restoration. Crack may not be apparent at first but removal of the restoration aids visualization. Positive response to vitality testing and pain can normally be elicited by getting the patient to bite with affected tooth on a cotton wool roll.

Rx: An adhesive composite restoration may be appropriate in teeth which are minimally restored but most cases will require a cast restoration with full occlusal coverage.

Periapical/periradicular pain

Progression of irreversible pulpitis ultimately leads to death of the pulp and pulpal necrosis. At this stage patient may experience relief from pain and thus may not seek attention. If neglected, however, the bacteria and pulpal breakdown products leave the root canal system via the apical foramen or lateral canals and lead to inflammatory changes and pain. Characteristically, the patient can precisely identify the affected tooth as the periodontal ligament, which is well supplied with proprioceptive nerve endings, is inflamed.

Pulpal necrosis with periapical periodontitis

Symptoms: Variable, but patients generally describe a dull ache exacerbated by biting on the tooth.

1 P. Dowell 1985 *BDJ* **158** 92.

255

Dental pain (*cont.*)

Signs: Usually no response to vitality testing, unless one canal of a multirooted tooth is still vital. The tooth will be TTP. Radiographically the apical PDL may be widened or there may be a periapical radiolucency (granuloma or cyst).

Rx: RCT or extraction.

Acute periapical abscess

Symptoms: Severe pain which will disturb sleep. Tooth is exquisitely tender to touch.

Signs: Affected tooth is usually extruded, mobile, and TTP. May be associated with a localized or diffuse swelling. Vitality testing may be misleading as pus may conduct stimulus to apical tissues. Radiographic changes can range from a widening of the apical PDS space to an obvious radiolucency. It is important to differentiate this condition from a periodontal abscess.

Rx: Drain pus and relieve occlusion. Drainage of pus can often be achieved by entering the pulp chamber with a high speed diamond bur. The tooth should be steadied with a finger to prevent excessive vibration. After drainage has been achieved it is preferable to prepare the canal and place a temporary dressing. Leaving the tooth on 'open drainage' should be avoided if possible, but if absolutely necessary for <24 hours, as after this time contamination of the root canal by anaerobic bacteria makes subsequent RCT very difficult. If a fluctuant swelling is present in the soft tissues then this should be incised to achieve drainage. Antibiotics should be prescribed if there is systemic involvement (pyrexia, lymphadenopathy) or if the infection is spreading significantly along tissue planes. When the acute symptoms have subsided then RCT must be performed or the tooth extracted.

Lateral periodontal abscess

Symptoms: Similar to periapical abscess with acute pain and tenderness and often an associated bad taste.

Signs: Tooth is usually mobile and TTP, with associated localized or diffuse swelling of the adjacent periodontium. A deep periodontal pocket is usually associated which will exude pus on probing. Radiographs normally show vertical or horizontal bone loss, and vitality testing is usually positive, unless there is an associated endodontic problem (perio-endo lesion).

Rx: Debride the pocket and achieve drainage of pus. Irrigate with a chlorhexidine solution. If there is systemic involvement or it is a recurrent problem then prescribe antibiotics (metronidazole or amoxycillin).

Non-dental pain

When no signs of dental or periradicular pathology can be detected then non-dental causes must be considered. Other causes of pain that can present as toothache include:

TMPDS (p. 478).
Sinusitis (p. 422).
Psychological disorders (atypical odontalgia) (p. 550).
Tumours (pp. 420 and 450).

Isolation and moisture control

Isolation is required to aid visibility, prevent contamination during moisture sensitive techniques, maintain an aseptic environment and protect the patient from caustic materials or aspiration of foreign material.

High volume suction, e.g. an aspirator.

Low volume suction, e.g. a saliva ejector. Many designs available—most useful are flanged metal type to keep tongue at bay when working in the lower arch, and disposable plastic type for working on upper teeth.

Compressed air This tends to re-distribute the moisture to somewhere else (e.g. your eye) rather than remove it. Should be used with care in deep cavities as prolonged use can cause pulpal damage.

Absorbents

- Cotton-wool rolls. Insert with a rolling action away from the alveolus. Moisten before removal to prevent tearing mucosa.
- Paper pads.
- Carboxymethylcellulose pads (Dry Tips). Very effective if inserted the correct way round with the impermeable plastic against the tooth.

Rubber dam provides effective isolation and also improves access to operating site. It is indicated where moisture control and airway protection is essential, e.g. RCT (RCT without rubber dam could be considered negligent), acid-etch technique. With practice rubber dam can be applied quickly and often saves time in the long run. The dam must be secured to the teeth, several methods are available:

- Rubber dam clamps. These consist of 2 metal jaws linked by 1 or more bows. Commonly used for posterior teeth.
- Floss ligatures.
- Wedges.
- Proprietary rubber bands (Wedjets) or pieces of dam, worked through contact points.
- Pinching dam between a tight contact point.

Types of dam: **1** Sheet grade, 6-inch square, which is supported with a frame. Moderate to thicker guages are preferable. **2** Mask type which is supported by a paper margin and is looped over ears with elastic.

Placement Several regimens have been described, the following is popular:

- Place cotton-wool roll in sulcus beside tooth for treatment.
- Mark position of centre of each tooth to be included with ball pen while dam held stretched.
- Punch holes in dam corresponding to tooth size.
- Try-in clamp (with floss tied to it).
- Apply brushless shaving foam to dam as a lubricant.

- Fit clamp into appropriate hole, with bridge distally and using forceps place clamp and dam on to tooth.
- Position dam on other teeth, using floss to ease through contact points.
- Secure dam anteriorly using one of the methods above.
- If frame required, position.
- Put napkin on patient's chin under dam. A saliva ejector may ↑ patient comfort.

Removal
- Take away clamps/ligatures, etc.
- Stretch dam, carefully cut interdental septa with scissors and remove.

Protection of the airway is mandatory when fitting crowns, bridges, inlays, and carrying out RCT. Best provided by rubber dam, but if this is not possible, can use a butterfly sponge, or gauze.

Gingival retraction ↓ gingival exudate and exposes subgingival preparations prior to impression taking. Some retraction cords are impregnated with substances such as adrenalin to ↓ bleeding. The cord should be packed into the gingival crevice with a flat plastic instrument (leaving a tag hanging out to aid removal) prior to impression taking. Braided cords are better than twisted. Bleeding from the gingival margin can be ↓ by applying alum/adrenalin solution.

Electrosurgery may be indicated where a margin extends subgingivally and gingival overgrowth is hampering restoration or impression taking.

Principles of cavity preparation

Why restore?
- To restore function.
- To prevent further spread of an active lesion which is not amenable to preventive measures.
- To preserve pulp vitality.
- To restore aesthetics.

However, these reasons need to be evaluated with regard to the patient and the rest of the dentition. For example, there is little point in attempting restoration of a non-functional 8̲.

Cavity design With caries prevalence declining, emphasis has changed from extension for prevention, to minimizing removal of tooth tissue. Cavity preparation should be based on the morphology of the carious lesion and the requirements of the filling material being used.

General principles of cavity preparation
- Gain access to caries.
- Remove all caries at ADJ junction (to prevent spread laterally).
- Cut away all carious and unsupported enamel.
- Extend margins so that they are accessible for instrumentation and cleaning.
- Shape cavity so that remaining tooth tissue and restorative material will be able to withstand functional forces (resistance form).
- Shape cavity so that restoration will be retained (retention form).
- Check cavity margins are appropriate for restorative material.
- Remove remaining caries unless indirect pulp cap to be carried out.
- Wash and dry cavity.

Helpful hints
- While care must be exercised not to overcut a cavity, do not skimp on access so that caries removal is compromised by poor visibility.
- Mark centric stops with articulating paper prior to cavity preparation and try to preserve if possible.
- Avoid crossing marginal ridges.
- In removing caries a tactile appreciation of the hardness of dentine is important, ∴ use slow speed instruments or excavators.
- The base of the cavity should not be flattened as this runs the risk of pulp exposure.
- Unless caries dictates, margins should be supragingival.
- Retention form is required to enable the restoration to withstand displacing forces. The adhesive materials allow more latitude in this respect, but where possible cavities for these materials should be retentive.

- All internal line angles should be rounded to ↓ internal stresses. Removing caries with a large diameter round bur automatically produces the desired shape.
- In a Class II box, margin should extend below the contact point. Failure to do so can result in ↑ risk of recurrent caries.

Amalgam[1] (see also p. 656).
- Amalgam is brittle, ∴ an amalgam margin of 90°, or at least 70°, is required to prevent ditching. Also avoid leaving amalgam overlying cavity margins and overcarving.
- Accepted minimal dimensions for amalgam are 2 mm occlusally and 1 mm elsewhere.
- Linings are required to protect against thermal injury.

Composite, p. 660.

Glass ionomer, p. 666.
- Wear precludes their use in load-bearing situations except for 1° teeth.

Gold
- Relies on minimally divergent walls and cement lute for retention.
- A preparation margin of >135° is advisable to give good marginal fit to restoration and to allow burnishing.

It would be foolish to think that experience in cavity preparation can be adequately assimilated from the written text. The purpose of the following pages is to give the reader some practical tips on how to do the procedures considered as well as to describe recent innovations and techniques.

Nomenclature

Class I   cavity in pits and fissures of molar/premolar.
Class II   cavity in approximal surface(s) of molar/premolar.
Class III cavity in approximal surface of incisor/canine not involving incisal edge.
Class IV as for Class III, but including incisal edge.
Class V   cavity in cervical third of buccal or lingual surface of any tooth.

1 P. B. Robinson 1985 *Dental Update* **12** 357.

Class I

Amalgam

Amalgam still remains the most widely used material for Class I cavities, probably because it is more forgiving of technique than some of the newer materials. If enamel margins are cut to an angle of 90° (or, if cusps steeply inclined, >70°) the resultant cavity will be adequately retentive.

Lining: shallow cavity—calcium hydroxide liner; average cavity—a base of e.g. modified ZOE, GI or polycarboxylate; deep cavity—subline with calcium hydroxide followed by a base as for average cavity.

Composite

The controversy surrounding posterior composites is dealt with on p. 266 and p. 660. A technique which has gained more widespread acceptance is the:

Preventive resin restoration Introduced by Simonsen (then by others as the minimal composite restoration!). Preparation is limited to caries removal and the resultant cavity restored using fissure sealant alone if small, or composite followed by sealant, if larger. Alternatively, GI can be used instead of composite. The rationale of this approach is that adjacent fissures are sealed for prevention. It is particularly useful for investigating any suspect areas of a fissure, a technique that is often referred to as an enamel biopsy (obviously coined by an academic). This involves exploring the area with a small bur, if no caries is found the 'cavity' can be aborted and sealant placed. If carious, a PRR can be carried out. It is often possible to complete preparation of a PRR without LA; however, if the cavity appears larger than originally thought LA can then be given. If the cavity extends significantly into load bearing areas a conventional cavity should be prepared and filled with amalgam (or posterior composite).

Technique for medium-sized cavity (1–2 mm)
- Assess whether LA required. If not, ask patient to signal if tooth becomes sensitive.
- Isolate tooth (preferably with rubber dam).
- Gain access to caries with a small bur at high speed.
- Use a small round bur run at slow speed to remove caries. Only remove as much enamel as required for access.
- Line dentine with calcium hydroxide.
- Etch cavity margins and occlusal surface, wash and dry.
- Place a thin layer of bonding resin (or sealant) in cavity and cure.
- Fill cavity with composite, but don't overfill.
- Paint sealant over whole of occlusal surface and cure.
- Check occlusion.

Where possible a related sealant and composite should be used to ensure a good bond.

Hints for composite restorations

- Put etchant gel in a disposable syringe with a wide-bore needle to aid placement.
- Additions are generally easy as new composite will bond to old.
- Avoid eugenol containing cements with composite restorations.
- For larger cavities composite must be cured incrementally.

Class II—amalgam and glass ionomer

▶ Avoid the creation of an overhang at the cervical margin and ensure a good contact point with adjacent tooth.

Amalgam Over the years the dimensions of the textbook cavity have diminished. In practice, cavity size is determined by the size of the carious lesion and extension beyond this should be minimal. The conventional Class II cavity comprises an approximal box and an occlusal dovetail for resistance against lateral displacement. Retention from occlusal forces is derived from a 2–5° divergence of the walls towards the floor in both parts of the preparation. The margins of the box should extend just outwith the contact area unless caries dictates a wider position. Amalgam restorations are prone to # at the isthmus ∴ sufficient depth must be provided in this area. The width of the isthmus should not be overcut (ideally $\frac{1}{4}$ to $\frac{1}{5}$ intercusp width) and the dovetail only minimally flared in order not to undermine support for the cusps. However, where these are weakened or missing they should be replaced with a pinned restoration (p. 308). A chisel can be used to plane away unsupported enamel from the margins of the completed preparation to produce a 90° butt joint. In molar teeth with mesial and distal caries it may be preferable to try and cut two separate cavities, but often a confluent MOD preparation is unavoidable.

Glass/cermet ionomer Although the resistance to wear and fracture toughness of cermet is superior to GI it is still inadequate for load-bearing conventional Class II cavities. Recently, a 'tunnel' approach to interproximal caries has been described.[1] Access to the caries is made through either the occlusal or buccal surfaces leaving the marginal ridge intact. This approach is only suitable for small lesions, as when preparation is completed at least 2 mm of marginal ridge must remain. The access cavity may need to be widened buccolingually in order to complete caries removal. A piece of mylar strip wedged into place will act as a matrix. A GI or cermet is used to fill the bulk of the cavity and the occlusal access cavity sealed with a posterior composite resin. Long-term results using this technique are still awaited.

1 J. W. McLean 1988 *BDJ* **164** 293.

Class II—composite and inlays

Composite Posterior composites have been shown to perform satisfactorily over a 3 yr period,[1] but the technique is more demanding taking approximately 50% longer. In addition, it is difficult to establish adequate contact points and occlusal stops Polymerization shrinkage can cause cuspal flexure, post-operative pain and marginal gaps. Posterior composites are ∴ best avoided in the following situations:

- Large restorations which include centric stops.
- Poor moisture control.
- Restorations with deep gingival extensions.
- Bruxism or heavy occlusion.

If a composite is to be used then a hybrid material with >75% filler is advisable. Pre-wedging aids creation of a contact point. Composite should be placed, and cured, incrementally. If possible, centric stops should be preserved on sound tooth tissue.[2]

Composite and porcelain inlays These new inlay techniques appear to overcome some of the problems associated with direct composite resin restorations. When used in conjunction with an acid-etch technique existing tooth tissue is reinforced. Curing composite outside the mouth overcomes polymerisation shrinkage and ↑ strength. As the inlays are bonded to the tooth with an adhesive, parallel walls are less important, but undercuts must be removed. In general, composite inlays are preferred to porcelain as they are a better fit and are less abrasive to opposing teeth. The long-term durability of the two techniques and their mode of cementation have yet to be assessed.

Composite inlays can be made: **1** directly at the chairside (e.g. the EOS system) on a silicone model obtained from an impression of the prepared cavity. Although this allows curing of the inlay from all sides, greater strength is achieved by additional heat curing (110 °C for 5 min). **2** Indirectly in the laboratory (where they are cured under heat and pressure).

Technique—preparation
- Prepare cavity with near parallel walls, rounded line angles and a slight bevel of the enamel margins. For onlays, a minimum 1.5 mm reduction of cusps is necessary.
- Place CaOH lining in deep part of the cavity.
- Cover all dentine with GI lining cement (minimum depth 0.5 mm), blocking out any undercuts.
- Take an impression of the preparation, the opposing arch and record an occlusal wafer.
- Choose shade.
- Make and place temporary (p. 306) with eugenol-free cement (indirect inlays only). Alternatively a proprietary resin-based temporary material (e.g. Fermit) may be used.

1 A. A. Robinson 1988 *BDJ* **164** 248.   **2** R. W. Bryant 1992 *Aust Dent J* 37 81.

Technique—cementation
- Place rubber dam.
- Remove temporary and clean tooth.
- Try-in inlay and carefully check marginal fit and adjust as necessary.
- Polish any adjusted areas.
- Remove inlay and clean with alcohol. For porcelain only, place layer of silane coupling agent on fitting surface.
- Etch enamel. Wash and dry.
- Place bonding agent on prep and fitting surface of inlay.
- Apply dual-cure composite to prep and inlay and carefully seat.
- Cure for 10 sec and then remove any excess composite.
- Complete light-curing (dual-cure composite will finish setting chemically under inlay).
- Trim any excess cement and polish.
- Check occlusion and adjust.

Class III

Glass ionomer and composite are now the most widely used materials for Class III restorations, the former being indicated where prevention of recurrent caries is more important than aesthetics. E.g. approximal caries in lower incisors is indicative of a high caries rate, ∴ GI cement is advisable.

Access should be gained from either the buccal or lingual aspect depending on the position of the lesion. As both materials are adhesive the cavity is just extended sufficiently to remove all peripheral caries, some unsupported enamel can be retained labially, but the margins should be planed with chisels to remove any grossly weakened tooth structure. Additional retention can be created with a cervical groove, but the incisal pit advocated for silicate restorations is better omitted, as it may undermine the incisal edge. Cavity preparation can be almost entirely completed with slow speed burs and hand instruments. A slight excess of material should be moulded into the cavity with a Mylar strip, wedged cervically. Once the material is set, the excess can be removed, though for GI, varnish or bonding resin should be placed first. After checking the occlusion the restoration can be polished using one of the proprietary products (e.g. Soflex discs, Enhance). For GI polishing should be delayed for 24 hrs and carried out under water spray.

Class IV

The restoration of choice is composite, the so-called 'acid etch tip' (p. 112); however, for large Class IV cavities in the adult patient, a PJC or porcelain veneer may give better retention and aesthetics.

Class V

Although Class V cavities are seen less frequently in younger patients, they are an ↑ problem in older age-groups with gingival recession. GI is the preferred material in this situation, but amalgam can be used for posterior teeth. Where aesthetics are important a GI/composite resin laminate restoration could be used.

Amalgam Once caries has been removed the margins should be trimmed so that no undermined enamel remains. Because of the prism orientation this results in the occlusal and gingival walls lying parallel to each other, whereas the mesial and distal walls diverge outwards. To ↑ retention, undercuts should be cut in the dentine gingivally and occlusally (but not mesially and distally).

Glass ionomer Cavity preparation is similar to above except that provision of undercuts is less important; however, for abrasion cavities it may be necessary to cut a small step 0.5 mm in depth gingivally, so that the GI can be finished to a butt joint. The best finish is achieved using a preformed matrix, which also helps to create a convex surface contour. The foil type are preferable (e.g. Hawes Neos) as they can be burnished to improve fit. For ease of handling tack the matrix to an instrument with sticky wax. Then pack the cavity to excess and position the matrix gently. Once the GI has set the matrix is removed and the restoration protected from moisture contamination (p. 668). Any excess is then trimmed away.

Composite resin In aesthetically important regions composite resin might be the material of choice, which should be used in conjunction with a dentine bonding agent when the cavity margins are not sited on enamel.

269

Other materials/techniques In abrasion cavities in the anterior part of the mouth a 'sandwich' technique using a layer of GI with a composite facing, has been suggested. Problems can arise with this type of restoration as the forces generated during polymerization of the composite are greater than the bond strength of the GI to dentine. If the aesthetics of a composite resin are required it would therefore seem more sensible to use a dentine bonding agent in preference to this technique. Newer types of tooth-coloured restorative materials have recently been introduced which hold promise for the restoration of aesthetically important Class V cavity. Some materials termed light-cured glass ionomers (e.g. Fuji 2LC), have aesthetics which approach the composites. Another type of material known as a compomer (e.g. Dyract), is half-way between a composite and a glass ionomer.

Root caries

As gingival recession is a prerequisite to root caries, it occurs predominantly in the 40+ age-group. Dentine is thus directly exposed to carious attack. It is sometimes seen secondary to ↓ saliva flow (which reduces buffering capacity and may alter dietary habits) caused by salivary gland disease, drugs or radiation. Long-term sugar-based medication may also be a factor. Rx requires, first, control of the aetiological factor, and for most patients this involves dietary advice and OHI. Topical fluoride mouthrinses may aid re-mineralization and prevent new lesions developing. However, active lesions require restoration with GI cement. See also rampant caries, p. 98.

Management of the deep carious lesion

Assessment
- Is the tooth restorable and is restoration preferable to extraction?
- Is the tooth symptomless? If not what is the character and duration of the pain?
- Test vitality and percuss the tooth (before LA!).
- Take X-rays to check extent of lesion and ? apical pathology.

Management depends upon a guesstimate of pulpal condition (p. 254). Irreversible pulpitis/necrotic pulp—Rx: RCT (p. 312) or extraction. Reversible pulpitis/healthy pulp—aim to maintain pulp vitality by removal of carious dentine without pulp exposure. If in doubt Rx as reversible pulpitis. Can always institute RCT later.

Indirect pulp cap Ideally, cavity preparation should involve the elimination of all caries, but where this would risk pulp exposure and the tooth is vital it may be more prudent to carry out an indirect pulp cap. This involves leaving a small amount of softened dentine at the base of a deep cavity with the aim of arresting further bacterial spread and maintaining pulp health.

Rationale
- Softening of dentine precedes bacterial invasion.
- Pulpitis does not occur until bacteria are within 0.5–1 mm of the pulp, ∴ if a vital tooth is asymptomatic the softened dentine closest to the pulp is unlikely to contain bacteria.
- Prognosis for continuing vitality of healthy pulp is better if exposure avoided.
- Materials with antibacterial properties help ↓ bacterial activity.
- Bacteria sealed under a restoration are denied substrate ∴ lesion arrests.

Sequence of treatment for vital pulp.
1 LA.
2 Apply rubber dam to ↓ risk of further bacterial contamination.
3 Prepare cavity outline, removing caries from ADJ and cutting back unsupported enamel.
4 Cautiously remove softened dentine from floor. If possible complete removal, but if likely to result in exposure and dentine only slightly softened, stop.
5 Apply calcium hydroxide cement (e.g. Dycal, Life) to floor.
6 Place a base (e.g. polycarboxylate/modified ZOE cement)
7 Adjust margins and restore.
8 Warn patient that some sensitivity should be expected initially, but if symptoms occur after that to return.
9 Follow-up for at least 1 yr.

If it is necessary to leave dentine that is probably infected, place calcium hydroxide and a ZOE dressing. Leave tooth for 3 months before re-entering to complete caries removal.

Exposure

- If traumatic, small (<1 mm), and uncontaminated, perform direct pulp cap with calcium hydroxide and restore.
- If carious exposure and continued pulp vitality is doubtful, RCT will be required. If time short can dress tooth with Ledermix and ZOE and extirpate pulp at next visit.

Pulpotomy is removal of coronal part of pulp in order to eliminate damaged or contaminated tissue. It is indicated for teeth with immature apices as continued vitality of apical pulp will allow root formation to proceed. Once the apex has closed conventional RCT can be carried out. The pulp is amputated to the cervical constriction, dressed with calcium hydroxide and the tooth temporarily restored.

271

Materials used in the maintenance of pulp vitality

Zinc oxide eugenol sterilizes underlying dentine, ↓ prostaglandin synthesis, aids re-mineralization and is hygroscopic. This makes it useful for indirect pulp capping. As it does not promote the formation of a calcific bridge it is C/I for direct pulp caps.

Calcium hydroxide has a pH of 11 which makes it bacteriostatic and promotes the formation of a calcific barrier. When calcium hydroxide comes into contact with the pulp a zone of pulpal necrosis is formed. This is subsequently mineralized with calcium ions from the pulp. It is the material of choice for direct pulp caps.

Ledermix is a mixture of triamcinolone acetonide (a steroid) and demethylchlortetracycline in a water-soluble base. It has anti-inflammatory and bacteriostatic properties, but also suppresses pulpal defences ∴ resulting in the rapid spread of any bacteria not affected by the antibiotic it contains. It is most useful for the management of irreversibly inflamed pulps where anaesthesia may be a problem, or when pulp extirpation has to be delayed.

Survival and failure of restorations

Survival of restorations

The results of Elderton's study into the durability of routine restorations placed in the General Dental Services in Scotland provided both a shock and a stimulus to the profession, as he found that 50% lasted for less than 5 yrs.[1] This led to debate over both clinical technique and the profession's readiness to replace restorations. It is interesting to note that those patients who change dentists frequently are more at risk of replacement restorations than those who are loyal to the same GDP.[2] In order to increase longevity we need to consider the reasons for the failure of restorations and diagnosis of secondary caries.

Reasons for failure of restorations

- Incorrect diagnosis and treatment planning, e.g. pulpal pathology; caries of another surface; extraction of tooth for another reason.
- Incorrect preparation of tooth, e.g. caries left at ADJ; incorrect margin preparation; inadequate retention; preparation too shallow; weakened tooth tissue left unprotected.
- Incorrect choice of restorative material, e.g. inadequate strength or resistance to wear for situation.
- Incorrect manipulation of material, e.g. inadequate moisture control; over or under contouring.

Before replacing a failed restoration it is important to identify the cause of failure and decide whether this can be dealt with by replacement or repair. When making this decision bear in mind that cavity size is ↑ on average by 0.6 mm each time a restoration is removed.[3]

Secondary caries

Unfortunately, placement of a restoration does not confer caries immunity upon a tooth. When caries occurs adjacent to a restoration it is called secondary or recurrent caries. While secondary caries is an accepted phenomenon, we as a profession have perhaps been a little too ready in the past to diagnose and treat it. Ditched amalgam margins are not a reason for replacement *per se* and active intervention is only required if caries can definitely be demonstrated. Secondary caries is difficult to diagnose, but careful observation rather than intervention is now advocated.

In order to prevent secondary caries it is important not only to educate the patient to reduce their caries rate, but also to examine our restorative technique, e.g. ensuring that adequate amalgam margin angles are prepared.

1 R. J. Elderton 1983 *BDJ* **155** 91.   **2** J. A. Davies 1984 *BDJ* **157** 322.   **3** R. J. Elderton 1979 *Proc Br Paedodont Soc* **9** 25.

Occlusion—1

In a book of this size it is (thankfully) not possible to consider all aspects of occlusion, therefore we will try to concentrate on the practical aspects and leave the more esoteric considerations to other texts. We also suggest that significant occlusal adjustment is only attempted by the experienced.

Definitions

Ideal occlusion Anatomically perfect occlusion—rare.

Functional occlusion An occlusion that is free of interferences to smooth gliding movements of the mandible, with absence of pathology.

Balanced occlusion Balancing contacts in all excursions of the mandible to provide ↑ stability of F/F dentures, not applicable to natural dentition.

Group function Multiple tooth contacts on working side during lateral excursions, but no contact on non-working side.

Canine guided occlusion During lateral excursions there is disclusion of all the teeth on the working side except for the canine.

Hinge axis The axis of rotation of the condyles during the first few millimetres of mandibular opening.

Terminal hinge axis The axis or rotation of the mandible when the condyles are in their most posterior, superior position in the glenoid fossa.

Retruded arc of closure The arc of closure of the mandible with the condyles rotating about the terminal hinge axis.

Intercuspal position (ICP) or centric occlusion Position of maximum interdigitation.

Retruded contact position (RCP) or centric relation Position of the mandible where initial tooth contact occurs on the retruded arc of closure. Occurs when condyles are fully retruded in the glenoid fossa. In approximately 20% patients RCP and ICP are coincident, remainder have forward slide from RCP to ICP.

Rest position The habitual postural position of the mandible when the patient is relaxed with the condyles in a neutral position.

Freeway space The difference between the rest and intercuspal position.

Centric stops The points on the occlusal surface which meet with the opposing tooth in ICP. Normally the cusp tips, marginal ridges and central fossae.

Supporting or functional cusps The cusps that occlude with the centric stops on the opposing tooth. Usually palatal on upper and buccal on lower.

Non-supporting cusps The cusps that do not occlude with the opposing teeth. Usually buccal on upper and lingual on lower.

Deflective contacts deflect mandible from natural path of closure.

Interferences are contacts that hinder smooth excursive movements of mandible.

Occlusal vertical dimension (OVD) Relationship between maxilla and mandible in ICP, i.e. face height.

Do occlusal factors play a role in temporomandibular dysfunction?[1]

TMPDS is recognized as being of multifactorial aetiology (p. 478). The evidence would suggest that occlusal interferences usually cause either subclinical or no dysfunction because they lie within the adaptive capacity of the patient's neuromusculature. However, this may be lowered by stress and emotional problems so that in susceptible patients occlusal interferences can result in muscle hyperactivity at certain times. It is important ∴ to ensure that iatrogenic intereferences are not introduced during restorative procedures.

1 R. W. Wassell 1989 *J Dent* **17** 101.

Occlusion—2

Occlusal examination

Prior to carrying out restorative treatment the dentist should examine the patient's occlusion. Occlusal contacts can be identified with a 10 μm metal foil (Shimstock) and marked using thin articulating paper (20 μm). Important features to look for are:
- Number and distribution of occluding teeth.
- Over-eruption, tilting, rotation, etc.
- Presence or absence of centric stops.
- The RCP and any slide between RCP and ICP.
- Anterior guidance—look for disclusion of posterior teeth on protrusion.
- Lateral excursions—? group function, ? canine guidance—check for non-working interferences.
- TMJs and muscles of mastication.

The clinical examination can only reveal a limited amount of information and in some circumstances (such as prior to crown and bridgework or in patients with TMPDS) a more detailed occlusal examination is required. This is called an **occlusal analysis** or a **diagnostic mounting** and is done by mounting study models on an adjustable articulator (see below) to facilitate the examination of the features above.

Occlusal considerations for restorative procedures

In most situations restorations are made to conform to the patient's existing occlusion and the main consideration is to prevent the introduction of iatrogenic occlusal interferences. This approach to treatment is known as the **conformative approach**. In some circumstances the conformative approach is not appropriate and a new occlusal scheme must be planned. This is often the case when extensive crown and bridgework is required such that the patient's existing occlusion will be effectively destroyed by the preparations. A new occlusion is established which is free of interferences and with the patient occluding in centric relation, which is the only reproducible position. This approach to treatment is called the **reorganized approach** and further consideration of this line of treatment is beyond the scope of this book.

For simple intracoronal restorations there is generally no need to employ any complex methods, but care must be taken to ensure the correct occlusal scheme is reproduced. Before preparing the cavity it is worthwhile marking the centric stops with articulating paper and trying to preserve them if possible. On completion of the restoration it must be checked in centric occlusion to ensure that it is not high, but also to ensure that it has recreated the centric stops, as if it is out of occlusion then overeruption will occur (which may produce interferences). The restoration should then be checked in all mandibular excursions to ensure that no interferences are introduced.

One or two units of extracoronal restorations can again be constructed in a relatively simple manner. They are usually constructed in the laboratory using hand-held models to reproduce the occlusion. Again, great care must be taken at the try-in stage to check the occlusion as above. This technique should be used with care when restoring the most distal tooth in the arch as it is very easy to introduce errors in this situation, it may be more appropriate to use an occlusal record (transfer coping technique—see later) and mount the models on an articulator.

More complex laboratory made restorations need to be constructed with the models mounted on an articulator. This allows the restorations to be constructed in harmony with the patient's occlusion in all mandibular positions, which should minimize the amount of time spent adjusting the restoration at the try-in stage. Also, if any changes to the patient's occlusion are planned, then they can be made on the articulator in a controlled fashion. An articulator is a device which holds the models in a particular relationship and simulates jaw movements. Numerous types of articulator are available but only certain types are appropriate for use in crown and bridgework, commonly used examples are the Denar mark 2 and the Dentatus, which are both semi-adjustable articulators. The articulator must accurately reproduce mandibular movements and to do this the casts must be mounted in the correct relationship to the 'TMJs', this is achieved by taking a facebow record, which relates the maxillary teeth to the hinge axis.

Occlusal records

Occlusal records are required to mount the models on an articulator in a particular position. Two positions are commonly used for the mounting, namely, the intercuspal position and the retruded contact position. A wax 'squash bite' has commonly been used to record ICP, however it is inaccurate as the mandible can be deviated as the teeth 'bite' through the wax. It is far better not to use any record and to mount the models to the position of 'best fit'. After the preparations have been carried out it can be difficult to locate the working model to this position of best fit, in this situation the **transfer coping technique** should be used. In this technique Duralay copings are constructed on the working dies which are taken to the clinic and seated in place on the preparations, they are then adjusted to ensure that they are clear of the occlusion. A further mix of Duralay is applied to the occlusal surface of the coping and the patient asked to close together which produces an indent of the opposing tooth in the resin and provides a very accurate occlusal registration.

When the models are to be mounted in RCP this is achieved by recording the position of the mandible on the retruded arc of closure just before tooth contact occurs. This is termed a **precentric record** and is generally registered with a relatively hard wax (Moyco Beauty Wax). The record is constructed on the maxillary model and trimmed flush with the buccal surfaces of the teeth. It is then softened and seated on the maxillary teeth and the mandible manipulated on to the retruded arc of closure to indent the wax, without allowing tooth contact to take place. The registration can be refined by using a low viscosity material (e.g. ZOE) in the wax record.

On occasions it may be necessary to record other positions of the mandible, such as a protrusive position or lateral positions. These are often necessary in order to set the angles on an adjustable articulator, and the records are usually taken in Moyco Beauty Wax similar to above.

Anterior crowns for vital teeth—1

▶ Defer preparation of any crowns until the patient can attain good OH. As not only will this help to ↑ their motivation, but healthy gingivae are necessary for correct placement of margins and accurate impressions.

Preliminary treatment
- Check vitality; if any doubt, institute RCT first.
- Take X-rays to establish size of pulp. If large, crown preparation C/I.
- Get study models and special tray made. A trial preparation and build-up in ivory wax on a duplicate model can be helpful (especially for the less confident operator). This helps to anticipate any complications and in addition can be used for fabrication of a temporary crown.
- Record the shade, so that it can be checked at subsequent visits.
- Examine the occlusion.

Porcelain jacket crown (PJC)

This gives best aesthetic result and can be used where occlusal forces are not too heavy. Newer ceramic systems (e.g. Dicor, Inceram, Empress) are available which offer ↑ strength.

Principles
- Sufficient tooth reduction to permit adequate thickness of porcelain for strength.
- Reduction should follow tooth contours. NB 2 plane reduction on labial face.
- Shoulder preparation: 1.0 mm labial (just into gingival crevice) and palatal (supragingival).
- 5° taper of opposing walls for retention.

Preparation
Interproximal: Use a long, tapered fissure bur (with flat end) to produce a shoulder. Walls should have 5° taper and converge lingually.

Labial: With the same bur first place three depth grooves and remove intervening tooth tissue. Extend 0.5 mm subgingivally.

Lingual: If possible carry out under direct vision. Continue interproximal shoulder round to form cingulum wall, supragingivally. Remainder of palatal surface should be prepared with a diamond wheel or flame-shaped bur to give 0.8 mm clearance from opposing teeth.

Incisal: A reduction of 2–3 mm is required angled towards palatal.

Finishing: An end-cutting fissure bur is helpful to define the shoulder. Finishing burs should be used to round off line angles.

Fabricate temporary crown (p. 306) next, as if time runs out impressions can be deferred, but a temporary crown cannot. Don't make too good a temporary or the patient may be tempted not to return.

Impressions (p. 672) Opposing arch can be recorded in alginate (cheaper) and don't forget a bite wafer. Check shade in natural and artificial light with help of DSA and patient. Fit temporary and arrange next appointment.

Fitting crown Protect airway. Remove temporary crown and try-in PJC. Check occlusion, if any adjustments are required either polish with fine abrasives or re-glaze. Make sure patient is happy before you cement crown.

Anterior crowns for vital teeth—2

Platinum bonded/McLean-Sced/twin foil crown is a modification of the PJC. A layer of platinum foil (coated with tin) is retained in crown to ↑ strength. Requires only 0.7 mm reduction which is helpful when a PJC is C/I by the occlusion, but where the reduction required for a bonded crown may jeopardize the pulp.

Porcelain fused to metal (PFM) crown ↑ strength compared to PJC, but ↑ expense, ↑ labial reduction and ↓ aesthetics. Preparation as for PJC, except that require 0.5 mm reduction of lingual surface with chamfered margin and labial reduction of 1.5 mm, with shoulder. Transition from shoulder to chamfer is on approximal surface. The junction between porcelain and gold must not be in an area of contact with the opposing teeth. Ideally, all occlusal surfaces in contact with opposing teeth should be finished in metal.

Common problems with anterior crowns

- *Preparation likely to expose pulp*. In young patient consider veneer as interim measure. In older patient, elective RCT.
- *Completed crown does not seat*. Check: (1) no temporary cement left on prep; (2) approximal contacts with floss, if too tight adjust; (3) ? distorted impression, check for undercuts on prep and repeat impressions; (4) die over-trimmed leading to over-extension of margin, cut-back crown margin.
- *Core material showing through crown*. Need to reduce prep so that sufficient bulk of enamel and dentine porcelain can be built up over core and remake.
- *Colour not right*. If technician handy can see if surface stains will give sufficient improvement. If not, re-choose shade and remake.

Removing old crowns (Protect airway)

A crown-removing instrument can be used to try and remove a crown without destroying it. If crown is to be replaced:

PJC: cut longitudinal groove in labial face of porcelain. Insert a flat plastic and twist.

PFM crown: cut through porcelain as above and then using a cross-cut tungsten carbide try to cut through metal, but not underlying dentine. Use a flat pastic to separate.

Anterior post and core crowns

In root-filled anterior teeth it is usually necessary to insert a post and core prior to the placement of a crown. Post and cores serve two functions: first, they provide retention for the restoration and secondly, they distribute stresses along the length of the root. As the placement of a post will preclude further orthograde endodontics it is important to check first that the root-filling and apical condition are satisfactory. If in doubt, repeat RCT and place gutta-percha filling.

Preliminary preparation The first step is to prepare the crown of the tooth to receive the appropriate coronal restoration, the author prefers to use a PJC rather than a PFM as it is a weaker restoration, and any subsequent trauma will probably lead to fracture of the crown rather than the root, which is likely to occur if a PFM is used. The appropriate reductions and margin preparations are carried out with the intention of retaining as much coronal dentine as possible. Grossly weakened tooth substance is removed, but the root face should *not* be flattened off. The retention of a core of tooth substance is important as it effectively increases the length of the subsequent post, obviously in some cases this will not be possible, e.g. if the tooth has fractured at gingival level. The coronal gutta-percha is removed with a heated instrument or Gates–Glidden bur, taking care not to disturb apical seal. The root canal is then prepared according to the particular technique being used. As a general guide the post should be as long as possible (without causing a perforation!) which gives maximum retention and gives the best stress distribution. A rule of thumb is that the post length should be at least equal to the anticipated crown height and a periodontal probe is helpful to check prepared canal length.

Types of post and core system

Many different types of post system are available and they can be classified in numerous ways:
• **Prefabricated or custom-made**—Prefabricated posts obviously have the advantage of being cheap and quick, however, they lack versatility and many of the systems require all coronal dentine to be removed. Custom-made techniques are preferred as they are more versatile, but they are more expensive and require an additional laboratory stage. The Wiptam technique is a hybrid of these two types where a custom-made core is cast on to a prefabricated post, however the technique is no longer popular as corrosion occurs between the dissimilar metals.
• **Parallel sided or tapered**—Parallel sided posts are generally preferred to tapered as they provide greater retention and do not generate as much stress within the root canal. Tapered posts, however, are less likely to perforate in the apical region. Again, hybrid versions exist between these two forms which are parallel in the coronal region and tapered towards the apex, e.g. Cytco.
• **Threaded, smooth or serrated**—Threaded posts provide greater retention than smooth sided, however, will ↑ stress within

the root canal. Serrated posts do not concentrate stress but simply increase the surface area for retention.

Other design features include antirotational components and cementation vents.

Examples

Custom-made The cast post and core is a popular choice. First of all the root canal is prepared using parallel sided twist drills, e.g. Panadrills and an antirotation groove is placed in the coronal dentine. The post and core can then be constructed either by a *direct technique* or by an *indirect technique*. In the direct technique a pattern is fabricated in the mouth using either inlay wax or a burn out resin (e.g. Duralay) which is then sent to the laboratory for casting. For the indirect technique an impression is taken using a spiral filler to get the material down the post hole, which is then supported with a length of wire. When using this technique it is generally inadvisable to have the post and subsequent crown constructed on the same impression. The Parapost system is basically a cast post but the kit provides a plastic burn out post which is serrated and has a cementation vent.

Preformed These are available in various different forms:

Parallel, smooth sided—e.g. Charlton post.

Parallel, threaded—e.g. Radix anchor, Kurer.

Tapered, threaded—e.g. Dentatus screw. These are the poorest design in terms of stress production and, in the authors' opinion, should not be used.

Some of these systems have a prefabricated core on the post whilst with others it must be built up around the neck using composite.

For the majority of crowns choice is one of personal preference. However, no one system will be versatile enough to cover every eventuality, so it is wise to be familiar with more than one method.

285

Anterior post and core crowns— practical tips

Some problems and possible solutions

- *Sub-gingival tooth loss*: either extrude tooth orthodontically or use cast post and core method, extending post into defect in the form of a diaphragm.
- *Insufficient space for separate core and PJC*: Construct post crown in one piece with porcelain bonded to labial face.
- *Extensive tooth loss and calcified canal* (e.g. dentinogenesis imperfecta, severe toothwear): Use dentine pins (plus dentine bonding agent) to retain pinned composite core for porcelain bonded crown.
- *Perforation of root by post*: Apical $\frac{2}{3}$—will need surgical approach to cut back excess post and seal perforation with amalgam. Coronal $\frac{1}{3}$—incorporate perforation into diaphragm preparation and make new cast post and core.
- *Loss of post*: Check (1) length adequate? If not, remake with ↑ length, (2) loose fit or too much taper? Correct and remake, (3) perforation? Take X-rays in parallax to check, and see above, (4) root # ? Extract.
- *Apical pathology*: If post and core crown satisfactory arrange apicectomy. If not remove and place orthograde root-filling and new post and crown.

Causes of failure in post and core crowns A recent survey of failed posts showed that most failed within 1 yr and that having survived satisfactorily for 3 yrs a post crown has a good chance of lasting for 10 yrs. Common causes of failure were caries, root # and mechanical failure of the post.[1]

Removing old posts and cores Unretentive posts may be removed by grasping with Spencer–Wells forceps and twisting. Post removers are available (but C/I for threaded posts) which work by drawing out post using root face as anchorage e.g. the Eggler post remover. Some proprietary kits, e.g. Masseran, can be used to cut a channel around the post to facilitate removal. In some cases an ultrasonic scaler tip can be used to vibrate the post loose.

1 R. Lewis 1988 *BDJ* **165** 95.

Veneers

Indications Mild discoloration (can ↑ success by bleaching first), hypoplasia, fractured teeth, toothwear lesions, closing space, or modifying shape (within limits). Veneers are particularly useful in adolescents where PJC preparation may risk exposure.

Contraindications Severe discoloration, cases where labial enamel has been lost or an edge-to-edge incisor relationship exists. Overlapping teeth, pencil-chewing or nail-biting are relative C/I.

Types

Acrylic laminate veneers (e.g. Mastique) Are bulky which results in ↑ gingival inflammation. Also liable to # and bond failure.

Composite resin Can be made directly (more commonly) or indirectly. Problems are shrinkage, staining, and wear. Average life-span approximately 4 yrs.[1]

Porcelain Clinical studies report better performance and aesthetics than acrylic or composite.[2] In addition, porcelain is less plaque retentive. Are made indirectly in the laboratory and roughened on their fitting surface by etching or sandblasting. This surface is treated with a silane coupling agent (e.g. Scotchprime) prior to bonding to the etched tooth enamel with composite.

Technique for porcelain veneers

Tooth preparation The veneers are usually 0.5 to 0.7 mm thick, ∴ unless deliberate overbuilding is required, the tooth needs to be reduced labially. To guide reduction depth cuts of 0.5 mm are advisable. A definite chamfered finishing line will make the technician's job considerably easier and this should be established first. If the tooth is discoloured the margin should be subgingival (but still in enamel), otherwise keep slightly supragingival. The finishing line is extended into the embrasures, but kept short of the contact points. Incisally, the veneer can be finished to a chamfer at the incisal edge or wrapped over on to the palatal surface (see opposite).

Any exposed dentine should be covered with a thin layer of GI cement. An impression of the preparation is taken using an elastomeric impression material in a lower stock tray and the shade taken with a porcelain shade-guide. Temporary coverage is not usually required, p. 307.

Try-in Careful handling is necessary so as not to contaminate the fitting surface of the veneer. The prepared tooth should be cleaned and isolated and then the veneer tried in wet (to ↑ translucency). Minor adjustments are best deferred until after cementation to ↓ risk of #. The effect of different shades of composite ± opaquers and tints can be tried prior to etching in

1 J. S. Clyde 1988 *BDJ* **164** 9. **2** S. M. Dunne 1993 *BDJ* **175** 317.

order to get the best colour match. If several veneers are to be fitted, check them individually and then together to work out the order of placement.

Placement The fitting surface of the veneer is cleaned with alcohol, dried and then coated with a thin layer of silane coupling agent followed by bonding resin. The tooth is re-isolated and cellulose acetate strips or a Contour-strip used to separate from adjacent teeth. After etching, washing, and drying, a film of bonding resin is placed and cured. Numerous cementation systems are available, e.g. Porcelite, Heliolink. The composite is placed thinly over the labial surface of the tooth and the veneer carefully positioned. Excess composite should be removed before curing. Adjustments are made with a flame-shaped diamond finishing bur before polishing. The patient should be instructed in the use of floss.

Porcelain slips are veneered corners or edges used to restore # incisors or close spaces by building out the tooth mesially or distally. Their long-term performance has yet to be established.

Incisal edge chamfer   Wrap around incisal edge

Porcelain veneer preparations

Posterior crowns

Posterior crowns are indicated as bridge abutments and for repair of tooth substance lost due to caries, wear or #. Tooth loss due to these causes should first be repaired using a pin retained amalgam restoration. Prior to preparation any doubtful restorations should be replaced.

Full veneer gold crown

Principles
- Remove enough tooth substance to allow adequate thickness of gold, i.e. 1.5 mm on functional cusp, 1 mm elsewhere, following original tooth contours.
- Wide bevel on the functional cusp (normally buccal—lower; palatal—upper) for structural durability.
- Convergence of opposing walls <10°.
- Height of axial walls as great as possible (without compromising occlusal reduction).
- Chamfer finishing line.
- Where possible, margins should be supragingival and on sound tooth substance.

Preparation
Occlusal: Using a short diamond fissure bur reduce the cusp height maintaining the original anatomy.

Bucco-lingual: With a torpedo-shaped bur eliminate undercuts, retaining 5° taper in cervical $\frac{2}{3}$rds, but usually remaining $\frac{1}{3}$rd will converge occlusally.

Approximal: Using a fine tapered diamond bur within the confines of the tooth, eliminate undercuts at an angle of 5°.

Finishing: Round axial line angles and cusps. Check no undercuts and smooth prep with fine diamonds.

Impressions, p. 672, and temporary coverage, p. 306.

Porcelain fused to metal crown is used where aesthetics is important.

The preparation is similar to the full veneer gold crown except that where porcelain coverage is required more tooth substance must be removed. The amount of porcelain coverage must be decided before the preparation is commenced, and the patient consulted at this stage to make sure that they are happy.
- **Occlusal reduction**—If it is acceptable to the patient it is better to provide an all-metal occlusal surface, as less tooth substance needs to be removed. Failing this, try to persuade the patient to accept metal centric stops. If the patient is adamant that an all porcelain occlusal surface is necessary then 2 mm will need to be removed from the supporting cusps and 1.5 mm from the non-supporting cusps which will clearly compromise retention in teeth with short clinical crowns.
- **Buccal reduction**—1.5 mm should be removed in order to provide enough room for the metal and porcelain.
- **Margins**—If it is acceptable to the patient it is better to produce a metal to tooth margin, which will necessitate a narrow

collar of metal around the gingival margin. In this case, the finishing line prepared should be a deep chamfer or a bevelled shoulder. If the patient insists on a porcelain to tooth margin then a 1.5 mm shoulder must be produced. Where no porcelain coverage is needed a chamfer finishing line is produced as for the full veneer gold crown.

Three-quarter gold crown

The preparation is as for full veneer gold crown except:
- Buccal surface is left unprepared.
- Retention grooves are placed on the mesial and distal surfaces—these must both be parallel to the line of withdrawal of the preparation.

- Proximal flares extend from the mesial and distal surfaces to the buccal surface to eliminate undercuts.
- A groove or 'occlusal offset' is prepared along the occlusal surface between the 2 retention grooves, just inside the tips of the buccal cusps. This serves to ↑ structural durability of the restoration which can be quite weak in this region.
- A minimal buccal overlap is prepared, just over the tips of the buccal cusps, which is then linked to the proximal flares.

Crowning root-filled posterior teeth

Single-rooted posterior teeth can be treated as anterior teeth, p. 284. In multi-rooted teeth the major problem is the divergence of the root canals. The two most commonly used methods of solving this problem are:

1 *Direct method*: Manufactured posts are cemented into each canal and then an amalgam, cermet or composite core built up around them. Composite has the advantage that the preparation can be completed at the same visit. A dentine bonding agent should be used to enhance retention.

2 *Indirect method*: A sprue is placed in the least divergent canal and a lubricated post (e.g. SS wire) positioned in the more divergent. A wax pattern (or Duralay) is then built up to form a core. The lubricated post is removed before the pattern is sent to the lab for casting. The cast post (for the least divergent canal) and core is cemented in position, then either Wiptam or SS wire is cemented through the hole in the casting into the more divergent canal.

With either technique a gold shell or porcelain bonded crown can be used for the final restoration.

Problems with post and core crowns in multirooted teeth

- Short or very curved canals: use a small Dentatus screw and build up core as in (1).
- Subgingival tooth loss: use pre-formed post(s) with amalgam core which is well-condensed into region of defect. Some corrosion between the dissimilar metals may be preferable to extending preparation and impressions below level of amalgam.

Bridges

Definitions

Bridge—A prosthetic appliance that is permanently attached to remaining teeth and replaces a missing tooth or teeth.

Abutment—A tooth which provides attachment for a bridge.

Retainer—The component that is cemented to the abutments to provide retention for the prosthesis.

Pontic—The artificial tooth that is suspended from the abutments.

Connector—The component that joins the pontic to the retainer. May be rigid or non-rigid.

Saddle—The area of edentulous ridge over which the pontic will lie.

Units—Number of units = number of pontics + number of retainers.

Retention—Prevents removal of the restoration along the path of insertion or long axis of the preparation.

Support—The ability of the abutment teeth to bear the occlusal load on the restoration.

Resistance—Prevents dislodgement of the restoration by forces directed in an apical or oblique direction and prevents movement of the restoration under occlusal forces.

Types of bridge

Fixed–fixed—The pontic is anchored to the retainers with rigid connectors at either end of the edentulous span. Both abutments provide retention and support. Both preparations must have a single line of draw.

Fixed–movable—The pontic is anchored rigidly to the major retainer at one end of the span and via a movable joint to the minor retainer at the other end. The major abutment provides retention and support whilst the minor abutment provides support only. This design allows some independent movement of the minor abutment and has the advantage that the preparations need not be parallel.

Direct cantilever—Pontic is anchored at one end of the edentulous span only.

Spring cantilever—The retainer and pontic are remote from each other and are connected by a metal bar which runs along the palate. Usually an upper incisor replaced from the premolars or a molar. It is useful where there is an anterior diastema or if the posterior teeth are heavily restored, however, they are often poorly tolerated.

Resin-retained—Retained by composite resin (p. 302).

Compound—Combination of more than one of above types.

Removable—Can be removed by dentist for maintenance.

Types of retainers

- Full coverage crown
- Three-quarter crown.
- Post-retained crown
- MOD Onlay
- Class 2 inlay

All of the above restorations have been used as retainers in conventional bridgework. They are listed in order from most retentive to least retentive. Wherever possible one of the first two should be used as the failure rate of the last three is much higher. Post crowns should be avoided if possible, and onlays or class 2 inlays should only ever be used as minor retainers in fixed–movable bridges.

Selection of abutment teeth

When selecting abutment teeth general factors must be taken into account, such as caries status and existing restorations, but there are two other considerations which specifically relate to bridgework. These are **retention** and **support**.

Assessment of retention

The factors that affect the amount of retention offered by a potential abutment tooth are the clinical crown height and the available surface area. Obviously, the larger teeth offer more retention and should be chosen in preference to the smaller ones. The teeth of both arches are listed below in order of the amount of retention offered (if a full coverage restoration is used).

	Greatest	→	→	→	→	→	Poorest
Maxilla	6	7	4	5	3	1	2
Mandible	6	7	5	4	3	2	1

Assessment of support

Three factors are important:

1 *Crown–root ratio*: ideally should be 2:3 but 1:1 is acceptable. As bone is lost so the lever effect on the supporting tissues is increased.
2 *Root configuration*: Widely splayed roots provide more support than fused ones.
3 *Periodontal surface area*: The more of the root that is attached to the bone via the PDL the greater the support offered. At one time great significance was applied to this factor and it formed the basis of Ante's law (1926) which states that '*the combined periodontal area of the abutment teeth must be at least as great as that of the teeth being replaced?*' Ante's law has no scientific basis and no longer has

Bridges (*cont.*)

a place in contemporary bridgework design. It does not take into account that we are dealing with a biological system—as the load is increased on the abutment teeth a biofeedback mechanism operates to cause a reduction in this load.

The teeth of both arches are listed below in order of the amount of support offered, assuming that the periodontal tissues are intact.

	Greatest	→	→	→	→	→	Poorest
Maxilla	6	7	3	4	5	1	2
Mandible	6	7	3	5	4	2	1

As a result of the considerations above, a rule of thumb of bridge design has arisen, which states that one should try to involve a molar or a canine if possible.

Taper and parallelism
- Opposing walls of abutments should have 5° taper.
- For most designs abutments should be prepared with a common line of draw.
- Checking parallelism: direct vision, with one eye; survey mirror with parallel lines inscribed.
- Management of tilted abutments, p. 296.

Types of pontic

Ridge lap—As name suggests, this type of pontic should make (minimal) contact with buccal aspect of ridge. Gives good aesthetics and is the most popular type.

Hygienic—Does not contact saddle ∴ easy to clean. Unaesthetic ∴ limited to molar replacement.

Bullet—makes point contact with tip of ridge.

Saddle—extends over ridge buccally and lingually ∴ difficult to clean. Should not be used.

Bridges—treatment planning and design

Treatment planning

First consider whether benefits of replacing missing teeth (improved aesthetics, occlusal stability, mastication, and speech) outweigh disadvantages (↑ oral stagnation, tooth preparation, cost). If replacement is indicated ? fixed or removable prosthesis. A number of factors affect this decision:

General	Local
Patient's motivation	OH and periodontal condition
Age	Number of missing teeth
Health	Position of missing teeth
Occupation	Occlusion
Cost	Condition of abutments
	Length of span

These factors need to be favourable if expensive and complex bridge-work is required. Removable prostheses are indicated if general or local factors are less than ideal (p. 334).

Designing bridges

- Assess prognosis of all teeth in vicinity to reduce risk of another tooth requiring extraction in near future.
- Assess possible abutment teeth (check restorations, vitality, periodontal condition, mobility, and take periapical radiographs).
- Select design of retainers, e.g. full or partial crown.
- Consider pontics and connectors.
- With this information compile a list of possible designs for bridge.
- Consideration of the advantages and disadvantages of each design combined perhaps with a diagnostic set-up for one or two designs should help to narrow down the choice. Where possible try the least destructive alternative first.

Specific design problems

1 *Periodontally involved abutments*: first control periodontal disease. Then ? bridge indicated. Fixed–fixed type of design preferable to splint teeth together.

2 *Pier abutments*: this is the central abutment in a complex bridge which supports pontics on either side, which are in turn anchored to the terminal abutments. In this situation the pier abutment can act as a fulcrum and when one part of the bridge is loaded the retainer at the other end experiences an unseating force which can lead to cementation failure. To overcome this a stress-breaking element msut be introduced, e.g. fixed–movable joint.

3 *Tilted abutments*: this occurs most commonly following loss of a molar. There are several approaches:
- Orthodontic treatment to upright abutments.
- Two-part bridge, e.g. fixed–movable.

- Telescopic crowns—placement of individual gold shell crowns on abutments, over which telescopic sleeves of bridge fit.
- Partial veneer preparations in which pins or slots are prepared to compensate for slight malalignment of abutments (least satisfactory).
- Precision attachments—a precision screw and screw tube can be incorporated into a two-part bridge. After cementation the screw is inserted which effectively converts the bridge to a fixed–fixed design.

4 *Canines*: The canine is often the keystone of the arch and is a very difficult tooth to replace. The adjacent teeth are poor in terms of the amount of retention and support that they offer and the canine is often subject to enormous stresses in lateral excursion (in a canine-guided occlusion). If a canine is to be replaced with a bridge then the occlusal scheme should be designed to provide group function in lateral excursion—never canine guidance.

Bridges—practical stages

1 Take a **history**—Why is a bridge necessary? When and why was tooth lost? Remember the past dental history, social history, and medical history.

2 Carry out a **clinical examination**—*extra-oral* and *intra-oral*. Extra-orally look for signs of TMJ dysfunction. Intra-orally look at the general condition of the mouth, the length of edentulous span (unit spacing), the condition and position of the potential abutment teeth. Carefully examine the occlusion and try to formulate some initial ideas on possible bridge designs.

3 **Special tests**—Radiographs of the potential abutments are mandatory. ? Vitality tests.

4 **Diagnostic mounting**—Take accurate impressions of both arches, a facebow record and have the models mounted on an adjustable articulator. The mounting can be carried out either in ICP (best fit) or in RCP, for which a precentric record will be necessary. If a reorganized approach to reconstruction is being considered or if the clinical examination has revealed significant occlusal interferences then an RCP mounting should be performed. Carefully examine the occlusion and consider what occlusal consequences the proposed restoration will have.

5 **Diagnostic waxing**—In effect, this is a mock up of the final restoration on the mounted models. Wax can be added to the teeth to simulate the effect that the restoration will have on the final occlusion and aesthetic result. In the anterior part of the mouth a denture tooth can be used. In addition to assessing aesthetics and occlusion the diagnostic wax up can serve as a template from which the temporary bridge can be constructed. An impression is taken of the wax up in a silicone putty and is saved for later. At this stage the design of the prosthesis must he finalized.

6 **Preparations**—Before the preparations are carried out any suspect restorations in the abutment teeth are replaced. Preparations are carried out in accordance with the basic principles (p. 260) and care is taken to ensure that a single path of insertion is established. When checking for parallelism one eye should be kept closed and the use of a large mouth mirror is very helpful. Custom-made paralleling devices can be used but they are very cumbersome.

7 **Temporary bridge**—This is normally constructed using the matrix which has been formed from the diagnostic wax up, in this way the temporary bridge should reproduce the aesthetics and occlusion of the final bridge (if the wax up was done properly!). The matrix is filled with one of the proprietary temporary crown and bridge resins (e.g. Trim) and seated over the preps. After it has set it is removed, trimmed, polished, and cemented with a temporary cement (e.g. Temp bond).

8 **Impressions**—An impression is taken using an elastomeric material, preferably in a special tray. Ideally all of the pre-

parations should be captured on one impression, but this can be very difficult if multiple preparations are involved. If difficulties are encountered in this respect they can often be overcome by using the transfer coping technique. In this technique acrylic (Duralay) copings are made on dies of the preparations for which a successful impression has been achieved. These are then taken to the mouth and seated on the appropriate tooth and the impression repeated to capture the other preps. On removal, the coping will be removed in the impression and the dies can be reseated in the copings and a new model poured around them.

9 **Occlusal registration**—Under most circumstances the models will be mounted in ICP in the position of best fit, and therefore an occlusal registration will not be necessary. Where numerous preparations have been carried out and it is difficult to locate this position, then some form of occlusal record will be necessary. The best technique again involves the use of transfer copings, and is described in the section on occlusion (p.277).

10 **Metal work try-in**—If a porcelain fused to metal bridge is being constructed then it is advisable to try in the metal work before the porcelain is added. At this stage the fit of the framework can be evaluated and the occlusion adjusted. On occasions it will be found that one retainer seats fully whilst the other does not. This can occur if there has been some minor movement of the abutments since the impression was taken. If this is the case the bridge should be sectioned, and it is hoped that both retainers will then seat. The two parts are then secured in their new position and sent back to the lab for soldering.

11 **Trial cementation**—The finished bridge is tried in and any necessary adjustments made. The bridge should then be temporarily cemented (with Temp bond) for a period of a month or so. The advantage of a trial cementation period is that if any further adjustments are necessary they can be carried out outside the mouth and the restoration repolished and reglazed. The patient is instructed in how to clean the bridge (use of Superfloss).

12 **Permanent cementation**—After the period of trial cementation the bridge is re-evaluated and the patient questioned to check that they are happy with it. If all is well the bridge is removed and cemented with a permanent cement (usually zinc phosphate or zinc polycarboxylate).

13 **Follow-up**—Arrangements are made to recall the patient to check that the bridge is still functioning satisfactorily.

Bridge failures

Most common reasons for bridge failure are:
- Loss of retention.
- Mechanical failure, e.g. # of casting.
- Problems with abutment teeth, e.g. secondary caries, periodontal disease, loss of vitality.

Management of failures Depending upon type and extent of problem.
- Keep under review.
- Adjust or repair *in situ*.
- Replace.

Before replacement of a bridge is embarked upon, a careful analysis of the reasons for failure is necessary. Minor problems in an otherwise satisfactory bridge should be repaired if at all possible. Fractured porcelain can be repaired with one of the specialized repair kits available, or alternatively use a hybrid composite plus a silane coupling agent. Secondary caries or marginal deficiencies, if small, can be restored with GI cement.

A recent survey of bridges placed by GDPs in Sweden found that 93.3% were still in service after 10 years.[1] The most common reason for failure was loss of vitality. This is not necessarily an indication for removal of the bridge because RCT can often be carried out through the retainer of the abutment.

Removing old bridges
- If wish to remove intact, try a sharp tap at cervical margin with a chisel or preferably a slide hammer. Orthodontic band removing pliers can also be used but these require a small hole to be cut in the occlusal surface. If one retainer only is loose, support bridge in position while trying to remove it so that it does not bind.
- Retainers can be cut through (p. 282), but this will destroy bridge.

1 S. Karlsson 1986 *J Oral Rehab* **13** 423.

Resin-bonded bridges

This technique involves bonding a cast metal framework, carrying the pontic tooth, to abutment teeth using an adhesive resin. The resin bonds to the abutment teeth using the acid-etch technique and to the metal framework by either mechanical or chemical means.

Classification by:
1 Position: anterior or posterior.
2 Retention: • Macromechanical: (a) perforated (Rochette)
                                             (b) mesh (Klettobond)
                                             (c) particular (Crystalbond)
             • Micromechanical: (a) etched (Maryland)
                                             (b) chemically etched
             • chemical: (a) sand-blasted
                                               (b) tin-plated

Chemical retention to a sandblasted metal surface is now used virtually exclusively. A dual affinity cement (Panavia Ex) is used which chemically bonds to both enamel and non precious alloys.

Indications	Advantages
• Short span	• Cheaper than conventional bridge
• Sound abutment teeth	• Minimal tooth reduction
• Favourable occlusion	• No LA required
• Younger patient	**Disadvantages**
	• Tendency to de-bond
	• Metal may show through abutments

Treatment planning as for conventional bridgework. If orthodontic treatment is needed to localize space or upright adjacent teeth, it is advisable to retain with a removable retainer for at least 3 months prior to bridge placement.

Tooth preparation is required to:
• Give a single path of insertion. Provide near parallel guiding planes eliminating undercuts, which allows coverage of maximal surface area for bonding.
• Provide space in occlusion to accommodate bridge. Need at least 0.3 mm for wings.
• Increase retention e.g. using a wrap-around design (covering > 180° of tooth circumference) to resist lateral displacement.
• Prevent gingival displacement. A minimal chamfer is sometimes used.
• Provide axial loading of the abutments—prepare cingulum or occlusal rests.

NB Tooth preparation must be confined to enamel and the framework should be designed with maximal coverage (to ↑ surface area available for bonding).

Technique (Chemical method using Panavia Ex)—Following tooth preparation an elastomeric impression of the abutment teeth is taken plus an alginate impression of the opposing arch. When taking the shade it is a good idea to place a metal instrument behind the abutment teeth and choose accordingly. At the try-in stage the bridge should be assessed for fit, aesthetics, occlusion, etc, and then returned to the lab to have the fitting surface sand-blasted. Contamination of the fitting surface with saliva must be avoided and cementation is best done under rubber dam. Following etching and washing of the abutments, the wings of the bridge are coated with Panavia Ex and the bridge seated into place and held firmly until set. Use of acetate strips and Superfloss at this stage will clear most of the excess cement and prevent it adhering to the adjacent teeth. The cement must then be covered with a substance known as Oxyguard which prevents oxygen inhibition of the surface layer. After 5 minutes or so the rubber dam is removed and any excess cement removed.

303

Problems
- *Dentine exposed during preparation*: use a dentine bonding agent.
- *Metal shining through abutments*: cut wings away incisally before cementation or use a more opaque cement. May have to consider conventional bridge.
- *Debonds*: if one flange only can usually detach other by a sharp tap with a chisel or by using ultrasonic scaler tips. If persistent problem consider conventional bridge.
- *Caries occurring under debonded wings*: remove bridge and repair.

Attrition, abrasion, and erosion

As these rarely occur individually, the term 'toothwear' is preferred. Also called 'tooth surface loss', but this is often an understatement!

▶ Some toothwear during life is inevitable, where it has resulted in an unsatisfactory appearance, sensitivity or mechanical problems, the condition warrants investigation and treatment.

Abrasion is physical wear of a tooth caused by an external agent. Classically, toothbrushes are blamed for the characteristic cervical notches, but it is now thought that other factors may also be operating.[1]

Attrition is physical wear caused by movement of one tooth against another. Affects interproximal and occlusal surfaces. ↑ in more abrasive diets and in bruxism. It is often assumed that attrition is greater in patients with reduced posterior support, but no evidence exists to support this.[2] Bruxism ↓ with ↑ age.

Erosion is loss of tooth substance from non-bacterial chemical attack. The incidence of erosion appears to be ↑, but this may be the result of an ↑ awareness of the problem. As the presence of acid results only in demineralization, for loss of tooth substance to occur erosion must act in conjunction with attrition or abrasion or both. Erosion will be enhanced if the buffering capacity of the saliva is ↓, e.g. in dehydration secondary to alcoholism. Classically, see smooth plaque-free surfaces with proud restorations whether the acid is industrial (rare), dietary, or gastrointestinal in origin. The latter may be caused either by gastric reflux, or by vomiting (e.g. bulimia, p. 551, or pregnancy). Such conditions warrant referral.

Diagnosis From the clinical picture and history. As toothwear may be due to a factor which no longer operates, it may be necessary (tactfully) to delve into the patient's past. In a proportion of cases, the aetiology will remain obscure and this will complicate prevention. It is important to establish whether the toothwear is ongoing. If teeth are sensitive, probably yes; however, sequential study models will provide definitive evidence.

Toothwear indices are only of value if reproducible, i.e. used regularly. If interested see paper by Smith.[3]

Management[4]
- Prevention requires an understanding of the aetiology. However, an explanation of possible exacerbating factors to the patient may help to limit loss even if the exact aetiology is unknown.
- Monitoring. Take study models and photos. Intervention indicated in cases with an unsatisfactory appearance, sensitivity, or mechanical problems.

1 B. G. N. Smith 1989 *Dental Update* **16** 204.   **2** A. F. Kayser 1985 *Community Dental Health* **2** 285.   **3** B. G. N. Smith 1984 *BDJ* **156** 435.   **4** E. A. M. Kidd 1993 *Dental Update* **20** 174.

- GI or composite restorations may help improve appearance and ↓ sensitivity, but if toothwear is progressive full-coverage crowns are preferable.
- If toothwear is excessive it may not be possible to provide aesthetic crowns without ↑ OVD. The patient's tolerance to this should be tested first with an acrylic splint. The majority of patients seem to cope with an increase of <5 mm.
- Overdentures may be indicated in some patients, but are aesthetically less satisfactory.
- Referral to a physician (GI problems), or psychiatrist (bulimic patient), or restorative specialist (complicated restorative problem).

Temporary restorations

Indications
- Protection of pulp and paliation of pulpal pain.
- Restoration of function.
- Stabilization of active caries prior to permanent restoration.
- Aesthetics.
- Maintenance of position of prepared and adjacent teeth.
- Prevent over-eruption of opposing teeth.
- Prevent gingival overgrowth.

Temporary dressings Choice of material depends upon the main purpose of the dressing, i.e. therapeutic or structural, but also must be capable of being readily removed. For paliation of pulpal pain ZOE is indicated. For caries control a calcium hydroxide base and modified ZOE. If the remaining tooth tissue requires support this can be gained using a copper ring or an orthodontic band. For a less 'temporary restoration' use GI cement. The interim seal during RCT is very important ∴ a relatively strong material which prevents microleakage (e.g. GI) should be used.

Temporary crowns 3 main types.

Pre-formed: **1** Polycarbonate crown (e.g. Directa) which is trimmed to correct shape and customized by lining with self-cure acrylic (e.g. Trim), NB epimine resin will not bond. **2** Soft metal alloy crowns (e.g. Ion).

Laboratory made Advisable if preparing multiple crowns or if temporary crown needs to last for several months while other aspects of treatment are completed. Preferred method for temporary bridges.

Chairside This is a versatile technique. Crowns are custom-made using an alginate impression of the tooth taken prior to preparation, as a mould. When the preparation is completed, any undercuts should be blocked out with carding wax to prevent the temporary crown material locking in the mouth. The material for the crown is then syringed into the impression around the preparation and re-seated in the mouth. When the initial set has been reached the impression is removed and the temporary left to finish curing before being polished. Suitable materials are Pro-temp (a bis-acryl resin) and Trim (poly-*n*-butyl methacrylate). An alternative technique is to make a duplicate model of the diagnostic waxing. This approach is useful for multiple crowns or where changes are being made to occlusion/aesthetics. Applicable to both anterior and posterior crowns.

If preparing several adjacent teeth consider linking temporary crowns to aid retention.

Temporary post and core crowns Some systems (e.g. Para post) come complete with temporary posts, otherwise they can be made at the chairside with a suitably sized piece of wire. The length of the post should be adjusted so that it protrudes 2–3 mm out of the canal without interfering with the occlusion. A one-piece temporary post and crown is made either by the chairside method described above or with a polycarbonate crownformer and acrylic.

Temporary bridges The best type is made in the laboratory in acrylic and re-lined at the chairside. Alternatively, make a chairside bridge using the diagnostic waxing.

Veneers Temporary coverage is usually not necessary, but if the patient complains of sensitivity tack a temporary composite veneer to the prepared surface by etching two small areas of enamel.

Temporary inlays A Scutan or Pro-temp temporary can be made, but if the permanent inlay is to be cemented with composite resin, an eugenol-containing cement is C/I. Recently, a light-cured temporary material (Fermit) has been introduced for this situation which purportedly does not require to be cemented as it swells *in situ*.

Temporary cements The choice depends upon how long the temporary coverage is required for and the closeness of the fit. For most purposes the proprietary cement Temp-bond is fine, but for longer periods consider using modified ZOE. NB Use of a ZOE temporary cement with a freshly prepared composite prep will decrease the strength of the final crown.

Pinned restorations

Extensive loss of the crown of a tooth leads to problems with retention of subsequent restorations. In posterior (and occasionally anterior) teeth these can be alleviated by the use of dentine pins. Pin retention may also be necessary to support the cores of full coverage crowns. Pins reduce the compressive and tensile strength of amalgam and ∴ should not be over-prescribed. A rule of thumb is one pin per missing cusp. There are three basic types of pin:

1 Self-threading (e.g. TMS, Stabilok), where the pin is carried to prepared channel in a handpiece. This type are self-shearing when the correct depth is reached (usually 2 mm).
2 Friction lock, where the pin is forced into a slightly under-sized prepared channel and held there by friction.
3 Cemented, where the pin is cemented into a prepared channel.

Retention can be ranked: self-threading >> friction lock > cemented.

▶ Care is required to avoid perforation into the pulp or PDL.

Technique (self-shearing threaded pin system)
• Complete cavity preparation and lining.
• Choose pin site: 1 mm away from amelo-dentinal junction and clear of bi- or trifurcation areas. An X-ray may be helpful.
• Use a small round bur to indent chosen site.
• Pin perforates into pulp: can try placing CaOH in pin-hole and monitoring vitality of pulp. Alternatively, can go straight to RCT which will allow use of post in canal to ↑ retention.
• Pin perforates into PDL: can monitor; or extend cavity margin to include defect and carefully re-site pin!; or try to smooth off if accessible in gingival crevice.
• Check occlusion and if necessary bend pin inwards. An old chisel with a groove cut in it is a suitable instrument.
• Fit matrix band or copper ring and condense amalgam (or composite).

Problems and their management
• Pin perforates into pulp: can try placing CaOH in pin-hole and monitoring vitality of pulp. Alternatively, can go straight to RCT which will allow use of post in canal to ↑ retention.
• Pin perforates into PDL: can monitor; or extend cavity margin to include defect and carefully re-site pin!; or try to smooth off if accessible in gingival crevice.

Pin retention in anterior teeth Following the introduction of the acid-etch technique pins are rarely required in anterior teeth. Gold veneer pinledge preparations are occasionally used as bridge abutments. They rely on pins (which are an integral part of the gold casting) for retention and avoid approximal involvement necessary for $\frac{3}{4}$ crown. Pins used to retain composite restorations may show through and detract from the aesthetics.

Bleaching

Bleaching provides a conservative solution to mild to moderately discoloured vital or root-filled teeth. Take photos as a pre-operative record or alternatively note shade of tooth, using a porcelain shade-guide.

Vital bleaching

Several methods have been described. Hydrogen peroxide (as a 30% solution) is used as the oxidizing agent. This is obtainable as Concentrated Hydrogen Peroxide Solution BP. It is caustic ∴ protective eyewear and care is required. Heat &/or UV light are commonly used to activate the hydrogen peroxide, but a dental curing light has been suggested.[1] A solution of 10% carbamide peroxide in a soft splint has been advocated for 'home bleaching', but is presently unavailable in the UK and efficacy is equivocal.

Technique
- Apply Orabase to gingivae prior to placing rubber dam over teeth to be treated.
- Polish teeth with pumice.
- Etch enamel, wash, and dry.
- Cut a piece of gauze to cover labial and palatal surfaces and soak with hydrogen peroxide.
- Apply light &/or heat for 30 min (Union Broach Light provides both).
- Remove gauze and wash teeth with copious amounts of water.
- Remove rubber dam and polish teeth.
- Advise patient to avoid tea, coffee, red wine, cigarettes, etc., for 1 week and that some sensitivity may occur.
- Can repeat up to 10 times.

Non-vital bleaching

This provides a conservative alternative to a post-retained crown for the discoloured root-filled tooth. However, usually only achieves an improvement in shade. There is a tendency for the discoloration to recur over time ∴ warn the patient and overbleach. Interestingly, it has been shown that the degree and duration of discoloration and the age of the patient do not affect the prognosis for a successful result.[2]

Technique (walking bleach)
- Place Orabase around gingivae of tooth to be treated and isolate with rubber dam.
- Open access cavity and remove root filling to 2 mm below gingival margin (a perio probe is a useful guide).
- Place thin layer of zinc phosphate cement over root filling.
- Remove any stained dentine within pulp chamber.
- Clean cavity with etchant on a pledget of cotton wool and then repeat with alcohol. Wash and dry.

1 R. J. Andlaw 1992 *A Manual of Paedodontics*, Churchill Livingstone. **2** R. A. Howell 1980 *BDJ* **148** 159.

- Mix together sodium perborate (Bocasan) and hydrogen peroxide to a paste and place in cavity.
- Seal with a pledget of cotton wool and GI (or a eugenol-free cement).
- Review patient after 1–2 weeks and repeat (up to 2 times) if necessary.
- Seal cavity permanently with white GP and a light shade of composite or GI.

Some authorities recommend omitting the hydrogen peroxide from the walking bleach technique, and using sodium perborate alone.

Thermocatalytic techniques where the hydrogen peroxide is heated within the pulp chamber should no longer be used, as associated with the development of cervical resorption lesions.

Root canal therapy

RCT involves the removal of pulpal remnants and cleaning and obturation of the resultant space, in order to prevent bacterial proliferation within the canal system. Apical lesions which appear radiolucent on X-ray are usually sterile and are caused by toxins from bacteria within the canal,[1] ∴ the first line of treatment should be an orthograde root filling.[2]

Indications

- Pulp irreversibly damaged &/or evidence of periapical disease.
- Restoration of crown of tooth requires retention from root canal.
- Crown of tooth requires extensive modification, e.g. bridge, overdenture.

▶ Those patients at risk from bacteraemia should be protected with antibiotic cover (p. 590) prior to endodontic treatment and for acute periapical conditions. Cover should be given for initial stages of RCT, but once working length has been determined (and all instrumentation confined within canals) it should not be necessary.

Anatomy

The apical foramina is usually sited 0.5–0.7 mm away from the anatomical and radiographic apex. The apical constriction usually occurs 0.5–0.7 mm short of the foramina. These distances increase with age due to deposition of secondary cementum. Root-filling to the constriction provides a natural stop to instrumentation, thus the working length should be established 1–2 mm from the radiographic apex.

Average working lengths (in mm):

	1	2	3	4/5	6	7
Maxilla	21	20	25	19	19	18.5
Mandible	19	19.5	24	20	19.5	18.5

Most canals are flattened mesio-distally, but become more rounded in the apical $\frac{1}{3}$. Lateral canals are branches of the main canal and occur in 17% to 30% of teeth.

Remember:

4—74% have >1 canal with >1 foramina.

5—75% have 1 canal with 1 foramina.

6, 7—Assume these teeth have 4 canals (2 MB; 1P; 1DB) until second MB canal cannot be found.

1, 2—>40% have 2 canals, but separate foramina are seen in only 1%.

4, 5—May have 2 canals, but these usually re-join to give 1 foramina.

1 L. I. Grossman 1982 *J Endodont* **8** 513. 2 British Endodontic Society: Guidelines 1983 *Int Endodont J* **16** 192.

$\overline{6, 7}$—Generally have 3 canals (MB; ML; D), but $\frac{1}{3}$ have 4 canals (2 in D root).

Assessment

- Check there is no doubt about which tooth requires treatment.
- Is tooth restorable following RCT?
- Good radiographs are essential. Minimum requirements are (1) pre-op assessment; (2) check working length; (3) check obturation and provide baseline for follow-up. In addition, an X-ray of the master-point prior to obturation is a wise precaution.

Root canal therapy—instruments

It is helpful to make up RCT kits containing the commonly used instruments, e.g. front surface mirror, extra-long probe, endo-locking tweezers, long-shanked excavator, flat plastic, root-canal spreaders and condensers, and a metal ruler. The whole kit can then be sterilized.

Broaches These are either smooth for exploring or barbed for pulp extirpation.

Reamers These are miniature twist drills made by twisting a tri-angular metal blank. They are used with a rotary action ($\frac{1}{4}$ to $\frac{1}{2}$ turn) in a clockwise direction and then withdrawn. Disadvantages of reamers include their inflexibility with ↑ size which can result in a wider channel being cut apically. They should never be used in the preparation of curved root canals.

Files These are more flexible than reamers and are used mainly with a longitudinal rasping action, although some types can also be rotated. The main types of file available are:
- *K-file*—Made by twisting a square metal blank. More twists than a reamer ∴ more cutting edges.
- *Hedstroem file*—Made by machining a continuous groove into a metal blank. More aggressive than K-file. Must never be used with a rotary action as liable to #.
- *K-flex file*—Similar to K-file but made by twisting a rhomboid shape blank → alternating blades with acute and obtuse angles. More flexible than K-file but becomes blunt more quickly.
- *Flex-o-file*—Looks similar to a K-file but is made from a tri-angular blank of a more flexible steel. The file also has a blunt tip which means that it is unlikely to create a false channel. This file is more flexible than K-types and is now becoming a popular replacement.

Files have traditionally been made from steel, but recently some have been introduced which are made from nickel titanium or titanium, these are much more flexible and appear promising.

Larger sized files and reamers can be sterilized and re-used more frequently than the smaller sizes. All instruments should be examined regularly, and discarded if there are any signs of damage.

Spiral root fillers are used to deposit paste materials within the canal, but are liable to #, ∴ the inexperienced operator should use them by hand. An alternative is to coat a file with the paste and spin it in an anti-clockwise direction to deposit the paste in the canal.

Gates–Glidden burs These are bud-shaped with a blunt end and are used, at slow speed, for preparing the coronal $\frac{2}{3}$rds of the canal.

Stops indicate the working length on RCT instruments. Some have a notch in to indicate the direction of a curvature.

Finger spreaders These are used to condense the cones of gutta-percha during canal obturation. They are sized to match the GP accessory cones.

Other equipment—Sterile cotton wool pledgets and paper points will be required for drying canals, and a syringe and needle for irrigating them.

Root canal therapy—materials

Irrigants are required to flush out debris and lubricate instruments. As the irrigant must be sterile and a syringe is necessary to dispense it, LA is a popular choice. However, sterile water or dilute sodium hypochlorite (e.g. Milton) are less expensive alternatives. Sodium hypochlorite is generally considered to be the best irrigant as it is bacteriocidal and dissolves organic debris, the normal concentration is 2.5% available chlorine. Chelating agents which soften dentine by their de-mineralizing action are particularly helpful when trying to negotiate sclerosed or blocked canals (e.g. EDTA paste, RC Prep).

Canal medication The emphasis in root canal therapy should be placed on thorough mechanical debridement rather than trying to sterilize their contents (which is impossible!). Canal medicaments therefore play a secondary role in RCT. Strong antiseptics, such as phenolic compounds, have been used in RCT, but they are potentially toxic and their effects appear short-lived—so their use is no longer recommended. For routine cases it is normal practice to leave the root canal empty but in certain situations a medicament is helpful. Two types are worthy of consideration:

1 *Non-setting calcium hydroxide paste* (see below). It can be very effective in treating an infected canal where there is a persistent inflammatory exudate from the periapical tissues.
2 *Antibiotic/steroid paste* (Ledermix) is useful if unable to achieve anaesthesia of a hyperaemic pulp. Dressing with Ledermix ↓ inflammation and allows pulp extirpation under LA next visit.

Filling materials Gutta-percha is a form of latex that is extracted from tropical trees. It comes the nearest to fulfilling the requirements of an ideal filling material. It is supplied in cones which come in two forms: *master cones*: sized according to ISO standards; and *accessory cones*: sized to match the finger spreaders. Paste filling materials are C/I as they are soluble in the long term and some (e.g. Endomethasone) contain paraformaldehyde which is highly toxic.

Sealers A wide variety of sealers are available. The calcium hydroxide materials (e.g. Sealapex) or the eugenol-based sealers (e.g. Tubliseal) are perhaps the safest choice. In anterior teeth, Kerr's sealer should be avoided as it contains silver which may cause staining. Paraformaldehyde containing sealers (e.g. Endomethasone or Spad) have been shown to cause tissue necrosis and should not be used.

Calcium hydroxide is considered separately because it has a wide range of applications in endodontics due to its antibacterial properties and an ability to promote the formation of a calcific barrier. The former is thought to be due to a high pH and also to the absorption of carbon dioxide upon which the metabolic activities of many root-canal pathogens depend. Indications for the use of calcium hydroxide include:

- To promote apical closure in immature teeth.
- In the management of perforations.
- In the treatment of resorption.
- As a temporary dressing for canals where filling has to be delayed.
- In the management of recurrent infections during RCT.

In RCT, a suspension of calcium hydroxide in carboxy-methyl cellulose, (e.g. Hypo-cal, Reogan Rapid) is the most useful.

Canal preparation—1

Aims of preparation The aims of canal preparation are twofold and are generally described as *cleaning* and *shaping*.

Cleaning aims to remove all bacteria and organic debris from the root canal.

Shaping aims to produce the ideal shape for the reception of the root filling material. As the filling material of choice is normally gutta-percha the shape should be a continuously tapering cone with its narrowest diameter at the apical constriction and its widest diameter in the coronal region.

Preparation of tooth Prior to starting RCT all caries must be removed from the crown of the tooth and an interim restoration placed. Grossly weakened posterior teeth may need to be temporarily restored with a strong cement in an orthodontic band or with an amalgam restoration. Large temporary dressings of, e.g. modified ZOE, in the posterior part of the mouth are not satisfactory. This initial preparation aids placement of rubber dam, prevents ingress of bacteria from the mouth, and provides a stable reference point for measurement of working length.

Isolation is required to maintain an aseptic environment, protect the patient from toxic materials and prevent inhalation of small RCT instruments. Use of rubber dam is mandatory.

Access should be designed to reduce the curvature required to negotiate the apical $\frac{1}{3}$rd of the canal and will involve removal of the entire roof of the pulp chamber including the pulp horns. The access cavity in anterior teeth should be midway between incisal edge and the cingulum and in posterior teeth will vary according to the anatomy of the pulp chamber. Lining up a bur with the pre-operative X-ray will help to gauge the depth of preparation. The turbine handpiece should be used to gain initial access, reverting to slow speed for removal of the roof of the pulp chamber and subsequent preparation. When completed the access cavity should have a smooth funnel shape. Long shanked burs are helpful.

Extirpation of the pulp is achieved using a barbed broach or fine files. Often, the pulp remnants are necrotic and copious irrigation is required to complete removal.

Working length[1] can be defined as the distance from a fixed reference position on the crown of the tooth to the apical constriction of the root canal. Remember that the apical constriction is normally 1–2 mm short of the radiographic apex of the tooth. There are two methods of establishing the working length: the first involves the use of radiography; and the second uses an electronic device known as an apex locator. The radiographic technique is the most commonly used and is described below. Apex locators work by measuring electrical impedance with an

1 C. Stock 1994 *BDJ* **176** 329.

electrode in the root canal, when electrode reaches the apical foramen it emits an audible or visible signal.

Common errors in canal preparation

- *Incomplete debridement*: working length short, missed canals.
- *Lateral perforation*: often occur as a result of poor access.
- *Apical perforation*: makes filling difficult.
- *Ledge formation*: can be very difficult to bypass.

Apical transportation (zipping)—A file will tend to straighten out when used in a curved canal and straightening can transport the apical part of the preparation away from the curvature.

Elbow formation—When apical zipping occurs a narrowing often occurs just above this in the canal such that the canal is hourglass in shape. This narrowing is termed an elbow.

Strip perforation—A perforation occurring in the inner or furcal wall of a curved root canal usually towards the coronal end.

Techniques of canal preparation

Numerous techniques have been described:

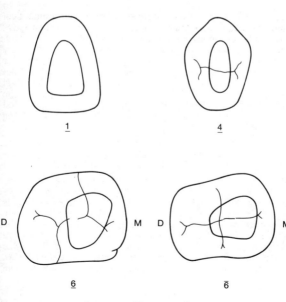

Access cavity preparations

Canal preparation—1 (cont.)

Standardized technique uses reamers or K-files with a rotary action to prepare the apical part of the canal to a round cross-section. Should not be used in curved canals as they will be straightened.

Stepback technique Popular technique, widely taught. The apical part of the root canal is prepared first and then the canal is then flared from apex to crown. Can get difficulties with blockage of canals using this technique and irrigation can be difficult.

Stepdown technique This technique (along with several others) prepares the coronal part of the canal before the apical part. This has numerous advantages (see later).

Balanced force technique This complex technique involves the use of blunt tipped files with an anticlockwise rotation whilst applying an apically directed force. It is a difficult technique to master but would be useful if preparing the apical part of severely curved canals.

Anticurvature filing This was developed to minimize the possibility of creating a 'strip' perforation on the inner walls of curved root canals. It is used in conjunction with other techniques or preparation and the essential principle is the direction of most force away from the curvature. Filing ratio 3:1 Outer wall:Inner wall.

In order to avoid confusion we have tried to provide a simplified guide to preparation of root canals. The recommended techniques are described on the next page.

321

Canal preparation—2

Technique for the preparation of large, straight root canals

Simple 'step back' technique

1 Obtain a good quality pre-operative radiograph. Identify the root canal.

2 Place a rubber dam and prepare the access cavity. Prepare the outline first (remembering your pulpal anatomy) with a high-speed bur and then change to a slow-speed rosehead. Remove the roof of the pulp chamber in its entirety leaving no ledges or overhangs.

3 Irrigate the pulp chamber, dry with cotton wool, and try to identify the root canal.

4 Go back to your pre-operative X-ray and estimate the distance (in mm) from an occlusal reference position to the apical constriction. Mark this distance on the shank of a file using a silicone stop. Insert the file into the canal to this length and take a radiograph with the file in place. If file will not go down to full length then try a smaller file, if this does not work then stop and take an X-ray to discover why not. Hopefully, the X-ray will confirm that the file is at the correct working length, if not, then the length can be adjusted as appropriate. Where the discrepancy is more than 2 mm it is better to repeat the working length X-ray.

5 The next step is to prepare the apical stop. Insert a file to the working length and turn it gently through about 30° whilst withdrawing it from the canal. Repeat this with successively larger files until the file that first binds in the apical region is found, use this file with a longitudinal filing motion until it is loose in the canal. Repeat this with at least the next two file sizes, and between each size reinsert the first one (recapitulation), and irrigate. When this has been completed using at least a size 25 file, and the filing debris consists of a clean dentine, the apical stop is complete. The last file used is referred to as the master apical file.

6 The rest of the canal is now flared by 'stepping back'. Take a file one size larger that the master apical file and insert it to a length 1 mm short of the working length. Work this file with a circumferential filing motion. Continue this enlarging procedure with successively larger size files, each 1 mm shorter than the previous, to complete the preparation. After each file is used it is important to reinsert the master apical file to the full working length (recapitulation) and irrigate thoroughly to ensure that the canal does not become blocked.

Technique for the preparation of fine and curved root canals

Modified 'step back' technique with 'orifice enlargement'

1 As for simple 'step back' technique.
2 As for simple 'step back' technique.

3 Irrigate the pulp chamber, dry with cotton wool, and try to identify the root canals using a fine file. Use the grooves in the floor of the pulp chamber as a guide to their location. Pass a fine instrument down the canals to ensure their patency, and if necessary enlarge them at this stage to allow the passage of a size 15 file, ensuring that you are filing short of the working length.

4 Go back to your pre-operative radiograph and, for each canal, estimate the distance (in mm) from an occlusal reference position, to the junction of the coronal and middle thirds of the root canal, which is usually around 16–18 mm. Take a size 15 Hedstroem file and mark this distance on the shank of the file using a silicone stop.

5 This step is known as 'orifice enlargement' and aims to improve the access to the root canal while at the same time reducing its effective curvature. Insert the Hedstroem file into the root canal to the depth determined in 4, and work it against the walls of the root canal with a longitudinal filing motion. The filing motions should be preferentially directed away from the furcation region (anticurvature filing) which helps prevent the formation of a 'strip' perforation in the furcal wall (which is very thin) and also straightens the coronal part of the canal. Repeat this process with several sizes of file. Copious irrigation must be used in order to prevent the canal from becoming blocked. With

Pre-operative

Orifice enlargement

Completed canal

Diagram of stages of canal preparation

experience this procedure can be speeded up by using Gates–Glidden burs in addition to the Hedstroem files.

6 The working length is now determined for each canal. It is important that this is not done before orifice enlargement, as the reference point on the crown will move as the coronal part of the canal is straightened. Remember to use the pre-operative X-ray to estimate the workig length prior to taking the radiograph.

7 The next step is to prepare the apical stop. This is done with files using a longitudinal filing motion and copious irrigation. It is permissible to use a quarter turn with the finer files at this stage in order to help engage the dentine. As a general guide the apical stop should be prepared two sizes larger than the first file that binds at the full working length, but in addition to this it must be at least a size 25 (preferably a 30). Remember that in curved canals the larger files (generally 25 and above) will need to be precurved, this is done by bending around a mirror handle. When using precurved files try to use very short cutting strokes—long ones will produce an incorrect canal shape. The *balanced force* technique can be very effective at this stage in severely curved canals, providing the operator is experienced in the technique.

8 The rest of the root canal is now flared by 'stepping back' in the conventional manner. This procedure is made much easier as the coronal part of the root has already been enlarged.

Advantages of orifice enlargement

Effectively decreases the curvature in the coronal part of the root canal allowing straighter access for files to the apical region. It therefore reduces the likelihood of apical transportation (zipping).

It allows improved access for the flow of irrigant solution within the canal.

It reduces the likelihood of apical extrusion of infected material as most of the canal debris is removed before apical instrumentation takes place. This is particularly important in view of the fact that the majority of bacteria in an infected root canal are located in the coronal region.

Canal obturation

Purpose To provide a three-dimensional hermetic seal to the root canal which will prevent the ingress of bacteria or tissue fluids which might act as a culture medium for any bacteria that remain in the root canal system. In the past, the critical factor was considered to be the achievement of an apical seal, but now it is realized that coronal seal is also important. For this reason the whole of the root canal must be filled, and techniques that only seal the apical region (e.g. silver points) have fallen from favour.

Techniques

Numerous techniques have been described, all of those mentioned here use gutta-percha:

Cold lateral condensation This is a commonly taught method of obturation and is the standard by which others are judged. The technique involves placement of a master point chosen to fit the apical section of the canal. Obturation of the remainder is achieved by condensation of smaller accessory points. The steps involved are:

1 Select a GP master point to correspond with the master apical file instrument. This should fit the apical region snugly at the working length so that on removal a small degree of resistance or 'tug-back' is felt. If there is no tug-back, cut 1 mm at a time off the tip of the point until a good fit is obtained. The point should be notched at the correct working length to guide its placement to the apical constriction.
2 Take a radiograph to confirm that the point is in correct position.
3 Coat walls of canal with sealer using a small file.
4 Insert the master point, covered in cement.
5 Condense the GP laterally with a finger spreader to provide space into which accessory points can be inserted until the canal is full.
6 Excess GP is cut off with a hot instrument and the remainder packed vertically into the canal with a cold plugger.

Warm lateral condensation As above but uses a warm spreader. Finger spreaders can be heated in a flame or a special electronically heated device (Endotec) can be used.

Vertical condensation In this technique the GP is warmed using a heated instrument and then packed vertically. A good apical stop is necessary to prevent apical extrusion of the filling, but with practice a very dense root filling can result.

Thermomechanical compaction This involves a reverse turning screw (e.g. McSpadden compactor or GP condenser) which softens the GP, forcing it ahead of, and lateral to the compactor shaft. This is a very effective technique, particularly if used in conjunction with lateral condensation in the apical region.

Thermoplasticized injectable GP (e.g. Obtura, Ultrafil) These commercial machines extrude heated GP (70 to 160 °C) into the canal. It is difficult to control the apical extent of the root filling and in addition some contraction of the GP occurs on cooling.

Coated carriers (e.g. Thermafil) These are cores of titanium or plastic that are coated with GP. They are heated in an oven and then simply pushed into the root canal to the correct length. The core is then severed with a bur. A dense filling results but again apical control is poor and extrusions common. They are very expensive.

Once the filling is in place the tooth will need to be permanently restored, provided the follow-up X-ray is satisfactory. Fillings that appear inadequate radiographically may be reviewed regularly, or replaced, depending upon the clinical circumstances.

327

Follow-up The tooth should be reviewed radiographically after 6 months and thereafter annually for up to 4 years. Failure of endodontic treatment may present as pain, swelling, discharge, or radiographically as an enlargement of a periapical radiolucency. Following effective RCT most radiolucent periapical lesions show signs of resolution within 2 years.

Some endodontic problems and their management

Acute periapical abscess Relief of symptoms requires drainage of the abscess and where possible this should be obtained through the tooth. Open the pulp with a diamond bur in a turbine handpiece whilst supporting the tooth to ↓ vibration. Regional anaesthesia and occasionally sedation may be required. Once opened the canal is irrigated with sodium hypochlorite and if at all possible re-sealed. It may be necessary to see the patient again in 24 h, but this is more labour saving long term than leaving the tooth on open drainage. Relieve any traumatic occlusion.

If a fluctuant abscess is associated with the tooth this should be incised. If drainage can be obtained through the tooth and there is no evidence of a cellulitis then antibiotics are not required (p. 408).

Pain following instrumentation This is usually due to instruments, irrigants or debris being forced into the apical tissues. Placement of a small amount of Ledermix in the canal may provide symptomatic relief, but care is required not to breach the apex. Occasionally, an acute flare-up of a previously asymptomatic tooth occurs following initial instrumentation—this is called a phoenix abscess. Loss of face is saved by warning patients that this can happen. Affected teeth should be opened and irrigated and if possible re-sealed. This may need to be repeated after 24–48 h.

Recurrent symptoms/intractable infection If thorough cleaning and dressing the canal with calcium hydroxide is unsuccessful, it may be necessary to do an apicectomy (p. 402).

Sclerosed canals As the incidence of pulp necrosis following canal obliteration is only 13–16%, elective RCT is not warranted.[1] However, where pulp death has occurred finding the canal orifice may be difficult.

If careful exploration with a small file is unsuccessful, investigation of the expected position of the canal entrance with a small round bur may help. Once the canal is found, a No. 8 or 10 reamer should be used to try and negotiate it, using EDTA or RC Prep as a lubricant and the canal prepared and filled conventionally. Success rates of 80% have been reported for canals that were hairline or undetectable on X-ray.[1] Occasionally, a total blockage of the canal is encountered, in which case the filling is placed to this level ± an apicectomy.

Pulp stones in the pulp chamber can usually be flicked out. If they occur in the canal use EDTA and a small file to try and dislodge them.

Fractured instruments Sometimes it is possible to get hold of the # 'd portion with a pair of fine mosquitos. If not, insertion of

1 J. O. Andreasen 1981 *Traumatic Injuries of the Teeth*, Munksgaard.

a fine file beside the instrument may dislodge it and should this be unsuccessful a Masseran kit (p. 286) may be required. Should the fractured piece be lodged in the apical portion of the canal it may be better to fill the canal below it and keep it under observation, resorting to an apicectomy, if indicated.

Immature teeth with incomplete roots, p. 122.

Removing old root-fillings If a single-point technique or a root-filling paste has been used removal is straightforward. If a well-condensed GP filling is present this may be softened, using a heated probe in order to gain purchase for a fine reamer to be inserted. Use of chloroform may aid softening and removal. If time is at a premium a pledget of cotton wool moistened with eucalyptus oil can be sealed in place for 1–2 weeks before removal is attempted. Apical silver points or amalgam will require an apicectomy in order to achieve a satisfactory apical seal.

329

Perforations can be iatrogenic or caused by resorption (p. 122). In the latter case, dressing with calcium hydroxide may help to arrest the resorption and promote formation of a calcific barrier. Management of traumatic perforations depends upon their size and position:

Pulp chamber floor: If small can cover with calcium hydroxide and fill with GP or amalgam, but if large hemisection or extraction may be necessary.

Lateral perforation: If this occurs near the gingival margin it can be incorporated in the final restoration of the crown, e.g. a diaphragm post and core crown. If in the middle third, the remainder of the canal may be cleaned by passing instruments down the side of the wall opposite the perforation. Then the canal can be filled with GP, using a lateral condensation technique to try and occlude the perforation as well. Larger perforations may require a surgical approach and in multirooted teeth hemisection or extraction may be unavoidable.

Apical $\frac{1}{3}$rd: It is usually worth trying a vertical condensation technique to try and fill both the perforation and the remainder of the canal. If this is unsuccessful an apicectomy will be required.

Ledge formation If this occurs, return to a small file curved at the apex to the working length and use this to try and file away the ledge, using EDTA or RC Prep as lubricants.

Perio-endo lesions, p. 246.

Four-handed dentistry

The introduction of the water-cooled turbine handpiece helped to force dentists to change from a standing to a seated position and resulted in the DSA taking a more active role intra-orally. The development of 4-handed dentistry, however is credited to J. Ellis Paul.

Advantages

1 ↑ efficiency
2 ↑ patient comfort
3 ↑ operator visibility
4 ↓ backache
5 ↑ professional satisfaction for the DSA

Seating the patient With the exception of the old and infirm or pregnant patient, a supine position is preferable. Remember to warn the patient that you are about to whisk them backwards.

Seating the dentist The aim of this is a relaxed comfortable position, with direct vision of the teeth to be treated. First adjust the position of the dentist's stool so that the dentist's thighs are parallel to the floor. Then position the dental chair so that with the operator's back straight, the patient's mouth is at the correct distance of vision. The usual operating position is between 10 and 12 o'clock relative to the patient, whose head is at the 12 o'clock position and feet at 6 o'clock.

Seating the DSA The DSA must also be seated comfortably with a straight back and positioned at a level about 10 cm above the dentist, so that vision is improved. The DSA's normal working environment is between 2 and 3 o'clock, within easy reach of the instruments and equipment to be used.

Role of the DSA

- Receive, seat and look after the patient.
- Place eye protection on patient.
- Position dental light.
- Retract patient's lips and protect their soft tissues.
- Pass instruments to the operator.
- Hold aspirator tip (usually held in right hand).
- Hold 3-in-1 syringe (usually held in left hand whilst aspirator being used and then transferred to right hand while cavity being filled).

Instrument transfer The transfer zone is just in front of, and slightly below the patient's mouth, not over their eyes. There are several techniques which enable the DSA to pass and receive instruments effectively from the dentist. Each technique has merits and disadvantages. All require practice. Each dental health team needs to choose, adapt, practise, and perfect a system that safely achieves instrument transfer.[1]

---

1 J. E. Paul 1973 *A Practical Guide to Assisted Operating*, BDA.

6 Restorative dentistry
Four-handed dentistry

7 Prosthetics and gerodontology

Treatment planning	334
Principles of partial dentures	336
Components of partial dentures	338
Partial denture design	342
Clinical stages for partial dentures	344
Immediate complete dentures	346
Principles of complete dentures	348
Impressions for complete dentures	350
Recording the occlusion for complete dentures	352
Trial insertion of complete dentures	354
Fitting complete dentures	356
Denture maintenance	358
Cleaning dentures	360
Denture problems and complaints	362
Candida and dentures	364
Denture copying	366
Overdentures	370
Gerodontology	372
Age changes	374
Dental care for the elderly	376

Relevant pages in other chapters: Bridges—treatment planning and design, p. 296; occlusion, p. 274; acrylic and other denture materials, p. 684; casting alloys, p. 676; impression materials, p. 672.

Principal sources: M. R. Y. Dyer 1989 *Notes on Prosthetic Dentistry*, Wright. J. N. Anderson 1981 *Immediate and Replacement Dentures*, Blackwell. J. C. Davenport 1988 *A Colour Atlas of Removable Partial Dentures*, Wolfe.

7 Prosthetics and gerodontology

Treatment planning

Reasons for prosthetic replacement of missing teeth
- Restore aesthetics.
- Increase masticatory efficiency.
- Improve speech.
- Preserve or improve health of the oral cavity by preventing unwanted tooth movements.
- Space maintenance.
- Prepare patient for complete dentures.

Disadvantages of prosthetic replacement
- ↑ oral stagnation.
- Damage to soft tissues and remaining teeth, due either to poor denture design or lack of patient care.

Treatment options for the partially dentate mouth
1 No replacement of missing teeth. If the benefits of a prosthesis do not outweigh the disadvantages, then replacement is C/I. An occlusion with first premolar to first premolar present in each jaw is usually functionally adequate. Poorly controlled epilepsy is a C/I to dentures.
2 Bridges (p. 292) are preferable for short bounded spans in well-motivated patients.
3 Removable partial dentures. Indicated for patients with satisfactory OH and whose remaining teeth have an adequate prognosis, or as a training/interim appliance prior to F/F.
4 Complete immediate dentures. These are indicated for patients who have already mastered wearing a partial denture and whose remaining teeth are of poor prognosis. Occasionally, a medical condition may require the extraction of the remaining teeth and provision of immediate complete dentures.
5 Extraction of the remaining teeth and provision of a denture after healing has occurred. Avoid if possible as considerable guesswork is involved in the subsequent denture and the chances of the patient coping successfully are ↓.

In the older, partially dentate patient it is important to assess whether the patient is likely to retain some functional teeth for the remainder of the life-span. If this is improbable, treatment should be aimed towards providing F/F dentures while the patient is still young enough to adapt.

Treatment planning for partial dentures
- ▶ It is important to enquire about previous denture history (just because a patient is not wearing a denture does not mean that they have not had one) and assess the reasons for failure or success. If a patient produces an extensive collection of unsuccessful dentures, unless there is an obvious and easily remedied fault, it is probably wiser to assume that you are unlikely to succeed where so many have failed and refer the patient for a specialist opinion.
- Relief of pain and any emergency treatment.

- History and exam, including a thorough clinical and radiographic assessment of remaining teeth and edentulous areas.
- Unless immediate dentures planned, extract any teeth of poor prognosis.
- OH and periodontal treatment.
- Preliminary design of partial denture.
- Carry out restorative treatment required.
- Modify design if necessary and commence prosthetic treatment (p. 344).

Treatment planning for complete dentures
- Relief of pain and any emergency treatment including temporary modification of existing dentures, if indicated.
- History and exam.
- Investigation and treatment of any systemic problems.
- Removal of pathological abnormalities (e.g. retained roots) and pre-prosthetic surgery if required.
- ? rebase (p. 358), copy (p. 366), or construct new dentures (p. 348).

▶ Discussion with the patient of the limitations of dentures prior to their construction is more likely to be viewed as explanation, whereas leaving it until after fitting the dentures will be seen as making excuses!

Principles of partial dentures

Definitions

Saddle that part of a denture which rests on and covers the edentulous areas and carries the artificial teeth and gumwork.

Connector joins together component parts of a denture.

Support is resistance to vertical forces directed towards mucosa.

Retainers are components which resist displacement of denture.

Indirect retention is resistance to rotation about clasp axis by acting on the opposite side to the displacing force.

Fulcrum axis is the axis around which a tooth- and mucosa-borne denture tends to rock when saddles are loaded.

Bracing resistance to lateral movement.

Guide planes two or more parallel surfaces on abutment teeth used to limit path of insertion and improve retention and stability.

Survey line indicates the maximum bulbosity of a tooth in the plane of the path of withdrawal.

Free-end saddle is an edentulous area posterior to the natural teeth.

Stress-breaker is a device which allows movement between saddle and the retaining unit of partial denture.

Gum-stripper a tissue-borne partial denture which can 'sink'.

Dysjunct denture has complete separation between tooth-borne and mucosa-borne parts.

Swinglock denture has a labial bar or flange which is hinged at one side of the mouth and locks at the other for additional retention. Popular in USA.

Two-part denture is made in two or more sections which are then fixed together with screws or other devices.

Classification

KENNEDY—describes the pattern of tooth loss:
 I bilateral free-end saddles
 II unilateral free-end saddle
 III unilateral bounded saddle
 IV anterior bounded saddle, only

Any additional saddles are referred to as modifications (except Class IV), e.g. Class I modification 1 has bilateral free-end saddles and an anterior saddle.

CRADDOCK—describes the denture type:
 I mucosa-borne
 II tooth-borne
 III mucosa and tooth borne

Acrylic versus metal dentures

Approximately 75% of the dentures provided in the UK have an acrylic connector and base. Although metal bases are generally

preferred, because the greater strength of metal permits a more hygenic design, an acrylic base is indicated for:
- Temporary replacement, e.g. following trauma or in children.
- Where there is inadequate support from the remaining teeth for a tooth-borne denture.
- When additions to the denture are likely in the near future.

However, where financial constraints C/I a metal base, attention to the following may avoid the production of a gum-stripper:
- Wide mucosal coverage to provide maximum support.
- Keep base clear of the gingival margins wherever possible.
- No interdental extensions of acrylic.
- Point contact and wide embrasures between natural and artificial teeth.
- Labial flanges for extra retention and bracing.
- Additional support from wrought SS rests.

337

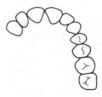

Kennedy Class I   Kennedy Class II

Kennedy Class III   Kennedy Class IV

Components of partial dentures

Saddles can be made entirely of acrylic or have a sub-framework of metal overlaid by acrylic.

Rests are an extension of the denture on to a tooth to provide support and/or prevent over-eruption. Occlusal rests are used on posterior teeth (over either the mesial or distal marginal ridge and fossa) and cingulum rests on anterior teeth. Rests may be wrought or cast, the latter is preferred for strength and fit.

Clasps provide direct retention by engaging the undercut portion of a tooth. The action of a clasp must be resisted either by a non-retentive clasp arm above the maximum bulbosity of the tooth or by a reciprocal connector. Clasps can be classified by their position (occlusally approaching or gingivally approaching) or by their construction and material.

Cast (cobalt chrome) Clasps are stiff, easily distorted and are liable to #. However, provided they are limited to undercuts of <0.25 mm, the advantage of being able to cast them as an integral part of a denture framework offsets these drawbacks.

Wrought Clasps are usually attached by insertion into the acrylic of a saddle. SS is the most commonly used alloy, but gold clasps are more flexible and easily adjusted (and distorted).

The stiffer the wire the smaller the undercut that can be engaged. This can be offset by reducing the diameter of the wire to ↑ flexibility (but ↑ the likelihood of #) or by increasing the length of the clasp arm (e.g. using reverse action clasp). Cast cobalt chrome can be too stiff for clasps on premolar teeth. The actual design used depends upon:

1 Depth of undercut: <0.25 mm—cast cobalt chrome; <0.5 mm—SS wire; <0.75 mm—wrought gold.
2 Position of undercut on tooth and relative to saddle, e.g.:
   • High survey line: ring clasp.
   • Diagonal survey line, (a) sloping down from saddle: re-curved or gingivally approaching clasp, (b) sloping up from saddle: 3-arm clasp.
   • Medium survey line: 3-arm clasp.
   • Low survey line: modify tooth shape, e.g. with composite.
3 Position of tooth. Gingivally approaching clasps are less conspicuous and are ∴ preferred for anterior teeth.
4 Material of denture base. Cast clasp arms are easily cast as part of the framework but for acrylic dentures wrought clasps are more usual.

Occlusally approaching 3-arm clasp
1 arm is the bracing reciprocal arm
1 arm is the retentive component
1 arm is the occlusal rest

Gingivally approaching T clasp

The two most commonly used types of clasp

Components of partial dentures(*cont.*)

Connectors
In addition to joining parts of the denture together, the connector can also contribute to support and retention.

P/- connectors

	Patient tolerance	Indirect retention	Support	Comments
Ant. pal. bar	−	+	−	useful for Kennedy IV
Mid. pal. bar	+	−	++	C/I torus
Post. pal. bar	++	+	−	need mucocompression
Ring	−	++	+	
Plate	+	++	++	less hygenic
Horseshoe	+	+	++	useful for multiple saddles

-/P connectors
- Lingual bar should only be used if there is >7 mm between floor of mouth and gingival margin to give 3 mm clearance from gingivae. Does not contribute to indirect retention. Usually cast. C/I if incisors are retroclined and for free-end saddles. If insufficient space can use sublingual bar.
- Sublingual bar lies horizontally in anterior lingual sulcus, but opinions differ as to patient tolerance.
- Lingual plate is well tolerated and provides good support, bracing and indirect retention, but covers gingival margins. Can be made of cast metal or acrylic.
- Continuous clasp is really a bar which runs along the cingulae of the lower anterior teeth and is usually used in conjunction with a lingual bar. Poorly tolerated.
- Dental bar is similar to continuous clasp, but of ↑ cross-sectional area and without lingual bar. Useful for teeth with long clinical crowns. Provides support and indirect retention.
- Buccal/labial bar is indicated when the lower incisors are retroclined.

Partial denture design

P/P design is carried out after assessment of the patient and with reference to any previous dentures. A set of accurately articulated study models is essential.

Surveying

Objectives:
1 Establish path of insertion.
2 Define those undercuts which may be used to retain denture.
3 Define those undercuts which require blocking out prior to finish.

If the path of insertion is at 90° to the occlusal plane insertion of the denture will be straightforward; however, where the teeth are tilted or few undercuts exist, an angulated path of insertion may be advantageous. Which provides more resistance to displacement during function, is controversial.

A survey line can then be marked on the teeth to indicate their maximum bulbosity in the plane of the path of withdrawal. Ideally, a proper dental surveyor should be used, but a steady hand and a propelling pencil can be substituted.

Design

1 **Outline saddles** Usually straightforward. If <½ tooth width or if in doubt of the need to replace a missing tooth, omit.

2 **Plan support** Support can be tooth only, mucosa only or both. Tooth borne support (occlusal and cingulum rests) should be used wherever possible as teeth are better able to withstand occlusal loading and support will not be compromised following resorption. Tooth and mucosa support is inevitable with large and free-end saddles and where plate designs are used. Tissue-only support should be utilized when no suitable teeth are available, and is less damaging in the upper than the lower arch, because of the palatal vault.

Need to assess the role of the denture, length of the saddles, the amount of support required (? denture opposed by natural or artificial teeth), and the potential of remaining teeth to provide support (root area in bone), before a final decision is made.

3 **Obtain retention** Retention can be:

(1) *Direct*, e.g. clasps, guide planes, soft tissue undercuts or precision attachments. Of these clasps are the most commonly used. The best arrangement is to use three clasps as far away from each other as possible. Guide planes help to establish a precise path of insertion and withdrawal. Need be only 2–3 mm in length.

(2) *Indirect*: This is derived by placing components so as to resist 'rocking' of the denture around direct retainers, e.g. by the position of clasps and rests and the type of connector. Particularly important with free-end and large anterior saddles.

4 Assess bracing required Bracing is provided by the connector, maximum saddle extension and the reciprocal arms of clasps. Elimination of occlusal interferences ↓ need for bracing.

5 Choose connector After consideration of above. Is there space in the occlusion to accommodate the chosen connector? Where possible the connector should be cut away from the gingival margins, but only for gaps >4 mm.

6 Re-assess ? as simple as possible ? aesthetic.

Instructions to technician Should include written details + diagram. Where some confusion may arise over the precise position of a component it may be helpful to mark this directly on the cast.

Some design problems

The lower bilateral free-end saddle (Class I) This presents a particular problem because of a lack of tooth support and retention distally, small saddle area compared to force applied and distal leverage on abutment tooth in function (which ↑ with resorption).[1] Possible solutions include:

- Maximize indirect retention by placing rests and clasps on mesial aspect of the abutment tooth and using plate design.
- Using a muco-compressive impression of saddle area to ↓ displacement in function. The altered cast technique.
- Use fewer and small teeth on saddle.
- Stress-breaker design (advantages more theoretical than practical).
- Use precision attachments.

Class IV Can sometimes avoid clasps anteriorly by the use of a flange. Possible designs include: anterior palatal bars with clasps and rests on 65|56, spoon denture which has no clasps and is held in place by tongue, horseshoe.

Multiple bounded saddles A horseshoe design, which utilizes guide planes for retention, may be indicated.

343

---

1 H. Devlin 1994 *BDJ* **176** 31.

Clinical stages for partial dentures

1 Assessment and treatment plan, p. 334.

2 Take first impressions These are usually taken using alginate in a stock tray. For free-end saddles modify the tray first with compound or silicone putty.

3 Record occlusion If ICP is obvious the occlusion can be recorded conventionally (p. 277) at the same visit as first impressions. If ICP is not obvious, wax record blocks will be required and a separate visit. Where there are no teeth in occlusal contact, the steps involved are the same as for recording the occlusion for F/F (p. 352). If there is an occlusal stop, but insufficient standing teeth to produce a stable relationship of the casts, the procedure is as follows:

- Determine the OVD and mark the position of 2 index teeth with pencil.
- Define the occlusal plane using the record block on which this is easiest, e.g. tooth to tooth, tooth to retromolar pad.
- Check the record blocks in the mouth, using the mark on the index teeth as a guide and adjust blocks if necessary.
- Record occlusion with bite-recording paste.
- Check the relationship of the index teeth on the articulated casts corresponds to that in the mouth.

4 Mounted casts are surveyed and denture designed (p. 342).

5 Tooth preparation may be required to:
- Accommodate rest seats. Rests need to be >1 mm for strength, ∴ if insufficient room in occlusion to accommodate this bulk, tooth reduction is required.
- Establish guide planes.
- Modify unfavourable survey line, e.g. ↓ bulbosity.
- Increase retention, e.g. by the addition of composite to create undercuts (NB use microfine type to ↓ abrasion of clasp).

6 Record second impressions using a special tray. Alginate is the most commonly used material, but elastomers are preferable for deep undercuts. Select tooth mould and shade.

7 Try-in
- Check extension, adaptation, and position of rests and clasps. If casting does not fit, use of correcting fluid on fitting surface may reveal which areas to relieve.
- Check upper and lower separately for OVD and occlusion, and then together.
- Check aesthetics with patient and only proceed when patient satisfied.
- Major faults: repeat second impressions &/or occlusion. Minor faults: adjust at finish.
- Prescribe post-dam, relief areas and management of undercuts.

8 Finish Once any fitting surface roughness is eliminated, the dentures are tried in separately, adjusting undercuts and contacts as required. The extension, occlusion, and articulation are then adjusted if necessary. Give the patient written and verbal instructions and a further appointment.

Rebasing P/P

Acrylic mucosa borne dentures can be rebased at the chairside with self-cure materials, but difficulty may be experienced in removing the denture in the presence of undercuts and the materials are generally inferior to the original denture base. Alternatively, P/P can be rebased in the laboratory by means of a technique similar to that used for F/F (p. 358). Alternatively, make a new denture. For cast metal dentures an impression can be recorded of saddle area using an elastomer or ZOE, whilst holding denture by the framework. In all cases care must be taken to avoid the introduction of occlusal errors, e.g. ↑ OVD.

Immediate complete dentures

When the remaining teeth have a poor prognosis management depends upon whether the patient is already a partial denture wearer or not.

Rx alternatives for patients with no previous denture experience
- Extraction of remaining teeth, wait 6 months for resorption to slow and then construct F/F dentures. A recipe for disaster!
- Extraction of majority of posterior teeth leaving sufficient only to maintain OVD and occlusal relationship, and then make immediate complete dentures when resorption has slowed.
- Provide partial denture and allow patient to adapt before progressing to an immediate complete denture. The best solution.

Rx alternatives for partial denture wearer

- A 'creeping partial denture' to which teeth are added as required. This allows a gradual progression towards edentulousness and is preferable for the elderly patient.
- Immediate complete denture. This has the advantage that the form and position of the natural teeth can be copied and is said to promote better healing and reduce resorption, but frequent adjustments and early replacement are necessary.
- Overdenture (p. 370).

Types of immediate complete denture
- *Flanged*: either full or part (extended 1 mm beyond maximum bulbosity of ridge).
- *Open face*: no flange, artificial teeth sit over (or just into) the socket of natural predecessor.

Flanged dentures are preferable as they afford better retention and make subsequent rebasing easier. However, where a deep labial undercut exists into which it would be impossible to extend a flange, the choice is either surgical reduction or an open-face denture. Most patients choose the latter.

Clinical procedures
1 *Assessment*. Warn the patient about the effects of resorption and the need for early rebasing/replacement.
2 *Primary impressions* (as for P/P, p. 344).
3 *Secondary impressions* in alginate or silicone (p. 344).
4 *Recording occlusion*. Where there are sufficient posterior teeth remaining, a wax wafer should suffice and this can be taken at the same visit as impressions are recorded. Otherwise, record blocks will be required.
5 *Try-in*. This will be limited to those teeth that are already missing. Check fit, extension, and stability, etc. In addition, need to prescribe:
   - Type of flange required.
   - Any proposed changes in position of anterior artificial teeth compared to natural teeth.
6 *Extraction* of remaining teeth as atraumatically as possible.

7 *Finish*. Repeated removal and insertion of the denture should be avoided, therefore adjustments should be limited to making the patient comfortable. They should be instructed not to remove the denture before the review appointment in 24 h.

- *Review*. The fitting and occlusal surfaces are adjused as required. If dentures are unretentive they will require temporary reline (see below).
- *Recall*. Regular inspection of immediate dentures is important as rapid bone resorption means that they will require rebasing early. However, this should be deferred, if feasible, for at least 3 months after the extractions. A possible regimen is 1 week, 1 month, 3 months, 9 months, and then yearly.

Laboratory procedures These are similar to F/F except that the plaster teeth are removed and the cast trimmed, before final processing.

Surgical procedures, p. 426.

Problems
- *Denture unretentive*. Use a temporary reline material (replaced regularly) to tide patient over initial 3 months and then re-line with heat-cure acrylic.
- *Gross occlusal error*. Adjust occlusal surface of one denture until even contact attained. This denture can then be replaced after initial resorption has occurred.

Principles of complete dentures

Retention The forces acting between the fitting surface and the denture bearing area to keep them in apposition. Mainly determined by (1) the anatomy of the denture-bearing area, particularly the ridge shape, (2) the viscosity and volume of saliva, (3) ability of the patient to avoid movements which result in denture displacement. As the area covered by the F/- is approximately twice that of -/F, retention is usually more of a problem with the latter.

Stability The forces acting on polished and occlusal surfaces which prevent displacement of the denture.

Neutral zone The area where the displacing forces are in balance.

Ways to optimize retention and stability

- Maximum extension of denture base (as far as the surrounding musculature will allow). The upper denture should extend distally over the tuberosities and onto the compressible tissue just anterior to the vibrating line on the palate. The lower denture should extend the full depth and width of the lingual pouch, and halfway across the retromolar pad. NB Overextension will result in a denture that is displaced in function.
- As close an adaptation of denture base to mucosa as possible, to maximize the surface tension effects of saliva.
- Placement of the teeth in the neutral zone. More important in -/F. The better retention of F/- often allows some latitude in this respect.
- Correct shape of the polished surfaces so that muscle action tends to re-seat the denture.
- A good border seal. This is achieved by ensuring that the flanges fill the entire sulcus width and by placing a post-dam on compressible tissue.
- Balanced occlusion free from interfering contacts.
- Weight of denture. F/- needs to be light, but if -/F heavy this will be advantageous.

Patient assessment This should include:
- Previous dental history. Including the age of the patient, when they became edentulous, number and degree of success of previous dentures, and their opinion of present F/F.
- EO examination of skeletal pattern and biological age.
- IO examination for signs of any pathology, and an assessment of ridge form, compressibility of mucosa, tongue size, tonicity of the lips, and the volume and viscosity of saliva.
- An evaluation of their present F/F. What to copy and what to correct.
- Personality.

Common denture faults These are, in order of prevalence:[1]
- Lack of freeway space.
- Failure to reproduce closely enough the features of previous successful dentures.

1 R. Yemm 1985 *BDJ* **159** 304.

- Occlusal errors.
- Incorrect adaptation and extension.

Impressions for complete dentures

▶ Tissues must be healthy before final impressions are recorded. If necessary use tissue conditioner in present F/F (p. 358).

Classically, two sets of impressions are recorded of the edentulous mouth. The purpose of the first is to record sufficient information for a special tray to be made in which to record the second or master impression. In practice, many use the first impression recorded in a stock tray for construction of the denture. With a careful technique this may suffice for some patients, but especially for those with retention problems, second impressions in a special tray are advisable.

First impressions

These are recorded using a (edentulous) stock tray, and alginate, elastomer (both preferable for undercut or flabby ridges), or composition. A line should be marked on the impression to indicate to the technician the desired extension of the special tray. In the upper, the posterior limit should be the hamular notches and the vibrating line, and in the lower the retromolar pads.

Special trays can be made in self-cure acrylic or shellac. The space left for the impression depends upon the material to be used: ZOE = 0.5 mm, elastomer = 0.5–1.5 mm (depending on viscosity), plaster = 2 mm, alginate = 3 mm. For trays with >1 mm space use greenstick stops to aid positioning.

Second impressions

These aim to record the maximum denture bearing area and develop an effective border seal. The special tray should be modified by reducing any over-extension and the peripheries adapted by the addition of greenstick tracing compound. Gently manipulate the patient's soft tissues and ask them to slightly protrude their tongue to imitate functional movements.

Muco-compressive versus muco-static A muco-compressive impression technique is advocated to give a wider distribution of loading during function and to compensate for the differing compressibility of the denture-bearing area, thus preventing # due to flexion. ZOE or composition are used. However, dentures made by this method are less well retained at rest, which is the greater proportion of time. Alginate and plaster are said to be more muco-static. Tissue adaptation following a period of use probably reduces the clinical difference between the two techniques.

Special techniques

Reciprocal denture impression technique This is used for recording the neutral zone in patients with limited natural retention for -/F.
• Record second impressions and occlusion.
• A fully extended acrylic baseplate is made on the lower cast,

with wire loops added which do not extend above occlusal plane.
- Silicone putty (with ↓ activator to extend working time) is pressed on to baseplate and around loops, and inserted.
- The patient is asked to swallow, purse lips and say 'Ooh' and 'Eee'.
- The impression is removed and trimmed down until it can be fitted on to the articulator to replace the lower occlusal rim.
- A mould of the impression is made into which wax is poured.
- The wax is cut away so that each denture tooth can be positioned within the zone recorded to make the trial denture. The polished surfaces should replicate the impression.

Flabby ridge classically occurs under a F/- opposed by natural lower teeth. If mild, then an impression recorded with alginate or elastomer in a tray perforated over the flabby area may suffice. For more severe cases a 2-stage technique is required, using a special tray with a window cut out over the flabby tissue. First, an impression is recorded in the tray with ZOE and the paste trimmed away from the flabby area. This is then re-seated and low-viscosity elastomer or impression plaster placed into the window to complete the impression.

Functional impression Tissue conditioner is placed inside the patient's existing denture. After several days of wear a functional impression is produced.

Common impression problems and faults
- A feather edge indicates under-extension. This can be corrected by the addition of greenstick to the tray and repeating.
- Tray border shows through impression material. The tray should be reduced in the area of over-extension and the impression repeated.
- Air blows. If small can be filled in with a little soft wax. If large, retake the impression.
- Tray not centred. This is often at least partially due to using too much material so that it is difficult to see what is where. Remember to line up the tray handle with the patient's nose (except for ex-boxers).
- Retching. A calm and confident manner is necessary for successful impressions. Gain the patients confidence by attempting the lower first and use a fast-setting, viscous material. Distraction techniques may help, e.g. wriggling the toes on the left foot and the fingers of the right hand at the same time (the patient, not the operator!).
- Patient with dry mouth. ZOE is C/I, use elastomer instead.
- Area where tray shows through in otherwise good impression. Can be overcome by prescribing a tin-foil relief when dentures being processed.

Recording the occlusion for complete dentures

When recording the occlusion the aim is to provide the technician with information for constructing trial dentures, including:

Vertical dimension The FWS is the space between the occlusal surfaces of the teeth when the mandible is in the rest position. In the majority of patients it is 2–4 mm. The OVD for an edentulous patient can ∴ be determined by measuring their resting face height and subtracting a FWS. Resting face height is assessed using:

- A Willis gauge, to measure the distance between the base of nose and the underside of the chin. Is only accurate to ± 1 mm.
- Spring dividers, to measure the distance between a dot placed on both the chin and the tip of the patient's nose. This method is less popular with patients and is C/I for bearded gentlemen (or ladies!).
- The patient's appearance and speech.

Position of the occlusal plane This should be placed so that about 1–2 mm (↓ with age) of tooth are visible below the patient's upper lip at rest. The occlusal plane should lie midway between the ridges parallel to the inter-pupillary and the ala-tragal lines. At rest the tongue should rise just above the lower occlusal plane posteriorly.

Horizontal jaw relationship Unless ICP can be copied from the remaining teeth (if constructing an immediate denture) or by copying existing successful dentures, then it is often easier to record the more reproducible RCP. In the natural dentition, ICP is about 1 mm forward of RCP, ∴ some prosthetists advise adjusting the finished dentures to allow the patient to slide comfortably between the two positions.

Position of the anterior and posterior teeth Ideally, the artificial teeth should lie in the space occupied by the natural dentition. The extent to which it is possible to compensate for a Class II or III malocclusion depends upon the retention afforded by the ridges. In the natural dentition the upper incisors lie about 10 mm anterior to the incisive papilla. With resorption this comes to lie on the ridge crest, ∴ the artificial teeth should be placed labial and buccal to the ridge, to give adequate lip support and a naso-labial angle of around 90°.

Mould and shade of artificial teeth Posterior teeth should be narrow to ↑ masticatory efficiency. Low cusped teeth are preferred, but cuspless teeth are useful for patients with poor natural retention or a 'wandering' ICP. When considering the colour, mould, and arrangement of the anterior teeth the patient's age, facial appearance, and most importantly their opinion, must be taken into account. If you disagree about the suitability of their choice, document it.

Type of articulator to be used for setting-up the teeth Most textbooks advocate adjustable or average value articulators for F/F dentures. However, most dentures are made on simple hinge articulators to the satisfaction of the majority of patients, probably because they are able to adapt to the occlusion that results. While hinge articulators are indicated for cuspless teeth or for patients who employ a chopping masticatory action, an average value type will give some degree of balanced articulation which can then be refined in the mouth and will avoid the provision of too deep an overbite.

Practical procedures

The occlusion is recorded using wax rims mounted on acrylic, shellac, or wax bases. A heated wax trimmer, or a plaster knife and a bunsen burner, are required to adjust the rims. As head posture can affect FWS, position the patient so that the Frankfort plane is horizontal.

1 Check fit of bases. If poor, can either repeat second impressions, or take a ZOE or low viscosity elastomer impression within the base (if acrylic or shellac) and proceed.
2 Adjust upper record block to give adequate lip support.
3 Trim occlusal plane of upper rim.
4 Trim lower record block to obtain correct lip support and bucco-lingual position of posterior teeth.
5 Adjust lower rim so that it meets upper evenly in RCP, with 2–4 mm of freeway space.
6 Mark centre lines.
7 Locate rims in RCP, e.g. with bite-recording paste or large office staples.
8 Prescribe mould and shade of artificial teeth for try-in.

Common pitfalls

• Inaccuracies caused by poorly fitting bases.
• Rims contacting prematurely posteriorly and flipping-up anteriorly or vice versa.
• Failure to provide adequate FWS. This is less likely to occur if the rest position is recorded with only the lower denture or rim in position.
• Attempting to correct too much when replacing old worn dentures and exceeding the adaptive capacity of the patient.

Trial insertion of complete dentures

Trial dentures are constructed by setting-up the prescribed teeth in wax on acrylic, shellac, or wax bases. Both the dentist and patient must be satisfied before the dentures are processed in acrylic.

Clinical procedures
Check the trial dentures
- On and off the articulator. Comparison with the patient's existing dentures is helpful to see if the features to be copied or modified have been successfully incorporated.
- Singly in the mouth. To check extension, stability, and the position of the teeth relative to the soft tissues.
- Together in mouth. Examine vertical dimension, occlusion, aesthetics and phonetics ('S' sound will be affected by an ↑ or ↓ FWS).

Seek the patient's opinion Some advocate getting patients to sign an acceptance slip before going to finish.

Prepare post-dam This should be placed just anterior to the vibrating line on the palate, which can be assessed by asking patient to say 'Aah'. The degree of compressibility of the tissues is assessed and the depth of the post-dam cut accordingly (usually about 1 mm). The post-dam is prepared on the upper cast with a wax knife in the shape of a cupid's bow.

Complete prescription to the technician This should include:
- Any changes in posterior tooth position or anterior tooth arrangement.
- For fibrous undercuts >4 mm and bony undercuts >2 mm, decide whether they are to be plastered out or the flange thickened for adjustment at finish.
- Tin-foiling for relief of hard or nodular areas, if required.
- Gingival colour and contour.
- Denture base material. This is usually heat-cure acrylic; however, metal bases are indicated for patients with a history of fractured dentures.
- Identification marker, which is preferably legible.

Common problems and possible solutions
- Over-extension of flanges. Reduce.
- Under-extension of flanges. Try a temporary wax addition to flange first, to check effect of extending it. If this is satisfactory a new impression is required.
- Teeth outwith neutral zone. Remove offending teeth and replace with wax which can be trimmed until correct.
- Incorrect OVD. If too small, can increase by adding wax to the occlusal surfaces of teeth, but if too large will need to replace lower teeth with wax and re-record OVD.
- Occlusal discrepancy or anterior open bite or posterior open bite. Replace lower posterior teeth with wax and re-record OVD.
- Too little of upper anterior teeth visible. Re-set anterior teeth

to correct position and ask lab to adjust occlusal plane accordingly.

- Too much of upper anterior teeth showing. The effect of reducing the length of the incisors can be judged by colouring incisal region with a black wax pencil and then indicating desired change in position to lab.
- Inadequate lip support. An increase in support can be assessed by adding wax to the labial aspect of the upper try-in.

A new try-in will be required if large errors are being corrected or if any doubt still exists about the occlusion.

355

Fitting complete dentures

Some adjustment of completed dentures is inevitable following processing. On average, a 0.5 mm increase in height occurs and a slight shift in tooth contact posteriorly. The main steps are:

Adjustment of fitting surface First, smooth any roughness and then gradually reduce the bulk of the flanges in areas of undercut until the denture can be easily inserted without compromising retention.

Check occlusion The vertical dimension of the dentures is maintained by contact between the upper palatal and lower buccal cusps, ∴ adjustment of these should be avoided if possible.

1 Get patient to occlude and check contact with articulating paper. If contact uneven, or heavy contacts seen, adjust the fossae.
2 For cusped teeth only, place articulating paper between occlusal surfaces and ask patient to make small lateral movements and adjust Buccal Upper and Lower Lingual (BULL rule) cusps only to remove any interferences.
3 Remove any interferences to protrusive movements.
4 If RCP position has been recorded and cusped teeth used, it will be necessary to adjust the distal slope of the upper buccal cusps and the mesial slope of the lower lingual cusps to allow a forward slide of approximately 1 mm. For crossbites this rule should be reversed!
5 Balancing contacts are desirable, but not essential unless they can be established easily by minor adjustments to working side contacts. Some authorities suggest providing even occlusal contact only at the time of fitting, allowing the patient to adapt to their new dentures before trying to achieve balanced articulation.

Advice to the patient Verbal and written instructions should be given.
- Most patients take some time to adapt to their new dentures. During this time a softer diet is advisable.
- If pain is experienced the patient should try to continue wearing their dentures and return for adjustment as soon as possible so that affected areas can easily be seen.
- Although patients should be encouraged not to wear their dentures at night, adaptation may be speeded up if they are worn full-time for the first 1–2 weeks.
- When the dentures are not being worn they should be stored in water to prevent them drying out and warping. Plastic denture boxes are cheap, and safer than a glass of water at the bedside.
- Cleaning, p. 360.

Review The patient should be seen 1–2 weeks after fitting to ease the dentures and adjust the occlusion. Localization of the cause of any irritation due to a flaw on the fitting surface can be helped by:

- Pressure relief cream which is painted on to the fitting surface of the denture.
- Indelible pencil, or denture fixative powder mixed with zinc oxide, which is applied carefully to area thought to be responsible and the denture inserted. On removal the mark will have been transferred to the adjacent mucosa and should correspond with the damaged area.

If there is no obvious cause relating to the fitting surface remember that occlusal faults can cause displacement and mucosal trauma and an excessive OVD is a common cause of generalized soreness under -/F (p. 362).

Stress the importance of regular review of all patients with dentures.

357

Denture maintenance

▶ Review patients with F/F annually. Regular maintenance will help prevent damage due to ill-fitting dentures and will ↑ the likelihood of early detection of oral pathology.

Problems caused by lack of aftercare of F/F As a result of resorption all dentures become progressively ill-fitting, leading to loss of retention and stability. Movement of dentures in function may result in:

• Resorption.
• Predisposition to candidal infection.
• Denture irritation hyperplasia, p. 426.
• Inflammatory papillary hyperplasia of the palate.

All of these are exacerbated by wear of the occlusal surfaces.

Rebasing

358

The terms rebasing and relining are commonly used interchangeably. Strictly speaking relining is replacement of the fitting surface (e.g. with a temporary material) and rebasing is replacement of most or all of the denture base.

Rebasing is indicated where the only feature of F/F that requires improvement is the fitting surface, otherwise consider replacement F/F using copy method. For rebasing the material of choice is heat-cure acrylic (p. 684), but this necessitates the patient being without their dentures while the addition is being made. Self-cure acrylic applied at the chairside appears attractive, but its properties are inferior. For a heat-cure rebase, a wash impression (ZOE or low viscosity elastomer) must be recorded inside the denture.

Technique: To avoid an ↑ in OVD, record the impression for one denture at a time.

• Check occlusion and adjust if required. Note OVD.
• Remove undercuts from fitting surface.
• Correct extension and place post-dam in greenstick.
• Apply impression material and insert in mouth. Get patient to close into contact with opposing denture. Check OVD and occlusion.
• Remove and examine impression, if unsatisfactory (or if in doubt) repeat.

An alternative method for inflamed tissues is to record a functional impression (p. 351) over several days with a tissue conditioner, in which case the resulting impression needs to be cast immediately.

Tissue conditioners

These are resilient materials which give a more even distribution of loading and thus promote tissue recovery. They are particularly useful where ill-fitting dentures have caused trauma, as it is important to allow the tissues to recover before impressions for replacement dentures or a rebase are taken.

Technique: Relieve any areas of pressure on the fitting surface and reduce any over-extension. A minimum thickness of 2 mm is required and the material should not be left for >1 week. Repeated applications may be necessary. Crushed Nystatin tablets can be incorporated in the powder of Visco-gel prior to mixing, if candidal infection is present. Ardee liner which is available as a sheet, is particularly useful for -/F.[1]

Soft linings are indicated for
- Older patients with a thin atrophic mucosa. Usually for -/F.
- Following prosthetic surgery.
- To utilize soft tissue undercuts for ↑ retention, e.g. following hemimaxillectomy, clefts.

It is wise to make a new denture first in acrylic and adjust the occlusion, before placing soft lining. A minimum thickness of 2 mm is required which may significantly weaken a lower denture necessitating placement of a metal strengthener on the lingual aspect. No material is ideal (p. 686).

359

1 G. R. Lewis 1988 *BDJ* **165** 288.

Cleaning dentures

When new dentures are fitted the importance of regular, thorough cleaning with soap, water and a brush to prevent the build-up of plaque, stain and calculus, should be emphasized. Unfortunately, few patients are sufficiently diligent, due in part to being conditioned by advertising to expect to use a denture cleaner.

Advise patients to clean their dentures over a basin of water to act as a safety net.

Formulation	Active ingredients	Problems
Powder	Abrasives, e.g. calcium carbonate	Abrasion
Paste	Abrasives + eugenol	Abrasion + crazing
(Dentu-creme)	Abrasives + phenol oil	Abrasion + sensitivity
Hypochlorite (Dentural)	Sodium hypochlorite	Can corrode metal
Effervescent (Steradent)	Dissolves to give alkaline peroxide solution	Not very effective
Dilute acids (Denclen)	3–5% hydrochloric acid or 5–10% phosphoric acid	Can corrode metal
Enzymatic	Proteolytic enzymes	Not widely available

Practical tips

Hypochlorite solutions are effective for acrylic dentures when used overnight, but if used with *hot* water are liable to cause bleaching ∴ warn patient.[1] The peroxide cleaners are popular but are ineffective if used for only 15–30 min as the manufacturers advise.

	Avoid	*Use*
Viscogel	Acids, alkaline peroxide	Hypochlorite
Molloplast	Acids, alkaline peroxide	Hypochlorite
Coe-comfort	Hypochlorite, alkaline peroxide	Soap + water
Metal denture	Hypochlorite	Alkaline peroxide
Any denture	Household bleach	

1 C. A. Crawford 1986 *J Dent* **14** 258.

Denture problems and complaints

The most common complaints are of pain &/or looseness which can be due to denture errors or patient factors. The latter should be foreseen and the patient warned in advance of the limitations of dentures. The wise prosthetist will tend to overestimate the difficulty of providing successful dentures. Unless the cause is immediately obvious, e.g. a flaw on the fitting surface, a systematic examination of the fitting, polished, and occlusal surfaces (including the jaw relationship) should be carried out.

Pain This can be due to a variety of causes, including roughness of the fitting surface, errors in the occlusion, lack of FWS, a bruxing habit, a retained root, or other pathology. Forward or lateral displacement of a denture due to a premature contact can lead to inflammation of the ridge on the lingual or lateral aspect respectively. With continued resorption bony ridges become prominent and the mental and lingual foramina exposed, which can lead to localized areas of specific pain.

Pain from an individual tooth on P/P
- Excessive load &/or traumatic occlusion.
- Leverage due to unstable denture.
- Clasp arm too tight.
- Inadequate lining under amalgam restoration failing to insulate against a galvanic couple with metal denture.

Looseness This more commonly affects the lower than the upper denture.

Denture faults	*Patient factors*
Incorrect peripheral extension	Inadequate volume or amount of saliva
Teeth not in neutral zone	Poor ridge form
Unbalanced articulation	↓ adaptive skills e.g. elderly patient
Polished surfaces unsatisfactory	

Burning mouth This can be due to (1) local causes: e.g. ↑ OVD, sensitivity to acrylic monomer, or be unrelated to the denture (e.g. irritant mouth washes), or (2) systemic causes, e.g. the menopause, deficiency states, cancerophobia, xerostomia.

Speech

Patient's complaint	*Possible cause*
Difficulty with f,v.	Incisors too far palatally
Difficulty with d,s,t	Alteration of palatal contour
	Incorrect overjet and overbite
s becomes th	Incisors too far palatally
	Palate too thick
Whistling	Palatal vault too high behind incisors
Clicking teeth	↑ OVD
	Lack of retention

Cheek biting Check first that teeth are in neutral zone. If satisfactory, ↑ buccal 'overjet', i.e. reduce buccal surface of lower molars (provided normal bucco-lingual relationship).

Retching
- Map out extent of sensitive area on palate using a ball-ended instrument and firm pressure, and check extension of denture.
- Palateless dentures may be a solution, but their retention is poor.
- Training dentures. These can take the form of a simple palate to which teeth are added incrementally starting with the incisors.
- Implants (p. 428) and a fixed prosthesis.

The grossly resorbed lower ridge Resorption is progressive with time, which is a good argument for avoiding rendering young patients edentulous. The mandible resorbs more quickly than the maxilla, which exacerbates the problem of retention for -/F. Management is dependent upon the severity of the problem and the patients biological age.
- Minimizing destabilizing forces upon the lower denture, e.g. (1) maximum extension of denture base, (2) ↓ number and width of teeth, (3) ↑ freeway space, (4) lowering occlusal plane.
- Reciprocal denture impression technique to record neutral zone (p. 350).
- Surgery (p. 426).
- Implants (p. 428).

Recurrent fracture Apart from carelessness, this is usually caused by fatigue of the acrylic due to continual stressing by small forces. Flexing of the denture can occur with flabby ridges, palatal tori, uneven occlusal contacts, and following resorption. Notching of a denture, e.g. relief for a prominent frenum, can also predispose to fracture. Treatment depends upon the aetiology, but in some cases provision of a metal plate or a cast-metal strengthener may be necessary.

Candida and dentures

Candida is a common oral commensal. It becomes pathogenic if the environment favours its proliferation (e.g. dentures, antibiotic alteration of the bacterial flora) or the host's defences are compromised.

Denture stomatitis

Also known as denture sore mouth, which is a misnomer because the condition is usually symptomless. Classically, seen as redness of the palate under a F/- denture, with petechial and whitish areas. 90% cases due to candida albicans, 9% due to other candida and <1% other organisms, e.g. klebsiella.

Incidence This is a common condition, having been reported in 30–60% of patients wearing F/F. It affects females more commonly than males, in a ratio of 4 : 1, and usually affects the upper denture bearing area only.

Aetiology is still not completely understood.
- Infection with candida.
- Poor denture hygiene.
- Night-time wear of dentures.
- Trauma is often cited as a contributing factor to denture stomatitis, BUT
  1 occurs more commonly under F/- rather than -/F
  2 it can affect patients wearing F/- only
  3 also found under well-fitting orthodontic appliances.
- Systemic factors can predispose to candida infection, e.g. iron and vitamin deficiency, steroids, drugs which cause xerostomia, and endocrine abnormalities.
- A high intake of sugar provides substrate for candida to multiply. It has been postulated that the reason that the upper denture-bearing area is more commonly affected is related to the more serous nature of saliva from the submandibular glands and the poorer fit of -/F which allows saliva to reach the underlying mucosa more easily.

Management
- Leave dentures out. This is not a realistic solution to most patients, but they should be encouraged to remove their dentures at night.
- Improved denture hygiene, e.g. brushing fitting surface and soaking in hypochlorite cleanser.
- Reduce sugar intake.
- Antifungals (p. 618). Nystatin suspension 100 000 units/ml, 1 ml qds (NB pastilles contain sugar) or Amphotericin suspension 100 mg/ml, 1 ml qds are the first choice. 2% Miconazole gel is more expensive and should therefore be reserved for patients with candida which is unresponsive to other agents, or associated with angular cheilitis.
- If suspect systemic factors exacerbating condition refer to GMP.

- Co-existing papillary hyperplasia of palate may need surgical reduction.

Angular cheilitis, p. 438.

Denture copying

Successful function with complete dentures depends to a marked degree upon the patient's ability to control them. This ability is learnt during a period of denture use. When replacement dentures become necessary, it is helpful if the new appliances require as little adaptation as possible to the existing skills. This is generally considered to be particularly important for the older patient. Not only may skills have been developed over a long period, but also the ability to relearn may be diminished. So-called denture copying techniques provide a more reliable method for provision of replacements.

Treatment planning

Before undertaking treatment it is essential to decide which features of the previous dentures are satisfactory, and which require modification, and by how much. Consider:

- Fitting surface—if this is the only feature that requires improvement, then rebasing is a possibility.
- Polished surface shapes.
- Occlusal surface, jaw relationship, occlusal vertical dimension (OVD). The effect of an increase in OVD can be assessed by self-cure addition to the existing dentures, but remember that this irreversibly alters them.
- Anterior tooth size, arrangement, relation to lips.
- Posterior tooth mould, and arch width (relation to tongue and cheeks).

Copying complete dentures

A number of techniques have been described. They vary in the materials used, and these in turn affect the acceptability in laboratory work, and the clinical freedom to incorporate 'corrections'. In general, copies of the old appliances are used as substitutes for record blocks, and as special trays.

A typical method involves the following stages:[1]

1 Clinic

- Correct under-extension with greenstick tracing compoud.
- Record impressions with silicone putty of polished surface and teeth, using large disposable tray. Complete mould with second mix of putty to record fit surface (use a separating medium—vaseline or, better, emulsion handcream).
- Open mould, clean dentures, and return to patient.
- Send putty moulds to laboratory.

2 Laboratory

- Fabricate shellac baseplates on the (heat-resistant) silicone model of the fit surface.
- Pour wax into remaining space.
- After cooling remove completed copy, cut off sprues, and polish.

1 R. Yemm 1991 *Int Dent J* **41** 233.

3 Clinic
- Employ the copies ('replica record blocks') to record required changes in denture shapes (see treatment planning), by adding or removing wax.
- Record working impressions in low viscosity silicone *with adhesive* to aid retention on base.
- Record jaw relationship with 'bite recording paste'.
- Select shade/moulds for new teeth.

4 Laboratory
- Cast impressions and articulate.
- Set up, cutting away modified replica rims to substitute new teeth (rather like setting up an immediate denture!)
- Wax-up, including borders defined by working impressions.

5 Clinic
- Try-in stage.

6 Laboratory
- Finishing stages as normal.

7 Clinic
- Normal insertion stage (and subsequent review).

Other methods use alternative materials (e.g. alginate for impressions of existing dentures, wax, and self-cure resin) to form the copy dentures. Choice will depend on acceptability to both clinic and laboratory. In no instance, however, is an all-wax copy regarded as being acceptable, since rigidity is inadequate for use as an impression tray.

Copying partial dentures for immediate dentures[1]

In patients with successful P/P, for whom extraction of the remaining teeth is planned, the transition to complete dentures can be facilitated by using a copy technique.

Clinic 1 Correct underextended flanges with greenstick and then take impressions of the dentures with putty in stock trays (see F/F technique). Record an alginate impression of the opposing arch, if no denture is planned for that arch.

Lab 1 Wax/shellac replica of partial denture is constructed.

Clinic 2 Use the replica denture to develop the prescription and then record a wash impression inside the base with a light-bodied silicone. Record the occlusion using bite recording paste. Finally, take an overall impression in a stock tray with the modified replica denture *in situ*.

Lab 2 The impression is cast and used as a base for articulating the wax replica with the cast of the opposing arch. The teeth

1 J. R. Drummond 1983 *BDJ* **155** 297.

Denture copying (*cont.*)

prescribed are then set up, and the wash impression retained in the replica, for the try-in.

Try-in and Finish as for complete immediate dentures, p. 346.

Overdentures

An overdenture (OD) derives support from one or more abutment teeth by completely enclosing them beneath its fitting surface. It can be a partial or complete denture.

Advantages

- Alveolar bone preservation around the retained tooth.
- Improved retention, stability, and support.
- Preservation of proprioception via PDL.
- Improved crown root ratio which ↓ damaging lateral forces.
- ↑ masticatory force.
- Additional retention possible using attachments.
- Aids transition from P/P to F/F.

Disadvantages

- RCT probably required.
- To avoid excessive bulk in region of retained tooth, denture base may need to be thinned which ↑ likelihood of fracture.
- ↑ maintenance for both patient and dentist.

Indications

- Motivated patient with good oral hygiene.
- Because of ↓ retention and stability of -/F and ↑ rate of mandibular resorption OD are particularly useful for -/F or free-end saddle.
- Cleft lip and palate.
- Hypodontia.
- Severe toothwear.

Choosing abutment teeth

- Ideally: bilateral, symmetrical with a minimum of one tooth space between them.
- Order of preference: canines, molars, premolars, incisors.
- Healthy attached gingiva, adequate periodontal support ($>\frac{1}{2}$ root in bone) and no mobility.
- Is RCT required and if so is it feasible?

Preparation of abutment teeth

Alternatives include:
1 Removal of undercuts only.
2 Preparation of crown for thimble/telescopic gold coping.
3 RCT, tooth cut to dome-shape and access cavity restored with amalgam.
4 RCT and gold coping over root face.
5 RCT and precision attachment.

Precision attachments[1] are useful for ↑ retention of dentures or bridges especially in cases with tissue loss (e.g. trauma or CLP), but they ↑ loading on abutment teeth, are expensive and difficult to rebase and repair. Usually of two parts which are matched to

---

1 H. W. Preiskel 1984 *Precision Attachments in Prosthodontics*, Quintessence.

fit together. One part is attached to the abutment tooth and the other to the denture. A variety of attachments are available including stud/anchor (e.g. Rotherman eccentric clip), bar (e.g. Dolder), and magnets.

Since precision attachments require the highest technical skill and are highly dependent on patient and professional maintenance, it is wise to first use a basic OD and then re-assess the need for additional retention. Hybrid dentures are partial dentures that utilize precision attachments (either intra- or extracoronal) on the abutment teeth for retention. Implants which are inserted in edentulous areas can be used with a precision attachment to increase retention.

Clinical procedures

1 Assessment (clinical examination, study models, radiographs, etc.).
2 RCT if required.

If abutment preparation is limited to crown reduction:

3 The steps involved are as for immediate dentures with the abutment teeth reduced less on cast than is planned clinically. At the visit the final dentures are to be fitted, the abutment teeth are prepared and the dentures re-lined with self-cure acrylic to improve their adaptation.

If precision attachments or copings to be used:

3 The teeth are prepared and an impression of the abutments, including post holes, is taken. In the lab, dies are prepared and transfer copings (usually metal) made. The transfer copings are tried on the abutments and if satisfactory an overall impression is recorded to accurately locate the copings to the remainder of the denture bearing area. Alternatively, a 2-stage technique using a special tray with windows cut out over abutments can be used.
4 Regular review (6-monthly) and maintenance is necessary for success.

Problems

The most important problems are:
• Caries of abutment teeth ∴ need good oral and denture hygiene and topical fluoride, e.g. toothpaste, applied to the fitting surface of the denture. Patients should be encouraged to leave their denture out at night.
• Periodontal breakdown.

Gerodontology

(or Gerodontics or Gerontology)

Definition Dentistry for the elderly. For those who have not reached pensionable age, the elderly is anyone over 65. Others suggest that is >75 yrs of age. Rather than arbitary cut-offs, biological age should be considered. This new speciality, however, is still in its infancy.

Epidemiology Two factors are mainly responsible for the increasing relevance of dentistry for the elderly, an increase in the proportion of the elderly in the population and the improvements in dental health which have resulted in more people keeping their natural teeth for longer. By 2001 the proportion aged >75 yrs will have ↑ by 22%. It is estimated that by 2001, 10% of adults will be edentulous, compared with 25% in 1983.

Problems The major overall problems are:
- Age changes, both physiological and pathological.
- Disease and drug therapy (Chapter 11).
- Delivery of care.

Restorative problems include:
- Root caries which can occur following exposure of root surfaces by gingival recession, in association with changes in diet, ↓ self-care, and ↓ salivary flow. Details of management are given p. 30. Prevention of root caries in susceptible patients is possible using either a topical fluoride mouthrinse or fluoride-containing artificial saliva, e.g. Luborant or Saliva Orthana.
- Tooth wear (p. 304) is especially prevalent when partial tooth loss has occurred.
- Pulpal changes including sclerosis (p. 328) and ↓ repair capacity.

Periodontal problems include:
- Reduced manual dexterity making OH procedures difficult.

Epidemiological studies of the periodontal needs of the elderly population are still sparse and some trends may be masked by a high rate of edentulousness. The available evidence suggests that although older patients develop plaque more quickly, periodontal treatment need increases up to middle age, and thereafter the majority can be maintained by regular non-surgical management.

Prosthetic problems include:
- Reduced adaptive capacity, ∴ if teeth are unlikely to last a lifetime, the transition to at least partial dentures should be made whilst the patient is able to learn the new skills necessary.
- Age changes in the denture-bearing areas including bone resorption and mucosal atrophy.

Age changes

Age changes are defined as an alteration in the form or function of a tissue or organ as a result of biological activity associated with a minor disturbance of normal cellular turnover.

In general

↓ microcirculation ↓ cellular reproduction, ↓ tissue repair, ↓ metabolic rate, ↑ fibrosis. Degeneration of elastic and nervous tissue. Results in reduced function of most body systems.

Dental

Oral soft tissues A ↓ in the thickness of the epithelium, mucosa and sub-mucosa is seen. Taste bud function ↓. With age, an ↑ occurs in the number and size of Fordyce's spots (sebaceous glands), lingual varices and foliate papillae. Recent evidence suggests that stimulated salivary flow rate does not fall purely as a result of age. However, medications or systemic disease can affect salivary output.

Dental hard tissues Enamel becomes less permeable with age. Clinically, older teeth appear more brittle, but there is no significant difference between the elastic modulus of dentine in old or young teeth. The rate of secondary dentine formation reduces with age, but still continues. Occlusion of the dentine tubules with calcified material spreads crownwards with age.

Toothwear is an age-related natural phenomenon. As it results in a more favourable crown root ratio, it can even be considered protective. However, excessive wear can be caused by environmental factors.

Dental pulp ↑ fibrosis and ↓ vascularity mean that the defensive capacities of the pulp reduce with age, ∴ pulp capping is less likely to succeed. Also ↑ secondary dentine and ↑ pulp calcification.

Periodontium ↑ fibrosis, ↓ cellularity, ↓ vascularity and ↓ cell turnover are found with ↑ age. Whether gingival recession is pathological or physiological is still hotly debated.

Systemic

Immune system A ↓ in cell-mediated response and ↓ in number of circulating lymphocytes leads to an ↑ incidence of autoimmune disease as well as ↓ in the older patient's defence against infection. Also an ↑ in neoplasia is seen. Steroid treatment for auto-immune disease may complicate dental treatment.

Nervous system Ageing involves both a physiological decline in function, and dysfunction associated with age-related disease (e.g. strokes, parkinsonism, trigeminal neuralgia). A dimunition in acuity compounds the problem.

Cardiovascular Hypertension and ischaemic heart disease worsen with age. Anaemia is more common in the elderly. In general, the greatest problems arise when a GA is required, or the practice is on the second floor.

Pulmonary system Lung capacity ↓ with age and chronic obstructive airways disease ↑ in prevalence.

Endocrine system Diabetes is more common.

Muscles ↓ bulk, slower contractions and less precision of control occur.

Nutrition Poverty, impaired mobility, ↓ taste acuity and ↓ masticatory function can result in nutritional deficiencies in the elderly. These can manifest as changes in the oral mucosa.

Mucosal disease which is more common with increasing age
- Oral cancer, p. 450.
- Lichen planus, p. 462.
- Herpes zoster is more common with ↑ age due to a dimunition of T-cell function. Neuralgia occurs more frequently after an attack in the elderly.
- Benign mucous membrane pemphigoid, p. 444.
- Pemphigus, p. 442.
- Candida is seen more frequently due to an ↑ proportion of denture wearers in the older age-groups and immune deficiencies.

This list is obviously not exhaustive.

375

Dental care for the elderly

The basis for delivery of care to the older patient has been, and is likely to continue to be, via the general dental practitioner.

General management problems

Medical and drug history (Chapter 11) It is wise to check any complicated medical problems with the patient's GMP or physician. Unfortunately, many doctors are only familiar with the dental treatment they have personally received, ∴ give details of what is proposed.

Communication with the elderly requires patience and understanding. Often older patients will try to cover up deafness, poor eyesight, and lack of comprehension, so it is better to err on the side of over-stressing an important point or instruction, but avoid sounding patronising. It is often helpful to enlist the assistance of a relative or friend of the older patient.

Oral hygiene may be compromised by arthritis and or a stroke. Advise an electric toothbrush or modifying the handle of an ordinary toothbrush to make it easier to grip, e.g. with elastoplast or bicycle handles. Double-headed toothbrushes are also available.

Delivery of care

Dental practice Consideration should be given to:
- Access for a wheelchair or Zimmer frame. This is easy to arrange when a purpose-designed practice is being built, but can pose considerable problems for the established practice. In the surgery it is important to allow sufficient space around the dental chair.
- Timing of appointments, e.g. for a diabetic patient these need to be arranged around meals and drug regimens, and early morning visits are probably C/I for arthritic patients as it may take them a couple of hours to 'get going'.
- Positioning of the patient. Many elderly patients are unhappy to be recumbent in the dental chair. In addition, this position is C/I for those with cardiovascular or pulmonary disease. Adjust dental chair gradually as rapid movement from a flat to upright position can result in postural hypotension.

Domiciliary care An estimated 12–14% of the elderly population are bedridden or housebound to such a degree that they cannot visit their GMP or GDP. Unfortunately, although the dental needs of this group are high the uptake of dental care is low, due to the low priority placed on such care and the difficulty experienced in obtaining it. Hygienists can now make domiciliary visits.
- *Site*: The kitchen is probably the most suitable room, with access to water and heat. However, it is wise to defer to the patient should they prefer another location for the improvised surgery. Care must be taken to protect the floor and work surfaces from any spillages.

- *Seating*: Where possible the patient should be seated in their wheelchair, or a straight-backed chair placed against a wall with a cushion for additional head support. When a patient is bedridden there is no choice, but the dentist should take frequent rests from bending over, to prevent back strain.
- *Equipment*: The amount of expenditure incurred on this will be determined by the volume of such work undertaken. Portable dental units are available, but costly. However, the outlay could be shared by several practices or FHSA. A Mothercare or fishing-tackle box is useful for holding small items.
- *After-care*: This is particularly important. It is the dentist's responsibility to ensure that the relatives or carers appreciate the need for good oral and denture care and how to carry it out. In the bedridden chlorhexidine can be applied to the teeth by cotton swabs.

▶ Take a chaperone.

377

Key points

- Can treatment be carried out successfully?
- Consider maintenance required by any proposed treatment. Elaborate procedures which fail may leave the patient worse off.
- The objective is to maintain optimum oral function. Sometimes retention of a few teeth can be disadvantageous.
- Medical crises (e.g. a period in hospital) can result in a very rapid change in a previously stable oral state, e.g. rapid caries attack, loss of denture-wearing skill through lack of use.
- Avoid sudden changes in occlusion. The shape/form of dentures should not be changed anteriorly. During restorative work refrain from introducing significant occlusal change. If necessary to extract teeth, do so a few at a time, with additions to existing dentures.

Some clinical techniques of particular value in elderly

- Adhesive restorations, e.g. glass ionomer for root caries.
- Acid-etch bridgework is less destructive to abutments and is ∴ more fail-safe.
- Gradual tooth loss, with additions to existing P/P, is less demanding of a ↓ adaptive capacity.
- Replacement dentures should be made with careful regard to existing appliances. Use of copying techniques again ↓ amount of adaptation required.
- If recording the occlusion proves difficult, use cuspless teeth.
- Mark dentures with the patient's name.

8 Oral surgery

Principles of surgery of the mouth   380
Asepsis and antisepsis   382
Forceps, elevators, and other instruments   384
The extraction of teeth   386
Complications of extracting teeth   388
Post-operative bleeding   390
Suturing   392
Dento-alveolar surgery: removal of roots   394
Dento-alveolar surgery: removal of unerupted
  teeth   396
Dento-alveolar surgery: removal of third
  molars   398
Dento-alveolar surgery: third-molar
  technique   400
Dento-alveolar surgery: apicectomy   402
Dento-alveolar surgery: other aids to
  endodontics   404
Dento-alveolar surgery: helping the
  orthodontist   406
Dento-facial infections   408
Biopsy   410
Cryosurgery   412
Non-tumour soft-tissue lumps in the mouth   414
Non-tumour hard-tissue lumps   416
Cysts of the jaws   418
Benign tumours of the mouth   420
The maxillary antrum   422
Minor preprosthetic surgery   426
Implantology   428
Lasers   430

Principal sources: Moore 1985 *Surgery of the Mouth and Jaws*, Blackwell Scientific. Seward, Harris, and McGowan 1987 *Outline of Oral Surgery*, Wright.

Additional background: *Pathology*: Soames and Southam 1993 *Oral Pathology*, OUP. *Third molars*: MacGregor 1985 *The Impacted Lower Wisdom Tooth*, OUP.

Principles of surgery of the mouth

The mouth is a remarkably forgiving environment in which to operate, because of its excellent blood supply and the properties of saliva. It is compromised less than could be expected by its teeming hordes of commensal organisms. This does not, however, constitute *carte blanche* for ignoring the basic principles of surgery, although these can and should be modified to suit the nature and the site of the surgery.

Asepsis and antisepsis, p. 382.

Analgesia Thankfully, nowadays all patients should expect and receive painless surgery, both perioperatively and *postoperatively*. Analgesia and anaesthesia, Chapter 13.

Anatomy and pathology are the interdependent building blocks of surgery. Know the anatomy and you can understand or even devise the operation. Know the pathology and you know why you are doing it, what can be sacrificed, and what must be preserved.

Access For all minor and certain major oral surgical procedures, access is through the mouth via intraoral incisions. Extra-oral surgery, Chapter 10.

Incisions For dentoalveolar surgery are full thickness, i.e. mucoperiosteal flaps; for mucosal and periodontal surgery split thickness flaps are raised (p. 236 and p. 240). For mucoperiosteal flaps, although the base does *not have* to be longer than its length, this design improves the blood supply to the flap and should be used where this is a concern. Improved access via a large flap, allowing minimally traumatic surgery, virtually always outweighs the trauma of additional periosteal stripping.

Always plan the incision mindful of local structures. *One cut*, at right angles, through mucoperiosteum to bone, is the aim. Do not split interdental papillae. Try to cut in the depth of the gingival sulcus. Raise the flap cleanly, working subperiosteally with a blunt instrument, p. 384.

Retraction of the raised flap should be gentle and precise. It is the assistant's duty to prevent trauma to the tissues by sharp edges, overheating drills, or bullish surgeons.

Bone removal by drills must be accompanied by sterile irrigation to prevent heat necrosis of bone, damage to soft tissues, and clogging of the bur. When using chisels to remove bone remember the natural lines of cleavage of the jaws and make stop cuts (chisel technique, p. 400).

Removal of the tooth/root is carried out using *controlled force*.

Debridement is removal from the wound of debris generated by both the pathology and the operation. It is as important as any other part of the operation.

Haemostasis and **Wound closure** are covered on p. 392.

Post-operative oedema is, to some degree, inevitable; it is minimized by gentle efficient surgery, which is more important than such measures as peri- or post-operative steroids.

Asepsis and antisepsis

Asepsis is the avoidance of pathogenic microorganisms. In practical terms, 'aseptic technique' is one which aims to exclude all microorganisms. Surgical technique is aseptic in the use of sterile instruments, clothing and the 'no touch' technique.

Antisepsis is an agent or the application of an agent which inhibits the growth of microorganisms while in contact with them. Scrubbing up and preparation of operative sites are examples of antisepsis. **Disinfection** is the inhibition or destruction of pathogens, whereas **Sterilization** is the destruction or removal of **all** forms of life. Pre-packaged sterile supplies and the use of an autoclave (121 °C for 15 min or 134 °C for 3 min) for resterilizable equipment are the only really acceptable techniques in dentistry. Disinfection using gluteraldehyde or hypochlorite are second choices, for use where true sterilization is not feasible. It is not possible to render the mouth aseptic and it is fruitless to try, there are, however, three basic techniques which are of value:

Avoid introducing infection This is achieved by *always* using *sterilized* instruments, and wearing gloves.

Avoid being infected yourself by the operative site—wear gloves.

Reduce the contaminating load to the site By pre-extraction cleaning of teeth, use of chlorhexidine mouthrinse, and prophylactic antimicrobials, when appropriate.

Cross-infection and its control

Much attention has been focused on this problem in recent years, first with hepatitis B and its related agents, and then with HIV. Although screening is possible in some instances, this is of little real value since the majority of individuals with Hep B and HIV infection *are asymptomatic* and hence not identifiable. Therefore safe practice mandates the use of sound cross-infection control as part of everyday practice on *all* patients.

Aerosols are easily created and are a potential source of cross-infection. Minimize wherever possible by high vacuum suction. Wear glasses and a mask if exposure to an aerosol cannot be avoided. Masks are routine in theatre, although of unproven value.

Cleaning and sterilizing Use disposable equipment when possible and **never** reuse. Clean instruments prior to sterilization. Use disposable or easily disinfected work-surfaces.

Gloves should be worn routinely. Sterile gloves for surgery.

Immunization against Hep B is available. Get it and get all staff with clinical contact to do likewise.

Waste disposal It is everyone's responsibility to ensure sharps are carefully placed in rigid, well-marked containers and disposed of by an appropriate service. Dealing with potentially contaminated impressions and appliances, p. 675. Treatment of the known high-risk patient, p. 734.

Needlestick injuries If this happens to you, rinse wound under running water and record date and patient details. In hospital, follow local policy; in practice, contact local public health laboratory.

Forceps, elevators, and other instruments

Extraction forceps come in numerous shapes and sizes. The choice of forcep is largely down to individual preference, or more frequently, availability. 'Universal' forceps are straight-bladed upper or lower forceps used to grip the roots of teeth to allow a controlled extracting force. 'Eagle beak' forceps are upper and lower molar forceps which engage the bifurcation of molar teeth allowing a buccally directed extraction force. 'Cowhorns' are designed to penetrate the molar bifurcation either to be used in a figure of eight loosening pattern or to split the roots. Most forceps come with a deciduous tooth equivalent.

Elevators are used to dilate sockets to facilitate extraction or to remove dental hard tissue by themselves. These are the instruments which should **always** be used to remove impacted teeth. They should be used with gentle (finger pressure) forces. The commonly used patterns are Couplands No. 1, 2, and 3, Cryers right and left and Warwick-James right, left, and straight.

Scalpel A Bard-Parker handle with a No. 15 blade is the usual.

Periosteal elevators A number are available, the Howarths, originally designed as a nasal rasparatory, is the favourite.

Retractors Tongue, cheek, and flap retractors are needed and are legion in number; Lack's tongue retractor, Kilner's cheek retractor, Bowdler-Henry's rake retractor, and the Minnesota flap retractor are favourites.

Chisels versus burs Depends upon your training. Generally, burs (No. 8 round T–C for bone removal, medium taper fissure T–C for tooth division) are kinder on the conscious patient and the best bet for the inexperienced. Chisels are more appropriate in theatre and are particularly useful (3 mm and 5 mm T–C tipped) for disto-angular third molars and upper third molars.

Curettes The Mitchells (no relation) trimmer is probably the most valuable instrument in this category.

Needle holders and sutures vary more than any of the above depending on your location. The usual suture size for intra-oral work is 3/0, the material may be non-absorbable (silk) or absorbable (catgut, Dexon, Vicryl).

Scissors Remember to keep dissecting scissors, e.g. McIndoe's, separate from suture cutting scissors, and keep both sets sharp.

Dissecting forceps are designed to hold soft tissue without damaging it, Gillies dissectors are popular. College tweezers are *not* dissecting forceps and are used to lift up sutures prior to removing them.

Aspirator Sterile/disposable suction tip small enough to get into the defect.

The extraction of teeth

The extraction of teeth must be viewed as a minor surgical procedure, ∴ the medical history will be pertinent, e.g. bleeding diathesis, at risk from bacteraemia, etc. More common and specific considerations are the sex, age, and build of the patient. Extractions in children are technically simple, it is the child who is most likely to be a problem, whereas stoical old men who may not bat an eye at the procedure often have teeth aptly described as 'glass in concrete'. Malpositioned teeth present problems of access and isolated teeth especially upper second molars tend to be ankylosed. Heavily restored and root-filled teeth tend to be very brittle. In all these cases a pre-extraction X-ray should be taken.

Extraction of teeth begins with *positioning*. After LA has taken, the patient is positioned supine at the height of the operator's elbow for upper teeth and sitting with the operator behind for (right-handed dentists) lower extractions on the right and in front for lower extractions on the left. The position is reversed for left-handers, but unfortunately the world seems biased against this group, and many dental chair systems seem to preclude comfortable positioning for them.

Common technique The socket is dilated either using an elevator between the bone of the socket wall and the tooth or by driving the forceps blades into the socket. The blades of the forceps are applied to the buccal and palatal/lingual aspects of the tooth and pushed either along the root of the tooth or, in certain molar extractions into the bifurcation. The tooth is then gripped in the forceps and, maintaining a consistent and quite substantial *vertical* force, the tooth is moved depending on its anatomy:

<u>1, 2, 3</u> have conical roots—rotate then pull.
<u>4, 5</u> have either two fine roots or a flattened root, move bucco-palatally until you feel them 'give', then pull down and buccally.
<u>6, 7</u> have three large divergent roots—these are moved buccally while maintaining upward pressure, but frequently need a variety of rocking movements before they are sufficiently disengaged to complete extraction.
<u>1, 2, 3</u> can usually be removed with a simple buccal movement, but sometimes need to be rocked or even rotated.
<u>4, 5</u> are rotated and lifted out.
<u>6, 7</u> are 2-rooted and can usually be removed by a controlled buccal movement. Remember to support the patients jaw.

Deciduous teeth are extracted using the same principles, but while permanent molars can be removed using forceps which engage the bifurcation, these should *not* be used on deciduous teeth.

Third molars (p. 398)
As with all operative techniques in dentistry, the doing is worth a thousand words. To become competent there are three golden rules: practice, practice, and practice.

Complications of extracting teeth

Access Small mouths present an obvious, but usually manageable problem. Crowded or malpositioned teeth may need trans-alveolar approach. Trismus, if due to infection is a C/I to out-patient GA and should be managed in hospital where facilities for external drainage and airway protection are available.

Pain Has the LA worked? Try further LA as regional block, infiltration or intraligamentary injection. Is it pain or pressure? If pressure reassure and proceed. If pain and other signs of adequate LA are present then acute infection is the most likely culprit. Can the extraction wait by using delaying tactics such as draining an abscess? The vast majority can and very few adult extractions really justify a GA.

Inability to move the tooth Don't worry; it happens to us all. Have you got an X-ray? If not, get one and look for: bulbous or diverging roots, very long roots, ankylosis or sclerotic bone. **Do not** press on regardless, it will work sometimes but shows lack of consideration and will cost you in time and goodwill in the long run. Most 'solid' teeth have an easily identifiable cause, e.g. diverging roots, and raising a flap and using a trans-alveolar procedure (p. 394) will quickly and easily remedy this.

Breaking the tooth is a common occurrence and may even assist extraction if, e.g. the roots of a molar are separated. More often, unfortunately, the crown fractures leaving a portion of root(s) *in situ*. It is quite acceptable to leave small (<3 mm) pieces of deeply buried apex, but provide antibiotics, tell the patient and review. Larger pieces of root must be removed as they have a high incidence of infective sequelae (p. 408).

# of alveolar and/or basal bone Breaking the alveolar bone is relatively common. If # only involves the alveolus containing the extracted tooth, remove any pieces of bone not attached to periosteum and close the wound. Rarely, the alveolus carrying other teeth will be involved in which case remove tooth by a trans-alveolar procedure and splint remaining teeth (p. 118). Basal bone # is rare; ensure analgesia (LA &/or systemic analgesics) and arrange reduction and fixation (p. 496).

Loss of the tooth Stop and look; in the mouth, is it under the mucoperiosteum or in a tissue space—these can usually be milked out. Look in the suction apparatus. Is it in the antrum (p. 422) or even the ID canal? Has it been swallowed or inhaled? Chest X-ray is mandatory if not found.

Damage to other teeth/tissues and extraction of the wrong tooth Prevent by confirming with the patient the teeth to be removed and making careful notes. Plan the operation, do not use inappropriate instruments or ones which you don't know how to use. If the wrong tooth is extracted, re-plant if feasible and proceed to remove the correct tooth. Tell the patient and make careful notes.

Dislocated jaw Reduce (p. 498); **Bleeding** (p. 390); **Pain, swelling**, and **trismus** are common sequelae and are discussed on p. 400.

Post-operative bleeding

Bleeding disorders are covered on p. 522.

Principles of management of post-operative bleeding.

1 Support the patient. If hypotensive and tachycardic, establish IV access and replace lost blood volume.
2 Diagnose each cause, nature, and site of blood loss.
3 Control the bleeding point.

Classically, post-op bleeding is described as: immediate (primary), reactionary, and secondary.

Immediate When true haemostasis has not been achieved at completion of surgery.

Reactionary Occurs within 48 h of surgery and is due to both general and local rise in BP opening up small divided vessels which were not bleeding at completion of surgery.

Secondary Occurs around 7 days post-op and is usually due to infection destroying clot or ulcerating local vessels. In practice, bleeding following removal of teeth is common and usually simple to diagnose. The patients are seldom shocked or hypotensive but are often very anxious and nauseated by the taste, smell, and sight of blood, and by blood in the stomach, which is irritant. Bleeding usually comes from one or all of three sources: (1) gingival capillaries; (2) vessels in the bone of the socket; and (3) a large vessel under a flap or in bone such as the inferior alveolar artery. The first two are by far the more common.

Management

Reassure the patient they won't bleed to death. Remove accompanying entourage and get the patient to an area with reasonable facilities. Take a drug history (anticoagulants?). *Wear gloves* and apron, they often vomit. (If patient has to wait to be seen they should bite firmly on a clean handkerchief or gauze, rolled to fit the area the bleeding seems to be coming from, and bite firmly on this.) In good light, with suction, clean the patient and their mouth, remove any lumps of clot and identify the source of bleeding. Is it from under a flap? If from a socket squeeze the gingivae to the outer walls of the socket between finger and thumb; if bleeding stops it is from a gingival vessel. In these cases, LA and suturing are needed. If bleeding continues it is from vessels in bone which need some form of pack.

Technique

Give LA if needed, have assistance with suction. If flap involved remove old sutures, evacuate clot, identify bleeding point and place a tight suture around it. Bleeding should be much reduced; if not, repeat until it is, then close wound and have the patient bite on a swab for at least 15 min. If it is a gingival bleed a tight interrupted or mattress suture will compress the capillaries, again followed by a swab to bite on. If bleeding from the depths of the socket, the clot may need to be removed and replaced by a pack or supported by a resorbable mesh (oxidized cellulose) ± agents such as bovine thrombin, adrenalin, or epsilon amino-caproic acid soaked into the mesh. If removing clot and packing the socket, remember this will delay healing and predispose to infection so use BIPP or Whitehead's varnish packs. If all else fails, all the above measures plus a pressure pack, analgesia, a sedative antiemetic, and a night in a hospital bed will do the trick. Patients requiring this degree of treatment should be investigated haematologically and for liver disease.

Suturing

Every dentist should master the basic skills of suturing.

Materials Most sutures are suture material fused to a needle although threaded reusable needles are used in some countries. Needles may be round or cutting, straight, curved, or J-shaped. Almost all intra-oral work is done with a 16–22 mm curved cutting or reverse cutting needle held in a needle-holder. Suture material may be resorbable (catgut, dexon, or Vicryl) or non-resorbable (silk, nylon, prolene, or Novafil). Monofilament suture (e.g. nylon) causes less tissue response than braided (e.g. silk).

Skin best closed with nylon, prolene, or Novafil. *Mucosa* and *Deep tissues* best closed with catgut, dexon, or vicryl. *Vessels* are tied off using resorbables, except major veins and arteries which are transfixed and tied with silk. Black silk sutures (BSS) can be and still are used for skin and mucosa. Suture strength is described as 0 (thickest) to 3/0 (commonest in use intra-orally) to 11/0 (thinnest for microvascular work).

Types of stitch are shown in the diagram.

Suture technique Closure of a wound or incision should, whenever possible, be without tension, by closing deep layers and over supporting tissue. Hold the needle in the needle holder around ⅔rds of the way from its tip. Suture from free to fixed tissue taking a bite of 2–3 mm on both sides. Leave the sutured wound edges slightly everted in apposition. Except when swaging tissue to bone, e.g. when arresting haemorrhage or when tying vessels do not overtighten the suture as wound margins become swollen, and you need to allow for this.

Knot tying The two most useful knots are the square (reef) knot and the surgeons knot (see diagram).

Instrument tying is easy to learn from a book but needs considerable practice to perfect. The knot is started by passing the suture once (square knot) or twice (surgeon's knot) around the tip of the needle holders, the knot is tightened and then locked by passing the suture around the needle holder in the opposite direction once. It is possible to control the suture tension by completing the knot in three loops instead of two.

Hand tying is invaluable for those wishing to develop surgical expertise or involved in major maxillofacial surgery. It takes a substantial amount of time and practice and is impossible to learn from a book. Get a sympathetic senior to demonstrate.

Suture removal is not someone else's job to be casually forgotten about. Do the stitches need to be removed? In inaccessible sites, difficult patients, or areas in which scar quality is less important, a resorbable suture should be used. This saves time and much worry for many patients. Facial skin sutures should be removed between 3–5 days. When removing sutures use *sharp* scissors (avoid 'stitch cutters' if you can), lifting up and

cutting a bit of suture that has been in the tissue, thus avoiding
dragging bacteria through the incision on removal.

393

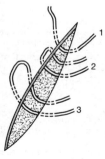

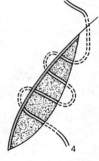

A Reef knot
B Surgeon's knot

1 Simple interrupted suture
2 Horizontal mattress suture
3 Vertical mattress suture
4 Continuous subcuticular suture

Dento-alveolar surgery: removal of roots

Does the root need to be removed? If large, being extracted for pulpal or apical pathology, is symptomatic, is an impediment to denture construction, or is in a patient in whom risk of minor local infection is not tolerable (e.g. immunocompromised or at risk from SBE) then the answer is yes.

Non-surgical methods The use of root forceps or elevators may allow simple removal of roots close to the alveolar margin. When using root forceps ensure the root can be seen to be engaged by the blades. Elevators can be used to direct a root along its path of withdrawal providing (a) one exists, and (b) an elevator can be introduced between bone and root. *Do not waste time persisting with non-surgical methods* if your initial attempt is unsuccessful.

Surgical methods Plan your operation. Do you know why the root cannot be delivered, exactly where it is, and any adjacent structures? If not, get an X-ray.

(a) The flap: if edentulous incise along the crest of the ridge, if dentate in the gingival margin. Flaps may be envelope, 2-sided, or 3-sided. Relieving incisions make reflection of the flap easier but must be avoided in the region of the mental nerve ($\overline{45}$) and are better avoided around the buccal branch of the facial artery (mesial root $\overline{7}$). Include an interdental papilla at either side of the flap and start vertical cuts $\frac{2}{3}$rds of the way distal to the included papilla. Big flaps heal as well as small ones, the important consideration is access.

(b) Identify obstructions to the path of withdrawal of the root, these are either removed or the root is sectioned to create another path of withdrawal, depending on which approach is the least traumatic.

(c) Remove the minimum amount of buccal bone compatible with exposing the maximum diameter of the root and a point of application for an elevator (No. 8 round T–C surgical bur).

(d) Elevate by placing an elevator between bone and tooth (remember to apply to the convex surface of curved roots) and direct the root along its natural path of withdrawal.

(e) Finally, debride and close the wound.

Special cases
1 Small apical fragments; use an apicectomy approach.
2 Multi-rooted teeth; always divide the roots, as this makes life much easier.
3 Cannot find the root; re-X-ray and look in the soft tissues. The root may have been displaced into another cavity or even be in the aspirator, look carefully but remember discretion is the better part of valour. Tell the patient.
4 Patient refuses operation; it's their body, record your advice and their decision.

Under GA Mallet and chisel can replace bur and the 'broken instrument' technique can be used. This involves using a straight instrument or elevator guided through the bone using the mallet, and either then being used as an elevator or being placed in contact with the tooth or fragment, which is delivered by a sharp blow. Use with care.

Outline of 2-sided flap in heavy shade.
× retained root or similar.
a line of additional incision to convert to a 3-sided flap.

Dento-alveolar surgery: removal of unerupted teeth

The teeth most commonly requiring removal other than third molars are maxillary canines and premolars, supernumaries and mandibular canines and premolars. Rarely, permanent or deciduous molars may be impacted or submerged.

Maxillary canines (p. 158) The canine may lie within or across the arch, buccally or most frequently palatally. Assessment requires a careful examination, palpation, and X-rays (either two films at 90° or the parallax technique, p. 160).

Techniques Buccal impactions are approached via a buccal flap, palatal via a palatal flap and cross or within arch impactions, need a combination of the two. Buccal flaps are as previously described (p. 394). Palatal flaps involve the reflection of the full thickness of the mucoperiosteum of the anterior hard palate, the incision running in the gingival crevice from 6 to 6 for bilateral canines or to the contralateral canine region for single impactions. The neurovascular bundle emerging from the incisive foramen is sacrificed, with no noticeable morbidity. **Never** incise the palate at 90° to the gingival crevice and **always** use an envelope flap, otherwise you will section the palatine artery. Remove bone over the bulge of the crown of the tooth until the maximum bulbosity of the crown and the incisal tip are exposed. If the root curvature and path of withdrawal are favourable, the entire tooth can be elevated out. If not, section at the cervical margin with a tapered fissure bur and winkle the pieces out separately. Debride the socket and close with vertical mattress sutures to minimize haematoma formation.

Mandibular canines Most lie buccally, and can be elevated or removed with root forceps. Unerupted, deeply impacted buccal or lingual canines rarely need to be removed. If necessary a degloving incision provides good access.

Maxillary premolars Most lie palatally. If partially erupted and conically rooted, they are simply elevated. Otherwise, a similar approach to that used for palatal canines is used. Premolars within the arch are approached buccally, sectioned, and removed piecemeal.

Mandibular premolars are often angled lingually. The 'broken instrument' technique can be invaluable. Otherwise, an extended buccal flap is raised to visualize and protect the mental nerve, buccal bone is removed, and the tooth sectioned at its cervical margin. The crown is then displaced downwards into the space created and the root elevated upwards.

Submerged deciduous molars If they need to be removed are approached buccally and sectioned vertically then elevated along the individual roots path of withdrawal.

Supernumeraries (p. 72) These are removed using the approach used for the tooth they impede or replace. This can be a surprisingly difficult operation, usually because of difficulty in finding and identifying the thing.

Dento-alveolar surgery: removal of third molars

Not all wisdom teeth need to be removed.[1] Those which have space to erupt into functional occlusion should be left to do so, and those which are deeply impacted and asymmptomatic are best left alone. Decisions about surgery vary widely.[2]

Aetiology As the last tooth to erupt, the third molar is most liable to be prevented from doing so in a crowded mouth. *Causes*: soft western diet doesn't create space by contact point abrasion, inherited tooth-jaw size incompatibility, and possibly an evolutionary tendency towards decreased jaw size.

Symptoms Pain, swelling, pericoronitis (p. 408), and sometimes a foul taste. Mostly due to localized infection, less commonly due to caries or resorbtion of second molar or cyst associated with third molar. Third molars covered by bone are very unlikely to become infected, whereas those where the crown has breached mucosa will almost inevitably do so. Infections in older, less vascular, bone are more difficult to treat.

398

Indications for removal Recurrent pericoronitis, unrestorable caries in $\overline{7}$ or $\overline{8}$, external or internal resorption, cystic change, periodontal disease distally in $\overline{7}$. The prevention of LLS crowding on its own is **not** an indication nor is vague TMJ pain.

Timing Symptoms are most common in the late teens and twenties and bone is soft, spongy and elastic in this age group, so this is the usual and most favourable time to operate. Prophylactic removal of third molars is justifiable only if symptoms in later life can reasonably be expected.

Choice of anaesthetic Bilateral impacted third molars are more kindly treated under GA, as are those where surgery may be technically more difficult, e.g. disto-angular $\overline{8}$. LA ± sedation is appropriate for most unilateral impactions not presenting particular difficulty. Consider medical history.

Assessment Look first, unerupted or partially erupted? Take pre-op X-ray (DPT is ideal).
Assess: angulation (vertical, mesio-angular, disto-angular, horizontal, or transverse); depth of impaction from the alveolar crest to the maximum diameter of the crown; degree of impaction (and against what); root shape; bone density; the relationship to the ID canal; and the presence of any other pathology or complicating factors.

Plan: the path of withdrawal. What impedes it? How much bone needs to be removed to provide a point of application for an elevator? How much to clear the path of withdrawal? Can this be done less traumatically by sectioning the tooth? In what direction?

Warn: the patient about pain, trismus, swelling, and the possibility of damage to the ID and lingual nerves.

1 J. Shepherd 1993 *BDJ* **174** 85.   **2** M. Brickley 1993 *BDJ* **175** 102.

399

Dento-alveolar surgery: third-molar technique

Technique

Mandible A buccal flap is raised along the external oblique ridge (lies well lateral to the arch) over the crest of the ridge if unerupted, or in the gingival margin if partially erupted. Extend to the distal aspect of the second molar and down into the buccal sulcus. Cut on to bone to raise a full thickness flap. If a mesio-angular or horizontal impaction extend the incision around $\overline{7}$ to the mesial border, but beware the buccal branch of the facial artery. Reflect and retract flap. If bone removal or tissue handling is from anywhere other than the buccal aspect, the lingual nerve should be protected by passing a smooth blunt instrument (e.g. Howarth's) *subperiosteally* and reflecting a lingual flap. Remove bone to provide a point of application for an elevator and clear obstruction to the path of withdrawal. This can be done with chisels or a bur. With a bur this is done by creating a disto-buccal gutter exposing the maximum bulbosity of the tooth. If needed the bur can then be used to section the $\overline{8}$ to provide an unimpeded path of withdrawal.

With chisels, place mesial and distal stop cuts around the tooth and split off a collar of buccal and distal bone to expose the bulbosity of the crown. This can then be extended into a lingual split, taking off a piece of lingual plate and allowing the tooth to be elevated lingually. Whichever technique is used, it is important to remember that if it has been done properly a minimum of directed force via an elevator will deliver the tooth or fragment out along its path of withdrawal. If this cannot be done, look carefully for another obstruction. Once the tooth is removed debride the socket, remove the follicle, and close the wound with loose sutures to allow for swelling. Achieve haemostasis with pressure.

Maxilla A flap similar to that in the mandible can be used for inaccessible wisdom teeth but many can be approached using a 'slash' incision (from disto-palatally on the tuberosity to disto-buccally at the second molar and into the buccal sulcus). Reflect and retract the flap. Bone removal can usually be effected by a hand-held chisel. The much softer bone rarely causes any problem with elevation, but care has to be taken to prevent displacement into the pterygoid space. The 'slash' incision often needs no closure whereas a conventional incision will need repositioning with sutures.

Post-operatively

Pain can be quite severe and responds best to NSAIDs. Peri-operative LA works well and leaves patients pain-free but numb for the duration. Trismus is due to pain and muscle spasm and can be reduced by adequate analgesia, p. 572. It has obvious but often forgotten connotations for meals post-op. Swelling can also be reduced by high-dose perioperative steroids. There is no evidence that ice-packs help, although they are often used. Haemorrhage can usually be controlled by biting on pressure packs. Rarely, it may be necessary to re-explore the wound.

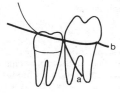
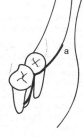

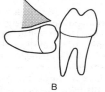

diagram—after Moore (1976)
a outline of incision for raising a third-molar buccal flap.
b modification to create an envelope flap.
A vertically impacted third molar.
B horizontally impacted third molar.
C mesio angular impaction of a third molar.
D disto angular impaction of a third molar.

Dento-alveolar surgery: apicectomy

There are four surgical aids to endodontics: apicectomy, root hemisection, endodontic (diodontic) implants, and removal of extruded endodontic paste.

Apicectomy

This is by far the commonest. It is a *second line* treatment after failure of, or as a supplement to, orthograde endodontics.

Indications
- impossible to prepare and fill apical $\frac{1}{3}$rd of tooth, e.g. pulpal calcification, curved apex, open apex
- broken instrument in canal
- post crown on tooth with apical pathology
- root perforation
- fractured and infected apical $\frac{1}{3}$rd
- persistent infection due to apical cyst

Assessment Intra-oral X-ray, best possible root filling *in situ*, free from acute infection.

Technique This operation is best performed under LA. Ensure an area of two tooth widths either side of the tooth being treated is anaesthetized, and give palatal infiltration. In the mandible, give a block plus infiltration to aid haemostasis. If associated with infection, prophylactic antibiotics, e.g. amoxycillin 1 g 1 h pre-op and 6 h post-op or metronidazole 400 mg 1 h pre- and 6 h post-op can help.

Flaps may be 2- or 3-sided, semilunar, or sublabial, the latter two avoid post-operative recession but give inferior access. Reflect and retract well above the apex (there is often a bulge or perforation of the cortical plate to aid location of the apex). A bony window is created to visualize the apex, which is often found sitting in a mass of granulation tissue. Excise the apical 1–2 mm and currette out the cystic and granulation tissue. Pack the cavity with bone wax or adrenalin-soaked ribbon gauze, identify the canal and prepare it with a $\frac{1}{2}$-round bur. Seal with amalgam or hot gutta-percha and debride. Close, using interrupted or vertical mattress sutures. An alternative is to perform a 'pull-through' procedure, where the canal is prepared for an orthograde root-filling at the same time as the apicectomy and forced through the apex at surgery.

Special points in apicectomy

Warn patients about post-operative swelling. Lower incisors present an access problem eased but not erased by a degloving incision and experience. Think twice about the mental nerve in lower premolar apicectomies. Think hard about alternatives to apicectomy of $\overline{6}$. Don't think at all about $\overline{7, 8}$. Remember the buccal and palatal roots in 4; section the buccal root low to see the palatal. Apicectomy of $\overline{6}$ is fine provided the palatal root can be treated by an orthograde approach, is hemisected or you are happy to deal with breaching the antrum.

diagram—after Moore (1976)

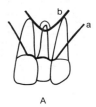

A   B

A an approach to apicectomy.
a outline of incision for 3-sided flap, good access best flap for the novice.
b outline of semilunar flap incision.

B an approach to apicectomy, a window is created in the buccal cortex to expose the apex, which is resected, leaving a smooth raw bony cavity.

Dento-alveolar surgery: other aids to endodontics

Root perforations are approached as for apicectomy; however, multirooted tooth perforations, unrootfillable roots or untreatable periodontal pockets may be dealt with by

Root hemisection This simply involves raising a flap around the tooth, identifying and horizontally sectioning the root and atraumatically elevating it out. The wound is closed and a cleanable undersurface sealed with amalgam left.

Endodontic (diodontic) implants have not received widespread acceptance. They can be used to secure an anterior tooth which lacks sufficient supporting bone after endodontics. A sterile alloy implant passes through the prepared root canal into periapical bone and is impacted into the bone transfixing the tooth. Such implants are being superseded by single-tooth implants, p. 428.

Removal of extruded paste Usually, all that is required is an apicectomy approach. However, careless use of 'paste only' techniques can result in paste in the floor of the nose, the antrum, or the ID canal. The nasal floor can be approached sublabially or intranasally, the antrum by standard methods (p. 422) and the ID canal by saggitally splitting the buccal cortex.

Prognosis Single-rooted tooth apicectomies should succeed >90% of the time. Multirooted teeth, revision apicectomy, and perforation repair have a much lower success rate.

Dento-alveolar surgery: helping the orthodontist

Many minor oral surgical procedures, e.g. extraction of 4s or removal of $\overline{8}$ are carried out at the instigation of an orthodontist. This page concerns itself with the specific procedures of; frenectomy, pericision, tooth exposure, and tooth repositioning.

Frenectomy This procedure is of value in closing a median diastema only if gentle traction on the upper lip and fraenum produces blanching in a palatal insertion around the incisive papilla. It follows that the excision of the frenum must include those fibrous insertions, which leaves a raw area of alveolus after excision—this can be dressed with surgicel, BIPP, or a periodontal pack. It is a different operation from preprosthetic fraenectomy and is performed for a different reason.

Pericision is simply incising supra-alveolar periodontal fibres to prevent relapse when derotating teeth.

Tooth exposure Orthodontic traction is the treatment of choice for malpositioned, unerupted canines and incisors if the apices are in good position for eruption. The essential aspect of the operation is to remove any sacrificable impediments to tooth movement. Whether brackets are bonded to teeth at the time of surgery or following removal of the pack is a matter for individual preference.

Technique Palatal teeth are exposed by a palatal flap. Remove bone carefully with chisels, expose the greatest diameter of the crown and the tip. (Moving the tooth is counter-productive, ∴ don't do it). Excise palatal mucoperiosteum generously; it grows back, bond a bracket if you're going to. Firmly pack the wound with, e.g. Whitehead's varnish and ribbon gauze and secure, or use an acrylic dressing plate with periodontal paste dressing. Close the remainder of the flap with vertical mattress sutures. Buccally located teeth are approached by a buccal flap, in order to preserve attached gingiva, and bonding should be done at operation. The flap can be repositioned coronally with the elastics tunnelling subgingivally. Teeth within the arch are approached buccally removing crestal bone as needed.

Tooth repositioning (transplantation) Although there are claims of success rates as high as 93%, few people match this and most would transplant only when exposure and orthodontic movement were rejected. The most commonly transplanted tooth is the maxillary canine. It is essential to measure the available space and compare this with the erupted contralateral tooth or a good X-ray estimation, as it is not acceptable to grind down healthy teeth at operation to accommodate the retrieved tooth. If the tooth appears to be too big for the available space then orthodontic Rx is required to create space. As this is often the reason the patient rejected exposure, an impasse is sometimes reached.

Technique The tooth is exposed by buccal or palatal flap, and once it is certain that it can be removed atraumatically, the deciduous tooth, if present, is extracted and a new socket is surgically prepared with a bur. The tooth is reimplanted without force, the flaps sutured, and a close-fitting but **not** cemented splint placed. Functional splinting is continued for 7–10 days and the tooth root-filled as soon as possible after surgery. Regular follow-up is essential to allow early detection of root resorption.

diagram—after Moore (1976)

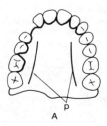

A

A the palatal flap, outline (in heavy print) of the incision for a palatal flap raised to expose a buried right maxillary canine.

p position of the palatine arteries.

Dento-facial infections

▶ Infection associated with teeth is rarely, if ever, treated *definitively* by antibiotics and analgesics.

The vast majority of infections in this area requiring surgical treatment are bacterial, usually arising from necrotic pulps, periodontal pockets, or pericoronitis. Rarely, can be life-threatening if allowed to progress, e.g. to the fascial spaces of the neck or the cavernous sinus, or as a focus for subacute bacterial endocarditis (p. 524).

Microbiology Culture of dentofacial infections usually produces several commensal organisms, of which anaerobes are the most important. The predominant species *Bacteroides* (anaerobe) and *Streptococci* (aerobe and anaerobe) are usually sensitive to the penicillins. Resistance is reported rarely. *Bacteroides* is nearly always sensitive to metronidazole.

Diagnosis is usually simple and clinical based on pain, swelling, temperature, and discharge.

Apical abscess Teeth with an apical abscess are TTP and non-vital. They may be discoloured or crowned and have a past history of trauma or RCT. Pain and tenderness to percussion are often diminished when the intrabony pus tracks through the soft tissues and discharges, usually in the buccal sulcus (exceptions are $\underline{2}$, and palatal roots of maxillary molars which discharge palatally, and $\overline{12}$ which often discharge on the chin).

Rx Drainage of pus either via the root canal, incision of any fluctuant abscess or extraction under LA or GA. Palatal or buccal abscesses can be drained quite simply under LA by infiltrating a small amount of LA *between* the abscess cavity and the overlying mucosa, then incising the abscess. Explore using blunt closed forceps and keep patent either by excising an elipse of tissue or inserting and suturing a small rubber drain; this is particularly important in the palate. Cover the procedure with 'best guess' antibiotics such as amoxycillin 500 mg PO continued tds for 5–7 days, or metronidazole 400 mg PO continued bd for 5–7 days, or both.

Periodontal abscesses arise in a pre-existing periodontal pocket (p. 220). Initial treatment involves incision and drainage, followed by elimination of the pocket, unless extraction is considered the only option.

Pericoronitis is inflammation and infection of a gum flap (operculum) overlying a partially erupted tooth, usually a $\overline{8}$ often traumatized by an overerupted $\underline{8}$. Rx: involves removal of the opposing $\underline{8}$, irrigation under the operculum with saline or chlorhexidine, and antibiotics (see above) if necessary. Nearly all third molars associated with pericoronitis need removal.

Dry socket is osteitis of a socket following tooth removal. Commonest in the mandible after removal of molars, especially $\overline{8}$. Predisposing factors are smoking, surgical trauma, LA, past history, bone disease, the pill, or immunodeficiency.

Diagnosis: pain onset after (usually 2–4 days) extraction, similar in nature but *worse* than the preceding toothache. The socket looks inflamed and exposed bone is usually visible. Rx is to gently clean the socket by irrigation and dress the exposed bone. Numerous preparations are available, e.g. ZOE packs (provide remarkable analgesia but are a horror to remove), and topical metronidazole. Chlorhexidine &/or hot salt mouthwashes may help. NSAIDs are the systemic analgesic of choice, p. 602. Prophylactic anaerobicidals such as metronidazole reduce the incidence of this condition.[1]

Actinomycosis (p. 202) persistent low grade infection, multiple sinuses. Rx: Drainage and up to 6 weeks amoxycillin 500 mg tds. Doxycycline 100 mg od is an alternative.

Staphylococcal Lymphadenitis especially seen in children, small occult skin or mucosal breach allows ingress. May mimic a 'slapped face' due to exotoxin. Drain and give flucloxacillin 125–500 mg qds (depending on age p. 128).

Atypical mycobacteria Lymphadenitis with no obvious cause. Cold nodes, non-febrile patient. Drain or excise. Culture for up to 12 weeks. Do **not** start antituberculous therapy as many atypical mycobacteria are resistant and side-effects are common and significant.

409

Biopsy

A biopsy is a sample of tissue taken from a patient for histopathological examination.

Types of biopsy Biopsies may be incisional or excisional. Examples of incisional biopsies are fine needle aspirate biopsies, punch biopsy, trephines, and 'true-cut' needle biopsy. The commonest technique by far, however, is to excise an ellipse of tissue including a portion of the lesion and surrounding normal tissue. Excisional biopsy provides after the fact information on the excised sample (reserve for lesions <1 cm).

What should be biopsied? Nearly everything that is worth excising is worth histological review, and so all excised specimens should be examined histopathologically. Any soft tissue lesion not amenable to accurate clinical diagnosis (by a reasonably trained eye) should be biopsied. All red lesions of oral mucosa and most white patches should be biopsied. If you think 'should I biopsy this?' then do it; you will always get some unpleasant surprises.

Special considerations Frozen sections are biopsy specimens taken during major surgery, either when the extent of the procedure will depend on the histological diagnosis of the lesion or to verify clearance of excision. It is essential to contact the pathology lab before the patient goes to theatre to warn them. Advance warning is also necessary for certain special tests, e.g. immunohistochemistry.

How it is done Tell the patient you need a piece of tissue to help make the diagnosis. LA or GA. For simple incisional biopsy, stabilize the tissue to be sampled. Transfixing with a 3/0 BSS helps, avoids crush artefact, and orientates the specimen. Cut an ellipse of tissue including lesion and normal surrounding tissue, lift up and dissect out, then close primarily with sutures.

Biopsy and oral cancer Incisional biopsy carries a (theoretical) risk of shedding malignant cells into the circulation. The alternative is to subject a patient to mutilating surgery before definitive diagnosis. If you suspect an oral malignancy, refer before biopsy because most consultants have a preferred approach, often involving geographic biopsy at the EUA stage.

Specimens are best laid out on paper if small or pinned out on a cork board if large. This allows orientation and decreases shrinkage artefact. Usual preservative is 10% formalin; ask if you are not sure, as some specimens are needed fresh. Consider a specimen for culture as well (e.g. lymph node biopsy). Make a diagram of the specimen on the pathology form to accompany the clinical details.

Cryosurgery

Cryosurgery is the therapeutic use of extreme cold.

Equipment The coolants (usually nitrous oxide or liquid nitrogen) act via a cryoprobe, a tubing system in which the coolant is not in direct contact with the tissues. The probe tip is applied to the lesion with an intervening layer of lubricant jelly, this gives rise to an 'iceball', which is essential for success. Liquid nitrogen can also be directly sprayed on to lesions and, rarely, bone can be immersed into liquid nitrogen prior to reuse as a framework for grafting.

Mechanism Cell death and subsequent tissue necrosis by cellular disruption, dehydration, enzyme inhibition, and protein denaturation follows application of extreme cold. Indirect effects include vascular stasis and an immune response.[1] There is a curious lack of infection and scarring following cryosurgery.

Indications Vascular malformations (haemangioma) respond very well. Areas of leukoplakia unsuitable for excision, and provided they have not already undergone malignant change, are often responsive. Occasionally, malignant change following cryosurgery of leukoplakia is reported, so some controversy surrounds the technique. Extensive hyperplastic lesions, e.g. palatal hyperplasia under F/-, may respond. Viral warts respond in most instances and some advocate its use for mucoceles. Superficial basal cell carcinoma is frequently treated with cryosurgery, usually liquid nitrogen spray, although its use in more aggressive malignancy is controversial, as is its use following enucleation of keratocysts. Intractable facial pain is one of the more accepted uses, the freezing of peripheral nerves being followed by a period of analgesia which extends beyond the original post-operative numbness.

Technique
- Warn the patient about the procedure, post-op oedema (which can be severe), and a slough which forms over frozen sites. There is sometimes depigmentation of skin lesions.
- LA for larger lesions or if biopsy needed. (LA may ↑ effectiveness of iceball.)
- Select a probe-tip suitable for the lesion, overlap ice zones if the lesion is large.
- Use 'KY' jelly to improve contact between probe and tissues.
- Usual freeze–thaw cycles are around 1 min, repeated at least twice.
- Do not remove probe until defrosting has occurred.
- Careful follow-up and check the histology of the lesion, except when using cryoanalgesia.

Simple analgesics and chlorhexidine mouthwash post-op often help.

The treatment of frank malignancy by cryosurgery remains controversial, but freeze–thaw cycles used must be in excess of the usual (up to 3 min) and tissue temperatures monitored.

1 V. Popescu *J Max-fac Surg* **8** 8.

412

Non-tumour soft-tissue lumps in the mouth

Abscess (p. 408). Generalized gingival swelling and gingivitis (p. 212).

Brown 'tumour' is not a tumour but a giant cell lesion sometimes found in soft tissue but more commonly within bone (p. 416). It occurs 2° to hyperparathyroidism, although this diagnosis is usually suggested after enucleation on finding giant cells in a fibrous stroma histologically. Check bone biochemistry (Ca^{2+}, $PO^{4+++}\downarrow$, alkaline, phosphatase $\uparrow$). If hyperparathyroidism is confirmed and treated, these lesions regress.

Dermoid cyst is a developmental cyst commonest at the lateral canthus of the eye, but next most often found in the midline of the neck above mylohyoid where it causes elevation of the tongue. Rx: conservative excision.

Congenital epulis by definition present at birth; usually presents as a pedunculated nodule. Histology reveals large granular cells. Rx: conservative excision.

Peripheral giant cell granuloma (giant cell epulis) is a deep red gingival swelling, probably caused by chronic irritation. Histology reveals a vascular lesion with multinuclear giant cells. Rx: excision with stripping of periosteum and curettage of underlying bone.

Pregnancy epulis An $\uparrow$ inflammatory response to plaque during pregnancy causes a lesion indistinguishable from a pyogenic granuloma. Onset usually in 3rd month. Rx: none (other than OHI) if possible, as it regresses after delivery. If very troublesome, simple excision; but it may recur.

Pyogenic granuloma is a red fleshy swelling, often nodular, occurring as a response to recurrent trauma and non-specific infection. Histology shows proliferation of vascular connective tissue, $\therefore$ bleeds easily. Rx: excision, debride area, good OH.

Fibroepithelial polyp is an over-vigorous response to low grade recurrent trauma. May be cessile or pedunculated and range from small lumps to lesions covering the entire palate. Excise with base. Histology shows dense collagenous fibrous tissue lined by keratinized stratified squamous epithelium.

Irritation (denture) hyperplasia is a very common hyperplastic response to repeated trauma e.g. following denture induced ulceration. Classically, seen as rolls of tissue in the sulcus related to a denture flange. Histology is similar to fibroepithelial polyp. Rx: Complete excision with temporary removal of dentures allows healing. Consider simple pre-prosthetic measures and replace F/F.

Mucoceles are usually mucous extravasation cysts, where saliva leaks from a traumatized duct and pools creating a compressed connective tissue capsule. Rarely, they are mucous retention cysts. Mostly affect lower lip—similar swellings in the upper lip are often minor salivary gland tumours (p. 506). Rx: excision with associated damaged glands and duct.

Ranula are mucoceles of the floor of the mouth, arising from the sublingual gland. Tends to recur if marsupialized. A plunging ranula crosses deep to mylohyoid and appears as a neck and floor of mouth swelling. Rx: excision of cyst and associated sublingual gland.

Granulomata Lumps characterized by the histological finding of granulomata may be caused by Crohn's disease (p. 464), sarcoidosis (p. 526), or implanted foreign bodies such as amalgam.

Haemangioma is a developmental lesion of blood vessels. Present at birth, they can grow with the child, remain static or regress. Blanch on pressure. Small lesions respond to cryotherapy. Do **not** biopsy.

Lymphangioma is a rarer developmental lesion, this time of lymphatics. May present as an enlarged tongue or lip. Rx is difficult; some can be beneficially excised.

Warts/squamous papillomata Main aetological factor is human papilloma virus (HPV). True warts are rare in the mouth and are usually transmitted from skin warts. ↑ in those with STD or AIDS.

Papillomas are common in the mouth, appear as multiply papillated pink or white asymptomatic lumps. Rx: excision biopsy (if on a stalk—ligate or diathermy base as they contain a prominent vessel).

Non-tumour hard-tissue lumps

Cysts (p. 418); benign tumours (p. 420); malignant tumours (p. 450).

Tori are bony exostoses found in both jaws. *Torus palatinus* is found in the centre of the hard palate. *Torus mandibularis* on the lingual premolar/molar region of the mandible. Rx: reassurance that these developmental anomalies cause no harm (they are *not* part of the Gardener syndrome, p. 744). Rarely, excision for denture construction is indicated.

Giant cell granuloma (p. 414) This can present as an intrabony swelling or symptomless radiolucency. Carefully enucleate.

Brown 'tumour' (p. 414) Again, imitates the giant cell granuloma; difference is in bone biochemistry.

Paget's disease of bone is relatively common over the age of 55 and affects the skull, pelvis, and long bones, as well as the jaws. Although aetiology is uncertain both the measles and respiratory syncytial virus have been implicated. The maxilla is more frequently affected than the mandible. Hypercementosis of roots makes extractions difficult in this group. There is a replacement of normal bone remodelling by a chaotic alternation of resorption and deposition, with resorption dominating in the early stages. Bone pain and cranial neuropathies can occur. X-rays show a 'cotton wool' appearance. Biochemistry shows an ↑ alkaline phosphatase and urinary hydroxyproline. Avoid GA, use prophylactic antibiotics and plan extractions surgically. Diphosphonates and calcitonin are used in treatment.

Fibrous dysplasia Areas of bone are replaced by fibrous tissue. Onset in childhood; ossifies and stabilizes with age. Jaw involvement usually presents as a painless hard swelling. Characteristic X-ray appearance is of 'ground glass' bone. Histology shows fibrous replacement of bone with osseous trabeculae which look like irregular 'Chinese characters'. **Rx** Skeletal resculpting after stabilization of growth ± orthognathic surgery/orthodontics.

Cherubism Hereditary, and presents between 2–4 yrs. Is a bilateral variant of fibrous dysplasia. In addition to the histological pattern for fibrous dysplasia there are also multinucleated giant cells. The treatment is similar.

Cysts of the jaws

Cysts are abnormal epithelium lined cavities which often contain fluid but only contain pus if they become 2° infected. Jaw cysts predominantly arise from odontogenic epithelium and grow by a means not fully understood but involving epithelial proliferation, bone resorption by prostaglandins, and variations in intracystic osmotic pressure.

Diagnosis

Many are detected as asymptomatic radiolucencies on X-ray, others present as painless swellings, almost always of the buccal cortex. Infected cysts present with pain, swelling, and discharge. Vitality test associated teeth. Take a DPT and a periapical film when possible to screen for size and coexisting pathology. Transillumination rarely helps, but aspiration is sometimes useful and can help distinguish some lesions. Rarely, cysts may present with a pathological # especially of the mandible.

Treatment

(a) Enucleation with 1° closure is commonest and generally the Rx of choice. It consists of removing the cyst lining from the bony walls of the cavity and repositioning the access flap. Any relevant dental pathology is treated at the same time, e.g. by apicectomy.

(b) Enucleation with packing and delayed closure, is used when badly infected cysts, particularly very large ones, are unsuitable for 1° closure. Pack with Whitehead's varnish or BIPP.

(c) Enucleation with 1° bone grafting. Rarely useful.

(d) Marsupialization. This is the opening of the cyst to allow continuity with the oral mucosa; healing is slower than with enucleation and a cavity persists for some time. It is useful to allow tooth eruption through the cyst or where enucleation is C/I.

▶ Always submit cyst lining for histopathology.

Types of cysts

Many classifications exist, few are helpful.

Inflammatory dental cysts are very common. Described as apical or lateral depending on position in relation to tooth root, or residual if left behind after tooth extraction. Necrotic pulp is the stimulus, and the epithelium comes from cell rests of Malassey. Rx: enucleation plus endodontics or extraction.

Eruption cysts, p. 68.

Dentigerous cysts form around the crown of an unerupted permanent tooth and arise from reduced enamel epithelium. May delay eruption. Rx: marsupialization or enucleation, depending on position.

Keratocysts are lined by parakeratinized epithelium derived from the remnants of the dental lamina, and have a fluid filling with a protein content <4 g/dL. Aspiration of samples for biochemistry and cytology for parakeratinized squames can be helpful. It is important to identify these cysts, as outpouching walls and 'daughter' or 'satellite' cysts make them more liable to recur. Their multiloculated appearance on X-ray may confuse them with an ameloblastoma (p. 420). Rx: consists of careful enucleation, +/− cryotherapy or aggressive curretage of the cavity. Rarely, excision is needed if recurrent.

Calcifying epithelial odontogenic cysts are rare and distinguished by areas of calcification and 'ghost cells' on histology. Enucleate.

Solitary bone cysts are usually an incidental finding on X-ray and are devoid of a lining, but may contain straw-coloured fluid. They probably arise following breakdown of an intraosseous haematoma, and are distinguished by a scalloped upper border on X-ray where the cyst pushes into cancellous bone between teeth but spares the lamina dura. Opening the cyst, gentle currettage, and closure heals these 'cysts', associated teeth need no Rx.

Aneurysmal bone cysts are expansile lesions full of vascular spongy bone. Presents as a symptomless swelling, unless traumatized, when bleeding causes pain and rapid expansion. Small ones can be carefully enucleated, but larger aneurysmal bone cysts need excision and possible reconstruction since they will recur if incompletely excised.

Fissural cysts are not associated with dental epithelium but arise from embryonic junctional epithelium. They are rare and include incisive canal cysts, incisive papilla cysts, and nasolabial cysts. Rx: by enucleation.

Benign tumours of the mouth

Non-odontogenic tumours

EPITHELIAL

Squamous cell papilloma (p. 415) resembles a white or pink cauliflower and is caused by papilloma virus. Usually presents on the palate. Does not undergo malignant change. Excise.

CONNECTIVE TISSUE

Fibroma Very rare. Is a benign fibrous tumour, usually pink and pedunculated. Excise with a narrow margin.

Lipoma A soft, smooth, slow growing, yellowish lump composed of fat cells. Enucleate or excise with narrow margin.

Osteoma Smooth, hard, benign neoplasm of bone. Usually unilateral and covered by normal oral mucosa. Not situated in the classical position of tori (p. 416). Gardener syndrome (p. 744).

Neurofibroma Rare tumour of the fibroblasts of a peripheral nerve. Usually affects the tongue; may be part of von Recklinghausen disease (p. 747). Can undergo sarcomatous change. Excise with a small margin.

Neurolemmoma (schwanomma) is a tumour composed of Schwann cells (cells of the axonal sheath).

Granular cell myoblastoma is a rare tumour of histiocyte origin which usually arises as a nodule on the tongue. Excise with a margin.

Ossifying fibroma may be a neoplasm or a developmental anomaly. It is a well-demarcated fibro-osseous lesion of the jaws. Presents as a painless slow-growing swelling, expanding both buccal and lingual cortices. X-ray shows a radiolucent area, circumscribed by a radiopaque margin. Histology is similar to fibrous dysplasia. Enucleation or conservative excision is curative. A faster growing but equally benign version occurs in children.

Odontogenic tumours

Many of these are fascinating (to some) rarities. Only the more important are discussed.

Ameloblastoma One of the commoner odontogenic tumours. Commonest in men and Africans, and in the posterior mandible. There are three basic types: unicystic, polycystic, and peripheral. The unicystic type is the least aggressive, the polycystic and peripheral types showing a tendency to invade surrounding tissue, whereas the unicystic expands it. Metastases are very rare. Histologically, 2 types are seen: plexiform and follicular. Rx: unicystic can be enucleated provided a rim of enclosing bone is removed as well, the other types require excision with a margin.

Adenoameloblastoma tends to occur in the anterior maxilla in females. Rx: conservative excision, as recurrence is not a problem.

Calcifying epithelial odontogenic tumour (Pindborg tumour)
Characteristically, a radiolucency on X-ray with scattered
radiopacities. Needs excision with a margin.

Myxoma can occur in both hard and soft tissues. Those arising
in the jaws are tumours of odontogenic mesenchyme. This is a
tumour of young adults arising within bone and can invade the
surrounding tissue extensively. Characteristically, has a 'soap
bubble' appearance on X-ray. Histology reveals spindle cells in a
mucoid stroma. These tumours need excision with a margin of
surrounding normal bone.

Ameloblastic fibroma Rare, affects young adults and appears as
a unilocular radiolucency on X-ray causing painless expansion
of the jaws. Enucleation is usually curative.

Odontomes are not true neoplasms but are malformations of
dental hard tissues. Classically, they are classified as *compound*
when they are multiple small 'teeth' in a fibrous sac, and
complex when they are a congealed irregular mass of dental
hard tissue. These are best regarded and treated as unerupted,
malpositioned, or impacted teeth and removed using standard
dento-alveolar techniques when required (p. 396).

421

Disturbances in tooth formation can lead to isolated abnormalities of enamel, dentine, and cementum. Cementomas are
worthy of mention because they create extreme difficulty in
tooth removal. Dens in dente, p. 76.

The maxillary antrum

These are the largest of the four paired paranasal air sinuses, lying in each half of the maxilla between the alveolus inferiorly, nasal cavity medially, and orbits superiorly.

Antral pathology often mimics symptoms attributable to maxillary teeth. Diagnosis is by exclusion of dental pathology, nasal discharge or stuffiness, tenderness over the cheeks and pain worse on moving the head. Occipito-mental X-rays (15° and 30°) may reveal antral opacity, fluid level, or # (p. 462). To define fluid level, repeat film with head tilted. Other X-rays: DPT for cysts and roots, lateral skull for roots, and CT scans for tumours and blow out #.

Extractions and the antrum The proximity of maxillary cheek teeth to the antral floor makes it easy for roots and even teeth to be displaced into the antrum. It also predisposes to # of the alveolar process during 6 7 8 extraction. Displaced roots can be retrieved either by an extended transalveolar approach similar to that for removing roots (useful when the roots are lying under the antral lining) or via a *Caldwell–Luc* approach.

Maxillary sinusitis

Acute sinusitis usually follows a viral URTI which is ↓ cilia activity, and is due to bacterial superinfection (usually mixed; anaerobes, haemophilus, staphylococci, and streptococci). Less commonly, due to foreign body, e.g. roots, water. Poor drainage via the osteum exacerbates the situation. Diagnosis as above and confirmed by proof puncture. Rx: erythromycin 500 mg PO qds or doxycycline 100 mg PO od or co-trimoxazole 960 mg PO bd. Decongestants, e.g. ephidrine 0.5–1% nasal drops, menthol inhalations. **Chronic sinusitis** may then develop, particularly if a foreign body or poor drainage is present. Mucosal lining hypertrophies and may form polyps. A post nasal discharge (drip) is often present. Rx: is aimed at reventilation of the sinus. Foreign bodies, if present, should be removed via an incision in the canine fossa (above the premolars) and creation of a bony window into the antrum (Caldwell–Luc). Ventilation is provided either by intranasal antrostomy or (ideally) by endoscopic enlargement of the osteum ± drainage of anterior ethmoids depending on cause (functional endoscopic sinus surgery—FESS).

Oro-antral fistula

Is the creation of a pathological epithelium-lined tract between the mouth and maxillary sinus. This most often occurs following the extraction of isolated molar teeth when the fistula tends to persist. Post-extraction reflux of fluids into the nose or minor nosebleeds are a diagnostic pointer. Confirm by getting patient to attempt to blow out against a closed nose; air bubbles through the fistula. Occasionally antral mucosa prolapses through the socket. Rx: Many small fistulae are asymptomatic and close spontaneously. Closure if diagnosis is made at time of

423

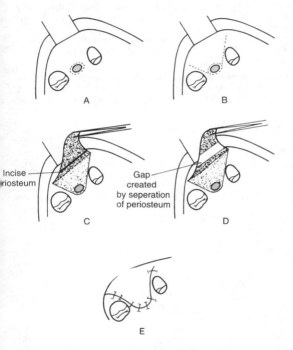

The buccal advancement flap (after Rehrmann).

Steps:
A excise the fistulous tract, easiest with a No. 11 blade.
B outline (dashed) of incision for a full thickness mucoperiosteal buccal flap.
C reflect full thickness mucoperiosteal flap and incise the
periosteal layer ONLY. This makes it possible to mobilize
the flap.
D 'stretch' the flap to assess the degree of elasticity once the
restraining effect of the periosteum is lost.
E the flap is advanced across the fistula and sutured to palatal
mucosa *over bone*.

extraction—close the socket by suture or buccal advancement flap, give antibiotics and decongestants as above, and advise not to blow nose. Closure if diagnosis is made >2 days after extraction—place on antibiotics, etc., for 2 weeks and review after 6. Many will have closed. If not, repair by:

1 Buccal advancement flap Excise fistula to prepare a line of closure over bone and raise a broad-based buccal flap. Incise periosteum to allow mucosa to stretch over the socket and close, **over bone**, with vertical mattress sutures. Use antibiotics, etc., remove sutures at 10 days. *Disadvantages*: thin tissue may break down, reduces sulcus depth, may need to remove another tooth to create space for flap.

2 Palatal rotation flap Excise fistula as above. Dissect a palatal mucoperiosteal flap based on palatine artery, rotate over socket, and suture in similar fashion. *Disadvantages*: bare bone left to granulate.

Sinus lift operation (p. 500)

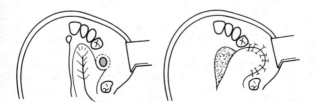

The palatal rotation flap (after Ashley).

A excise fistula, outline palatal flap based on greater palatine artery.
B mobilize and rotate palatal flap, suturing its leading edge to buccal mucosa *over bone*. Leave donor site to granulate under surgical dressing or pack.

Minor preprosthetic surgery

When teeth are extracted alveolar bone resorbs, ∴ should aim to preserve alveolar bone whenever possible, either by not extracting teeth (overdentures, p. 370) or by using a minimally traumatic technique.

At time of extraction of remaining teeth Careful extraction, compress the sockets, remove only small unattached pieces of bone, cover any exposed areas of bone with gingival flaps and surgically remove roots ONLY when necessary (infected, loose, $> \frac{1}{3}$rd root length). Consider interseptal alveolotomy if ridge is prominent and heavily undercut (e.g. Class II). This consists of creating a labial osteomucosal flap by dividing the septae and extending bone cut at the 3 region through the buccal plate and collapsing in the bone flap. Prominent fraena should simply be excised. Attempts at ↓ the rate of ridge resorption have been made by leaving roots under mucosal flaps and by implanting hydroxyapatite or bioconal cones into extraction sockets.

Problems in denture wearers

▶ Only use surgery when denture faults and psychogenic disorders have been excluded. Screen jaws with DPT.

Retained roots and bone sequestrae are removed using standard transalveolar technique except in the maxilla where buried canines may be removed using an osteoplastic flap (where bone is raised on a mucoperiosteal hinge).

Small bony irregularities can be smoothed with a bur but consider ridge augmentation if extensive.

Fibrous (flabby) ridges Reduce by raising a flap of attached gingiva to repair the defect, excise remaining soft tissue ridge, and repair with flap raised first. Fibrous tuberosities can be dealt with similarly.

Fibrous bands and irritation hyperplasia should be excised. Results are improved if palatal mucosal grafts are used to repair the defects and minimize scarring.

Tori can be reduced with a bur under a local flap. Mandibular tori can be removed with a chisel—this is more hazardous with palatal tori.

Muscle attachments to the mylohyoid ridge or genial tubercles can be displaced by resecting the bone from the mandible with a chisel and dissecting away the muscles. Genioglossus and geniohyoid should be re-attached to the labial sulcus.

Ridge augmentation (p. 500) The use of subperiosteally injected porous hydroxyapatite as an out-patient procedure under LA has brought some respectability to augmentation, which fell into disrepute due to the rapid resorption of onlay bone grafting. In this technique, a subperiosteal tunnel is raised along the crest of the ridge and filled with a hydroxyapatite/ saline sludge. It is very dependent on the shape of the ridge, and works best with concave ridges as opposed to the more often seen feather-edge ridge. *Problems*: migration of particles after periosteal elevation. Biocoral is replacing particulate hydroxyapatite for this purpose.

Sulcus deepening (p. 500) When adequate vertical and horizontal basal bone exists but there is a shortage of ridge +/or attached gingiva, these procedures can help. Depends on: (a) dissecting away non-attached mucosa to leave a raw 'new' sulcus; (b) lining this new sulcus with skin or mucosa; (c) *securing the new depth with a 'stent'*—a denture or baseplate lined with tissue conditioner or impression compound, which is held in place by nylon sutures for 10–14 days, then replaced immediately by a new denture with a soft lining extended to the new sulcus and worn continually for the first 3 months.

Implantology

The 'screw-in tooth' has arrived!

History Numerous procedures for oral implants have been described, including subperiosteal, endosseous, and submucosal. All have been strongly advocated by bands of enthusiasts and have one thing in common—failure. Genuine advances in the discipline have come about, thanks to a major contribution by Bränemark—'osseointegration'. All current implants are based on this now well-accepted concept.

Osseointegration is the direct abutment of bone to implant surface such that osteoblasts can be seen on electronmicrographs to be growing on the implant surface. In addition, a tight fibrous/epithelial attachment above the crestal bone between gingiva and implant is essential. Finally, the implant must be designed to resist displacement and evenly dissipate occlusal loads.

Types of implant Large range. Materials are titanium or hydroxyapatite-coated titanium. Bioceramic or gold are less common. May be inserted transmucosally or in two stages (see below).

Indications Edentulous mouths unable to retain dentures, partially dentate for bridge abutments, single anterior tooth replacement, and maxillofacial prostheses post cancer surgery or trauma. If implants survive first 2 yrs there is a 98% success rate with dramatic improvement in all functional parameters.[1]

Techniques Joint planning between oral surgeon and restorative dentist is essential for success. Conventional denture modification should have been tried, a balanced occlusion should be creatable, and a high standard of OH is mandatory. The surgical procedure is highly equipment-dependent, and the surgeon needs to be trained in the particular technique used. Most commonly, this involves:

(a) Fixture installation. A gingival-mucosal flap is raised, based lingually and a receiving channel is prepared in bone, using matched spiral drills. The entrance to the fixture site is countersunk, and depending on the type of implant, it is either pressed into place or if the channel is threaded, a fixture is screwed in. In 2-stage procedures the implant is covered by the flap at the end of the procedure. It is helpful when placing multiple implants that a direction indicator is used to achieve parallelism. Bone over-heating must be avoided by constant irrigation. In 2-stage procedures, a healing period of 4 months in the mandible and 6 months in the maxilla is recommended. In 1-stage procedures a connecting bar can be fitted within 2 weeks but load-bearing or retentive studs should be avoided for 4–6 months.

(b) In 2-stage procedures, abutment connection is then carried out by punch excision of mucosa overlying the implants,

1 R. Adell 1983 *J Prosthetic Dent* **49** 251.

removal of cover screws, and insertion of the abutment. A post-operative surgical pack is usually used, prosthetic procedures starting about 2 weeks after abutment connection.

Implant salvage/bone augmentation The principles of GTR (p. 238) can be used to allow bone formation around osseo-integrated implants either to ↑ bone height or bulk, or to cover an exposed side of the implant.

Craniofacial Implants (p. 514)

Transmandibular Implants (p. 500)

Lasers

Definition light amplification by the stimulated emission of radiation. Light consists of packets of photons transmitted in electromagnetic waves (visible light 400–700 nm). Laser energy is produced by light stimulation of active media to generate collimated light energy at a specific frequency. The active media determines the characteristics of the laser.

Clinical lasers

These consist of two main groups, *hard* and *soft* lasers.

Hard lasers work principally by thermal effect although certain benefits such as ↓ scarring and pain are thought to be due to the photochemical effects of the laser beam.

Carbon dioxide laser is a hard laser in common use in the hospital service. Principle role is as a cutting beam which seals small vessels as it cuts. Also used to evaporate benign white patches of oral mucosa. Used as continuous wave or pulsed beam at 10–20 watts of energy.

Argon laser produces a light beam which is selectively absorbed by haemoglobin and melanin, therefore particularly useful for pigmented and vascular lesions.

Neodymium-Yttrium aluminium garnet (Nd-Yag) laser Originally marketed as a hard laser with a relatively low power output, now available as a soft dental laser.

Tuneable dye laser Expensive variable frequency laser.

Soft lasers These are thought to work by stabilizing cell membranes by a non-thermal photochemical process, increase cellular metabolism by a minor thermal change, and possibly induce endorphin release.

Helium-Neon The red aiming beam on hard lasers, classroom pointers and the 'Terminators' weaponry. Part of the soft laser group it has no cutting effect but seems to act photochemically on cells.

Neodymium-Yag A system using this active media is now marketed for use in dentistry and is purported to increase cell turnover, decrease inflammatory response, inhibit oedema, increase rate of cell regeneration e.g. peripheral neurones, and decrease wound scarring. All this without any recognized side-effects. Many of these claims have yet to be widely validated.

Summary

Lasers are available for a range of uses in dentistry, they are expensive and require special safety precautions. Some benefits can be achieved in other ways and some have yet to be proven. They are a tool not a magic wand.

9 Oral medicine

Bacterial infections of the mouth 434
Viral infections of the mouth 436
Oral candidosis (candidiasis) 438
Recurrent apthous stomatitis (ulcers) 440
Vesiculo-bullous lesions—intraepithelial 442
Vesiculo-bullous lesions—subepithelial 444
White patches 446
Premalignant lesions 448
Oral cancer 450
Abnormalities of the lips and tongue 453
Salivary gland disease—1 454
Salivary gland disease—2 456
Drug-induced lesions of the oral mucosa 458
Facial pain 460
Oral manifestations of skin disease 462
Oral manifestations of gastrointestinal
   disease 464
Oral manifestations of haematological
   disease 466
Oral manifestations of endocrine disease 468
Oral manifestations of neurological disease 470
Oral manifestations of AIDS 472
Cervico-facial lyphadenopathy 474
An approach to oral ulcers 476
Temperomandibular pain—dysfunction/facial
   arthromyalgia 478

Principal sources: Much of the skill of oral medicine lies in the clinical recognition of lesions, ∴ a colour atlas of oral mucosal disease is invaluable, e.g. W. R. Tyldesley 1980 *A Colour Atlas of Oral Medicine*, Wolfe, C. Scully 1993 *Colour Guide—Oral Medicine*, Churchill Livingstone. Textbooks include W. R. Tyldesley 1989 *Oral Medicine*, OUP. J. Soames 1993 *Oral Pathology*, OUP.

Bacterial infections of the mouth

Caries (p. 28); periodontal disease (p. 208); dento-facial infections (p. 408). This page refers primarily to mucosal infections.

Scarlet fever This is a paediatric infection caused by an erythrotoxin produced by certain strains of beta-haemolytic streptococci. Usually causes an URTI. The sore throat is accompanied by a skin rash, general malaise, and fever. The oral mucosa is reddened and the tongue undergoes pathognomonic changes, the dorsum develops a white coating through which white oedematous fungiform papillae project—the 'strawberry tongue' of scarlet fever. Later the white coating is shed and the dorsum becomes smooth and red with enlarged fungiform papillae—'raspberry tongue'. Rx is directed towards the systemic condition with high dose penicillin. The oral manifestations resolve within 14 days.

Tuberculosis Oral involvement with either *Mycobacterium tuberculosis* or the atypical mycobacteria is extremely rare and when it does occur it is usually 2° to open pulmonary infection or coexisting HIV. The oral lesion presents as a deep painful ulcer gradually increasing in size. Any part of the oral mucosa may be involved, although the posterior aspect of the dorsum of the tongue is the commonest site. Histopathology shows necrotizing granuloma with giant and epitheliod cells and a Ziehl–Neilson stain reveals mycobacteria. Refer to a chest physician for management.

Syphilis All phases 1°, 2°, and 3° affect the mouth.

1° lesion A chancre (a firm ulcerated nodule) develops at the site of innoculation, usually the lips or tongue. This lesion is highly infectious and *Treponema pallidum* can easily be isolated. There is usually marked cervical lymphadenopathy which resolves spontaneously in 1–2 months.

2° lesion Develops 2–4 months after the 1° with a cutaneous rash and ulceration of the oral mucosa. This oral involvement occurs regardless of the site of 1° infection, with superficial grey, sloughy ulcers known as mucous patches or snail-track ulcers. These are also highly contagious and *T. pallidum* can be easily isolated. Syphilis serology is positive at this stage. The ulcers generally clear up within a few weeks, although there may be recurrences.

3° lesion Develops several years later and is marked by gumma formation. This is a necrotic granulomatous reaction usually affecting the palate or tongue, which enlarges and ulcerates and may lead to perforation of the palate. In the past, syphilis of the tongue was associated with malignant change, often presenting as a leukoplakia. It is not entirely clear whether it was the condition or its treatment which caused this.

Congenital syphilis Classical appearance of saddle nose, frontal bossing, Hutchinsonian incisors (peg-shaped with notch), and mulberry (Moon) molars.

Gonorrhoea This is a result of oro-genital contact with an infected partner and presents as a non-specific stomatitis or pharyngitis with frequent persisting superficial ulcers. Swabs may reveal gram −ve intracellular diplococci. Rx is with high-dose penicillin.

Viral infections of the mouth

Herpes simplex Although oral infection with herpes simplex types 1 and 2 have been described, type 1 remains the dominant pathogen. Antibodies indicating past infection are virtually ubiquitous in adults. There are two oral manifestations:

Primary herpetic gingivostomatitis varies widely in severity (↑ with age). In infancy is often mistakenly attributed to 'teething'. Presents as a widespread stomatitis with vesicles which break down to form shallow painful ulcers; enlarged, tender cervical lymph nodes; fever and a general malaise for 10–14 days. Although generally self-limiting, rare complications include herpetic encephalitis. *Diagnosis* is based on the clinical features and history, although the virus can be grown in cell culture. Microscopically ballooning degeneration of epithelial cells with intranuclear viral inclusions 'Lipshutz bodies' are seen. A four-fold ↑ in convalescent phase antibodies is also diagnostic. Rx: topical and systemic analgesia (Xylocaine viscous, paracetamol), a soft or liquid diet with ↑ fluid intake and prevention of 2° infection (chlorhexidine mouthwash) is usually adequate in healthy patients. Severely ill or immunocompromised patients should receive systemic acyclovir.

Secondary herpes (herpes labialis, cold sore) is a reactivation of the 1° infection which is believed to lie dormant in the trigeminal ganglion. Precipitating factors include trauma and immuno-suppression and, less commonly, exposure to sunlight, stress, or other illness. Usually recurs on the skin of lip or nose supplied by one branch of Trigeminal nerve, classically at the muco-cutaneous junction, or rarely as an intra-oral blister. Lesions may respond to topical acyclovir 5% cream. Should consider systemic acyclovir (p. 618) in the immunosuppressed.

Herpes (varicella) zoster causes chickenpox as a 1° infection and shingles as a reactivation.

Chickenpox Classically an itchy, vesicular, cutaneous rash which may rarely affect the oral mucosa.

Shingles is confined to the distribution of a nerve, the virus staying either in the dorsal root ganglion of a peripheral nerve or the trigeminal ganglion. Always presents as a unilateral lesion never crossing the midline. Facial or oral lesions may arise in the area supplied by the branches of the Trigeminal nerve. *Diagnosis*: pre-eruption pain, followed by development of painful vesicles on skin or oral mucosa which rupture to give ulcers or crusting skin wounds, in the distribution outlined above. These usually clear in 2–4 weeks, but apparent resolution is often followed by severe post-herpetic neuralgia which may continue for years. Rx: symptomatic relief for chickenpox. There is some evidence to suggest that aggressive early treatment of shingles with acyclovir ↓ the incidence and severity of post-herpetic neuralgia. Refer to an ophthalmologist if the eye is involved.

Coxsackie is an RNA virus causing 2° oral mucosal conditions.

Herpangina Caused by Coxsackie (range of A+B virus types) is confined to children and presents with widespread small ulcers on the oral mucosa with fever and general upset. May be preceded by sore throat and conjunctivitis. Can also be mistaken as 'teething'. Self-limiting in 10–14 days. Fairly rare.

Hand, foot, and mouth disease Caused by Coxsackie (usually A16) and is also confined to children. A papular, vesicular rash appears on the hands and feet in conjunction with nasal congestion and oral mucosal vesicles. These break away, leaving painful superficial ulcers, particularly on the palate. The gingivae are rarely involved. It is self-limiting in 10–14 days. Rx as herpetic stomatitis. Quite common.

Measles The prodromal phase of measles may be marked by small white spots with an erythematous margin on the buccal mucosa, known as Koplik spots. A few days later the maculo-papular rash of measles appears.

Glandular fever (infectious mononucleosis) is seen mostly in children and young adults and spread by infected saliva. It varies widely in severity and presents with sore throat, generalized lymphadenopathy, fever, headaches, general malaise, and often a maculo-papular rash. There may be hepatosplenomegaly. Oral manifestations may mimic 1° herpetic gingivostomatitis, with widespread oral ulceration and in additional petechial haemorrhages and bruising may be present. The cause is usually Epstein–Barr virus and less commonly cytomegalovirus or toxoplasmosis can give a similar picture. *Diagnosis*: initially monospot test, Paul–Bunnell test to exclude Epstein–Barr virus and acute and convalescent titres for CMV and toxoplasmosis. Be aware that early HIV infection can mimic this condition. **Rx** symptomatic as for 1° herpes, except toxoplasmosis which responds to co-trimoxazole 960 mg bd 10 days. NB Ampicillin should not be given to patients with a sore throat who may have glandular fever as it inevitably produces an unwanted response, ranging from a rash to anaphylaxis.

Reiter syndrome Causative agent unknown. Consists of urethritis, arthritis, conjunctivitis ± oral ulcers or erosions. Predominantly affects young males and has a strong association with HLA B27.

Oral candidosis (candidiasis)

Candida albicans is a common fungal commensal in the human mouth. Overt infection occurs when there are local or systemic predisposing factors, ∴ the prime tenet of management is to look for and treat these factors. Candidosis is conveniently divided:

Acute candidosis

Acute pseudomembraneous candidosis (thrush) Commonest in infancy, old age and the immunosuppressed or debilitated (e.g. radiotherapy, cytotoxics, steroids, diabetes, cancer, HIV, and haematological malignancy). *Diagnosis*: appears as creamy lightly adherent plaques on an erythematous oral mucosa, usually on the cheek, palate, or oropharynx. Occasionally symptomless, but more commonly cause discomfort on eating. These plaques can be gently stripped off, leaving a raw under-surface and, with gram staining, show candidal hyphae. In infancy, widespread oral candidosis can be associated with a livid facial rash and an associated nappy rash. Colonization of a breast-feeding mother's nipples can lead to mutual recoloniza-tion. Rx: nystatin sugar free suspension 100 000 units rinsed then swallowed qds 10 days or nystatin pastille 1 qds, are cheap and usually effective. Chlorhexidine mouthwash is an effective adjunct to treatment. Amphotericin and miconazole are more expensive (and mutually antagonistic) alternatives. Fluconazole 50 mg od is the systemic drug of choice.

Acute atrophic candidosis is an opportunistic infection follow-ing the use of broad spectrum antibiotics, and sometimes inhaled steroids. It is painful and is exacerbated by hot or spicy foods. The oral mucosa has a red, shiny, atrophic appearance and there may be coexisting areas of thrush. Rx: eliminate cause (if due to inhaled steroids rinse mouth with water after inhaling), otherwise as above.

Chronic candidosis

Chronic atrophic candidosis (denture stomatitis) (pp. 126, 364).

Angular cheilitis is a combined staphylococcal, streptococcal, and candidal infection, involving the tissues at the angle of the mouth, often with an underlying precipitating factor, e.g. iron deficiency anaemia. Therefore, haematological deficiency should be investigated with a FBC. Anecdote suggests that an inadequate OVD can also predispose, but correction of this alone will not resolve the condition. Often associated with chronic atrophic candidosis. Clinically, see red, cracked, macerated skin at the angles of the mouth, often with a gold crust. Infecting organisms can be identified on culture of swabs of the area, although it is usual to make a clinical diagnosis and Rx with miconazole cream which is active against all three infecting organisms. Treatment needs to be prolonged, up to 10 days after resolution of the clinical lesion and carried out in conjunction with elimination of any underlying factors.

Median rhomboid glossitis is no longer considered to be an anatomical abnormality but a form of chronic atrophic candidosis affecting the dorsum of the tongue. Seen in patients using inhaled steroids and smokers. Rx only if symptomatic as discomfort can be improved with topical antifungals, but the appearance cannot.

Chronic hyperplastic candidosis (candidal leukoplakia) presents as dense, white, adherent, keratotic patches on the oral mucosa which may be multifocal and is often confused with 'leukoplakia'. Although there is a slightly ↑ risk of malignant change, the initial approach *after ensuring the diagnosis microbiologically and histopathologically* is to eradicate the candidal infection. Candidal hyphae can be seen in the superficial layers of the epidermis and this is one reason its eradication is so difficult. Rx: systemic antifungals such as fluconazole and ictraconazole, while expensive, are indicated in an attempt to remove the infecting organism. Often associated with iron and B12 deficiency, and smoking, which should be corrected.

Chronic mucocutaneous candidosis is a rare syndrome complex with several subgroups including: **candidal endocrinopathy** where skin and mouth lesions occur in conjunction with endocrine abnormalities, **granulomatous skin candidosis**, a **late onset predominantly male group** and an **AIDS associated group**. Rx: ketoconazole, fluconazole, and ictraconazole (p. 618).

439

Histoplasmosis This and other rare fungal infections have occasional oral manifestations.

Recurrent aphthous stomatitis (ulcers)

This is the term given to a fairly well defined group of conditions characterized by recurrent oral ulceration. There are three subgroups, any of which may occur in conjunction with genital or conjunctival lesions as part of the Behçet syndrome.

Minor aphthous ulcers This very common condition (*c.* 25% of population) affects all age groups and F>M. Usually appears as a group of 1–6 ulcers at a time of variable size (usually 2–5 mm diameter). They last about 10 days and heal without scarring. Mainly occur on buccal or labial mucosa, floor of mouth or tongue, and extremely rarely on the attached gingiva or hard palate. Prodromal discomfort may precede painful ulcers. Exacerbated by stress, local trauma, menstruation, and may be an oral 'marker' of iron, B12, or folate deficiencies. In some cases are a manifestation of Crohn's disease, ulcerative colitis, or gluten enteropathy. Aetiology, although not fully understood, is almost certainly autoimmune. Rx prevent superinfection with chlorhexidine mouthwash and relieve pain (simple analgesics, benzydamine mouthrinse). Topical tetracycline and steroid preparations are sometimes useful (p. 606). It is important to look for and treat any underlying deficiency or coexisting pathology.

440

Major aphthous ulcers A more severe variant with fewer, but larger ulcers (up to 10 mm) which may last 5–10 weeks. They are associated with tissue destruction and scarring and any site in the mouth and oropharynx may be affected. There is an even higher association between major aphthae and gastrointestinal and haematological disorders. They are also seen in AIDS. Seldom a cyclical pattern. Rx: as for minor apthae and in addition topical or systemic steroids (p. 606).

Herpetiform ulcers This is a descriptive term, as these ulcers have nothing whatsoever to do with infection with the herpes virus. Manifests as a crop of small but painful ulcers which usually last for 1–2 weeks, the commonest site being the floor of the mouth, lateral margins, and tip of tongue. They heal without scarring and may occur on both keratinized and non-keratinized surfaces. Rarely, merge to form a large ulcer which heals with scarring. Rx: as for minor apthous ulceration.

Behçet syndrome Any of the above types of oral ulceration may appear in the Behçet syndrome, which affects the mouth, skin, genital mucosa, eyes, heart, blood vessels, chest, joints, and nervous system. It is a severe and dangerous multisystem condition, commoner in males, which cannot be diagnosed on the basis of the oral presentations. Athough there is an association with certain HLA types there are **no** diagnostic tests for this condition, ∴ diagnosis is by exclusion of other conditions. Rx the severity of the condition and the nature of the multisystem disease usually necessitates systemic immunosuppression with corticosteroids, azathioprine, cyclosporim A, or dapsone. Thalidomide has been used with some success in this condition (prescribed on a named-patient-only basis with a strict contraceptive protocol).

Oral ulcers See also p. 476.

Vesiculo-bullous lesions—intraepithelial

Vesicle is a small blister a few millimetres in diameter.

Bulla is a larger blister.

Intraepithelial blisters are caused by loss of attachment between individual cells (acantholysis).

Subepithelial blisters separate the epithelium from the underlying corium.

Ulcer is a breach in the mucous membrane.

Erosions are shallower than ulcers.

Because the vesiculo-bullous lesions constitute a defined group with examples from several different pathological processes, they are a favourite examinations topic. One method of classifying this group is into intraepithelial and subepithelial according to where the blisters form.

Pemphigus is a chronic skin disease which also affects mucous membranes. Autoimmune in aetiology, there are circulating autoantibodies to epithelial intercellular substance. Acantholysis causes separation of epithelium above the basal cell layer, and oedema into this potential space produces a superficial, easily burst, fluid-filled bulla. Rupture leaves a large superficial, easily infected ulcer. The first identifiable lesions are quite often found in the mouth, especially on the palate, although these are usually seen as ulcers because the bullae break down rapidly. It is mainly a disease of middle age (F>M) and has a ↑ incidence in Jews. Rarely, it may be drug-induced. *Diagnosis*: stroking the mucosa produces a bulla (Nikolsky's sign), but this is inducing pathology for the sake of diagnosis. Other methods are by direct or indirect immunofluorescent techniques (biopsy samples need to be fresh frozen). Rx: systemic steroids +/− azathioprine.

Benign familial chronic pemphigus differs from the above by having a strong family history with onset of the disease in young adults.

Viral infections, p. 436.

443

Vesiculo-bullous lesions—subepithelial

Benign mucous membrane pemphigoid Commonest in females >60. Presents as mucous membrane bullae which rupture and heal with scar formation. Rare to see skin bullae. Conjunctiva may be affected and if scarring occurs can lead to loss of vision, therefore regard oral signs as a warning to prevent ocular damage. The natural history is of a long lasting disease which persists with periods of activity and inactivity alternating and may be quiescent for several years. More common than pemphigus and pemphigoid. *Diagnosis*: again direct and indirect immunofluorescence is used the antibodies being found at the level of the basement membrane. The bullae are blood-filled and tense and may be found in conjunction with atrophic gingivitis (p. 212). Rx: topical steroids (p. 606).

Pemphigoid affects >60 yr age-group. Subepithelial bullae form which are firm and less likely to break down than those in pemphigus. The oral mucosa is only affected in around 20% of patients. May be an external 'marker' of internal malignancy.

Dermatitis herpetiformis is a rare chronic condition of unknown aetiology but often associated with small bowel disease. Commoner in middle-aged men; it affects both skin and mucous membranes, bullae in the mouth break down to leave large erosions. Rx: dapsone may be used both diagnostically and therapeutically.

Lichen planus affects both skin and mucous membranes. Bullous lichen planus is a rare variant in which subepithelial bulla form and break down, leaving large erosions. See also p. 462.

Epidermolysis bullosa This is a rare skin disease which exists in numerous different forms. The dystrophic autosomal recessive form is most likely to present with oral manifestations and appears shortly after birth. Associated with bullae formation after minor trauma to skin or mucosa, these break down leaving painful erosions. Healing with scarring, resulting in difficulty in eating, speaking, and swallowing as scar tissue limits movement. Skin involvement can lead to destruction of extremities and may be overtaken by carcinomatous change. Prognosis varies widely depending on type. Phenytoin and steroids may help some varieties.

Erythema multiforme This is a group of signs and symptoms of multifactorial aetiology, the most severe form is known as Stevens Johnson syndrome (p. 747). Affects skin and mucous membranes with an acute onset usually in young adult males and is probably due to deposition of immune complexes. It is associated with exposure to certain drugs or infecting organisms in a susceptible individual. *Diagnosis*: is from clinical features which include 'target lesions', concentric rings of erythema on the palms, legs, face, or neck. The oral mucosa is covered in bullae which break down, the lips and gingivae becoming crusted with painful erosions. There is usually a fever. It is a self-

limiting condition in 3–4 weeks but patients will need symptomatic support, ranging from simple analgesics and antiseptics to
intravenous rehydration and systemic steroids. If the causative
agent is found e.g. tetracycline the patient should **not** be re-
exposed to it diagnostically.

White patches

Numerous conditions manifest as white patches of the oral mucosa; some of these are transient, such as thrush (p. 438) or chemical burns (e.g. aspirin). More are persistent, and there exists some confusion over the terminology applied to these white patches.

White spongy naevus This is a rare, benign, familial disorder inherited as a simple dominant. It appears as a widespread soft, uneven thickening of the superficial layer of the epithelium which characteristically has no definite boundary and may affect any part of the mouth. Histology shows hyperplastic epithelium with gross intraepithelial oedema. Rx: neither exists nor is required.

Frictional keratosis This is a white patch due to hyperplastic hyperkeratotic epithelium induced by local trauma. It is managed by removal of the source of the friction which will generally allow complete resolution of the lesion. If this doesn't happen, biopsy is indicated.

Smokers' keratosis Characteristically a white patch affecting *exposed* areas of the palate with multiple red papular swellings protruding through it (inflamed mucous glands). Probably due to a combination of low-grade burn and the chemical irritants of smoke. It is seen particularly in pipe smokers. There is little evidence that these patches are premalignant and they resolve on stopping smoking.

'Syphilitic leukoplakia' A white patch on the dorsum of the tongue is one of the classical appearances of tertiary syphilis (p. 434). Active disease must be treated; however, this **will not** resolve the area of leukoplakia, which has a propensity to undergo malignant change.

Chronic hyperplastic candidosis/candidal leukoplakia, p. 438.

Lichen planus, p. 462.

Lupus erythematosis, p. 462.

Leukoplakia, p. 448.

Hairy leukoplakia, p. 472.

Panoral leukoplakia is where the entire oral mucosa appears to be undergoing hyperplastic field change. Has sinister connotations.

Oral carcinoma Occasionally, oral cancer may appear as a white patch as distinct from a leukoplakia becoming malignant.

Skin grafts may appear as a white patch in the mouth—and are a trap for the unwary in exams.

Renal failure can produce soft oval white patches which resolve on Rx of renal failure.

447

Premalignant lesions

There exists a group of conditions which have an ↑ risk of malignant transformation of the oropharyngeal mucosa. Although a great deal of attention has been paid to these pre-malignant conditions, it should be remembered that only a small number of oral cancers are preceded by them and also that the designation 'premalignant' does not necessarily imply certain malignant transformation. Indeed, the majority of patients with so-called premalignant lesions will not go on to develop oral cancer. The ↑ risk of progression to carcinoma necessitates accurate diagnosis, treatment if indicated, and long-term follow-up in an attempt to pre-empt life-threatening disease.

Leukoplakia 'A white patch on the oral mucosa which cannot be wiped off and is not susceptible to any other clinical or histo-logical diagnosis' (after Pindborg).

The histopathology of these lesions varies widely from the essentially benign to carcinoma-*in-situ*. They are usually characterized by a thick surface layer of keratin with thickening of the prickle cell layer of the epithelium, acanthosis, and in-filtration of the corium by plasma cells; however, the most important variable is cellular atypia amongst the epithelial cells. Pointers to look for are: nuclear hyperchromatism, an ↑ nuclear/cytoplasmic ratio, cellular and nuclear pleomorphism, ↑ &/or atypical mitoses, individual cell keratinization, and focal disturbance in cell arrangement and adhesion. The degree of cellular atypia is one of the most important factors to be con-sidered in the management of a leukoplakia. The second major consideration is the site, e.g. floor of mouth and ventral surface of tongue are more likely to undergo malignant change than most. Thirdly, relation to cause, e.g. buccal leukoplakia, in the preferred site for a betel quid is at high risk if the habit is not discontinued.

On average 5% of leukoplakias progress to carcinomas.

Erythroleukoplakia (speckled leukoplakia) This is basically leukoplakia with areas of erythroplakia. Exhibits an ↑ risk of malignant transformation.

Erythroplakia is generally a well-demarcated, red velvety patch of the oral mucosa which histologically shows marked cellular atypia, no surface keratinization, and a degree of atrophy of the surface layer. Most of these lesions are carcinoma-*in-situ* or frank carcinoma.

Erosive lichen planus is a comparatively rare variant of lichen planus, which some authorities believe to be premalignant. The common forms of lichen planus have no premalignant potential.

Submucous fibrosis is a condition found particularly in those of Indian extraction, and is thought to be a tanning of the oral mucous membrane induced by betel nut or quid chewing *with-out* the addition of tobacco (some cultures use a quid of betel leaf, nut, and slake lime; others add tobacco to this; the first seems to produce submucous fibrosis, the second carcinoma).

The mucosa is pale, with constraining fibrous bands, and fibrosis of the submucosa occurs, making the lips and cheeks immobile and resulting in trismus. Histology shows hyalinization and acellular dense fibrous tissue with narrowed blood vessels and a lymphocytic infiltrate. There is epithelial atrophy and cellular atypia. Rx: stop habit, intralesional steroids/ exercise, surgery with flap reconstruction.

Patterson-Brown–Kelly syndrome (Plummer–Vinson syndrome), p. 746.

Management of premalignant lesions Record site, preferably photographically. Consider site, histology, age, and health of the patient, in conjunction with aetiological factors, before deciding on long-term observation or active intervention. Stop patient from smoking completely and 60% will disappear.[1] Observation may consist of clinical examination with repeated cytology (although cytology has generally been disappointing), or biopsy if change is seen. Rx options: cryotherapy, surgical excision, topical bleomycin, after removal of any identifiable aetiological factors. Follow-up at 3-monthly intervals.

▶ It is impossible to predict the behaviour of a patch of leukoplakia with precision.

449

1 J. Pindborg 1980 *Oral Cancer and Precancer* John Wright Bristol.

Oral cancer

Cancer of the mouth accounts for about 2% of all malignant tumours in northern Europe and the USA, but around 30–40% in the Indian subcontinent. >90% of these are squamous cell carcinomas.

Site The floor of the mouth is the commonest single site, and when combined with lingual sulcus and ventral surface of tongue, creates a horseshoe area, accounting for over 75% of carcinomas seen in European or American practice. M>F, although this difference is less marked than it has been in the past, possibly due to changes in smoking habits between the sexes. It is an age related disease with 98% of patients >40 years.

Aetiology Main aetiological factor in cancer of the lip is exposure to sunlight, as with skin cancer. It is estimated that the risk of developing lip cancer doubles every 250 miles nearer the equator. Excessive alcohol and tobacco use are important factors in the aetiology of cancer of the mouth. Perhaps the most clear-cut aetiological factor is the chewing of tobacco and betel quid. Late-stage syphilis is now an exceedingly rare risk factor.

Clinical appearance Most often seen as a painless ulcer, although may present as a swelling, an area of leukoplakia, erythroleukoplakia or erythroplakia, or as malignant change of longstanding benign tumours or rarely in cyst linings. Pain is usually a *late* feature when the lesion becomes superinfected or during eating of spicy foods. Referred otalgia is a common manifestation of pain from oral cancer. The ulcer is described as firm with raised edges with an indurated, inflamed, granular base and is fixed to surrounding tissues.

Staging The **TNM** classification is most commonly used;

T primary tumour	**N** cervical nodes
T1 <2 cm diameter	**N0** no nodes
T2 2–4 cm diameter	**N1** single node <3 cm
T3 >4 cm diameter	**N2** multiple ipsilateral nodes
T4 massive, invading beyond mouth	or single node 3–6 cm
	N3 bilateral cervical nodes or ipsilateral node > 6 cm

M distant metastases
M0 absent
M1 present

Survival is dependent on site, stage, and patient age. The presence or absence of metastatic cervical nodes is the most important single prognostic factor.

Histopathology Almost always squamous cell carcinoma. Characteristically shows invasion of deep tissues with cellular pleomorphism and ↑ nuclear staining. The presence of a lymphocytic response may have prognostic value as does the manner of invasion (pushing or spreading). Spread can occur by local infiltration, lymphatic (cervical nodes) and late spread via blood stream.

Verrucous carcinoma is a distinctive exophytic, wart-like lesion which grows slowly, is locally invasive, and is regarded as a lower grade squamous cell carcinoma, characterized by folded hyperplastic epithelium and a lower degree of cellular atypia. Surgical excision +/− radiotherapy is the treatment. Inadequate radiotherapy has been reported to induce more aggressive behaviour.

Other tumours Malignant connective tissue tumours (sarcomas) are rare in the mouth, but fibrosarcoma and rhabdomyosarcoma are seen in children. Osteosarcoma of the jaws has a slightly better prognosis than when found in long bones.

Salivary gland tumours, p. 506. Malignant melanoma, p. 546.

Management of oral malignancy, p. 510.

Abnormalities of the lips and tongue

Although many diseases of the oral mucosa will involve the lips and tongue, there are a number of conditions specific to these structures due in part to their highly specialized nature. The tongue is a peculiar muscular organ covered with specialized sensory epithelium and the lips form the interface between skin and mucosa.

The tongue

Ankyloglossia (tongue tie) is the commonest of the developmental variations of the tongue and may be associated with microglossia. Rx: frenectomy.

Macroglossia Congenital; Down syndrome, Hurler syndrome, Benign tumours (e.g. lymphangioma), or acquired; acromegaly, amyloidosis, p. 520.

Fissured tongue Deep fissuring of the tongue is not pathological in itself (affecting 3% of tongues) but may harbour pathogenic microorganisms. Different fissure patterns are identified by various delightful names such as scrotal tongue. The Melkerson–Rosenthal syndrome is a deeply fissured tongue in association with recurrent facial nerve palsy and swelling.

Hairy tongue is a peculiar condition of unknown aetiology probably due to elongation of the filiform papillae which may or may not be accompanied by abnormal pigmentation. Sometimes responds to podophylin paint, a thorough scrape or surgical shave.

Median rhomboid glossitis, p. 438.

Geographic tongue (benign migratory glossitis, erythema migrans) This peculiar condition involves the rapid appearance and disappearance of atrophic areas with a white demarcated border on the dorsum and lateral surface of the tongue, which gives it the appearance of moving around the tongue surface. It is due to temporary loss of the filiform papillae. Several clinical variants exist. It is self-limiting and entirely benign.

Depapillation of the tongue also appears in a number of haematological and deficiency states and in severe cases may also appear lobulated.

Sore tongue (glossodynia) may occur in the presence or absence of clinical changes; however, it should be remembered that even the presence of glossitis may not explain the symptoms of sore or burning tongue. Main causes of glossitis are; iron deficiency anaemia, pernicious anaemia, candidosis, Vitamin B group deficiencies, and lichen planus. Sore, but clinically normal, tongue is an ↑ common problem and is often psychogenic in origin; however, the first line of treatment is to exclude any possible organic cause, e.g. haematological deficiency states and unwanted reactions to self-administered or professionally administered medicines or mouthwashes.

Lips

Granulomatous cheilitis (orofacial granulomatosis) This is characterized by swelling of the lips and is histologically similar to Crohn's disease (granuloma being found on biopsy). Intralesional steroids, e.g. triamincinolone, 40 mg into affected lip, may help.

Persistent median fissure This may be found as a developmental abnormality but is usually secondarily infected which is extremely difficult to eradicate. May be associated with granulomatous cheilitis.

Sarcoidosis A chronic granulomatomous condition which can affect any body system. Lip swelling, gingival, and palatal modules occur. Heerfordt syndrome p. 744. Biopsy reveals granuloma with Schoumann inclusion bodies. CXR-hilar lymphadenopathy. Serum angiotensin—converting enzyme ↑. Ask on opthalmologist to exclude uveal tract involvement. Rx: steroids, intralesional or systemic.

Actinic cheilitis Sun damage to the lower lip causes excessive keratin production and ↑ mitotic activity in the basal layer. Is premalignant. Advise sun blocks.

Exfoliative cheilitis Similar to above but of unknown aetiology.

Dry sore lips Except when accompanied by frank cheilitis, this is usually entirely innocent and can be treated symptomatically with lip salves. Common causes are lip licking, exposure to wind or sunlight. It is also a manifestation of viral illness. Rx: lip salves.

Peutz–Jegners syndrome, p. 746.

Herpes labialis, p. 436.

Mucocele, p. 414.

Allergic angio-oedema Severe type I allergic response affecting lips, neck, and floor of mouth. Usually an identifiable cause. Rx: Mild—antihistamine PO; Severe—as anaphylaxis, p. 562.

Hereditory angiooedema Defect of C_1-esterase inhibitor. Lip, neck, floor of mouth swelling, and swelling of feet and buttocks. Precipitated by trauma. Rx: acute attacks—fresh frozen plasma (contains C_1-esterase inhibitor), prophylaxis, stanozalol.

453

Salivary gland disease—1

The salivary glands consist of the major glands, the paired sublingual, submandibular and parotid glands and the minor salivary glands present throughout the oral mucosa, but particularly dense in the posterior palate and lips.

Xerostomia Dry mouth can be both a sign and a symptom. Note that some patients complain of a dry mouth when, in fact, they have an abundance of saliva. True xerostomia predisposes the mouth, pharynx and salivary glands to infection and caries. Common causes include irradiation of the head and neck, drugs (e.g. tricyclic antidepressants), anxiety states, and Sjögren syndrome. Rx is aimed at the underlying cause. Symptomatic relief with carboxymethylcellulose saliva, e.g. glandosane spray or Saliva-Orthana, helps. Optimal OH is essential.

Sialorrhoea is rare, although apparent sialorrhoea can occur with drooling due to inflammatory conditions of the mouth, or neurological disorders which inhibit swallowing. Rare causes include mercury poisoning and rabies. Problems with swallowing may be overcome by surgical re-positioning of the major salivary ducts.

Sialadenitis is inflammation of, usually, the major salivary glands. **Acute bacterial sialadenitis** presents as a painful swelling usually with a purulent discharge from the duct of the gland involved. It may also develop as an exacerbation of **chronic bacterial sialadenitis** which often exists as a complication of duct obstruction. Both conditions are almost always unilateral and common infecting organisms are oral streptococci, oral anaerobes, and *S. aureus*. Rx: exclusion or removal of an obstructing calculus (p. 508). Culture of pus from the duct and aggressive antibiotic therapy (amoxycillin/clavulanate and metronidazole). Stimulation of salivary flow by chewing or massage. Rarely, loculated pus collection within the gland necessitates incision and drainage. Once the acute symptoms have resolved sialography is indicated to define duct structure and may prove therapeutic. Other treatments include irrigating the gland with antibiotics &/or steroids. Rarely recurrent chronic sialadenitis is an indication for removal of the gland (p. 508). **Viral sialadenitis**, commonly mumps, an acute, infectious viral disease which primarily affects the parotid. It is transmitted by direct contact with droplets of saliva and usually affects children and young adults with sudden onset fever, pain, and parotid swelling. Classically, one gland is affected first, although bilateral swelling is the norm. In adults, the disease is more severe with multisystem problems such as orchitis. Protection is now conferred by the measles, mumps, rubella vaccine.

Rarely, sialadenitis can occur as a manifestation of allergy to various drugs, foodstuffs, or metals.

Sialolithiasis, p. 508.

Salivary duct and salivary gland fistulae Communications between the duct or gland and the oral mucosa or skin may occur post-traumatically or post-operatively. Duct repair or gland excision may be needed. Propantheline 15 mg tds before food can ↓ salivary flow and dry up small fistulae.

Mucocele, Ranula, p. 414.

Salivary gland disease—2

Sialosis is a non-inflammatory, non-neoplastic swelling of the major salivary glands, usually the parotids and usually bilateral. It is of unknown aetiology, although linked with endocrine abnormalities, nutritional deficiencies, and alcohol abuse. Sialography is essentially normal. Histology is of serous acinar cell hypertrophy. Treatment is aimed at any aetiological factors.

Sjögren syndrome (secondary Sjögren syndrome) is the triad of xerostomia, keratoconjunctivitis sicca, and a connective tissue disorder, usually rheumatoid arthritis. When the connective tissue component is absent the condition is called **primary Sjögren syndrome (sicca syndrome)** The aetiology is probably autoimmune and there is an ↑ risk of malignant lymphomatous transformation of the affected gland. Diagnosis is by Schirmer's test for lacrimal flow rate, parotid flow rate, sialography (looking for sialectasis), labial gland biopsy (looking for lymphocytic infiltrate of minor glands), and autoantibody serology. As there is no established method of disease control, treatment involves elimination of candidal superinfection, synthetic saliva (p. 622), synthetic tears, meticulous OH, and regular review in view of risk of lymphomatous change. See also p. 746.

Salivary gland tumours 80% are benign, 80% occur in the parotid, and 80% of these are in the superficial lobe. The majority are **pleomorphic adenomas** which have a mixed cellular appearance on histopathology. Although benign, the cells lie within the capsule of the tumour and satellite cells may lie outwith the capsule creating a tremendous propensity for recurrence if simply enucleated. Any tumour in the superficial lobe of the parotid should ∴ be removed by superficial parotidectomy taking a safe margin of normal tissue (p. 508). **Lymphangiomas and haemangiomas** are the commonest tumour found in salivary glands in children. **Adenolymphoma** is found almost exclusively in the parotid and **adenoid cystic carcinoma** is more commonly found in the minor than the major salivary glands. Tumours of the submandibular, sublingual and minor salivary glands are more likely to be malignant than those found in the parotid. Pointers to malignant change in salivary gland tumours are: fixation to surrounding tissues, nerve involvement (particularly the facial nerve in parotid tumours), pain, rapid growth, and lymphadenopathy. For management, see p. 508.

Miscellaneous Lymphoepithelial lesion (Miculicz disease) is essentially an aggressive form of the Sjögren syndrome without the eye or connective tissue component. NB Miculicz syndrome is salivary enlargement of known cause.

Frey syndrome, p. 744.

British Sjögren Syndrome Society, c/o 7 Winchells, Bennetts End, Hemel Hempstead HP3 8HZ

Drug-induced lesions of the mouth

One way of thinking of these reactions is to divide them into local and systemic effects.

Local reactions

Chemical burns, e.g. from an aspirin tablet being held against the oral mucosa beside a painful tooth is still seen and makes one despair of the public at times. The burns are superficial necrosis of the epithelium and can appear as a transient white patch. Rx: re-education and removal of the irritant. The mucosa will spontaneously heal. Iatrogenic causes include trichloracetic acid and phenol.

Interference with commensal flora Prolonged or repeated use of antibiotics, particularly topical antibiotics can lead to the overgrowth of resistant organisms, especially candida. Corticosteroids can cause a similar problem by immunosuppression.

Oral dyaesthesia A sore but normal-appearing tongue can be caused by certain drugs (e.g. captopril).

Systemic effects

Depressed marrow function There are a wide range of drugs which will depress any or all of the cell lines of the haemopoetic systems and these in turn can affect the oral mucosa, e.g. phenytoin. Long-term use can result in folate deficiency and macrocytic anaemia which can produce severe apthous stomatitis. Chloramphenicol and certain analgesics can induce agranulocytosis, leading to severe oral ulceration. Chloramphenicol can also induce aplastic anaemia, which affects haemostasis, although spontaneous oral purpura and haemorrhage is a rare presentation.

Immunosuppression Steroids and other immunosuppressants predispose to viral and fungal infections.

Lichenoid eruptions Classically associated with the use of gold in the treatment of rheumatoid arthritis (p. 540).

Erythema multiforme (Stevens–Johnson syndrome), p. 747.

Fixed drug eruptions are recurrent, sharply circumscribed lesions at the same site occurring on exposure to a specific drug. Extremely rare in the oral mucosa.

Exfoliative stomatitis is simply an oral manifestation of the very dangerous drug reaction known as exfoliative dermatitis in which the skin and other membranes are shed. Again, gold has been implicated.

Gingival hyperplasia is common in patients on phenytoin, and cyclosporin A, nifedipine, and certain other calcium channel-blockers. It is characterized by progressive fibrous hyperplasia and, while improved by OHI, it will occur even in the presence of meticulous OH.

Oral pigmentation Black lines in the gingival sulcus are described as being a sign of heavy metal poisoning. Chlorhexidine causes a black or brown discoloration of the dorsum of the tongue and some antibiotics can also do this.

Xerostomia, p. 454.

Allergic reactions Penicillin is a common offender.

There are a host of conditions affecting the oral mucosa, which may be ascribed to the use of drugs. Recognition of these is of course important, providing the drug can be withdrawn, but one has to pay attention to the reason the drug was given in the first place, and it may be that minor oral symptoms have to be tolerated when the drug is essential for the overall well-being of the patient.

Facial pain

Pain is a complex and multifaceted symptom and several other sections of this book are relevant. The commonest source of pain in the region of the jaws and face is the tooth pulp. Pain not directly related to the teeth and jaws is dealt with here.

Trigeminal neuralgia is an excruciating condition affecting mainly the >50s. It presents as a shooting 'electric shock' type of pain of rapid onset and short duration which is often stimulated by touching a trigger point in the distribution of the trigeminal nerve. Patients may refuse to shave or wash the area which stimulates the pain although, strangely, they are rarely woken by pain. In the early stages of the disease there may be a period of prodromal pain not conforming to the classical description and it may be difficult to arrive at a diagnosis, patients often have multiple extractions in an attempt to relieve the symptoms. It is thought to be a sensory form of epilepsy. Diagnosis is by the history and carbamezipine is useful both therapeutically and diagnostically, with an 80% response rate. Injection of LA can break pain cycles and be useful diagnostically. Cryotherapy can induce protracted analgesia, but sectioning the nerve rarely helps.

Glossopharyngeal neuralgia is a similar condition to trigeminal neuralgia, but less common. Affects the glossopharyngeal nerve, causing a sharp shooting pain on swallowing. There may be referred otalgia. Again, carbamezipine is the drug of choice.

▶ Patients under the age of 50 presenting with symptoms of cranial nerve neuralgia require full neurological examination and investigation, as this may be the presenting symptoms of an intracranial neoplasm.

Temporal arteritis (cranial arteritis) is a condition affecting older age groups. The pain is localized to the temporal and frontal regions and is usually described as a severe ache, although it can be paroxysmal. The affected area is tender to touch. Major risk is involvement of retinal arteries with sudden deterioration and loss of vision; underlying pathology is inflammatory arteritis. Tongue necrosis following lingual artery involvement has been described. *Diagnosis*: pulseless temporal arteries, classical distribution of pain, and an ↑ESR. Rx: aim is to relieve pain and prevent blindness and involves systemic steroids, guided by symptoms and ESR.

Migraine, p. 543.

Periodic migrainous neuralgia (cluster headache) Similar aetiology to migraine but with different clinical presentation; periodic attacks of severe, unilateral, pain lasting up to 30 min, located around the eye. Attacks often occur at a particular time of night (early morning 'alarm clock' wakening). Nasal stuffiness and lacrimation are associated complaints. Attacks tend to be closely concentrated over a period of time followed by a longer period of remission. Most sufferers describe alcohol intolerance. Rx: responds to ergotamine. Pizotifen is used prophylactically.

Pain associated with herpes zoster, p. 436.

Glaucoma gives rise to severe unilateral pain centred above the eye, with a tense, stony hard globe. Due to raised intraocular pressure. Acute and chronic forms are recognized. The acute form presents with pain and responds to acetazolamide. Will need ophthalmological referral.

Myocardial infarction and angina may on occasion radiate to the jaws.

Multiple sclerosis May mimic trigeminal neuralgia or cause altered facial sensation. Eye pain (retrobulbar neuritis) is associated. *Diagnosis*: depends upon finding multiple focal neurological lesions, disseminated in time and place.

Atypical facial pain This constitutes a large proportion of patients presenting with facial pain. Classically, their symptoms are unrelated to anatomical distribution of nerves or any known pathological process, and these patients have often been through a number of specialist disciplines in an attempt to establish a diagnosis and gain relief. This diagnosis tends to be used as a catch-all for a large group of patients, with the connecting underlying supposition that the pain is of psychogenic origin. There may be a florid psychiatric history or undiagnosed depression; alternatively, the patient may simply be over-reacting to an essentially minor discomfort or recently noticed anatomical variant as part of a general inability to cope with life. Pointers to a psychogenic aetiology include imprecise localization, often bilateral pain or 'all over the place'. Bizarre or grossly exaggerated descriptions of pain. Pain is described as being continuous for long periods of time with no change, and none of the usual relieving or exacerbating factors apply. Sleeping and eating are not obviously disturbed, despite continuous unbearable pain. Most analgesics are said to be unhelpful although many will not have tried adequate analgesia. No objective signs are demonstrable and all investigations essentially normal. After exclusion of any possible organic cause the introduction of an antidepressant, e.g. dothiepin, may (or may not) produce dramatic improvement in the pain.

Bell's palsy Although the main symptom is facial paralysis, pain in or around the ear often radiating to the jaw, precedes or develops at the same time in about 50% of cases. Rx: prednisolone 20 mg tds for 3 days; half-dose for 3 days; half again, then stop, is one advocated regimen. None is proven.

461

Oral manifestations of skin disease

Lichen planus is a disease involving both the skin and mucous membranes. The oral mucosa is frequently involved either in isolation or preceeding the skin manifestations. Aetiology is unknown although there is probably an autoimmune component and 'lichenoid eruptions' may be an unwanted reaction to some drugs. Affects adults and F>M. It has an insidious onset commonly involving the buccal mucosa, but also the tongue, lips, gingiva, floor of mouth, and palate. Commonest oral lesion is as white papules with a lacy, reticular pattern; variants include a coalesced plaque like lesion, a bullous form, and an erosive form. Skin lesions affect the flexor surfaces of arms and wrists and the legs, particularly the shin, as purple papules with fine white lines (Wickham's striae) overlying them. Histology shows hyperparakeratosis with elongated rete ridges, a prominent granular cell layer, acanthosis, and basal cell liquefaction. There is usually a dense band of lymphocytes directly beneath the epithelium. It is chronic disease, lasting for months or years, and can usually be differentiated clinically (and always histologically) from leukoplakia. It is a benign condition. Rx: distinction from leukoplakia and reassurance is often all that is needed. If drugs are implicated, these should be avoided if possible. Prevention of superinfection is helpful in erosive lichen planus. Occasionally, topical steroids may be needed for particularly irritating patches and systemic steroids or surgery for some cases of erosive lichen planus, which has premalignant potential.

Dyskeratosis congenita A rare condition, characterized by oral leukoplakia, dystrophic changes of the nails, and hyperpigmentation of the skin; the oral lesion is prone to malignant change.

Vesiculous-bullous lesions, p. 442.

Oral manifestations of connective tissue disease

Ehlers–Danlos syndrome is a rare inherited condition affecting skin and other connective tissues which manifests with elasticity and fragility of the skin, easy bruising, and bleeding. Oral features include pulp stones and occasionally hypermobility of the TMJ. Bleeding during surgery, weak scars, sutures 'pulling through', and difficulty with RCT are the main practical points.

Rheumatoid arthritis (p. 540) Main associations are Sjögren syndrome and rheumatoid of the TMJ (10% cases), which may cause pain, swelling, and limitation of movement. Rarely, pannus formation within the joint may occur.

Systemic lupus erythematosus (SLE) is a systemic multisystem disease of unknown aetiology, associated with the presence of antinuclear factor. Gives rise to skin lesions, classical malar 'butterfly' rash, and oral mucosal lesions in 10–20%, which include ulceration and purpurae. Antinuclear antibodies are present.

Chronic discoid lupus erythematosus The lesions of this condition are limited to skin and mucosa. May present as disc-like white plaques in the mouth and can progress to SLE, although more likely to remain as a chronic and recurring disorder. Lip lesions in women may be premalignant. Rx: SLE—systemic steroids; DLE—topical steroids.

Systemic sclerosis A chronic disease characterized by diffuse sclerosis of connective tissues. F>M It has an insidious onset and is often associated with Raynaud's phenomenon (painful reversible digital ischaemia on exposure to cold). Classically,the face has a waxy mask-like appearance. Eating becomes difficult due to immobility of underlying tissues, and dysphagia occurs due to oesophageal involvement. Autoantibodies are present. Rx: penicillamine may help.

Polyarteritis nodosa Characterized by inflammation and necrosis of small and medium-sized arteries; necrosis at any site may occur and is seen as ulceration in the mouth.

Dermatomyositis Inflammatory condition of skin and muscles, 15% are associated with internal malignancy. Tenderness, pain, and weakness of the tongue may be an early finding.

Reiter syndrome, p. 541.

Oral manifestations of gastrointestinal disease

Patterson-Brown–Kelly syndrome, p. 746.

Coeliac disease This common form of intestinal malabsorption may present with oral ulceration as the only symptom in adults. Although children also present with ulceration, they are more likely to show weight loss, weakness, and failure to thrive. Other findings are glossitis, stomatitis, and angular cheilitis.

Ulcerative colitis Rarely, pyostomatitis vegetans, a papilliferous, necrotic mucosal lesion can occur; more commonly ulcers indistinguishable from apthae are seen.

Crohn's disease Chronic inflammatory disease affecting any part of the gut from mouth to anus. Primarily affects the terminal $\frac{1}{3}$ of the ileum although about 1% of cases will present with oral ulceration predating any other symptoms. These tend to affect the gingiva, buccal mucosa, and lips with purplish-red non-haemorrhagic swellings, linear long-standing ulcers, and granulations. Granulomatous cheilitis is probably a variant. Painful oral lesions seem to respond well to simple excision but treatment is aimed at the systemic disease.

Gardener syndrome, p. 744.

Peutz–Jeghers syndrome, p. 746.

Cirrhosis Glossitis occurs in about 50% of patients. Sialosis is another association.

465

Oral manifestations of haematological disease

Anaemia The nutritional deficiencies associated with anaemia; iron, B12 and folate are all associated with **recurrent oral ulceration** (p. 440) and specific deficiencies may be present, even in the absence of a frank anaemia. **Atrophic glossitis** was formerly the commonest oral symptom of anaemia but is less often seen now. Red lines or patches on a sore but normal looking tongue may indicate B12 deficiency. **Candidosis** (p. 438) may be precipitated or exacerbated by anaemia, particularly iron deficiency, and **angular cheilitis** is a well-recognized association. The sore, clinically normal tongue (**burning tongue**) is sometimes a manifestation or even precursor of anaemia. **Patterson-Brown–Kelly syndrome** (p. 746).

Leukaemia This and other haematological malignancies are associated with a reduction in resistance to infection. The mouth may be involved either secondarily to this ↑ tendency to infection or as a direct consequence of infiltration of the oral tissues. The oral lesions of leukaemia are painful and can lead to difficulty in swallowing. Prevention of superinfection with chlorhexidine mouthwashes and aggressive appropriate treatment of infections which arise, is of real help. There is an ↑ tendency to bleed manifested as fine petechial haemorrhages or bruising around the mouth, and the gingiva may bleed heavily in the presence of only negligible trauma. Management of the bleeding is aimed at the underlying disorder; local techniques include improving OH, avoiding extractions, and using local haemostatic methods (p. 390). Spontaneous gingival bleeding may be controlled, using impressions as a made-to-measure pressure dressing.

Cyclical neutropenia This condition may manifest as oral ulceration, acute exacerbations of periodontal disease, or AUG.

Myeloma Macroglossia is an occasional finding.

Purpura is due to platelet deficiency. Commonest as ideopathic thrombocytopenic purpura (ITP) in children. Palatal petechiae or bruising may be seen. Palatal petechiae are also seen in glandular fever, rubella, HIV, and recurrent vomiting.

'Angina bullosa haemorrhagica' Oral blood blisters which are irritating but of no known significance.

467

Oral manifestations of endocrine disease

Acromegaly Oral signs of acromegaly include enlargement of the tongue and lips, spacing of the teeth, and an ↑ in jaw size, particularly the mandible resulting in a Class III malocclusion. Jaw pain is sometimes described, which can respond to treatment of the growth-hormone-secreting pituitary tumour responsible for the disease.

Addison's disease (adrenocortical hypofunction) Classically, causes melanotic hyperpigmentation of the oral mucosa, commonly of the cheek. May also be part of the endocrine–candidosis syndrome.

Cushing syndrome The appearance of a 'moon face' and oral candidosis are the common head and neck manifestations of this syndrome.

Hypothyroidism Congential hypothroidism is associated with enlargement of the tongue, with puffy enlarged lips and delayed tooth eruption. In adult hypothroidism, puffiness of the face and lips also occurs, but there are no particular oral changes.

Hyperthroidism Not associated with any particular oral changes.

Hypoparathroidism may be a component in the endocrine–candidosis syndrome, but there are no other specific changes.

Hyperparathyroidism Rare. Caused by hyperplasia or adenoma of the parathyroids. ↑ parathormone causes ↑ plasma calcium liberated from bone. Appears in the jaws as loss of lamina dura, a 'ground glass' appearance of bone and cystic lesions (often looking multilocular), which contains dark-coloured tissue, 'brown tumour' histologically indistinguishable from a giant cell granuloma (p. 416).

Diabetes No *specific* oral changes, although manifestations of ↓ resistance to infection can be seen if poorly controlled (e.g. severe periodontal disease). Xerostomia and thirst are prominent features of ketoacidosis. Sialosis is sometimes seen as a late feature of diabetes. Burning mouth may be a presenting feature of diabetes, and oral or facial dysaesthesia may reflect the peripheral neuropathies seen in diabetics. There is a tendency to slower healing following surgery.

Sex hormones There is a well-recognized ↑ in the severity and frequency of gingivitis at puberty and in pregnancy. Some females have recurrent apthae clearly associated with their menstrual cycle, and several symptoms, usually burning tongue or mouth or general soreness of the tongue or mouth, have been described during the menopause. It should be remembered, however, that there are profound psychological changes at this time of life in many women, and these symptoms may be a manifestation of atypical facial pain rather than a directly hormonally mediated effect. Hormone replacement does not seem to help.

469

Oral manifestations of neurological disease

Examination of the cranial nerves p. 542 and general concepts of neurological disease on p. 544. Of the cranial nerves, the trigeminal and facial nerves contribute most to disorders affecting the mouth, face, and jaws.

Trigeminal nerve *Ophthalmic* lesions result in abnormal sensation in the skin of the forehead, central nose, upper eyelid, and conjunctivae. *Maxillary* lesions affect skin of cheek, upper lip and side of nose, nasal mucosa, upper teeth and gingiva, palatal and labial mucosa. The palatal reflex may be lost. *Mandibular* lesions affect skin of lower face, lower teeth, gingivae, tongue and floor of mouth. Lesions of the motor root manifest in the muscles of mastication. Taste sensation is *not* lost in such lesions, although other sensations from the tongue are. Testing is performed by having the patient close his eyes and report on sensations experienced, in comparison to each other, while the areas of superficial distribution of the nerve are stimulated by light touch (cotton wool) and pin-prick (probe or blunt needle). The motor branch is tested by moving the jaws against resistance. A blink should be elicited by stimulating the cornea with a wisp of cotton wool (corneal reflex).

Facial nerve is motor to the muscles of facial expression and stapedius, secretomotor to the submandibular and sublingual salivary glands, and relays taste from anterior $\frac{2}{3}$rds of tongue via the chorda tympani. Tested by having the patient raise eyebrows, screw eyes shut, whistle, smile, and show their teeth. Upper and lower motor neurone lesions can be distinguished because the forehead has a degree of bilateral innervation and is relatively spared in upper motor neurone lesions. Taste is tested, using solutions which are sour, salt, sweet, and bitter. If taste is intact, flow from the submandibular duct can be assessed by gustatory stimulation. Test hearing to assess stapedius.

Neurological causes of facial and oral pain, p. 460.

Neurological conditions causing altered sensation may be: **1** Local causes e.g. third molar surgery, local neoplasms or rarely viral infections, e.g. H. zoster. **2** Distant causes, e.g. multiple sclerosis (may cause altered facial sensation) or intracranial tumours (remember these in differential diagnosis of trigeminal neuralgia). Rarer causes include systemic lupus erythematosus, polyarteritis nodosa, and dermatomyositis.

Neurological causes of facial paralysis May be upper motor neurone or lower motor neurone; of the former, strokes are the commonest. Combination lesions can be caused by amyotrophic lateral sclerosis of the cord. Lower motor neurone paralysis can be caused by Bell's palsy (p. 461), trauma, infiltration by malignant tumours, Ramsey Hunt syndrome, Guillain–Barré syndrome (post-viral polyneuritis, may even appear to be bilateral). Apparent paralysis may occur in myaesthenia gravis where abnormally increased fatigue of striated muscle causes ptosis and diplopia. Therapeutic paralysis may be induced for facial spasm, using botulinus toxin injected locally.

Neurological causes of abnormal muscle movement Tetanus is an obvious cause. Muscular dystrophy may present with ptosis and facial paralysis. Hemifacial spasm and other tics may be caused by a tumour at the cerebello-pontine angle. Oro-facial dyskinesia can be a manifestation of Parkinson's disease or an unwanted effect of major tranquillizers. Phenothiazines and metoclopramide are notorious for causing dystonic reactions in young women and children. Bizarre attacks of trismus due to masseteric spasm have been ascribed to metoclopramide.

Oral manifestations of AIDS

AIDS is the terminal stage of infection with the Human Immunodeficiency Virus HIV, which is recognized as undergoing a number of mutations. It is discussed in general terms on p. 552. The underlying severe immunodeficiency leads to a number of oral manifestations which, although not pathognomonic, should raise the possibility of HIV infection. They have been classified.[1]

Group I Strongly associated with HIV.

Candidosis (p. 438). Erythematous, hyperplastic, thrush, and angular cheilitis in young people.

Hairy leukoplakia Bilateral white non-removable corrugated lesions of the tongue unaffected by antifungals.

HIV gingivitis unusually severe gingivitis for the general state of the mouth.

Acute ulcerative gingivitis (p. 218) in young otherwise healthy mouths.

HIV periodontitis Severe localized destruction out of place with oral hygiene.

Kaposi's sarcoma One or more erythematous/purplish macule or swelling frequently on the palate.

Non-Hodgkin's lymphoma Similar to the above.

Group II Less strongly associated with HIV.

Atypical oropharyngeal ulceration

Ideopathic thrombocytopenic purpura

Salivary gland disease

Wide range of common viral infections

Group III Possible association with HIV.

Wide range of rare bacterial and fungal infections

Cat scratch disease

Neurological abnormalities

Osteomyelitis/sinusitis/submandibular cellulitis

Squamous carcinoma

Clearly, the conditions in *Group III* are likely to be seen in patients who are HIV negative at least as often as in patients who are HIV positive.

Persistent generalized lymphadenopathy Otherwise inexplicable lymphadenopathy >1 cm persisting for 3 months, at 2 or more extra-inguinal sites. Cervical nodes particularly commonly affected. May be prodromal or a manifestation of AIDS.

1 S. Challacombe 1991 *BDJ* **73** 305.

Rx for AIDS AIDS is currently incurable, but then again so is life. Zidovudine may prolong and improve quality of life in those with active AIDS but has otherwise proved dissappointing. A range of symptomatic Rx for associated conditions, e.g. co-trimoxazole for pneumocystis pneumonia and fluconazole for candidosis, exist and can both prolong and improve the quality of life for these individuals.

Dental treatment for patients with AIDS This group present two risks with regard to dental treatment. **1** To personnel carry-ing out the treatment. Affected patients carry an infectious disease for which there is no known cure and which is trans-mitted by blood and blood products. As it is impossible to adequately identify all such patients, routine cross-infection control is now a necessity. **2** As these patients are immuno-compromised, any treatment with a known risk of infective complications, e.g. extractions, should be covered with anti-septic and antimicrobial prophylaxis, and any surgery should be as atraumatic as possible. There may also be a slight ↑ tendency to bleeding in these patients, and local haemostatic measures (p. 392) may be needed.

473

Cervico-facial lymphadenopathy

You cannot palpate a normal lymph node, ∴ a palpable one must be abnormal. The most important distinction to make is whether this is part of the lymph nodes' physiological response to infection or whether it is undergoing some pathological change.

Investigations Routine extra-oral and intra-oral examination (p. 12) to exclude the common causes: apical and periodontal abscesses, pericoronitis, tonsillitis, otitis, etc. The fundamentals are history and palpation.

History Ask about pain or swelling in the mouth, throat, ears, face, or scalp. Was there any constitutional upset when the lump appeared? Has it been getting bigger progressively or has it fluctuated? Is it painful, and how long has it been present?

Palpation Fully expose the neck and palpate from behind, with the patient's head bent slightly forward to relax the neck. Examine systematically, feeling the submental, facial, submandibular, parotid, auricular, occipital, the deep cervical chain, supraclavicular and posterior triangle nodes. Differentiating between the submandibular salivary gland and node can be a problem and is made simpler by palpating bimanually; the salivary gland can be felt moving between the external and internal fingers. Supraclaviular nodes are more liable to be due to occult tumour in the lung or upper gastrointestinal tract, whereas posterior triangle nodes are more liable to be haematological in origin.

If a node is palpable, note its texture, size, site, and whether it is tender to touch or fixed to surrounding tissues.

Nodes which are acutely infected tend to be large, tender, soft, and freely mobile.

Chronically infected nodes are soft to firm and less liable to be tender.

Metastatic carcinoma in nodes tends to be hard and fixed.

Lymphomatous nodes are described as rubbery and have a peculiar firm texture.

Supplementary investigations Examine axillary and inguinal nodes, liver, and spleen. Carry out a FBC to look for leukocytosis and a monospot test for glandular fever. Once infection is excluded it is essential to exclude, as far as possible, an occult primary malignancy of the head and neck; the best way to do this is by EUA, panendoscopy, and chest X-ray.

If the diagnosis has still not been established it is reasonable to proceed to excision biopsy of the node, which should be cultured for mycobacteria as well as examined histologically. Occasional examinations include ultrasound, CT or MRI scans, and fine needle aspiration cytology (this is useful *only* when skilled cytology services are available and the technique is used regularly).

Common causes Dental abscesses, pericoronitis, tonsillitis, glandular fever, lymphoma, metastatic deposits.

Rare causes Brucellosis, atypical mycobacteria, TB, AIDS, toxoplasmosis, actinomycosis, sarcoidosis, cat scratch fever, syphilis, drugs (e.g. phenytoin). Other neck lumps, see p. 512.

An approach to oral ulcers

Oral ulceration is probably the commonest oral mucosal disease seen; it may also be the most serious. It is important, therefore, to have an approach to the management of oral ulcers firmly established in your mind.

Duration How long has the ulcer been present?

▶ If > 3 weeks, referral for biopsy is mandatory.

If of recent onset, ask whether it was preceded by blistering. Are the ulcers multiple? Is any other part of the body affected and have similar ulcers been experienced before? Then look at the site &/or distribution of the ulcer(s).

Blistering Preceeding the ulcer suggests a vesiculo-bullous condition (p. 442) such as herpetic gingivostomatitis. Blistering with lesions elsewhere in the body suggests erythema multiforme, or hand, foot, and mouth disease.

Distribution Limited to the gingivae suggests AUG (p. 220). Unilateral distribution suggests herpes zoster (p. 436). Under a denture or other appliance suggest traumatic ulceration.

Recurrence of the ulcers after apparent complete resolution is characteristic of recurrent apthae (p. 440).

Pain The presence or absence of pain is not particularly useful diagnostically, although the character of the pain may be of value. Pain is often a late feature of oral carcinoma and the fact that an ulcer may be painless *never* excludes it from being a potential cancer.

For most ulcers of recent onset, and a few present for an indeterminant period of time, a trial of therapy is often a useful adjunct to diagnosis. This is especially useful in recurrent oral ulceration, viral conditions (where treatment is essentially symptomatic), and lesions probably caused by local trauma (treatment being removal of the source of trauma and review after 1 week).

Ulcers which need early diagnosis include:

Herpes zoster As early aggressive treatment with acyclovir may reduce post-herpetic neuralgia.

Erythema multiforme In order to avoid re-exposure to the antigen.

Erosive lichen planus As this may benefit from systemic steroids.

Oral squamous cell carcinoma For obvious reasons.

477

Temporomandibular pain—dysfunction/ facial arthromyalgia

The allocation of this section to a speciality chapter created problems in just the same way as the referral of the 'TMJ' patient to a specialist is fraught with confusion. We have selected this site because we feel it is important to look at these patients from a physician's approach rather than a dental or surgical one.

What is it? The problem being addressed is pain in the pre-auricular area and muscles of mastication with trismus, with or without evidence of internal derangement of the meniscus. Conditions which can otherwise be classified as facial pain syndromes or other forms of joint disease are excluded and can be found in the relevant pages (pp. 460, 498).

Prevalence Affects around 40% of the population at some time in their lives. F>M.

Aetiology Idiopathic. Multiple theories put forward regarding occlusion, trauma, stress. To date, the concept of stress-induced parafunctional habits (bruxism, clenching) causing pain and spasm in the masticatory apparatus coupled with a ↓ pain threshold has seemed the most reasonable. This is compatible with the observed high association with back pain, headaches, and migraine. It does not explain the cause in those patients who can identify no different levels of stress in their lives nor does it help explain the high incidence of internal derangement of the meniscus. The discovery of a biochemical marker (tyramine sulphate in the urine) in non-depressed 'TMJ' patients has suggested that these patients are somehow biochemically sensitive to both mediators of damage in the joint, such as neuropeptides, and centrally (via serotonin) resulting in a lowered ability to cope with the local discomfort. The neuropeptide release can explain both joint pain and internal derangement (see diagram 1).[1]

Clinical features Pain, clicking, locking, crepitus, and trismus are the classical signs and symptoms. Some patients may be clinically depressed but most are not. Clicking commonly occurs at 2–3 mm of tooth separation on opening and sometimes closing. This is due to the meniscus being displaced anteriorly on translation of the head of the condyle and then returning to its usual position (the click). A lock is when it does not return.

Management Success has been claimed for a wide range of treatments reflecting confusion over diagnosis and the multi-factorial and self-limiting nature of the condition. Simple conservative treatment within the range of every dentist is successful in up to 80% of cases (see diagram 2).

1 *Reassurance and explanation* Advice as to the nature of the problem and its benign and frequently self-limiting course is all that many patients require. Do not create a problem where

1 M. Harris 1993 *BDJ* **174** 129.

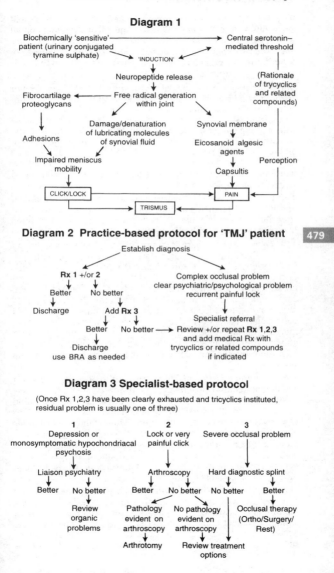

Diagram 1

Biochemically 'sensitive' patient (urinary conjugated tyramine sulphate) → Central serotonin-mediated threshold

'INDUCTION'

Neuropeptide release

(Rationale of trycyclics and related compounds)

Free radical generation within joint → Fibrocartilage proteoglycans

Fibrocartilage proteoglycans → Adhesions

Free radical generation → Damage/denaturation of lubricating molecules of synovial fluid

Free radical generation → Synovial membrane → Eicosanoid algesic agents → Capsultis

Adhesions → Impaired meniscus mobility

Perception

CLICK/LOCK — PAIN

TRISMUS

Diagram 2 Practice-based protocol for 'TMJ' patient

Establish diagnosis

Rx 1 +/or 2 → Better → Discharge

Rx 1 +/or 2 → No better → Add Rx 3 → Better → Discharge use BRA as needed

Add Rx 3 → No better → Review +/or repeat Rx 1,2,3 and add medical Rx with trycyclics or related compounds if indicated

Complex occlusal problem clear psychiatric/psychological problem recurrent painful lock → Specialist referral → Review +/or repeat Rx 1,2,3 and add medical Rx with trycyclics or related compounds if indicated

Diagram 3 Specialist-based protocol

(Once Rx 1,2,3 have been clearly exhausted and tricyclics instituted, residual problem is usually one of three)

1
Depression or monosymptomatic hypochondriacal psychosis → Liaison psychiatry → Better / No better → Review organic problems

2
Lock or very painful click → Arthroscopy → Better / No better → Pathology evident on arthroscopy → Arthrotomy / No pathology evident on arthroscopy → Review treatment options

3
Severe occlusal problem → Hard diagnostic splint → No better / Better → Occlusal therapy (Ortho/Surgery/Rest)

Temporomandibular pain—dysfunction/ facial arthromyalgia (*cont.*)

there is none! This is also the time to make a gentle but thorough social and family history to identify clinically depressed patients or those with significant stress.

2 *Simple analgesia, rest, gentle heat, and remedial exercises* Whether these are given by dentist, physiotherapist (in the form of short-wave diathermy and ultrasound) or ancillary staff is unimportant as the crucial part is that carried out by the patient at home by taking the analgesics and performing the exercises as instructed.

3 *Splint therapy* Upper/lower, hard/soft have all been used with varying success. The initial aim of a splint is: (a) to show something is being done (placebo); (b) ↓ bruxism and joint load; and (c) by ↑ the gap between condyle and fossa the disc may be freed. A simple full coverage upper or lower splint should be worn as often as possible, nights and evenings especially, and reassessed after 4 to 6 weeks.

These three simple measures should relieve symptoms in around 80% of patients and identify those needing referral to a specialist. **Do not** persist with ineffective treatment if symptoms have not improved within 3 months.

4 *Drug therapy* There is a natural reluctance among many patients to taking drugs, particularly those associated with psychiatry. There is also a misconceived reluctance among clinicians to use the tricyclics and related compounds. They are non-addictive and the commoner side-effects of weight gain, constipation, and dry mouth can be overcome. Benzodiazepines are not recommended but nortriptyline, dothiapine, and related compounds have been demonstrated to have analgesic and muscle relaxant effects independent of their anitdepressant effect and probably work via the central biochemical sensitivity identified by the presence of urinary conjugated tyramine sulphate.

5 *Occlusal adjustment* There are a group of patients where a significant occlusal problem exists. In these cases a hard diagnostic occlusal splint can be constructed for the mandible (Tanner) or for the maxilla (Michigan) and should be made to give multiple even contacts in centric relation and anterior guidance. The patient attempts to wear this full-time for up to 3 months. If pain is abolished while wearing the appliance, returns when it is removed, and is abolished on reinstitution, then occlusal adjustment by orthodontic, surgical or restorative means is a reasonable option.

6 *Surgery for internal derangement* If pain can be abolished by other methods and the patient continues to be bothered by a painful click particularly with recurrent locking, treatment aimed specifically at the meniscus is justified. The first line which may be useful diagnostically and improve pain due to capsulitis is arthroscopy. This is examination and irrigation of the upper joint space by a rigid endoscope through which lysis and lavage of adhesions and synovial inflammatory mediators can be performed. Menisci damaged beyond the scope of arthroscopy can be repositioned at open arthrotomy. Consensus dictates that the minimum of interference to the articulatory surfaces and the avascular meniscus is carried out.

Advanced trauma life support—ATLS 484
Primary management of maxillofacial
  trauma 486
Assessing head injury 488
Mandibular fractures 490
Mid-face fractures 492
Nasal and malar fractures 494
Treatment of facial fractures 496
Surgery and the temporomandibular joint 498
Major preprosthetic surgery 500
Clefts and craniofacial anomalies 502
Orthognathic surgery 504
Salivary gland tumours 506
Surgery of the salivary glands 508
Oral cancer 510
Neck lumps 512
Flaps and grafts 514

In general, maxillofacial surgery is a postgraduate subject which has evolved from oral surgery, with foundations in medicine, dentistry, and surgery. It is included here as an introduction to students, an *aide-mémoire* for junior hospital staff, and a guide for those who will be referring patients.

Principal sources: N. Rowe 1985 *Maxillofacial Injuries*, Churchill Livingston. J. R. Moore 1985 *Surgery of the Mouth and Jaws* 1985, Blackwell Scientific. A. Maran 1993 *Head and Neck Surgery*, Butterworth–Heinemann.

Advanced trauma life support—ATLS

ATLS is a system providing one safe way of resuscitating a trauma victim. It was first conceived in Nebraska, subsequently developed by the American College of Surgeons, and has now reached international acceptance. It is not the only approach but it is one which works. It is highly recommended.

Trauma deaths have a trimodal distribution. The first peak is within seconds to minutes of injury. The second is within the first hour, the 'golden hour', and is the area of main concern. The third is days to weeks later but may reflect management within the golden hour.

The core concept behind ATLS is the **primary survey** with **simultaneous resuscitation** followed by a **secondary survey** leading to definitive care.

Primary survey

This uses the mnemonic **ABCDE** on the basis of identifying and treating the most lethal injury first.

A *Airway with cervical spine control*—establish a patent airway and protect the cervical spine from further injury. Chin lift, jaw thrust, oral airway, nasopharyngeal airway, intubation, surgical airway as needed, coupled with manual inline immobilization of cervical spine or rigid cervical collar, sandbags, and tape..

B *Breathing and ventilation*—inspect, palpate, and ausculate the chest. Count respiratory rate. Give 100% oxygen. Chest decompression by needle puncture in 2nd interspace mid-clavicular line if indicated for tension pneumothorax.

C *Circulation with haemorrhage control*—assess level of consciousness, skin colour, pulse and BP, manual pressure control of extreme haemorrhage. Establish 2 large venous cannulae, take blood for X-match and baseline studies. Give 2 litres of prewarmed Hartmann's solution. Establish ECG,.

D *Disability*—(neuro evaluation) quick test is AVPU, is the patient Alert, responding to Vocal stimuli, responding to Painful stimuli, or Unresponsive. Establish urinary catheter unless urethral transection is suspected and naso- or oro-gastric tube.

E *Exposure*—remove all clothing to allow full assessment of injuries. Establish monitoring; respiratory rate, BP, pulse, arterial blood gases, pulse oximetry, and ECG prevent hypothermia.

If all these can be established and monitored parameters are stable the patient's chances of living are optimized.

X-rays at this stage obtain a cross table cervical spine, chest, and pelvis film in the resuscitation room.

Reassess the ABCs

If all is stable move to the **secondary survey** which is a head-to-toe examination of the patient. It is at this stage *only* that specific non-immediately life-threatening conditions should be identified and dealt with in turn.

Maxillofacial injuries other than those with a direct effect on the airway, cervical spine, or causing exsanguinating haemorrhage should not be dealt with until the **ABC**s have been completed and this question should be asked of all referring doctors prior to accepting responsibility for a patient. Remember to exclude intracranial, visceral, and major orthopaedic injuries.

While ATLS is designed primarily for front-line physicians, modified courses for dental graduates are available and are recommended to anyone who may undertake the care of the trauma patient.

Primary management of maxillofacial trauma

The first consideration is whether the patient has suffered severe trauma, which may have multiple and life-threatening ramifications, or, as is more commonly the case, simple trauma confined to the face. In the former the prime concern is keeping the patient alive, and the maxillofacial injuries can await treatment (p. 484). Remember that isolated facial injuries rarely cause sufficient bleeding to induce hypovolaemic shock.

Airway The brain can tolerate hypoxia for only around 3 min. Most conscious patients can maintain a patent airway if the oropharynx is cleared. Give all traumatized patients oxygen initially. Oral airways are seldom tolerated. Nasopharyngeal tubes are only of value if they can be inserted safely and kept patent. If the patient is unconscious and the airway is obstructed they should be intubated. If this is not possible, an emergency airway can be maintained by cricothyroid puncture with a wide-bore cannula. Long-term security of the airway can be achieved by tracheostomy. This should not be carried out by the inexperienced as an emergency, but on a prepared patient as an urgent planned operation.

Cervical spine injuries Until these are excluded by lateral C-spine X-ray the patient should wear a semi-rigid cervical collar reinforced by sandbags and tape, immobilizing the cervical spine.

Bleeding As above. Gunshot wounds and lacerated major vessels are exceptions which can cause extensive bleeding. Specific techniques to control naso- and oropharyngeal bleeding are: bilateral rubber mouth props to immobilize the maxilla, bilateral balloon catheters passed into the post-nasal space, inflated then pulled forward and bilateral anterior nasal packs.

Head injuries These are the main cause of death and disability in patients with isolated maxillofacial trauma. Assessment, p. 488.

CSF leaks Facial and skull # can create dural tears, leading to CSF rhinorrhoea (from the nose) or otorrhoea (from the ear). Although controversial, prophylactic antibiotics are usually recommended. A penicillin in high dose *plus* either sulphadiazine 500 mg qds IM or PO or rifampicin 600 mg od PO, as penicillins do *not* cross the healthy blood–brain barrier in adequate levels.

Tetanus immunity If in doubt give tetanus toxoid.

Analgesia May not be needed. Avoid opioids. If needed, use diclofenac 75 mg IM bd or codeine phosphate 60 mg IM 4 hrly.

Patients to admit Any question of danger to the airway, skull #, history of unconsciousness, retrograde amnesia, bleeding, middle $\frac{1}{3}$rd #, mandibular # (except when very simple), malar # if positive eye signs, children, and those with domestic or social problems. If in doubt *admit*. Place on, at least, initial hourly neurological observations (most will need $\frac{1}{4}$ hrly obs initially). Consider IV access. If not admitting, give the patient a head-injury card.

If teeth have been lost ensure they are not in the chest or soft tissues.

X-rays, p. 490, p. 492, p. 494.

Assessing head injury

▶ All patients with recent facial trauma warranting hospital admission need at least initial assessment for head injury.

Change in the level of consciousness is the earliest and most valuable sign of head injury.

A combination of the following is generally used.

Glasgow coma scale (GCS)

Eyes open
- spontaneously   4
- to speech   3
- to pain   2
- do not open   1

Best verbal response		**Best motor response**	
• orientated	5	• obeys commands	6
• confused	4	• localizes pain	5
• inappropriate	3	• normal flexion	4
• incomprehensible	2	• abnormal flexion	3
• none	1	• extension	2
		• none	1

Pulse and BP ↓ pulse and rising BP is a late sign of ↑ intracranial pressure.

Pupils Measure size (1–8 mm) and reaction to light in both pupils.

Respiration ↓ rate is a sign of ↑ intracranial pressure.

Limb movement
Indicate normal
mild weakness
severe weakness
extension
no response

for arms and legs (record right and left separately if there is a difference).

CT scan is the definitive investigation. However patients must *never* be transferred before correcting hypoxia and hypovolaemia.

Using Glasgow coma scale (GCS)

▶ Severe head injury +/or deterioration = call for help.

One accepted method of categorizing head-injured patients by severity using GCS is:[1]

Severe   <8
Moderate 9–12
Minor   13–15

1 *ATLS Core Course Manual* 1993, American College of Surgeons.

In addition:
A severe head injury is present if there are:
- Unequal pupils (except traumatic mydriasis).
- Unequal limb movement (except orthopaedic injury).
- Open head injury (i.e. compound to brain).
- Deterioration in measured parameters.
- Depressed skull #.

Subtle signs of deterioration include:
- Severe &/or worsening headache.
- Early unilateral pupillary dilatation.
- Early unilateral limb weakness.

A GCS of <6 in the absence of drugs has a dismal prognosis.

Mandibular fractures

Mandibular # are the commonest # of the facial skeleton. Most are the result of fights and road traffic accidents (the former appears to be ↑ whereas the latter is ↓ as a result of seat belts, etc.). Rarely, they may be comminuted with hard and soft tissue loss, e.g. gunshot wounds.

Classification The most useful is based on site of injury; dento-alveolar, condyle, coronoid, ramus, angle, body, symphyseal, or parasymphyseal. Further subclassification into unilateral, bilateral, multiple, or comminuted aids treatment planning. In common with all # they can be grouped into simple (closed linear #), compound (open to mouth or skin), pathological (# through an area weakened by other pathology), or comminuted; again, this influences treatment.

Muscle pull Pull on # segments render the # favourable or unfavourable depending on whether or not the # line resists displacement. This is of less importance than recognizing the # and its associated injuries.

Common # Condylar neck # are the commonest and usually need the least invasive Rx. Often found with a # of the angle or canine region of the opposite side of the jaw. Rarely, bilateral condylar # is found with a symphyseal # 'guardsman's #' from falling on the point of the chin. Angle # usually occurs through wisdom-tooth socket and body # commonly through canine socket.

Diagnosis History of trauma. Ask if the patient can bite their teeth or dentures together in the manner which is normal to them. Inability to do this and a lingual mucosa haematoma is pathognomonic of a mandibular # . Look at the face; there is usually bruising and swelling over the # site and sometimes lacerations. If the # is displaced, there may be gagging on the posterior teeth and the mouth hangs open. The saliva is usually bloodstained. The patient may complain of paraesthesia in the distribution of the IDB. Gentle palpation over the mandible will reveal step deformities, bony crepitus, and tenderness.

Examination of the mouth may reveal broken teeth or dentures which should be removed. Rinse the mouth to clean away blood clots prior to examining both the buccal and lingual sulcus. Look for step deformities in the occlusion, and examine the teeth. Palpate for steps in the lingual and buccal sulcus. If diagnosis is uncertain it is sometimes worth trying to elicit abnormal mobility across the suspected # site, using gentle pressure. In cases where you are very unsure, place one hand over each angle of the mandible and exert gentle pressure; this will produce pain if there is a # , even if it is only a crack # . Never perform this if you have proved otherwise that there is a # .

X-rays OPG and PA mandible are the essentials. Right and left lateral obliques if OPG is unobtainable. Rotated PA (helpful for # between the symphysis and canine region), IO periapicals, occlusal, high OPG, or reverse Townes for condyles, are all second-line investigations which may help.

Preliminary Rx (p. 484) Most patients will be admitted to hospital, nursed sitting up and leaning forward, as this is the most comfortable position. Barrel bandages are a waste of time. Keep nil by mouth and maintain hydration by IV crystalloid. Compound # need prophylactic antibiotics.

Mid-face fractures

The mid-facial skeleton is a complex composite of fine fragile bones, which rarely # in isolation. It forms a detachable framework which protects the brain from trauma. Severe trauma can move the entire mid-face downwards and backwards along the base of the skull, lengthening the face and obstructing the airway (clot and swelling exacerbates this). Most *conscious* patients, however, can compensate for this. # of the cribiform plate of the ethmoids can lead to dural tears and CSF leak (p. 486).

Orbit The globe and optic nerve are well-protected by the bony buttress of the orbit. Most # lines pass around the optic foramen; however, swelling can cause proptosis. Late changes can include tethering and enopthalmos. *Retrobulbar haemorrhage* is an arterial bleed behind the globe following trauma. It presents as a painful, proptosed eye with decreasing visual acuity, and is a surgical emergency. The clot must be decompressed and evacuated. Medical management with mannitol 20% 2 g/kg, acetazolamide 500 mg and dexamethosone at least 1 mg/kg all IV may help while theatre is being arranged.

Bleeding Severe bleeding is rare, but severe mid-facial trauma may lacerate the 3rd part of the maxillary artery, resulting in profuse bleeding into the nasopharynx which requires anterior and posterior nasal packs and possibly direct ligation.

Classification Mainly based on the experimental work of René Le Fort. Le Fort I # detaches the tooth-bearing portion of the top jaw via a # line from the anterior margin of the anterior nasal aperture running laterally and back to the lower $\frac{1}{3}$ of the pterygoid plate. Le Fort II detaches the true mid-face in a pyramidal shape (see diagram). Le Fort III detaches the entire facial skeleton from the cranial base, as diagram.

Diagnosis Le Fort I may occur singly or associated with other facial #. The tooth-bearing portion of the upper jaw is mobile, unless impacted superiorly. There is bruising in the buccal sulcus bilaterally, disturbed occlusion, and posterior 'gagging' of the bite. Grasp the upper jaw between thumb and forefinger anteriorly, place thumb and forefinger of other hand over the supraorbital ridges and attempt to mobilize the upper jaw to assess mobility. Spring the maxillary teeth to detect a palatal split. Percussion of the upper teeth may produce a 'cracked cup' sound. Le Fort II and III # produce similar clinical appearances; namely, gross oedema of soft tissues, bilateral black eyes (panda facies), subconjunctival haemorrhage, mobile mid-face, dish-face appearance, and extensive bruising of the soft palate. Look for a CSF leak and assess visual acuity. Le Fort II # may also show infra-orbital nerve paraesthesia and step deformity in the orbital rim. Peculiar to Le Fort III # are tenderness and separation of the frontozygomatic suture and deformity of zygomatic arches bilaterally and mobility of entire facial skeleton.

X-rays Occipito-mental 10° and 30°, submento-vertex, lateral skull, postero-anterior skull only. If C-spine is confirmed to be intact. Otherwise secure C-spine and await CT scan as the definitive imaging technique.

Preliminary Rx as p. 484, 486; **Definitive Rx**, p. 496.

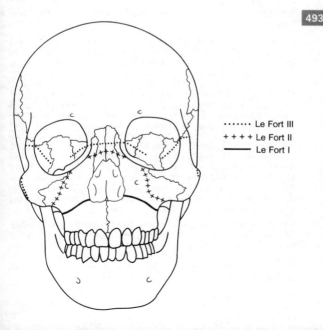

······ Le Fort III
+ + + + Le Fort II
——— Le Fort I

Nasal and malar fractures

Malar (or zygoma) # is a common and easily missed injury, usually the result of a blow with a blunt object (e.g. fist). The malar forms the cheekbone and resembles a 4-pointed star on occipito-mental X-ray. The 'star' points to the maxilla (orbital margin and lateral wall of antrum), frontal bone, and temporal bone. # can involve the arch alone or the whole malar, which may or may not be displaced. The nature of the displacement is of value in planning the treatment.

Diagnosis From history, examination, and X-ray. Bruising around the eye with subconjunctival haemorrhage (unilateral). Diplopia, a step deformity of the orbital rim, sometimes paraesthesia of the infra-orbital nerve, limitation of lateral excursion of the mandible or mouth opening and unilateral epistaxis. There is often tenderness on palpation of the zygomatic buttress intra-orally and usually some flattening of the cheek prominence.

Orbital floor # Main signs are those of the malar # (or middle $\frac{1}{3}$rd # if that is the presenting injury). Late signs are enopthalmos and tethering of inferior rectus, causing diplopia in upwards gaze. Also known as orbital blow-out #, fat and muscle herniates through the thin orbital floor (a similar injury can occur to the medial wall). Classically, seen on X-ray as 'hanging drop sign' (radiolucent fat hanging into antrum). Confirm with coronal CT scan. Lateral wall and/or roof # are much less common.

Nasal # Simple nasal trauma is seen by a number of different specialities and is considered rather trivial. This is unfair to the patients, as long-term results of nasal # leave a lot to be desired. Nasal # are frequently associated with deviation or crumpling of the septum, obvious nasal deformity, epistaxis, and a degree of nasal obstruction. Rx: often consists of simply manipulating the nasal bones with the thumb and splinting with plaster of paris. This leaves the septum untreated and contributes to poor long-term results, many needing rhinoplasty later.

Nasoethmoid # consists of nasal bones, frontal process of maxilla, lacrimal bones, orbital plate of ethmoid, and displacement of the medial canthus of the eye. These # require accurate reduction, stabilization, and fixation of the medial canthus.
Diagnosis: bilateral black eyes, obvious nasal deformity (particularly depression of the nasal bridge), septal deviation, epistaxis, and obstruction. Look for CSF leak.

Septal haematoma is a comparatively uncommon complication of nasal trauma which demands immediate evacuation as, if ignored, it can lead to septal necrosis.

Treatment of facial fractures

Essentially involves reduction, fixation, and immobilization of the # segments. In mandibular and maxillary #, this has traditionally been achieved by intermaxillary fixation (IMF), i.e. immobilizing the jaws in occlusion. Wisdom teeth, grossly broken down or periodontally infected teeth in the # line should be removed. Otherwise, provision of prophylactic antibiotics and adequate reduction and immobilization with good follow-up (including endodontics as a secondary procedure) often allows preservation of teeth in the # line. Minimum period of immobilization is 3 weeks, though children may need less. Add 1 week for: tooth in the # line, # of symphysis, age >40.

Open reduction and internal fixation (ORIF) of facial bone # s has revolutionized Rx. # sites are exposed usually via mucosal incisions and reduced under direct vision, the teeth are placed in temporary IMF +/− a wire around the tooth on either side of the # (tension band or bridle wire) and small bone plates of titanium, vitallium or stainless steel are placed to fix the reduction. IMF can then be released for recovery and elastics placed instead on the ward when the patient is fully conscious. Rarely interosseous wires replace plates or IMF may be used on its own.

Edentulous mandible The absence of teeth in occlusion created problems when relying on IMF, and modified dentures called Gunning splints were used. This technique has been superseded by the use of bone plates in the majority of cases.

Condylar # Can usually be managed conservatively with a soft diet and analgesics, provided the occlusion is not deranged. In intra-capsular # *do not* immobilize the joint. # dislocation often require ORIF for best results.

Mid-face # Various methods are used. Either:

External fixation, e.g. Levant frame, Royal Berkshire halo, which fix the mid-face to the cranium, *or*

Internal fixation Interosseous wiring, plating, transfixation with Kirschner wires. Plating to recreate the pyriform and zygomatic buttress system is currently most widespread.

Most of these techniques are used in conjunction with IMF, although this can be minimized if plating techniques are used.

Malar # Are elevated by a temporal approach (Gillies) or bone hook percutaneously and supplemented if needed as above.

Nasal bones Are manipulated and splinted. Some benefit from mini submucous resection of the septum.

Nasoethmoidal # are openly reduced and wired or microplated to reposition the medial canthi and restore anatomy.

Fractures in children are considerably rarer. Plates and pins tend to be avoided because of the risk to unerupted teeth. Patients <10 yrs, usually require a form of Gunning splint which fits over the mixed dentition. # are usually firmly united within 3 weeks. Microplates can sometimes be used.

Post-op care Prophylactic antibiotics and scrupulous OH are required. The main problems are presented by IMF, and these patients need to be 'specialed' post-operatively (p. 486). IMF creates a need for a liquidized diet of at least 2500 calories (2000 for F) and 3 litres of fluid daily. Do not discharge until this can be maintained at home. ORIF requires a soft diet for 3 weeks.

Complications *Mandible* Infection, paraesthesia, damage to teeth, TMJ pain, malunion, delayed union, non-union, bony sequestration, plate and wire infection. *Maxilla* post-operative deformity, epiphora, diplopia, late enophthalmus, anosmia. Failure of union is very rare, although mal-union is a problem with poor reduction or late referral. *Malar* Diplopia, retro-bulbar haemorrhage, enophthalmos, paraesthesia. *Nasal* deformity, nasal obstruction.

497

Surgery and the temporomandibular joint

TMPDS (p. 478).

Ankylosis May be true or false. *True ankylosis* is restriction of movement caused by a pathological condition of the joint. Usually due to trauma (intracapsular # in childhood) or infection. Extreme limitation of movement and X-rays will confirm degree of bony union. In fibrous union, exercises are of value. In bony union, interpositional arthroplasty or condylectomy with reconstruction is needed. Post-operative exercise is crucial.

False ankylosis restriction of movement imposed by extra-articular abnormalities is very rare. Rx depends upon cause. *Trismus* (limitation of movement due to spasm of muscles of mastication) may be confused with ankylosis but does not affect the joint. It is much more common and complicates many oral surgical procedures; may follow trauma, infection, or may be a manifestation of occult malignancy.

Trauma Condylar #, p. 490.

Intracapsular # Occlusal disturbance may be due to a traumatic synovitis without #. Most condylar trauma is managed conservatively. If #, joint must be kept mobile to prevent ankylosis.

Dislocation may occur in normal joints due to exceptional circumstances, or in lax joints where dislocation is recurrent. May be unilateral or bilateral. Condyle can be palpated anterior to articular eminence, X-rays confirm position, mouth is gagged open. Rx: Immediate reduction: the vast majority can be performed with LA around the dislocated joint ± sedation. Place thumbs over molar teeth and exert downward and backward pressure (if LA is used, less force is needed and thumbs can be placed in buccal sulcus, avoiding the risk of being bitten). Advise jaw support when yawning, etc.; this is usually enough, and avoids IMF. In chronically recurring dislocations, patients can be taught to self-reduce. Exercises are of limited benefit but may avoid surgery. Sclerosant injections are unpleasant and no longer used. Operations are legion; capsular plication, pins to augment articular eminence, Dautrey downfracture of zygomatic arch, obliteration of upper joint space, eminectomy, high conylectomy. No one operation has gained pre-eminence.

Condylar hyperplasia Rare. Rx: high condylectomy.

Tumours Rare. Rx: surgery; add radiotherapy if malignant.

Arthritides Rheumatoid, psoriatic, and gouty arthritides all manifest in the TMJ but in <15% of patients with the systemic disease. The signs and symptoms are joint stiffness, pain, tenderness, and crepitus. *Diagnosis*: knowledge of systemic disease and local signs. Rx: treat systemic condition. Symptomatic Rx of joint with appliances, physiotherapy, exercises and intra-articular steroids. Main long-term problem is ankylosis.

Osteoarthrosis is a distinct entity of the TMJ, with a different clinical course from that seen in other joints. Appears to be a degenerative condition of articular cartilage. *Diagnosis*:

crepitus, limitation, and pain on movement, tenderness over condyle, often with X-ray evidence of condylar erosions. Most have symptoms for around 1 yr which gradually ↓ over the next 2 yrs. X-rays show condylar remodelling to a flat smooth surface. Patients then enter a long period of remission. Rx is ∴ aimed at pain relief, using standard TMPDS measures (p. 478). Those remaining unresponsive may benefit from intra-articular steroids (p. 606) which probably accelerates natural remodelling. A small group will remain with pain 3–4 months after steroids, usually with X-ray abnormalities. They should be considered for high condylar shave or high condylectomy.

Surgery and TMPDS Approximately 20% of patients with TMPDS remain unresponsive to conservative Rx. They may benefit from arthrography or arthroscopy both as an investigation and a treatment (distension of the joint space-releasing adhesions). Surgery for pure TMPDS in patients unresponsive to the above may be of benefit (probably by cutting the nerve supply to the joint), although clicking can certainly be eliminated by meniscal plication &/or pterygoid myotomy, and irrevocably damaged menisci can be removed.

499

Major preprosthetic surgery

Minor procedures, p. 426; implantology, p. 428. The aim is to enable an edentulous patient to live comfortably with functioning dentures, therefore surgery without liasion with an understanding and competent prosthodontist is pointless. All procedures for improving the denture-bearing area of the jaws are dependent on the use of a denture with a modified fitting surface (stent) which is placed at operation and must be worn virtually permanently for up to 8 weeks post-op, the fitting surface being modified at intervals with a soft acrylic lining material. There are many who would claim these procedures do not work, and indeed there is little scientific support for them.

▶ Warn all patients undergoing lower jaw procedures about post-op mental nerve damage.

Epithelial inlay Basically a skin graft to the alveolar surface of the jaw requiring a deeper sulcus. Important points: excise all areas of scarred or hyperplastic tissue, dissect off any strands of mentalis in the lower jaw, ensure preservation of the alveolar periosteum, ensure the stent extends to the new sulcus depth, and ensure the skin-mucosa junction lies on the labial surface.

Combination mucosal flap and epithelial inlay Usually suffices in the maxilla due to its ↑ potential for denture retention.

Mental nerve repositioning Mental nerve compression by a denture flange following alveolar resorption is a common problem, producing a sensation like an electric shock but with a background ache which worsens during the day and eases overnight. Leaving the -/F out for several days relieves the pain. The mental nerve is repositioned by creating a new foramen under its present position or laterally transposing the entire nerve into the soft tissues in grossly atrophic mandibles.

Alveolar augmentation Problems with osseous donor site morbidity and length of procedure in elderly patients put these operations out of favour. The use of 'sandwich' procedures (where the augmenting material is literally sandwiched between horizontally osteotomized jaw), coupled with effective bone substitutes and tunnelling procedures (p. 426), have made augmentation, even in the comparatively frail elderly, a far better propostion. Younger, fit patients with severe jaw atrophy or those with incipient or actual pathological # may still benefit from split rib grafting.

Sinus lift ↑ popular procedure combined with simultaneous or delayed implants. After raising subperiosteal buccal flaps a window is created to expose antral lining. Lining of the floor and walls is elevated intact and this space is filled with bone from the iliac crest to provide retention for implants.

Transmandibular implant A box frame construction placed in the mandible from a submental incision. Reputed to ↑ bone deposition.

Clefts and craniofacial anomalies

▶ 20% of cases of congenital facial malformation are accompanied by a 2nd or 3rd systemic malformation.

Cleft lip and palate

See also p. 198.

▶ The aim is to replace anatomical structures in their correct position, the price is scarring, which will restrict growth to some degree. It is important to recognize that the stigmata of surgery increase due to growth of the patient. At least as much deformity has been caused in the past by poor surgery as has been caused by the cleft.

Lip closure Two main philosophies: (1 '*plastic*' *approach*. Performed neonatally or up to 3 months, flaps transgress skin boundaries and use supraperiosteal dissection. Gives good early aesthetic results (e.g. Millard); (2 '*functional*' *approach*. All skin boundaries are respected, uses subperiosteal dissection. Immediate aesthetics are less good due to pout caused by muscle repair but ↑ function and growth potential (e.g. Delaire).

Palatal closure Extensive surgery restricts maxillary growth ∴ minimal simple repair which recreates palatal muscle sling (e.g. Von Langenbeck or Delaire).

Alveolus Vomer flap to close anterior alveolus and gingivoperiosteoplasty are primary procedures advocated by some and decried by others.

Ears Pre-school audiology. Many will benefit from grommets.

Nasal deformity is perhaps the greatest surgical challenge. 1° functional lip/nose repair (Delaire) may alter this.

Secondary surgery Lip revision (simple or complex). Sometimes required pre-school or at time of alveolar bone graft.

Alveolar bone graft, p. 199.

Speech, nasendoscopy, and pharyngoplasty All cleft palate patients have impaired speech. Fibre-optic nasendoscopy allows visualization of the palate during speech and aids assessment. Pharyngoplasty narrows the velopharyngeal opening to reduce hypernasal speech.

Orthognathic surgery, p. 504.

Craniofacial anomalies

These are a broad group of conditions involving the craniomaxillofacial region. A simple classification is:

Congenital
- Orbital malformations (hypertelorism, orbital dystopia)
- Craniosynostoses (fusion of cranial sutures)

Acquired
- Tumours (benign or malignant)
- Dysplasias (fibro-osseous)

- Craniofacial synostoses
  (Crouzon, Apert syndromes,
  p. 742)
- Encephaloceles
- Others
  (Treacher–Collins, hemifacial
  microsomia, hemifacial
  hypertrophy and atrophy)

- Neurofibromatosis

- Post-trauma deformity

These patients need craniofacial teams (minimum: craniofacial surgeon, neurosurgeon, and anaesthetist). Bicoronal flaps to deglove the face are the mainstay of access, followed by craniotomy and osteotomy as required. Main risks are cerebral oedema, infection, damage to optic nerves and vessels; and, in neonates and children, paediatric fluid balance.

Orthognathic surgery

This is the surgery of facial skeletal deformity and merges with cleft and craniofacial surgery. Prime indications are functional: speech, eating. Secondary indications are aesthetic.

▶ Patients may not see these as quite such separate issues. Their reasons and motivation for seeking surgery must be understood and the limitations of surgery made absolutely clear before embarking on protracted and complex treatment.

Diagnosis and treatment planning, p. 196.

Mandibular procedures

Involve ramus, body, alveolus, or chin.

Intra-oral vertical subsigmoid osteotomy Used to push back the mandible. EO approach used when suitable equipment for IO not available. The IO procedure is straightforward, performed via an extended third-molar type incision. Bone cuts are made with a right-angled oscillating saw from sigmoid notch to lower border. Technique is very instrument-dependent.

Sagital split osteotomy Can move mandible backwards or forwards. IO incision similar to above. Bone cuts made from above lingula, across retromolar region, down buccal aspect to lower border. Split saggitally with osteotome. Main complication is paraesthesia of IDB.

Inverted L- and C-shaped osteotomy Usually EO approach. Rarely used; can lengthen ramus if used with bone graft.

Body ostectomy Shortens body of mandible. Need to gain space orthodontically or remove tooth. Watch mental nerve.

Subapical osteotomy Used to move dento-alveolar segments. Technically more difficult than it looks. Risk to tooth vitality.

Genioplasty The tip of the chin can be moved pretty much anywhere, the secret is to keep a sliding contact with bone and a muscle pedicle. Fixation should be kept away from areas of muscle activity as this leads to bone resorption.

Maxillary procedures

Segmental Can be single-tooth or bone and tooth blocks, e.g. Wassmund procedure which involves tunnelling incisions in buccal sulcus and palate to move 3+3. Problems are finding space for bone cuts and avoiding damaging teeth.

Le Fort I Mainstay procedure. Standard approach is the 'down-fracture' with horseshoe buccal incision, bone cuts at Le Fort I level, and segment pedicled on the palate. The freed maxilla can be moved easily up, down, or forward. In cleft palate cases, concern over the adequacy of the palatal blood supply has led to tunnelled buccal incisions to make the bone cuts, thus preserving some of the buccal blood supply to the maxilla.

Le Fort II Usually used for mid-face advancement. Bilateral canthal and vestibular incisions allow bone cuts at the Le Fort II level.

Le Fort III Really a subcranial craniofacial operation using a bicoronal flap plus vestibular and orbital incisions to move the entire mid-face and malar complex.

Malar osteotomy Often used for post-traumatic defects. Approach via bicoronal incisions. Risk to infraorbital nerve from maxillary bone cut.

Rhinoplasty is the correction of isolated nasal deformity. Usually done intranasally, supplemented by tiny incisions over the nasal bones to allow bone cuts. 'Open rhinoplasty', which involves degloving the nasal skeleton, is becoming popular.

Stability ↑ use of mini-plates in the fixation of the osteotomized segments has ↓ the reliance on bone grafts. Pre-surgical orthodontics (p. 196), makes a significant contribution to ↓ the rate of relapse. IMF +/or elastic traction remains vital for long-term success.

Salivary gland tumours

Diseases: see also p. 454.

Classified by the WHO into epithelial tumours, non-epithelial tumours, and unclassified tumours.

Common benign tumours *Pleomorphic adenoma* (p. 456). *Monomorphic adenoma* (adenolymphoma) affects M>F, rare <50 yrs. Bilateral in 10%. Does not undergo malignant change. Feels soft and cystic.

Carcinomas Rare. *Adenoid cystic carcinoma* Has a characteristic 'Swiss cheese' appearance histologically. Spreads locally particularly along perineural spaces. Can be compatible with long-term survival, although the disease is rarely, if ever cured. *Adenocarcinomas* There are a wide range of these malignant tumours, ranging from the highly aggressive to the relatively indolent. It is essential that adequate histological diagnosis is made early (refer pathology to Salivary Gland Tumour Panel). Other carcinomas are rare, but may arise in a pre-existing pleomorphic adenoma or *de novo*. 5 yr survival about 30%.

Other epithelial tumours include *acinic cell tumour* and *mucoepidermoid tumour*. Both have variable, unpredictable behaviour, can recur locally, and metastasize, and can occur at any age. On average, both compatible with around 80% survival.

Non-epithelial tumours include haemangioma, lymphangioma, and neurofibroma. Account for 50% of salivary tumours in children.

Unclassified group includes lymphomas, secondaries, lipomas, and chemodectomas.

Parotid History and examination are the prime diagnostic tools. Long history, no pain, no facial nerve involvement, suggests benign tumour. Facial palsy, pain, rapid growth suggests malignancy. The feel of many tumours are characteristic. CT, fine needle aspiration cytology (FNAC), and sialograms are sometimes helpful; however Rx almost always involves parotidectomy and most investigations simply delay this.

Submandibular Tumours are less common. Pleomorphic adenoma remains most common. Malignant tumours account for up to 30%. Treatment of most is gland excision via a skin crease in the neck. Radical neck dissection may be needed for tumours spreading beyond the gland.

Sublingual and minor glands >50% of tumours are malignant (mostly adenoid cystic) and require extensive surgery.

Surgery of the salivary glands

Surgery of the major salivary glands is primarily for tumours, obstruction and less commonly inflammatory conditions. The minor glands are most commonly removed for mucoceles (p. 414) and more rarely for tumours. With the exception of lymphomas *all* salivary gland tumours should have surgery initially, patient permitting.

Parotidectomy Principles are complete excision of tumour with margin of healthy tissue and preservation of facial nerve. Clinically benign tumours in superficial lobe have superficial parotidectomy; and in deep lobe, total conservative parotidectomy. In possibly malignant tumours, frozen section may help to decide whether facial nerve can be preserved. Malignant tumours require radical excision with nerve sacrifice and reconstruction ± radiotherapy.

Salivary duct calculi History is of recurrent pain and swelling in the obstructed gland, particularly before and during meals. Plain X-rays (lower occlusal for submandibular, cheek for parotid) reveal radioopaque stones, but do not exclude radiolucent stones and mucous plugs. Sialography, reveals a stricture or obstruction. Commonest in the submandibular duct. Rx: *Submandibular duct* for calculi lying anterior in duct—remove by passing a suture behind the calculus to prevent it slipping further down the duct, dissect the duct from an intra-oral approach and lift out stone; either marsupialize the duct or reconstruct. Posterior calculi—excise gland and duct. Rx: *Parotid duct* expose duct via intra-oral approach for anterior stones or via a small skin flap on to a probe in the duct for more posterior calculi. Otherwise, superficial parotidectomy is the only safe approach.

Recurrent sialadenitis Severe recurrent infection of the parotid or submandibular glands leads to dilatation and ballooning of the ducts and alveoli called sialectasis. Sialography is the investigation of choice and is often therapeutic, inducing long remissions between episodes of infection. *Conservative* Rx consists of irrigation of gland with tetracycline solution. When remission periods are short or intolerable or the patient requires *definitive* Rx, submandibular gland excision or total conservative parotidectomy with removal of 90% of the duct.

Surgery for drooling In severe cases it is possible to re-site the parotid ducts into the hypopharynx ± bilateral submandibular gland excision or duct repositioning to control drooling without impairing lubrication for swallowing and oral health.

Mucoceles, p. 414.

Oral cancer

Aetiology, epidemiology, diagnosis, and staging, p. 450. Neck lumps, p. 512. Salivary tumours, p. 506.

Various parameters affect the choice of treatment for patients with oral cancer; not least among these, but often forgotten, are the patients themselves, their general health, understanding of their disease, geographical location, social and domestic commitments. Classically, broad treatment principles are based on tumour staging, using TNM classification, p. 450, and the patient's fitness for surgery (which tends to imply that if they can't cope with surgery they can cope with travelling for radiotherapy, which is not always the case). In many instances of advanced cancer, combined surgery and radiotherapy constitutes optimal Rx.

Suggested management plan (this will vary according to the surgeon concerned).

1 Establish provisional diagnosis. History, examination, and get to know patient. FBC, ESR, U&Es, LFTs (including albumin), bone biochemistry, VDRL, blood group, chest X-ray, ECG, OPG. Think about bone scan.
2 Arrange tissue diagnosis. Best achieved by EUA with geographic biopsy and tattooing of the tumour margins (for future reference if pre-op radiotherapy is given). Perform pharyngo-oesophagoscopy, laryngoscopy, (bronchoscopy if facilities available), and examine post-nasal space. This is to exclude synchronous 1° tumour in upper aerodigestive tract (present in at most 15% cases). Palpate the neck for nodes. Stage tumour (TNM).
3 Unless patient has made it very obvious they do not want to know the diagnosis, inform first them, then relatives, fully.

- T1 N0 Surgery or radiotherapy offer equal cure rate.
- Tumour close to bone having radiotherapy; safest to remove associated teeth to prevent osteoradionecrosis.
- T2 and T3 >50% of patients will have occult metastases; consider whether to watch and wait, prophylactic radiotherapy (works for occult but not for obvious bulky nodes), or prophylactic functional or radical neck dissection.
- For large tumours radiotherapy may be given pre- or post-operatively.
- Resections should be in continuity; if nodes are involved this is called a 'commando' (resection plus neck dissection).
- Vastly improved access is obtained for resection by osteotomizing the mandible (position plate beforehand).
- Anterior floor of mouth cancer may spread to bilateral lymph nodes; bilateral *radical* neck dissection must be staged.

5-year survival 90% for T1 N0 but 30% for T2/3 N1 and worse for T4; however, this does *not* mean that extensive combination therapy and reconstruction is pointless in those with extensive oral cancers. Death comes in many ways, and a fungating uncontrolled cancer of the head and neck is one of the less pleasant. Attempted surgical 'cure' which alleviates local disease and symptoms and allows the patient a few more years of life and a gradual demise due to carcinomatosis or another disease, is still worthwhile from all viewpoints.

Chemotherapy There is, as yet, no proven role for cytotoxics in oral cancer, other than rare forms of palliation.

Neck lumps

▶ Do not leave chronic cervical lymphadenopathy undiagnosed.
▶ A 1° head and neck malignancy must be excluded before biopsy.

Children are an exception; inflammation is common, and tumour is rare, so a watch and wait policy is reasonable.

Diagnosis Listen to the story, look at the patient and the lump, and palpate it. If needed, carry out a full head and neck examination. Most diagnoses will be made by then.

Investigations Ultrasound, aspiration cytology, biopsy.

Causes Think (a) anatomy, (b) pathology, (c) oddity.

Skin Lesions lie superficially.
Sebaceous cyst Look for punctum, is within skin. Excise.
Lipoma Soft, often yellowish. Excise.
Sublingual dermoid cyst Lies in floor of mouth often under mylohyoid. Arises from trapped epithelium during embryonic fusion, contains keratin. Rx: total excision.

Lymph nodes Deep to skin.
Infection (p. 484) Nodes are large and tender. Causes: viral (e.g. glandular fever, HIV), bacterial (e.g. mycobacterial (which can calcify), actinomycosis) or reactive to other head and neck infection.

Malignancy Either metastatic from head and neck 1° (hard, rock-like nodes) or lymphoma/leukaemia (large rubbery), p. 484.

Glandular Think anatomically.
Salivary (p. 508) Submandibular/lower pole of parotid—abscess, sialadenitis, obstruction, sarcoidosis, Sjögren syndrome. Sublingual—ranula, tumour.
Thyroid Benign and malignant tumours, goitre, thyroglossal cyst (may lie anywhere between foramen caecum of tongue and thyroid, tract goes behind hyoid bone, moves with swallowing).

Arterial Don't biopsy!
Carotid aneurysm (pulsatile).
Carotid body tumour Found anterior to upper $\frac{1}{3}$rd sternomastoid. Usually firm, not hormonally active, 5% malignant. Rx: excise if symptomatic.

Pharynx
Diverticulum (or pharyngeal pouch). Fills on swallowing.

Larynx
Laryngocele Rare. Mainly M >60. 80% unilateral. Excise.

Sternomastoid

'Sternomastoid tumour' (congenital ischaemic fibrosis causing torticollis).

True muscle tumours Rare.

Bone

Cervical rib, prominent hyoid bone

Infections (see also p. 408) Ludwig's angina, submasseteric abscess, retropharyngeal abscess, parapharyngeal cellulitis, collar stud abscess (TB), infected cysts or pouches.

Oddities

Branchial cyst Either a remnant from 2nd and 3rd branchial arches or degeneration of lymphoid tissue. Is an epithelial lined cyst which presents as a deep-seated swelling lying anterior to sternomastoid at or above the level of the hyoid. Prone to infection. Rx: total excision.

Branchial fistula A fistula from the tonsillar fossa to the skin overlying anterior lower $\frac{1}{3}$rd of sternomastoid. Present at birth, discharges intermittently. Rx: total excision of tract.

Cystic hygroma Presents in infancy and is a form of lymphangioma which appears as endothelium-lined multilocular cysts containing lymph. Found behind lower end of sternomastoid. May suddenly ↑ in size if bled into or ruptured. Rx: total excision (in practice, excise as much as is possible) as soon as child is fit for operation.

513

Flaps and grafts

A **graft** is transferred tissue which is dependent on the donor site capillaries for its survival. A **flap** is transferred tissue which is, at least initially, independent of the donor site capillaries for survival.

It is the possibility of functional reconstruction of the head and neck in conjunction with the potential for cure that justifies the mutilation of radical surgery for oral cancer; however, head and neck reconstruction is not used only post oncology surgery, and its use in other aspects of maxillofacial surgery is increasing.

Mucosal grafts, p. 240. **Mucosal flaps**, p. 241.

Skin grafts may be split thickness or full thickness. Split thickness (taken by knife or dermatome from thigh or inner arm) take quickly and become wettable in the mouth. They are 'quilted' in place with sutures. Full thickness (supraclavicular or post-auricular) provide a good colour match when repairing skin defects of the face. Full thickness donor sites are closed 1°.

Free bone grafts From rib, iliac crest or calvarium. Rib, which is partially split at 1 cm intervals, can be bent to conform to the shape of the mandible. Iliac crest supplies cortical or cancellous bone and can be cut to a template, but ↑ risk of DVT. Various synthetic mesh containers as a mould for bone exist.

Nasolabial flap Random pattern pedicle flap based above and lateral to the upper lip; useful small local flap. Requires division later.

Tongue flaps Random pattern pedicle flap for lip and palate repair. Requires division later.

Forehead flap Based on anterior branch of superficial temporal artery and is a very safe flap. Entire length of forehead can be raised, quickly, but leaves bad donor site. Requires division later.

Masseter muscle flap Limited in size, can be used intra-orally.

Temporalis flap Inferiorly based on deep temporal branches of maxillary artery. For intra-oral reconstruction, limited use.

Deltopectoral flap Based on perforating internal mammary vessels. Thin skin suitable for skin or mucosa repair. Usually divided and inset after 3 weeks.

Pectoralis major myocutaneous flap Also described with bone, but the bone is really a free rib graft. Based on acromiothoracic axis. Usually tunnelled after neck dissection. Very bulky flap, of great value following 'commando'.

Latissimus dorsi myocutaneous flap Very bulky flap based on thoracodorsal vessels. Needs to be tunnelled through axilla; can, however, be raised as a free flap.

Free tissue transfer by microvascular re-anastomosis has been the biggest recent advance in reconstruction. The following are useful and commonly used flaps.

Radial forearm flap This is a fasciocutaneous flap based on the radial artery. The skin available is thin and supple and can conform to the complex anatomy of the mouth (skin is hairless which is a big bonus). A thin segment of up to 10 cm of radius can also be transferred for bony reconstruction.

Deep circumflex iliac flap Based on the deep circumflex iliac artery, although the superficial circumflex iliac should also be taken. This flap has potential for sufficient bone transfer to reconstruct the entire mandible; technically demanding with major donor site problems.

Free fibula flap 25 cm of fibula can be excised within a muscle cuff supplied by the peroneal vessels. Excellent length and thickness of bone for mandibular reconstruction. Disadvantage is short pedicle. Minimal donor site morbidity.

Free rectus abdominus Bulky skin/muscle flap based on inferior epigastric vessels. Useful for massive facial defects.

Craniofacial implants Prosthetic eyes, ears, and noses can be securely fixed to the facial skeleton with implants using techniques similar to oval implants.

11 Medicine relevant to dentistry

Medical theory
Anaemia 518
Haematological malignancy 520
Other haematological disorders 522
Cardiovascular disease 524
Respiratory disease 526
Gastrointestinal disease 528
Hepatic disease 530
Renal disorders 532
Endocrine disease 534
Endocrine-related problems 536
Bone disease 538
Diseases of connective tissue, muscle, and
    joints 540
Neurological disorders 542
More neurological disorders 544
Skin neoplasms 546
Dermatology 548
Psychiatry 550
The immunocompromised patient 552

Emergencies
Useful emergency kit 554
Fainting 556
Acute chest pain 558
Cardiorespiratory arrest 560
Anaphylactic shock 562
Collapse in a patient with a history of
    corticosteroid use 564
Fits 566
Hypoglycaemia 568
Acute asthma 570
Inhaled foreign bodies 572
If in doubt 574

In-patients

Management of the dental in-patient	576
Venepuncture and arterial puncture	578
Intravenous fluids	580
Blood transfusion	582
Catheterization	584
Enteral and parenteral feeding	586
Pain control	588
Antibiotic prophylaxis	590
Management of the diabetic patient undergoing surgery	592
Management of patients requiring steroid supplementation	594
Common post-operative problems	596

Principal sources: C. Scully 1987 *Medical Problems in Dentistry*, Wright; R. Hope 1993 *Oxford Handbook of Clinical Medicine*, OUP; Various 1988 *Procedures in Practice*, BMJ; Various 1993 *Prescribers Journal* **33** 6.

517

Anaemia

Anaemia is a ↓ in the level of circulating haemoglobin below the normal reference range for a patients age and sex. It indicates an underlying problem, and as such the cause of the anaemia should be diagnosed *before* instituting treatment.

▶ Never rush into transfusing patients presenting with a chronic anaemia. Elective surgery in patients with an Hb <10 g/dl is rarely appropriate.

Clinical features Of anaemia are notoriously unreliable, but beloved of examiners and include: general fatigue, heart failure, angina on effort, pallor (look at conjunctivae and palmar creases), brittle nails ± spoon-shaped nails (koilonychia), oral discomfort ± ulceration, glossitis, and classically angular cheilitis. Syndromes, p. 712.

Types of anaemia

Microcytic (MCV <76 fl) Iron-deficiency anaemia is by far the commonest cause. Causes: inadequate diet, chronic blood loss (G-I or menstrual). FBC and biochemistry shows microcytic, hypochromic anaemia with a low serum iron and a high total iron binding capacity (TIBC). ↑ RBC zinc protoporphyrin is a fast and sensitive early test. Thalassaemia and sideroblastic anaemia are rare causes of microcytosis.

Normocytic Commonly, anaemia of chronic disease. Other causes: pregnancy, haemolytic anaemia, and aplastic anaemia. Once pregnancy is excluded the patient needs investigation by an expert. The TIBC is usually lowered.

Macrocytic (MCV >96 fl) Low B12 +/or low folate are the common causes. B12 is lowered in pernicious anaemia (deficit of intrinsic factor), alcohol abuse, gut disease, and chronic exposure to nitrous oxide. Low folate is usually dietary, but may be caused by illness or drugs, such as phenytoin, methotrexate, and co-trimoxazole, coeliac disease and skin disease.

Management In all cases the cause must be sought, this may necessitate referal to a haematologist. Drugs used in iron deficiency are ferrous sulfate 200 mg tds. Transfusion of packed cells under frusemide (40 mg PO) indicated rarely for severe microcytic anaemia. Lifelong IM hydroxycobalamin 1 mg 3-monthly is used to treat B12 deficiency and folic acid 5 mg od for folate deficiency.

▶ Never use folate alone to treat 'macrocytosis' unless it is proven to be the only deficiency. NB Folic acid is **not** the same as folinic acid.

Note on sickle cell anaemia A homozygous hereditary condition causing red cells to 'sickle' when exposed to low oxygen tensions. This results in infarctions of bone and brain. In sickle cell trait (heterozygous form) the cells are less fragile and sickle only in severe hypoxia. *Management*: perform sickledex test on all black patients planned for GA. Avoid anaesthetics in practice.

Haematological malignancy

Leukaemias are a neoplastic proliferation of white blood cells. Acute leukaemias are characterized by the release of primitive blast cells into the peripheral blood and account for 50% of childhood malignancy. Acute lymphoblastic leukaemia, the commonest childhood leukaemia, now has a >50% 5 yr survival. Acute myeloblastic leukaemia is the commonest acute leukaemia of adults, but although an 80% remission rate is possible this is rarely maintained. Chronic leukaemias have cells which retain most of the appearance of normal white cells. Chronic lymphocytic leukaemia is the commonest and has a 5 yr survival of >50%. Chronic myeloid leukaemia is characterized by the presence of the Philadelphia chromosome, a fact beloved by examiners. It affects the >40s, and while remissions are common a terminal blast crises supervenes at some stage.

Myeloproliferative disorders are proliferation of non-leukocyte marrow cells, and have a wide range of presentation and behaviour.

Monoclonal gammopathies such as multiple myeloma, are B-lymphocyte disorders characterized by production of a specific immunoglobulin by plasma cells. Multiple myeloma is also a differential diagnosis of lytic lesions of bone, particularly the skull.

Lymphomas are a group of solid tumours arising in lymphoid tissue; they are divided into Hodgkin's or non-Hodgkin's lymphomas, with the latter carrying a poorer prognosis. Lymphoma should always be considered in the differential diagnosis of neck swellings.

Cytotoxic chemotherapy has been the mainstay of treatment for these disease with supplemental radiotherapy for masses or prior to bone marrow transplant. It is essential to remember that any patient receiving these drugs will be both immunocompromised and liable to bleed.

Hints In haematological malignancy, anaemia, bleeding, and infection are the overwhelming risks. Look for and treat anaemia. Avoid aspirin, other NSAIDs, trauma, and IM injections. Prevent sepsis, and if it occurs treat very aggressively with the locally recommended broad-spectrum antibacterials and antifungals, e.g. Azlocillin 5 g and Gentamycin 80 mg IV tds plus fluconazole up to 200 mg daily.

Amyloidosis is characterized by deposits of fibrillar eosinophilic hyaline material in a wide range of organs and tissues. Divided into 1° amyloidosis (AL amyloid) which is an immunocyte dyscrasia, signs and symptoms include peripheral neuropathy, cardiomyopathy, and macroglossia. Rx: immunosuppression sometimes helps. 2° amyloidosis (AA amyloid) reflects an underlying chronic disease: infection, rheumatoid, neoplasia. May respond to Rx of underlying disease. Diagnosis—biopsy of rectum or gingivae—stain with Congo red.

Other haematological disease

For the practical management of a bleeding patient, see p. 390.

Bleeding disorders

Platelet disorders These may present as nosebleeds, purpura, or post-extraction bleeding. They include diseases such as Von Willebrand disease; idiopathic thrombocytopenic purpura (ITP); virally associated (especially HIV) thrombocytopenic purpura; thrombocytopenia secondary to leukaemia, cytotoxic drugs, or unwanted effects of drugs, notably aspirin and chloramphenicol. *Management*: Ensure platelet levels of $>50 \times 10^9$/L, preferably 75×10^9/L for anything more than simple extraction or LA. If actively bleeding, use a combination of local measures (p. 390), tranexamic acid, and platelet transfusion. Platelet transfusions are short-lived and if used prophylactically must be given immediately prior to surgery. The quality of preparation varies by locality.

Coagulation defects Present as prolonged wound bleeding +/or haemarthroses. Causes include the haemophilias, anticoagulants, liver disease, and Von Willebrand disease.

Others Less common causes include: hereditary haemorrhagic telangectasia, aplastic anaemia, chronic renal failure, myeloma, SLE, disseminated intravascular coagulation, and isolated deficiency of clotting factors.

Haemophilia A (factor VIII deficiency) is the commonest clotting defect. Inherited as a sex-linked recessive it affects males predominantly, although female haemophiliacs can occur. All daughters of affected males are potential carriers. Usually presents in childhood as haemarthroses. Following trauma, bleeding appears to stop, but an intractable general ooze starts after an hour or so. Severity of bleeding is dependent on the level of factor VIII activity and degree of trauma.

Haemophilia B (factor IX deficiency) Clinically identical to Haemophilia A, also known as Christmas disease.

von Willebrand disease is a combined platelet and factor VIII disorder. It affects males and females. Mucosal purpura is common; haemarthroses are less so. Wide range of severity. May improve with age +/or pregnancy.

Management The haemophilias and Von Willebrand disease should always be managed at specialist centres. Check the patient's warning card for the contact telephone number.

Anticoagulants

Heparin is given IV or high-dose SC for therapeutic anticoagulation. Its effect wears off in around 8 h although it can be reversed by protamine sulphate in an emergency. Measure in kaolin–cephalin clotting time (KCT).

Warfarin is given orally; effects take 48 h to be seen. Normal range is an International Normalized Ratio (INR) of 2–4.

Extractions are usually safe at a level of <2.5. Avoid attempts to reverse warfarin with Vitamin K unless *in extremis*. Use fresh frozen plasma if needed, but consider why the patient is anti-coagulated in the first place.

Cardiovascular disease

This is the commonest cause of death in the UK.

Clinical conditions

Heart failure is the end result of a variety of conditions, not all of them cardiovascular. Basically, the heart is unable to meet the circulatory needs of the body. In right heart failure, dependent oedema and venous engorgement are prominent. In left heart failure, breathlessness is the principal sign. The two often co-exist. There is an ever-present risk of precipitating heart failure, even in treated patients, by increasing the demands on the heart, e.g. by fluid overload or excessive exertion.

Ischaemic heart disease is ↓ of the blood supply to part of the heart by occlusion of the coronary arteries, usually by atheroma, causing the pain of angina pectoris. If occlusion becomes complete a myocardial infarction occurs (p. 558).

Hypovolaemic shock is collapse of the peripheral circulation due to a sudden reduction in the circulating volume. If this is not corrected there can be failure of perfusion of the vital organs resulting in heart failure, renal failure, and unconsciousness ending in death.

Hypertension is a consistently raised BP (> 160/95 for an adult male) and is a risk factor for ischaemic heart disease, cerebro-vascular accidents, and renal failure. Up to 80% of hypertension has no definable cause—essential hypertension. 20% is secondary to another disease such as renal dysfunction or an endocrine disorder.

Murmurs are disturbances of blood flow which are audible through a stethoscope. They may be functional or signify structural disorders of the heart. They are of great relevance to dentists as their presence warns of the potential for colonization of damaged valves by blood-borne bacteria. Such a bacteraemia is often caused by dental procedures. Instrumentation liable to do this should be covered by antibiotic prophylaxis (p. 590). Colonization of the valves may lead to a potentially fatal illness—Subacute Bacterial Endocarditis (SBE).

Dental implications

SBE prophylaxis as above. Patients with a past history of rheumatic fever are very likely to have some damage to a heart valve, usually the mitral valve. They should receive antibiotic prophylaxis unless valvular damage is excluded by a cardiologist. The risk of precipitating heart failure or myocardial infarction in patients with compromised cardiovascular systems is ever-present. Prevent by avoiding GA, especially within 6 months of an MI, using adequate LA with sedation if necessary and avoid excessive adrenalin loads. Consider potential drug interactions (p. 624) and remember some of these patients will be anticoagulated.

Exclusion of septic foci may be requested in patients at high risk from bacteraemia, e.g. heart transplant recipients, those with severe valvular damage, or those with a history of SBE. It is prudent to err on the side of caution with these individuals and some will need dental clearances.

Respiratory disease

Disease of the chest is an everyday problem in developed countries. The principal symptoms are cough, which may or may not be productive of sputum, dyspnoea (breathlessness), and wheeze. The coughing of blood (haemoptysis) mandates that malignancy be excluded.

Clinical conditions

Upper respiratory tract infections include the common cold, sinusitis, and pharyngitis/tonsillitis (which may be viral or bacterial), laryngotracheitis, and acute epiglotitis. *All* are C/I to elective general anaesthesia in the acute phase. Sinusitis (p. 422). Penicillin is the drug of choice for a streptococcal sore throat. Avoid amoxycillin and ampicillin, as glandular fever may mimic this condition and these drugs will produce a rash, of varying severity, in such a patient. Epiglotitis is an emergency, and the larynx should never be examined unless expert facilities for emergency intubation are to hand.

Lower respiratory tract infections Both viral and bacterial lower tract infections are debilitating and constitute a C/I to GA for elective surgery. Bear in mind tuberculosis and atypical bacteria, e.g. legionella, mycoplasma, and coxiella. Open TB is highly infectious and cross-infection precautions are mandatory (p. 734).

Chronic obstructive airways disease (COAD) A very common condition usually caused by a combination of bronchitis (excessive mucus production, persistent productive cough >3 months per year for 3 yrs) and emphysema (dilation and destruction of air spaces distal to the terminal bronchioles). Smoking is the prime cause and must be stopped for treatment to be of any value.

Asthma Reversible bronchoconstriction causes wheezing and dyspnoea. Patients complain of the chest feeling tight. More than 2% of the population are affected and there is often an allergic component. Beware prescribing in asthmatics, as penicillin and aspirin allergies are more common. Management of acute asthma, p. 570.

Cystic fibrosis is an inherited disorder whereby viscosity of mucus is ↑. Patients suffer pancreatic exocrine insufficiency and recurrent chest infections. Diagnosis by history and sweat sodium measurement.

Bronchogenic carcinoma causes 27% of cancer deaths. Principal cause is smoking. ↑ in females. Symptoms are recurrent cough, haemoptysis, and recurrent infections. 5 yr survival is only 8%. Mesothelioma is an industrial disease caused by asbestos exposure.

Sarcoidosis Most commonly presents as hilar lymphadenopathy in young adults. Oral lesions can occur. Erythema nodosum common.

Dental implications

Avoid GA. Use analgesics and sedatives with caution as many depress respiratory drive. Advise your patients to stop smoking; give up yourself if you have developed the habit. Refer if suspicious, especially in the presence of confirmed haemoptysis.

Gastrointestinal disease

The mouth and its mucosal disorders and disorders of the salivary glands are covered in Chapter 9.

The oesophagus presents symptoms which can be confused with those originating from the mouth, the most important being *dysphagia*. Difficulty in swallowing may be caused by conditions within the mouth (e.g. ulceration), the pharynx (e.g. foreign body), benign or malignant conditions within the oesophagus, compression by surrounding structures (e.g. mediastinal lymph nodes), or neurological causes. It is a symptom which should be taken seriously and investigated by at least CXR, barium swallow, +/or endoscopy. Reflux oesophagitis is a common cause of dyspepsia. It is not, however, a cause of oral disease.

The stomach has two major and relatively common pathologies in peptic ulceration and gastric carcinoma. The former usually responds to medical treatment, while the latter requires surgery and has a poor prognosis. It is essential ∴ that epigastric symptoms are investigated by endoscopy prior to starting treatment. Non-malignant ulcers may be duodenal (epigastric pain worse at night, helped by eating) or gastric (epigastric pain exacerbated by eating and worse during the day); both may be complicated by vomiting, haematemesis, or melaena. NSAIDs are a potential cause of peptic ulceration and this condition is a strong relative C/I to their use. The mainstay of treatment are the H2 antagonists, cimetidine and ranitidine, ± antacids and alginates. Omeprazole, a proton pump inhibitor, provides a second line of treatment.

The small bowel has a multitude of associated disorders which tend to present in a similar manner; namely, malabsorption syndromes, diarrhoea, steatorrhoea, abdominal pain, and chronic deficiencies. Coeliac disease and Crohn's disease are the best known conditions. Coeliac disease is a hypersensitivity response of the small bowel to gluten and is treated by strict avoidance. A number of oral complaints are related. Crohn's disease may affect any part of the G-I tract but has a preference for the ileo-caecal area. It is a chronic granulomatous disease affecting the full thickness of the mucosa and may result in fistula formation. Ulcerative colitis is often mistaken for Crohn's disease initially, but affects the colorectum only.

Large bowel Diverticular disease is a condition with multiple outpouching of large bowel mucosa which can become inflamed causing diverticulitis. The irritable bowel syndrome is a condition associated with increased colonic tone causing recurrent abdominal pain; there may be some psychogenic overlay.

Colonic cancer is common in older patients; it may present as intestinal obstruction, tenesmus (wanting to defecate but producing nothing), abdominal pain, or anaemia. It is treated surgically with a 30% 5 yr survival. Familial polyposis coli is associated with the Gardener syndrome (p. 744). Antibiotic-induced colitis results from overgrowth of toxigenic *Clostridium difficile* after use of antibiotics such as ampicillin and clinda-mycin. It responds to *oral* vancomycin or metronidazole.

The pancreas Malignancy has the worst prognosis of any cancer and most treatment is essentially palliative.

Acute pancreatitis is often a manifestation of alcohol abuse. Aetiology is not entirely clear. Causes acute abdominal pain. Amylase levels are a guide but not infallible. Patients need aggressive rehydration and observation of electrolytes with analgesia.

Hepatic disease

The main problems presented by patients with liver disease are: the potential for increased bleeding, inability to metabolize and excrete many commonly used drugs, and the possibility that they can transmit the hepatitis B, C +/or D (Hep A and E are spread by faecal–oral route). The liver is also a site of metastatic spread of malignant tumours. Patients in liver failure needing surgery, especially under GA, are a high-risk group who should have specialist advice on their management.

Jaundice is the prime symptom of liver disease. It is a widespread yellow discoloration of the skin (best seen in good light, in the sclera). It is caused by the inability of the liver to process bilirubin, the breakdown product of haemoglobin. This occurs either because it is presented with an overwhelming amount of bilirubin to conjugate (acholuric jaundice), or it is unable to excrete bile (cholestatic jaundice). Cholestatic jaundice in turn may be either intrahepatic or extrahepatic. *Intrahepatic cholestasis* represents hepatocyte damage; this is reflected by increased aspartate transaminase levels on liver function tests, and results in impaired bile excretion, as indicated by increased plasma bilirubin and liver alkaline phosphatase levels. Causes include: alcohol and other drugs, toxins, bacterial and viral infections. A degree of hepatitis is present with these causes, whereas primary biliary cirrhosis and anabolic steroids cause a specific intrahepatic cholestasis without hepatitis. *Extrahepatic cholestasis* is caused by obstruction to the excretion of bile in the common bile duct by gallstones, tumour, clot, or stricture. Carcinoma of the head of the pancreas, or adjacent lymph nodes may also compress the duct.

Surgery in patients with liver disease
- Ascertain a diagnosis for the cause. If Hep B/C/D check HIV status if life-style indicates. Cross-infection precautions (p. 734).
- Do coagulation screen. May need correction with Vitamin K or fresh-frozen plasma.
- **Always** warn the anaesthetist, as it will affect the choice of anaesthetic agents.
- If a jaundiced patient must undergo surgery, ensure a good perioperative urine output by aggressive IV hydration and mannitol diuresis to avoid hepato-renal syndrome (see OHCM).
- Do not use IV saline in patients in fulminant hepatic failure, as there is a high risk of inducing encephalopathy.

Liver disease patients in dental practice
- Know which disease you are dealing with. If Hep B employ strict cross-infection control (p. 734).
- Be cautious in prescribing drugs (consult the BNF/DPF) and with administering LA.
- **Do not** administer general anaesthetics.

- Take additional local precautions against post-operative bleeding following simple extractions (p. 390). A clotting screen should be obtained for anything more advanced, and in all patients with severe liver disease.

Renal disorders

The commonest urinary tract problems, infections, are of relevance only to those who manage in-patients. Rarer conditions such as renal failure and transplantation are, surprisingly, of more general relevance because these patients are at ↑ risk from infection, bleeding, and iatrogenic drug overdose during routine treatment.

Urine This is tested in all in-patients. 'Multistix' will test for glycosuria (diabetes, pregnancy, infection), proteinuria (diabetes, infection, nephrotic syndrome), ketones (diabetic ketoacidosis), haematuria (infection, tumour), and bile as bilirubin and urobilinogen (cholestatic jaundice).

Urinary tract infections are a common cause of toxic confusion in elderly in-patients especially females. Send a mid-stream urine (MSU) for culture and sensitivity then start Trimethoprim 200 mg bd PO or Ampicillin 250 mg qds PO and ensure a high fluid intake. Minimal investigations of renal function are U&Es and ionized Ca^{2+}.

Nephrotic syndrome is a syndrome of proteinuria (>4 g/day), hypoalbuminaemia and generalized oedema. Facial oedema is often prominent. Glomerulonephritis is the major precipitating cause and investigations should be carried out by a physician with an interest in renal medicine.

Acute renal failure (ARF) is a medical emergency causing a rapid rise in serum urea and K^+. It may follow surgery or major trauma and is usually marked by a failure to pass urine. *Remember* the commonest causes of failing to PU post-op are under-infusion of fluids and urinary retention. Rx: ↑ IV fluid input and catheterize (p. 584). If ARF is suspected get urgent U&Es, ECG, and blood gases. Obtain assistance from a physician. Control of hyperkalaemia, fluid balance, acidosis, and hypertension are the immediate necessities.

Chronic renal failure is basically the onset of uraemia after gradual, but progressive, renal damage commonly caused by glomerulonephritis (inflammation of the glomeruli following immune complex deposits), pyelonephritis (small scarred kidneys due to childhood infection, irradiation, or poisoning), or adult polycystic disease (congenital cysts within Bowman's capsule). It has protean manifestations, starting with nocturia and anorexia, progressing through hypertension and anaemia to multisystem failure. Continuous ambulatory peritoneal dialysis and haemodialysis are the mainstays of treatment.

Main problems relevant to dentistry
- ↑ risk of infection, worsened by immunosuppression.
- ↑ bleeding tendency.
- ↓ ability to excrete drugs.
- Veins are sacrosanct, **never** use their A-V fistula.
- Bone lesions of the jaws (renal osteodystrophy, 2° hyperparathyroidism).

- Generalized growth impairment in children.
- Risk of infection and potential carriage of Hep B, HIV.

Renal transplantation is an increasingly common final treatment of renal failure and when successful renal function may reach near normal levels. They are, however, immunosuppressed and at greatly increased risk from infection. They may share the problems associated with chronic renal failure depending on the level of function of the transplant.

Hints
- Take precautions against cross-infection (p. 734).
- Treat all infections aggressively and consider prophylaxis.
- Use additional haemostatic measures (p. 390).
- Caution with prescribing drugs (p. 600).
- Never subject these patients to GA in dental practice.
- Remember veins are precious.

Endocrine disease

Addison disease is primary hypoadrenocorticism. Atrophy of the adrenal cortices causes failure of cortisol and aldosterone secretion. Secondary hypoadrenocorticism is far commoner, due to steroid therapy or ACTH deficiency (p. 594).

Conn syndrome Primary hyperaldosteronism causes hypokalaemia and hyponatraemia with hypertension.

Cushing disease and Cushing syndrome These conditions are due to excess corticosteroid production, the disease refers to secondary adrenal hyperplasia due to ↑ ACTH, whereas the syndrome is a 1° condition, usually due to adenoma. Classical features are obesity (moon face, buffalo hump) sparing the limbs, osteoporosis, skin thinning, and hypertension.

Diabetes insipidus Production of copius dilute urine due to ↓ antidiuretic hormone (ADH) secretion or renal insensitivity. May occur temporarily after head injury.

Diabetes mellitus Persistent hyperglycaemia due to a relative deficiency of insulin (p. 568).

Gigantism/acromegaly are excess production of growth hormone, before and after fusion of the epiphyses respectively.

Goitre is a large thyroid gland of whatever cause.

Hyperthyroidism Symptoms of heat intolerance, weight loss, and sweating occur. Signs are tachycardia (may have atrial fibrillation), lid lag, exophthalmos, and tremor. Commonest cause is Graves disease (p. 744). Functioning adenomas are another cause.

Hypothyroidism can be primary due to thyroid disease, or secondary to hypothalamic or pituitary dysfunction. 1° disease is often an autoimmune condition. Symptoms are poor tolerance of cold, loss of hair, weight gain, loss of appetite, and poor memory. Signs are bradycardia and a hoarse voice.

Hyperparathyroidism Primary is caused by an adenoma. Secondary is a response to low plasma Ca^{2+}, e.g. in renal failure, and tertiary follows on from secondary when the parathyroids continue ↑ production even if Ca^{2+} is normalized.

Hypoparathyroidism is usually secondary to thyroidectomy. Plasma Ca^{2+} ↓, resulting in tetany. Chvostek's sign is +ve if spasm of facial muscles occurs after tapping over the facial nerve.

Hypopituitarism can lead to 2° hypothyroidism or 2° hypoadrenocorticism.

Inappropriate ADH secretion Caused by certain tumours (e.g. bronchogenic carcinoma), head injury, and some drugs. Hyponatraemia, overhydration, and confusion occur.

Lingual thyroid May be the only functioning thyroid the patient has; do not excise lightly. Do pre-op isotope scan.

Phaeochromocytoma is a very rare tumour of the adrenal medulla, secreting adrenalin and noradrenalin. Symptoms are recurring palpitations and headache with sweating. Simultaneous hypertension with a return to baseline on settling of symptoms, is a good marker.

Pituitary tumours may erode the pituitary fossa (seen on lateral skull X-ray) and can cause blindness via optic chiasma compression.

Endocrine-related problems

▶ Always ask yourself 'is she, or can she be, pregnant?'

Pregnancy is a C/I to elective GA, the vast majority of drugs (p. 600), and non-essential radiography (the most vulnerable period being in the first three months). Elective treatment is best performed in the mid-trimester.

Menopause is the end of a woman's reproductive life and her periods. It is often associated with hot flushes and other relatively minor physical problems. Emotional disturbances may coexist, and the incidence of psychiatric disorders increases at this time.

Related problems

Suxamethonium sensitivity Around 1 in 3000 people have an inherited defect of plasma cholinesterase. These families are absolutely normal in every respect except in their ability to metabolize suxamethonium. This leaves them unable to destroy the drug which, normally wearing off in 2–4 minutes, produces prolonged muscle paralysis. This then requires ventilatory support until the drug wears off, which, in the homozygote may be as long as 24 hours.

Malignant hyperpyrexia is 'anaesthesia's own disease' when hyperthermia follows a trigger, usually an anaesthetic agent. Dantrolene sodium and cooling are life-saving.

Rare endocrine tumours

Glucagonoma Secretes glucagon causing hyperglycaemia.

Insulinomas Secrete insulin.

Gastrinomas Secrete gastrin causing duodenal ulcers and diarrhoea (Zollinger–Ellison syndrome).

Multiple endocrine neoplasia (MEN) syndromes Are a rare group of endocrine tumours. MEN 3 (2b) is medullary thyroid cancer, phaeochromocytoma, and oral mucosal neuromas.

Bone disease

Pathology of the bones of the facial skeleton is covered in Chapter 8.

Osteogenesis imperfecta (brittle bone disease) is an autosomal dominant type 1 collagen defect. Multiple # following slight trauma with rapid but distorted healing is characteristic. Associated with blue sclera, deafness, and dentinogenesis imperfecta (p. 75). The jaws are *not* particularly prone to # following extractions.

Osteopetrosis (marble bone disease) There is an ↑ in bone density and brittleness, and a ↓ in blood supply. Prone to infection which is difficult to eradicate. Bone pain, # and compression neuropathies may occur. Anaemia can complicate severe disease. Facial characteristics are frontal bossing and hypertelorism.

Achondroplasia is an inherited defect in cartilaginous bone formation, usually autosomal dominant. Have a 'circus dwarf' appearance with skull bossing; many have no other problems.

Cleidocranial dysostosis is an inherited defect of membraneous bone formation, usually autosomal dominant. Skull and clavicles are affected. Multiple unerupted teeth with retention of 1° dentition is characteristic.

Disorders of bone metabolism

Rickets/osteomalacia is failure of bone mineralization in, respectively, children and adults. Can be caused by deficiency, failure of synthesis, malabsorption or impaired metabolism of Vitamin D, and also hypophosphataemia or ↑ calcium requirement in pregnancy.

Osteoporosis is a lack of both bone matrix and mineralization. Important causes are steroid therapy, post-menopausal hormone changes, immobilization, and endocrine abnormalities. Hormone replacement therapy in post-menopausal women appears helpful.

Fibrous dysplasia is replacement of a part of a bone or bones by fibrous tissue with associated swelling. It usually starts in childhood and ceases with completion of skeletal growth. Termed monostotic if one bone is affected, polyostotic if >1 bone, and Albright's syndrome if associated with precocious puberty and *café au lait* areas of skin hyperpigmentation.

Cherubism is a bilateral variant of fibrous dysplasia.

Paget's disease of bone is a common disorder of the elderly, where the normal, orderly replacement of bone is disrupted and replaced by a chaotic structure of new bone, causing enlargement and deformity. The hands and feet are spared. Complications include bone pain and cranial nerve compression, or, more rarely, high output cardiac failure or osteosarcoma.

Diseases of connective tissue, muscle, and joints

Connective tissue diseases

These are mainly vasculitidies (inflammation of vessels).

Cranial arteritis (temporal arteritis) is a giant cell vasculitis of the craniofacial region. Presenting symptoms are unilateral throbbing headache. Signs are high ESR with a tender, pulseless artery. Major complication of temporal arteritis is optic nerve ischaemia causing blindness so start high dose steroids (60 mg prednisolone PO od) and monitor using ESR. Biopsy confirms.

Polymyalgia rheumatica is a more generalized vasculitis affecting proximal axial muscles. Responds to steroids, gradual improvement with time.

Diseases of muscles

Muscular dystrophy is a collection of inherited diseases characterized by muscle degeneration. Most are fatal in early adulthood.

Myotonic disorders are distinguished by delayed muscle relaxation after contraction. They are genetically determined in a complex fashion.

Polymyositis is a generalized immune-mediated inflammatory disorder of muscle. If a characteristic rash is present the condition is known as **dermatomyositis** and has an association with occult malignancy.

Joint disease

Osteoarthritis is inflammation of weight-bearing or otherwise traumatized joints, resulting in pain and stiffness. Osteophyte formation and subchrondral bone cysts, which collapse leading to deformity, are characteristic. Physiotherapy, weight loss, and analgesia are the mainstays of treatment.

Rheumatoid arthritis is an immunologically mediated disease where joint pain and damage are the most prominent symptoms. Morning pain and stiffness in the hands and feet, usually symmetrical, is characteristic. There may be systemic upset. Ulnar deviation of the fingers is pathognomonic. Maintenance of mobility by a range of means, including anti-inflammatory drugs, is the mainstay of treatment. Dry eyes and mouth may be associated with rheumatoid arthritis (Sjögren syndrome, p. 456). TMJ symptoms are rare in rheumatoid arthritis, although up to 15% of patients have radiographic changes in the joint.

Juvenile rheumatoid arthritis (Still's disease) is a rarer form of the disease affecting children. It can be much more severe than the adult condition and can cause TMJ ankylosis.

Psoriatic arthritis is associated with the skin condition and affects the spine and pelvis. It is milder than rheumatoid arthritis and has no serological abnormalities. The TMJ can be affected, but symptoms are usually mild despite some isolated case reports to the contrary.

Gout Urates are deposited in joints, causing sudden severe joint pain, often in the great toe. Affected joints are red, swollen, and very tender. Gout secondary to drugs, radiotherapy, or haematological disease is commoner than that caused by an inborn error of metabolism.

Ankylosing spondylitis affects the spine, usually in young men. Inflammation involves the insertion of ligaments and tendons. It is associated with HLA-B27.

Reiter syndrome is arthritis, urethritis, and conjunctivitis, usually in response to an infection. Oral lesions are often present.

Perthes disease is osteochondritis of the femoral head in, mainly, boys aged 3–11 years. No systemic implications.

Neurological disorders

Cranial nerves

I Olfactory Sense of smell is rarely tested, although damage is quite common following head &/or mid-face trauma.

II Optic Examine the pupils for both direct and consensual reflex; assess the visual fields and examine the fundus with an opthalmoscope (p. 12).

III Oculomotor is the motor supply to the extraocular muscles *except* lateral rectus and superior oblique. It supplies the ciliary muscle and the constrictor of the pupil. A defect ∴ causes impairment of upward, downward, and inward movement of the eye, leading to diplopia, drooping of the upper eyelid (ptosis), and absent direct and consensual reflexes.

IV Trochlear Supplies superior oblique, paralysis of which causes diplopia, worst on looking downward and inward.

V Trigeminal is the major sensory nerve to the face, oral, nasal, conjunctival and sinus mucosa, and part of the tympanic membrane. It is motor to the muscles of mastication. Sensory abnormalities are mapped out using gentle touch and pin-prick. Motor weakness is best assessed on jaw opening and excursion.

VI Abducens Supplies lateral rectus. A defect causes paralysis of abduction of the eye.

VII Facial is motor to the muscles of facial expression. Supplies taste from the anterior ⅔rds of tongue (via chorda tympani) and is secretomotor to the lacrimal, sublingual, and submandibular glands. It innervates the stapedius muscle in the middle ear. The lower face is innervated by the contralateral motor cortex, whereas the upper face has bilateral innervation. Assess by demonstrating facial movements.

VIII Vestibulocochlear is sensory for balance and hearing. Deafness, vertigo, and tinnitus are the main symptoms.

IX Glossopharyngeal Supplies sensation and taste from the posterior ⅓rd of the tongue, motor to stylopharyngeus, and secretomotor to the parotid. Lesions impair the gag reflex in conjunction with

X Vagus Has a motor input to the palatal, pharyngeal, and laryngeal muscles. Impaired gag reflex, hoarseness, and deviation of the soft palate to the unaffected side are seen if damaged. The vagus has a huge parasympathetic output to the viscera of the thorax and abdomen.

XI Accessory is motor to sternomastoid and trapezius, causing weakness on shoulder shrugging and on turning the head away from the affected side.

XII Hypoglossal Motor supply to the tongue. Lesions cause dysarthria (impaired speech) and deviation towards the affected side on protrusion.

Headache

The vast majority of headaches are benign, the secret is to pick out those which are not; read on.

Tension headache Commonest type. Due to muscle tension in occipitofrontalis. Usually worse as the day progresses, may feel 'band-like'. Responds to reassurance, anxiolytics, and analgesics.

Migraine is a distinct entity characterized by a preceding visual aura (fortification spectra). Severe, usually unilateral headache with photophobia, nausea, and vomiting. Thought to be due to cerebral vasoconstriction, followed by reflex vasodilation (the latter is the cause of the pain). Ergotamine abolishes an attack if used early and pizotifen is used prophylactically. F>M, the oral contraceptive being a contributing factor. There are many variants of classical migraine.

Migrainous neuralgia is rarer than migraine and causes localized pain, usually around the eye, with associated nasal stuffiness. M>F. There is a typical time of onset, often in early morning which recurs for several weeks—'clustering'. Alcohol is a common precipitant. Ergotamine and pizotifen help.

Raised intracranial pressure is a cause of headache demanding further investigation. Pointers are headache, worse on waking, irritation, decreasing level of consciousness, vomiting, sluggish or absent pupillary reflexes, bulging of the optic disc (papilloedema), *Rising BP and slowing pulse* are late premorbid signs of ↑ICP.

543

More neurological disorders

CNS infections

Bacterial meningitis Must be considered in the differential diagnosis of headache with photophobia and neck stiffness. Organisms are *Haemophilus influenzae*, *Neisseria meningitidis* (meningococcus), *Streptococcus pneumoniae*, and *Neisseria gonorrhoeae*. In children, the meningococcus is especially important and is classically associated with a purpuric rash. This is one of the very few indications for instituting immediate blind antibiotic therapy (parenteral penicillin).

Viral meningitis is usually mild and self-limiting. It is distinguished from bacterial meningitis by lumbar puncture.

Herpetic encephalitis is a rare manifestation of infection with the herpes simplex virus. Can be distinguished from drunkenness or dementia by history and rapid onset. Parenteral acyclovir can be curative.

CNS tumours Most brain tumours are secondary deposits. Although both benign and malignant primary tumours are found, they are rare. Despite this, they are the commonest cause of cancer death in children after leukaemia.

Epilepsy is an episodic outflow from the brain causing disturbances of consciousness, motor, and sensory function. Most causes are idiopathic but those with onset in adult life must be investigated for local or general cerebral disease. Major or *grand mal* epilepsy is characterized by an aura, loss of consciousness, and seizure followed by a tonic and clonic phase. Incontinence is a good guide to a genuine seizure. The fit rarely lasts >5 min, if it does the patient has entered status epilepticus (p. 566).

Petit mal Are epileptic attacks usually confined to children, taking the form of a short absence when movement, speech, and attention cease.

Temporal lobe epilepsy is characterized by hallucinations of the special senses.

Localized (Jacksonian) epilepsy Affects limbs in isolation. Patients with established epilepsy (once any treatable cause has been excluded) must be maintained on adequate levels of antiepileptic drugs.

Febrile convulsions are fits, usually in children <5 yrs old, secondary to pyrexia.

Cerebrovascular accidents (strokes) These are a very common cause of death in the elderly. A stroke is basically death of part of the brain following cerebral ischaemia, either due to bleeding into the brain or occlusion of vessels. It is often clinically difficult to distinguish the different types of stroke, namely subarachnoid haemorrhage, cerebral haemorrhage, thrombosis, and embolism. CT scanning is sometimes of value when treatment is to be attempted and cerebral angiography defines subarachnoid bleeds.

Multiple sclerosis This disorder is characterized by demyelination in multiple 'plaques' throughout the CNS. Symptoms are multiple and disseminated in both time and place. It is the commonest neurological disease of young adults and has neither cure nor specific diagnostic test. Progress, although relentless, is widely variable.

Myasthenia gravis is muscle weakness due to inadequate response to, or levels of, acetyl choline. Extraoccular muscles are often first affected. It is diagnosed, using the 'tensilon' test.

Parkinson's disease is a disease affecting the basal ganglia associated with a decrease in the local levels of dopamine. It is characterized by a 'pill-rolling' tremor, 'cog-wheel' rigidity, and a shuffling gait.

Skin neoplasms

▶ The skin of the face is the commonest site of curable skin cancers, so look and think.

Basal cell carcinoma (epithelioma, rodent ulcer) is an indolent skin cancer which very rarely metastasises. If it kills it does so by local destruction. Sunlight is a major aetological factor. There are various forms, the commonest being an ulcerated nodule with raised pearly margins and a telangiectatic surface. Rx: excision—(micrographic or conventional), radiotherapy (especially electron beam), cryotherapy, curretage, and electro-dessication.

Squamous cell carcinoma (SCC) of the skin is surprisingly indolent in comparison to SCC of mucosa. Presents as an ulcerated lesion with raised edges. Keratin horns may be present, and it may arise in areas of previously sun-damaged skin or in gravitational leg ulcers. Surgical excision or radiotherapy are treatments of choice.

Malignant melanoma This condition is being increasingly diagnosed, with a doubling of the incidence in the last 20 yrs. The prognosis is dependent primarily on the depth of the tumour (Breslow thickness) as the thicker the lesion the poorer the prognosis. Early metastasis is common. Sunlight is a major aetiological factor. Suspect if a pigmented lesion rapidly enlarges, bleeds, ulcerates, or changes colour. Prompt referral for specialist management is needed.

Naevi are areas of skin containing a disproportionate number of melanocytes.

Lentigo simplex is a freckle.

Dysplastic naevi are premalignant lesions often found in patients with malignant melanoma. They should be excised and patients advised to use high-factor sunscreens.

Lentigo maligna is a premalignant pigmented lesion of the elderly.

Carcinoma-*in-situ* (Bowen's disease) Presents as a scaly, red plaque. It is basically a squamous carcinoma which has not yet penetrated beyond the basal layer.

Actinic keratosis is persistent sun-damaged areas of skin in which cancer may arise.

Mycosis fungoides is a lymphoma of the skin which resembles psoriasis in appearance.

Metastatic deposits to the skin occur most frequently from breast, kidney, and lung, but skin secondaries from oral cancer are being seen increasingly. Kaposi's sarcoma is a purplish tumour seen in skin and mucosa in AIDS.

Dermatology

Psoriasis is a common relapsing proliferative inflammatory skin disease. It appears as a red plaque with silvery scale, chiefly on extensor skin of knees and elbows, although any area can be affected. It can be associated with systemic disease particularly arthropathy (p. 540). Rx is chiefly topical; steroids, coal tar, dithranol &/or UVB radiation, or psoralen-sensitized UVA radiation (PUVA). Rarely, methotrexate can be used.

Eczema Also called dermatitis. Has several variants according to aetiology.

Atopic eczema Starts in the first year of life with a red symmetrical scaly rash. Emulsifying ointments help prevent fissuring, although steroids are sometimes needed. Up to 90% grow out of it by age 12.

Exogenous eczema Can be produced in anyone exposed to a sufficient irritant. The hands are the usual target, with blistering, erythema, and cracking of skin.

Allergic contact eczema is a genuine allergic response, e.g. to nickel.

Seborrhoeic eczema is a fungal infection mainly affecting the scalp ('cradle cap' in neonates).

Skin infections Fungal infections are particularly common, causing angular cheilitis, athlete's foot, paronychia, vaginitis, etc. Furuncles are staphylococcal boils. Erysipelas is a streptococcal cellulitis. Viruses cause herpes zoster and simplex infections, molluscum contagiosum, and warts.

Infestations of the skin bring a shudder to most people, but they are also a hazard of working closely with patients! Head lice respond to malathion. Flea bites, as well as being unpleasant, can spread plague among other serious diseases. Scabies is an infestation with a mite which creates a characteristic itchy burrow in the finger webs.

Acne Acne vulgaris is characterized by the blackhead (comedo), and is an inflammatory condition caused by increased sebum secretion. It is hormone-dependent, although superinfection with the acne bacillus is a contributing factor. Tends to scar. After proprietary lotions, low-dose tetracyclines help.

The skin and internal disease

The skin, like the mouth, acts as an outside indicator for many internal diseases.

Erythema nodosum Painful, red, nodular lumps on the shins.

Erythema multiforme Circular target lesions.

Erythema marginatum Vanishing and recurring pink rings. These are all non-specific markers for a variety of diseases.

Vitiligo is an autoimmune hypopigmentation and is associated with other autoimmune conditions.

Pyoderma gangrenosum Blue-edged ulcers, especially on the legs, is associated with ulcerative colitis and Crohn disease.

Granuloma annulare Subcutaneous circular thickening and **necrobiosis lipoidica** yellow plaques on the shins are associated with diabetes.

Dermatitis herpetiformis Vesicular rash of knees, elbows, and scalp. Is associated with coeliac disease.

Pretibial myxoedema Red swellings above the ankle. Is associated with hyperthyroidism.

Skin diseases associated with malignancy are **acanthosis nigricans** (rough, pigmented, thickened areas of skin in axilla or groin) and **thrombophlebitis migricans** (tender nodules within blood vessels which move from site to site).

Psychiatry

One way of getting to grips with a new subject—and to virtually all dentists psychiatry as opposed to psychology is new—is to categorize. The major adult psychiatric diagnoses are listed in order of severity. This is known as the 'hierarchy of diagnosis'.

Organic brain syndromes

Acute organic reaction (delirium, toxic confusion). Clouding of consciousness and disorientation in time and place are major symptoms. Mood-swings are common, and visual hallucinations, rare in other psychiatric conditions, are often present.

▶ There is an underlying, frequently treatable, cause to this condition (infection, drugs, dehydration, alcohol withdrawal, etc.). Rx: find the cause and correct it, using sedation until the cause is identified and treatment has taken effect.

Chronic organic reaction (dementia) is a global intellectual deterioration highlighted by a worsening short-term memory. *Never* label someone as demented until all other possible causes, including depression, have been excluded by a psychiatrist. Alzheimer's disease and multi-infarct dementia are the commonest causes. There is no cure, although effective support services can improve the quality of life considerably.

Mental handicap, p. 52.

Psychosis

Contact with reality is lost and normal mental processes do not function. There is loss of insight. If an organic condition is excluded, the diagnosis is one of three.

Schizophrenia is a disorder where the victims live in an incomprehensible world full of vivid personal significance. First rank symptoms are a good guide to diagnosis: thought insertion, broadcasting and withdrawal, passivity feelings, auditory hallucinations.

Affective disorders are mania, hypomania, manic-depressive psychosis, and depression. Mania and hypomania are characterized by euphoria, hyperactivity, overvalued ideas or grandiose delusions, and pressure of speech. They differ only in degree. Cyclical mania and depression is known as bipolar or manic-depressive psychosis. Rx is with major tranquillizers and prophylaxis with lithium carbonate.

Depression May be either psychotic or neurotic. Markers of major depressive illness are anhedonia (failure to find pleasure in things which once did please), anorexia especially with weight loss, early morning wakening, tearfulness, inability to concentrate, feelings of guilt and worthlessness, and suicidal ideation.

Paranoid states are psychoses where paranoid symptoms predominate and despite lack of insight, other diagnoses do not apply.

▶ The commonly abused drugs can all mimic or precipitate psychotic states, as can giving birth—puerperal psychosis.

Neuroses

A neurosis is a maladaptive psychological symptom in the absence of organic or psychotic causes of the symptom and after exclusion of a psychopathic personality. Insight is present.

Anxiety neurosis frequently coexists with depression. These patients often have physical symptoms for which there is no physical explanation.

Obsessional neurosis are intrusive thoughts or ideas which the subject recognizes as coming from within themselves, but resents and is unable to stop. May be associated with *compulsive behaviour* where repeated purposeless activity is carried out due to an inexplicable feeling that it must be done.

Phobia is the generation of fear or anxiety out of proportion to the stimulus. Numerous stimuli exist, including dentists.

Anorexia nervosa/Bulimia nervosa

The development of weight reduction as an overvalued idea. Associated with weight loss of >25% ideal body weight and obsessive food avoidance. Commonest in females, it is also associated with amenhorrea. Binge-eating followed by vomiting &/or laxative abuse can occur. Bingeing without weight loss is bulimia nervosa. Dental effects, p. 304.

Personality disorders

These are not illnesses but extremes of normal personality traits, e.g. obsessional, histrionic, schizoid (cold, introspective). The most important is the psychopathic (sociopathic) individual who has no concept of affection, shame, or guilt, and is characterized by antisocial behaviour. They are often superficially personable, highly manipulative, and totally irresponsible. They have insight and are responsible for their own actions (bad not mad).

The immunocompromised patient

There are a group of individuals who present special problems because of defects in or suppression of their immune system. The condition with the highest profile among these is the Acquired Immune Deficiency Syndrome (AIDS).

The chief effect of being immunocompromised is an ↑ susceptibility to infection, often due to opportunistic organisms. Anything which changes the host environment in favour of opportunistic pathogens (e.g. surgery, broad-spectrum antibiotics) can lead to potentially fatal infection with rare or otherwise innocuous organisms.

Drugs which suppress the immune response: corticosteroids, cyclosporin A, azathioprin, cytotoxics, etc., are now in common use therapeutically. Cross-infection, p. 734. Aggressive Rx of infections and antimicrobial prophylaxis are needed, p. 590.

Congenital immunodeficiency states There are at least 18 of these. The commonest is selective IgA deficiency, which affects around 1 in 600 and has a wide spectrum of severity but may remain asymptomatic.

Acquired immunodeficiency

Autoimmune disease, e.g. systemic lupus erythematosus, rheumatoid arthritis, carry a minor ↑ risk of infection.

Chronic renal failure (p. 532) Moderately ↑ risk.

Deficiency states, e.g. anaemia. Carry a minor ↑ risk.

Diabetes mellitus is common and carries a moderate ↑ in risk of infection.

Infections Severe viral infections, tuberculosis, AIDS (specific defect).

Neoplasia All haematological malignancies severely ↑ risk of infection.

AIDS

An increasingly common disease caused by the human immunodeficiency virus (HIV-1, HIV-2) CD_4. T-lymphocyte defect ensues with failure of (mostly) cell-mediated immunity. Although HIV exposure produces antibody response, the virus remains infective in the presence of antibody, it must therefore be regarded as a marker of infectivity. Absence of HIV antibody *does not*, however, guarantee that that person is not infected with HIV. HIV antibody positive patients are at risk from developing AIDS usually after a prolonged latent period during which CD_4 cells ↓ in number. AIDS-related complex which includes cervical lymphadenopathy, oropharyngeal candidiasis, and 'hairy leukoplakia' is precursor to full-blown AIDS. Infections characteristic of AIDS are pneumocystis carinii pneumonia and disseminated mycobacterial infection. Kaposi's sarcoma is the tumour most often associated with the condition. The mode of transmission, for those visiting from another planet, is anal or vaginal sex or as a recipient of contaminated blood or blood products or mother to fetus transmission. The main risk groups, in the developed world, are therefore male homosexuals (although transmission through the heterosexual population is increasing), IV drug abusers, Transfusion recipients, and haemophiliacs who were at risk prior to screening of blood products now have a minimal risk.

In the developing world, heterosexual spread is common and mother to fetus transmission is creating a huge ↑ in HIV +ve children. While there is no cure or vaccination, numerous symptom-reducing and life-prolonging treatments are available with mixed results. Psychological and social support is the most helpful option after preventive advice. Prophylaxis consists of screening blood products, and avoidance of unprotected sexual activities and shared needles. Oral manifestations of AIDS, p. 472. Practical procedures for control of cross-infection, p. 734.

Useful emergency kit

Every practice should possess apparatus for delivering oxygen, or at least air. In addition, the facility to deliver nitrous oxide and oxygen mixture, e.g. via an anaesthetic or relative analgesia machine, can be invaluable.

The following should be readily available:
- Oral airway, preferably with a bag system, e.g. ambu-bag.
- High vacuum suction.
- Disposable syringes (2, 5, and 10 ml), needles (19 and 21 G) and a tourniquet. Butterfly needles and IV canulae are great assets to those familiar with their use.

Drugs:

- Adrenaline, 1 in 1000 solution (1 mg adrenaline in 1 ml saline).
- Hydrocortisone (as sodium succinate or sodium phosphate) (100 mg and water for injection).
- Benzodiazepine, diazepam in lipid emulsion, or midazolam (10 mg ampoules).
- Glucose, as dextrose 20% or 50% solution.
- Chlorpheniramine 20 mg injection.
- Flumazenil 100 microgram/ml 5 ml ampoule.

Practices equipped for GA will be required to meet specific criteria, including those indicated on p. 648.

Fainting

Fainting (vaso-vagal syncope) is innocuous providing it is recognized. It is easily the most common cause of sudden loss of consciousness, with up to 2% of patients fainting before or during dental treatment. The possibility of vaso-vagal syncope while under GA, and hence failure to recognize the condition and correct cerebral hypoxia, is the major reason for recommending the supine position.

Predisposing factors are pain, anxiety, fatigue, relative hyperthermia, and fasting. Characteristic signs and symptoms are: a feeling of dizziness and nausea; pale, cold and clammy skin; a slow, thin, thready pulse which rebounds to become rapid, and loss of consciousness with collapse, if unsupported.

A faint may mimic far more serious conditions, most of which can be excluded by a familiarity with the patient's past medical history. These include strokes, corticosteroid insufficiency, drug reactions and interactions, epileptic fit, heart block, hypoglycaemia, and myocardial infarction.

Prevention
- Avoid predisposing factors.
- Treat patients in the supine position unless specifically contraindicated (e.g. heart failure, pulmonary oedema).

Management
- Lower the head to the level, or below, of the heart. Best achieved by laying the patient flat.
- Loosen clothing (in the presence of a witness!)
- Monitor pulse. If recovery does not occur rapidly, then reconsider the diagnosis.
- Determine the precipitant and avoid in the future.

Acute chest pain

Severe, acute chest pain is usually the result of ischaemia of the myocardium. The principal differential diagnosis is between angina and myocardial infarction. Both exhibit severe retrosternal pain described as heavy, crushing, or band-like. It is classically preceded by effort, emotion, or excitement, and may radiate to the arms, neck, jaw, and occasionally to the back or abdomen. Angina is rapidly relieved by rest and glyceryl trinitrate (0.5 mg) given sublingually, which most patients with a history of angina carry with them.

Failure of these methods to relieve the pain, and co-existing breathlessness, nausea, vomiting, and loss of consciousness with a weak or irregular pulse, suggest an infarct.

Management depends on your immediate environment, but always ensure the patient is placed in a supported *upright* position, as the supine position increases pulmonary oedema and hence breathlessness.

Management

In dental practice—Summon help (ambulance). Administer analgesia; the most appropriate form available here will be nitrous oxide/oxygen mixture. Don't panic. Be prepared should cardiac arrest supervene. Give aspirin 75–150mg PO.

In hospital—Nurse upright. Give oxygen. Establish IV access and give an opioid analgesic if available (2.5–10 mg of diamorphine is most useful). Summon help; in dental units integrated into general or teaching hospitals this may best be achieved by contacting the resident medical officer via the switchboard. Remember to give aspirin.

Cardiorespiratory arrest

▶ Don't await 'expertise'. **ACT**.

90% of deaths from cardiac arrests occurring outside hospital are due to ventricular fibrillation (VF). This condition is potentially reversible. The commonest underlying cause is ischaemic heart disease, but other causes, especially in younger people, may exist, such as acute asthma, anaesthesia, drug overdose, electrocution, immersion, or hypothermia. In certain instances properly performed cardiopulmonary resuscitation (CPR) can sustain life for up to an hour while a precipitating condition is being treated.

Diagnosis Sudden loss of consciousness, absent breathing, absent pulse.

Management
- Ensure immediate expert help is called for.
- In *witnessed* or monitored arrests, a single sharp blow over the heart (precordial thump) is worthwhile.[1]
- **Airway** Extend neck, support mandible, and clear oropharynx.
- **Breathing** Look for chest movement. Listen for breath sounds and feel for breath on the back of your hand. If none, begin ventilation by a method you are familiar with. Basic ventilation starts with mouth to mouth; if you wish to use equipment such as airways, face masks, bag valve masks, etc., ensure they work and that you can use them. Begin with two slow expirations which cause the chest to rise.
- **Circulation** No palpable carotid pulse for 5 sec, then begin external cardiac massage. The positioning of your hands on the sternum and technique of compression is vital and will not be learned from a book. Have someone demonstrate the position and practise on a manikin.

Rates of compression/ventilation.
- Ventilation only (good cardiac output) 12–15 breaths/min.
- CPR single rescuer 15 compressions to 2 ventilations.
- CPR two rescuers 5 compressions to 1 ventilation. Aim for around 60–80 chest compressions per minute.

Unfortunately, few dentists are trained in the skill of endotracheal intubation, and your first cardiac arrest is not the best time to learn! It is ∴ safer to continue basic CPR and await assistance. For those in hospital practice and fully equipped surgeries where endotracheal tubes, monitors, and defibrillators are available, the recent recommendations of the Resuscitation Council (UK) are presented.

1 D. A. Chamberlain 1989 *BMJ* **299** 446.

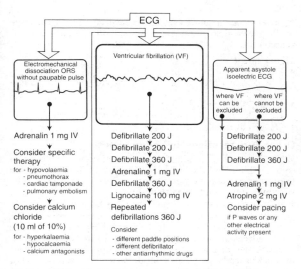

ECG

Electromechanical dissociation ORS without paupable pulse

Adrenalin 1 mg IV

Consider specific therapy

for - hypovolaemia
    - pneumothorax
    - cardiac tamponade
    - pulmonary embolism

Consider calcium chloride
(10 ml of 10%)

for - hyperkalaemia
    - hypocalcaemia
    - calcium antagonists

Ventricular fibrillation (VF)

Defibrillate 200 J
Defibrillate 200 J
Defibrillate 360 J
Adrenaline 1 mg IV
Defibrillate 360 J
Lignocaine 100 mg IV
Repeated defibrillations 360 J

Consider
    - different paddle positions
    - different defibrillator
    - other antiarrhythmic drugs

Apparent asystole isoelectric ECG

where VF can be excluded | where VF cannot be excluded

Defibrillate 200 J
Defibrillate 200 J
Defibrillate 360 J

Adrenalin 1 mg IV
Atropine 2 mg IV
Consider pacing
if P waves or any other electrical activity present

Continue CPR for up to 2 min after each drug. Do not interrupt CPR for more than 10 s, except for defibrillation. If an IV line cannot be established, consider giving double doses of adrenalin, lignocaine, or atropine via an endotracheal tube.

PROLONGED RESUSCITATION:	POST RESUSCITATION CARE
Give 1 mg adrenalin IV every 5 minutes. Consider 50 mmol sodium bicarbonate (50 ml of 8.4%) or according to blood gas results.	Check - arterial blood gases - electrolytes - chest X-ray Observe monitor and treat patient in an intensive care area.

Anaphylactic shock and other drug reactions

Penicillins are the commonest offender, but it is worthwhile remembering that there is a 10% cross-over in allergic response between penicillins and cephalosporins.

An anaphylactic reaction is not an all-or-nothing response, and grades of severity are seen. Generally, the reaction starts a few minutes after a parenteral injection and not immediately as does a simple faint. Some caution should be exercised, though as the quicker the onset of an anaphylactic reaction the more severe it is likely to be.

Principal symptoms are facial flushing, itching, numbness, cold extremities, nausea, and sometimes abdominal pain. Signs include wheezing, facial swelling and rash, cold clammy skin with a thin thready pulse. Loss of consciousness may occur, with extreme pallor which progresses to cyanosis as respiratory failure develops.

Angio-oedema is sudden onset, severe face and neck allergic swelling. The airway is at risk and ∴ should be managed as for anaphylaxis.

Management
- Place patient supine with legs raised.
- 1 ml of 1 in 1000 adrenalin IM or SC. Repeat every 15 min until improved.
- Up to 500 mg of hydrocortisone IV.
- Up to 20 mg of chlorpheniramine slowly IV (if available).
- Oxygen by mask.

Other drug reactions and interactions

While there is a multitude of drug interactions which the dental surgeon should be aware of as a prescriber, the drugs most liable to present an emergency problem to the dentist are those which we administer as local anaesthetics.

Although it is possible to achieve toxic levels of lignocaine, adrenalin, prilocaine, or felypressin without intravascular injection, this generally requires a particularly cavalier attitude to the administration of LA. Commonly, this effect is due to intravascular injection of a substantial proportion of a cartridge of local anaesthetic. Confusion, drowsiness, agitation, fits, or loss of consciousness may occur. Do not use more than 10 × 2.2 ml cartridges of lignocaine/adrenalin (500 mg lignocaine). In practice, you will rarely consider coming near this amount.

Management
- Place supine.
- Maintain airway, give oxygen.
- Await spontaneous recovery (in around 30 min) unless tragically a serious event such as myocardial infarction supervenes, in which case treat as indicated.

Collapse in a patient with a history of corticosteroid use

The use of corticosteroids therapeutically or otherwise for whatever cause may suppress the adrenal response to stress. The longer the course of treatment and the higher the dose used the more likely this is to occur.

The prime aim is to prevent the occurrence of stress-induced collapse; therefore, if patients have received steroids in the past year or are on steroids at present, cover any stressful procedure, anaesthetic, infection, or episode of trauma with 100 mg hydrocortisone IM 30 min prior to elective stress. In patients presenting acutely, treat immediately. If collapse occurs in such a patient, diagnosis is established by pallor, rapid, thin pulse with a profound and sudden drop in blood pressure, and loss of consciousness.

Management
- Place in supine position.
- Up to 500 mg hydrocortisone IV immediately.
- Maintain airway and supplement with oxygen if available.
- Ensure help (i.e. an ambulance) is requested.
- Exclude other causes of collapse.

Fits

The majority of epileptic fits do not require active intervention as the patient will usually recover spontaneously. All that is needed is sensible positioning to prevent the patient from damaging himself. Fits may be precipitated in a known epileptic by starvation, flickering lights, certain drugs such as methohexitone, tricyclics or alcohol, or menstruation. They may also follow a deep faint.

Diagnosis Many epileptics have a preceding aura followed by sudden loss of consciousness with a rigid extended appearance and generalized jerking movements. Frequently, they are incontinent of urine and may bite their tongue. There is a slow recovery with the patient feeling sleepy and dazed.

▶ Should the fitting be repeated, the patient has entered the state of status epilepticus. This is an emergency and requires urgent control.

Management In a simple major fit the patient should be placed in the recovery position when practicable and allowed to recover. If they enter status epilepticus, IV diazepam (preferably as a lipid emulsion—Diazemuls) usually aborts the fit in a dose of 10–20 mg, although more may be needed. Assess cardiorespiratory function; clear and maintain airway, and give oxygen if available. It is worthwhile considering placing an IV cannula or a butterfly in any epileptic patient with less than perfect control, as stress is an important precipitant. Status epilepticus should not be allowed to continue for more than 20 min as the mortality rate and chance of permanent brain damage increase with the length of attack.

Management in hospital After giving IV benzodiazepines and maintaining an airway, give an IV bolus of up to 50 ml of 20–50% dextrose. Establish a 0.9% saline infusion and repeat the benzodiazepines if necessary. If the fits are not controlled it may be necessary to use a phenytoin infusion or induce anaesthesia with thiopentone, an inhalational agent, and ventilate. Before this stage is reached help should have been requested.

Hypoglycaemia

Hypoglycaemia is the diabetic emergency most likely to present to the dentist. It is an acute and dangerous complication of diabetes and may result from a missed meal, excess insulin, or increased calorific need due to exercise or stress. Most diabetics are expert in detecting the onset of hypoglycaemia themselves; however, a small number may lose this ability, particularly if changed from porcine to human insulin. Recognition of this state is essential and **an acutely collapsed diabetic should be assumed hypoglycaemic until proven otherwise**, e.g. by BM sticks or blood-glucose levels.

Diagnosis—disorientation, irritability, increasing drowsiness, excitability, or aggression in a known diabetic suggests hypoglycaemia. They often appear to be drunk.

Treatment
- If conscious, give glucose orally in any available form.
- If unconscious, protect airway, place in recovery position, establish IV access and give up to 50 ml of 20–50% dextrose. If available, 1 mg of glucagon IM may be used. Ensure help is requested.

Acute asthma

An acute asthmatic attack may be induced in a patient predisposed to bronchospasm by exposure to an allergen, infection, cold, exercise, or anxiety. Characteristically, the patient will complain of a tight chest and shortness of breath. Examination will reveal breathlessness, with widespread expiratory wheezing. The accessory muscles of respiration may be used to support breathing. If the patient is unable to talk, you are dealing with a potentially fatal episode.

Management Make use of the patient's own anti-asthmatic drugs, such as salbutamol inhalers. Ideally, this should be administered in the form of a nebulizer using 24% oxygen and nebulized salbutamol. A DIY nebulizer can be fabricated from the patient's own inhaler pushed through the base of a paper cup. Repeated depressions of inhaler plunger will create an aerosol inside the cup which the patient can inhale. This will relieve most reversible airways obstruction. Steroids should be administered either as oral prednisolone, if the patient carries these with them, or as IV hydrocortisone up to 200 mg IV. This combination of salbutamol, steroids, and oxygen will often completely resolve an attack; however, in individuals who do not respond, an urgent hospital admission is required. Patients who are only partially responsive must have underlying irritants such as a chest infection either excluded or treated.

Management in dental practice
- Keep the patient upright.
- Administer salbutamol either by inhaler or by DIY nebulizer.
- Give oxygen.
- Give steroids.

If a complete response takes place it is reasonable to allow the patient to return home. If there is any doubt arrange for the patient to be seen at the nearest casualty department.

Management in hospital
- Nurse patient upright.
- Give nebulized salbutamol 2.5 mg up to 2 hrly.
- Establish IV access and give up to 200 mg hydrocortisone IV.
- Monitor peak expiratory flow and arterial blood gases.
- Obtain a chest X-ray.
- If incomplete response obtain expert help rather than embarking on alternative treatments such as aminophyline or ipratropium.

Inhaled foreign bodies

The combination of delicate instruments and the supine position of patients for many dental procedures inevitably increases the risk of a patient inhaling a foreign body. Two basic scenarios are likely, depending on whether or not the item impacts in the upper or lower airway.

Upper airway This will stimulate the cough reflex, which may be sufficient to clear the obstruction. A choking subject should be bent forward to aid coughing. **Do not** slap on back. If the object is not cleared, employ the Heimlich manoeuvre by encircling the victim with your arms from behind, and deliver a maximally forceful sudden squeeze by pulling them towards you. The force is directed upwards and backwards and creates an 'assisted cough' which expels the obstruction. If this fails after repeated attempts, a cricothyroid puncture using a wide-bore cannula or similar instrument may preserve life if the obstruction lies above this level.

Lower airway As only a segment of the lungs will be occluded this presents a less acute problem. It is also easier to miss. Classically, this involves a tooth or tooth fragment slipping from the forceps and being inhaled. With the patient in a semi-upright position the object ends up in the right posterior basal lobe. Should this happen, inform the patient and arrange to have a chest X-ray taken as soon as possible. If the offending item is in the lungs, removal by a chest physician by fibre-optic broncho-scopy is indicated, as this is inevitably followed by collapse and infection distal to the obstruction.

If in doubt

When presented with a suddenly collapsed patient the first thing to assess is your own response. **Don't panic**. You are only of value to the patient if you can function rationally. If presented with a case of sudden loss of consciousness, in the absence of an obvious diagnosis, the following steps should be followed.

- **Place in supine position.** If the patient has simply fainted they will recover virtually immediately.
- **Are they breathing?** if not begin artificial respiration, p. 560.
- **Maintain the airway** and provide oxygen if available.
- **Feel for the pulse**. If it is absent, there has been a cardiac arrest (p. 560). If it is present:
- **Establish IV access and give up to 20 ml 20–50% dextrose IV.**
- **Give hydrocortisone up to 200 mg IV.**

These measures will usually resolve most cases of sudden, non-traumatic loss of consciousness.

If the patient is acutely distressed and breathless they should be treated in an upright position and given oxygen while you try to differentiate between an acute asthmatic attack and heart failure and treat as indicated.

Always ensure that someone has requested assistance in the form of an ambulance or, if in hospital, the appropriate staff have been contacted.

Immediately after resolution of an emergency there tends to be a period of numb inactivity amongst the staff involved. Use this period to review your management of the situation and carefully document what happened. If the patient has been transferred to hospital or another department a brief, legible account of proceedings must accompany them. Include drugs used, their dosages, and when they were given. Try to ensure that a friend or relation of the patient is aware of the situation.

Management of the dental in-patient

The vast majority of in-patients will experience considerable anxiety on being admitted for operation, including those procedures which are in themselves 'routine'. In addition, as dentists have little training in medical clerking and ward work, there is a substantial risk of compounding an already stressful situation by being overly stressed yourself. Minimize this by preparation. Learn about the ward(s) you will work on before taking up a post. Never be afraid to ask nursing staff if you are unsure, and try to know a day in advance what cases are coming in.

Pre-operation

All patients attending as in-patients for operation must (a) be examined and 'clerked', and (b) have consented to surgery. In addition, many will require a variety of pre-operative investigations, these vary widely from consultant to consultant, so get to know the local variations. Common investigations and their indications are listed below. Sampling techniques, p. 578; samples, p. 16.

(a) Clerking This basically consists of taking a complete medical and dental history from the patient, including any drugs which they are taking at present (which you must remember to continue while in hospital by writing up in the drug 'kardex'), a family history for inherited disease, and a social history for problems related to smoking, alcohol, drug abuse, and ability to cope at home post-operation. This is followed by a systematic clinical examination (p. 10). Any special investigations are then arranged, and the results of these should be seen before the patient goes to theatre. Any problems uncovered should be relayed to the anaesthetist, who is the only person capable of saying whether or not the patient is 'fit for anaesthesia'. Any required pre-, peri-, or post-operative drugs are written up, p. 600.

(b) Consent All patients undergoing GA must give written, informed consent. It is often advised that patients receiving heavy sedation also do so. Every hospital has its own surgical consent form which you complete. After having the procedure and its likely potential risks explained, the patient also signs. It is essential that you are happy in your own mind that you understand what the operation entails; if in doubt ask. No one should wake up with their teeth wired together or a tracheostomy without being previously warned.

Investigations (p. 16)

Full blood count Elderly patients. Any suspicion of anaemia.

Sickle cell test All Afro-Caribbeans for GA.

Urea and electrolytes All patients needing IV fluids, on diuretics, or with renal disease. Have a low threshold for doing this test.

Coagulation screen All major surgery, any past history of bleeding disorders, liver disease, or history of ↑ alcohol, anticoagulants.

Liver function tests Liver disease, alcohol, major surgery.

Group and save/cross-match Major surgery, trauma, shock, anaemia.

ECG Heart disease, all major surgery, most patients >50.

Chest X-ray Trauma, active chest disease, possible metastases.

Hepatitis B and HIV markers Varies; usually at risk groups only, pretest counselling for HIV.

Post-operation

Immediately post-operation patients are resuscitated in a recovery room adjacent to theatre, with a nurse monitoring cardiorespiratory function. Once recovered, unless they are to be monitored in the intensive care unit they will be returned to the ward. In all patients, ensure a patent airway and consider:

Analgesia may take the form of LA (should be given post-anaesthetic/pre-surgery), oral or parenteral NSAIDs, or oral or parenteral opioids. Immediately post-op, analgesia is best given parenterally. Antiemetics such as 10 mg metoclopramide IM or IV or prochlorperazine 12.5 mg IM should be given if nausea or vomiting is present.

Antimicrobials are given in accordance with the selected regimen (p. 590). Certain patients may benefit from corticosteroids pre- and post-operatively to reduce oedema; regimens vary.

Nutrition is a problem principally for patients undergoing major head and neck cancer surgery (p. 510), but it is worthwhile reminding nursing staff to order soft diets for all oral surgery patients who can feed by mouth and to have liquidizers available for patients in IMF.

Fluid balance is covered on p. 580. Special consideration needs to be given to patients in IMF and those with tracheostomies. Although the use of immediate post-operation IMF is ↓ thanks to the introduction of mini-plating systems, it is still required. These patients need to be specially attended by a nurse looking after that patient only, for the first 12–24 h ('specialed'). Lighting, suction, and the ability to place the patient head-down if they vomit, is mandatory. Wire cutters and other tools to remove the IMF should also be available, although removal of IMF is actually unlikely to be quick enough to save a life. Other techniques include tongue suture, nasogastric intubation, prolonged retention of a naso-pharyngeal airway, elective prolonged intubation, and tracheostomy. Avoiding this situation, see p. 496. Tracheostomy is a great aid to secure airway management but is enormously inconvenient for patients and has its own complications. If indicated it is essential to care for the tube by suction and humidification to ensure patency.

Venepuncture and arterial puncture

Venepuncture To become proficient in the skills of vene-puncture you must practise the art in all its forms. To develop the skill of placing IV cannulae, cultivate a sympathetic anaesthetist, as anaesthetized patients are venodilated and will not feel pain! When carrying out your first few cannulations and arterial punctures on patients in the ward, a drop or two of 2% plain lignocaine deposited SC with a fine needle will aid both your peace of mind and the patient's comfort.

Tools of the trade Tourniquet, alcohol wipes, cotton wool. Green (21G) needles and butterflies are most commonly used. Sometimes finer needles or butterflies are needed, e.g. blue (23G). Patients who are difficult to cannulate can have fluids and certain drugs through fine (20G) or even 22G IV cannulae; most have 18G. Shocked patients or those needing blood should have at least a 17G and preferably a 16G or 14G cannula inserted. Note that gauges and colours are not consistent between needles and cannulae.

Sites of puncture

For sampling—First choice is the cubital fossa. Inspect and palpate; veins you can feel are better than those you can only see. Insert the needle at a 30–40° angle to the skin and along the line of the vein. If no veins are found in the cubital fossa try the back of the hand with a butterfly and use a similar approach. The veins of the dorsum of the foot are a last resort before the femoral vein lying just medial to the femoral artery in the groin.

For infusion—Single bolus injections, use a 21G butterfly in a vein on the back of the hand.

- For IV fluids or multiple IV injections place a 18G IV cannula in a straight segment of vein in the forearm, hand, or 'anatomical snuffbox'. Try to avoid crossing a joint as the cannula tissues more quickly if subjected to repeated movements. When inserting the cannula ensure the skin overlying the vein is fixed by finger pressure, pierce the skin, and move the stillete along the line of the vein until it enters the vein and blood flows into the cannula. As soon as you enter the vein, pull the stillete back into the cannula to minimize the risk of going through the vein. Insert the full length of the cannula into the vein and secure. Keep patent with heparinized saline.

Arterial puncture Whenever possible, obtain an arterial sampling syringe. Use LA unless you are very good. The syringe and needle must be flushed with heparin. Use radial, brachial, or femoral arteries. Palpate, prepare area with alcohol wipe, and insert needle at 30–60° to skin. When the needle enters the artery, blood pulsates into the syringe. Only 1–2 ml are needed. Remove needle and place in a bag with ice, contact the bio-chemist, and treat as an urgent specimen. The puncture site needs to be firmly pressed on for 2–3 min to prevent formation of a painful haematoma.

Intravenous fluids

Principles The maintenance of daily fluid requirements plus replacement of any abnormal loss by infusion of (usually) isotonic solutions. Normal requirements are around 2.5–3 l in 24 h. This is lost via urine (normal renal function needs an absolute minimum of 30 ml/h, but aim for 60 ml/h), faecal loss, and sweating. Where possible replace with oral fluids; IV fluids are a second best.

Common IV regimens 1 l normal saline (0.9%) and 2 l 5% dextrose in 24 h, or 3 l dextrose/saline solution in 24 h. Add 20 mmol potassium chloride per litre after 36 h, unless U&Es suggest otherwise.

Increase the above in the presence of abnormal losses, burns, fever, dehydration, polyuria, and in the event of haemorrhage or shock.

Special needs
(a) For burns use plasma.
(b) For fever use saline.
(c) For dehydration or polyuria use 5% dextrose, unless hyponatraemic. Exception is ketoacidosis—use saline.
(d) Haemorrhage demands replacement whole blood if available; packed cells are second best, and shock needs plasma, colloid, or blood. Be guided by the pulse, BP, urine output, haemoglobin, the haematocrit, and the U&Es.

Decrease the above in heart failure, and avoid saline. Shock and dehydration are rare complications of maxillofacial trauma or any other condition principally presenting to the dentist, therefore in their presence consider damage to other body systems and seek appropriate advice.

Polyuria Post-operatively is usually due to over-transfusion. Review anaesthetic notes, and if this is the case simply catheterize and observe.

Oliguria Post-operatively, is usually due to under-transfusion or dehydration pre-operatively. First, catheterize the patient to exclude urinary retention and allow close monitoring of fluid balance. Then increase rate of infusion of fluid (**max** 1 l/h) **unless** the patient is in heart failure or is bleeding. The former needs specialist advice, the latter needs blood. In an otherwise healthy post-operative patient, if this does not produce a minimum of 30 ml/h of urine, a diuretic (20–40 mg frusemide PO/IV) may be tried. Take care when using fluid challenges, as if failure to PU is renal it is possible to fluid overload the patient quickly. Review the fluid balance over several days for an overview and consider using albumin or plasma protein fraction to provide circulating protein.

Blood transfusion

Blood may be required for patients in an acute (e.g. traumatized) situation, electively (e.g. perioperatively) during major surgery or to correct a chronic anaemia (p. 518). In practice, the former two are much more commonly encountered by junior dental staff.

Whole blood is indicated in patients who have lost >20% blood volume, exhibit signs of hypovolaemic shock, or in whom this appears inevitable.

▶ Remember maxillofacial injuries *alone* only rarely result in this degree of blood loss (p. 486).

Always take blood for grouping in severely traumatized patients, and proceed to cross-match as indicated by the clinical signs or the Hb. Always use cross-matched blood except in utter extremis when ORhesus −ve blood can be used. Massive transfusions create problems with hyperkalaemia, thrombocytopenia, and low levels of clotting factors. Therefore, in patients with severe haemorrhage, simultaneous fresh frozen plasma (4–6 units) and platelets (6 units) will be needed.

Autologous blood is blood donated by a patient prior to elective surgery for use only on themselves. It avoids risks of cross-infection but requires a specially interested haematology department, so check locally.

Packed cells are used for the correction of anaemia if too severe for correction with iron, or if needed prior to urgent surgery. This reduces the fluid load to the patient, but elderly individuals and those in heart failure should have their transfusion covered with 40 mg Frusemide PO or IV.

Useful tips
- Cross-match one patient at a time and be sure you are familiar with local procedures.
- Nurses will perform regular observations of temperature, pulse, respiration, and urine output during transfusion—look at them!
- Don't use a giving set which has contained dextrose, as the blood will clog.
- Except in shock, transfuse slowly (1 unit over 2–4 h; >4 h the cannulae start to clog).
- 1 unit of blood raises the Hb by 1 g/dl = 3% haematocrit.

Complications
- ABO incompatibility. Causes anaphylaxis; manage accordingly (p. 562).
- Cross-infection.
- Heart failure can be induced by over-rapid transfusion.
- Milder allergic transfusion reactions are the commonest problem, Rx by slowing the transfusion. If progressive or temperature >40 °C stop transfusion. IV hydrocortisone 100 mg and chlorpheniramine 10 mg are useful standbys.

- Citrate toxicity is a hazard of very large transfusions and can be countered by 10 ml of calcium gluconate with alternate units.

Catheterization

This is not a topic normally covered by the dental syllabus, but the dental graduate may find him or herself confronted with a patient needing urethral catheterization if working on an oral and maxillofacial surgical ward. The only procedure they are likely to need to be able to perform is temporary urethral catheterization. This is indicated for urinary retention (almost always post-operatively), for precise measurement of fluid balance, or rarely in order to avoid use of bedpan or bottle. **Avoid** catheterization if a history of pelvic trauma is present, as an expert is needed. Catheters are associated with a high incidence of urinary tract infections, and presence of an UTI is a relative C/I.

Equipment Most wards have their own; ask for it and ensure you have an assistant available. It should contain a tube of local anaesthetic gel, a dish with some aqueous chlorhexidine, swabs, a waterproof sheet with a hole in the middle, sterile gloves, a 10 ml syringe filled with sterile water and a drainage bag. In most cases a 14–16 French gauge Foley catheter will suffice. Use a silicon catheter if you anticipate it being *in situ* for more than a few days.

Procedure Explain what you are going to do and why. If catheterizing for post-operative fluid balance, do it in the anaesthetic room after intubation (ask the anaesthetist first!). The procedure can be made both aseptic and aesthetic by wearing two pairs of disposable gloves, one of which is discarded after achieving analgesia, which is done by instillation of lignocaine gel (also acts as a lubricant). Find the urethral opening and, using the nozzle supplied, squeeze in the contents of the tube. In the female, finding the opening is the only significant problem and most women are catheterized by female nursing staff. In the male, it is necessary to massage the gel along the length of the penis and leave *in situ* for several minutes before progressing. Once analgesia is obtained, the female urethra is easily catheterized, if you have been unable to find the meatus to instil LA you should not proceed but get help. In the male, hold the penis upright and insert the tip of the catheter, pass it gently down the urethra until it reaches the penoscrotal junction, pull the penis down so that it lies between the patient's thighs, thereby straightening out the curves of the urethra, and advance the tip into the bladder until urine flows.

Once urine flows, the catheter balloon can be inflated. This should be painless; if not, deflate the balloon and reposition the catheter. Connect the drainage bag and remember to replace the foreskin over the glans to avoid a paraphimosis.

▶ If in the presence of adequate analgesia you are unable to pass the catheter **do not** persist, but get expert assistance.

Enteral and parenteral feeding

The main role for this type of feeding, as far as dental staff are concerned, is in the care of debilitated patients and those undergoing surgery for head and neck cancers. In view of the latter, it is worthwhile having a basic understanding of the subject.

Enteral feeding is providing liquid, low residue foods either (1) by mouth, or (2) more commonly by fine-bore nasogastric tube. A range of proprietary products are available for these feeds. Be guided by the local dietitian who has expertise in this area which you will lack. The major problem is osmotic diarrhoea, which can be ↓ by starting with dilute solutions.

Fine-bore nasogastric tubes are the mainstay in enteral nutrition for patients undergoing major oropharyngeal surgery. In many institutions, nurses are not allowed to pass these tubes, although they are allowed to pass standard nasogastric tubes for aspirating stomach contents. You can therefore find yourself asked to pass such a tube by someone who is more proficient but is forbidden to do so.

Technique Wear gloves. Explain the procedure to the patient and have them sit upright with the chin close to the chest. Place a small amount of lignocaine gel in the chosen nostril after ensuring it is patent. Keep a small drink of water handy. Select a tube; check the guide wire is lubricated and does not protrude. If a tracheostomy tube is present the cuff should be deflated to allow passage of the tube. Lubricate the tube and introduce it into the nostril; pass it along the nasal floor. There is usually a little resistance as the tube reaches the nasopharynx; press past this and ask the patient to swallow (use the water if this helps). The tube should now pass easily down the oesophagus, entering the stomach at 40 cm. Secure the tube to the forehead with sticky tape and only now remove the guide wire, being careful to shield the patient's eyes. Inject air into the tube and listen for bubbling over the stomach. Confirm position with a chest X-ray.

Problems
- *Nasal patency*. Select the least narrow nostril, use a lubricant and, if necessary, a smaller tube and topical vasoconstrictor.
- *Gagging/vomiting*. Press ahead; all sphincters are open.
- *Tube coiling into mouth*. Cooling the tube makes it more rigid and often helps. As a last resort, topical anaesthesia and direct visualization with a laryngoscope while an assistant passes the tube, may be necessary.
- *Tube pulled out by patient*. If they must have the tube, consider a septal suture or stitch to soft tissue of the nose.

Parenteral feeding is extremely hazardous and expensive. It requires a central venous line and considerable expertise if your patient is to survive. Avoid if at all possible.

Pain control

LA (p. 632); RA (p. 636); IV sedation (p. 640); cryoanalgesia (p. 412); and TENS (p. 633).

The aim of pain control is to relieve symptoms while identifying and removing the cause, the exception to this being when the cause is not treatable, as in terminal care. Then, the approach is to deal with the symptoms in order to enable the patient to have a quality to the end of their life.

Acute and post-operative pain May often be well controlled by LA (p. 630); it is, however, often necessary to use systemic analgesics. Useful analgesics in this situation include aspirin (900 mg PO 4 hrly), and paracetamol (1000 mg PO 4 hrly). Both these drugs can be combined with codeine (8–30 mg). Also, NSAIDs such as ibuprofen (600 mg PO 6 hrly) and diclofenac (50 mg PO 8 hrly or 75 mg IM perioperatively, followed by one further dose of 75 mg IM). These drugs are all simple analgesics and, with the exception of paracetamol, have the advantage of an anti-inflammatory action which is at least as important as their central analgesic effect. In *post-operative* pain, opioid analgesics are helpful when used parenterally and short-term. Papaveretum (10–20 mg IM/IV), morphine and diamorphine (5–10 mg IM/IV) combined with an antiemetic such as metoclopramide (10 mg IM/IV) are useful analgesics in the immediate post-op period.

Pain following maxillofacial trauma is a problem, as these patients must be assessed neurologically for evidence of head injury which C/I opioids. The addition of codeine phosphate (60 mg IM) to a NSAID is often effective and does not significantly interfere with neurological observations. Alternatively, PR or IM diclofenac sodium may suffice.

Facial pain not of dental or iatrogenic origin is covered on p. 460. These conditions often require the use of co-analgesics such as antidepressants, antiepileptics, and anxiolytics.

Pain control in terminal disease This is an important subject in its own right. Points to note include: aim for continuous control using oral analgesia; use regular, **not** on demand, analgesia titrated to the individual; diagnose the cause of **each** pain and prescribe appropriately (e.g. steroids for liver secondaries or raised intracranial pressure, NSAIDs &/or radiotherapy for bone metastases, co-analgesics for nerve root pain, etc.). Remember that psychological dependence is very rare in advanced cancer patients using long-term opioids, and analgesic tolerance is slow to develop. When starting patients on opioid analgesia, always consider using an antiemetic (p. 610) and a laxative. Oral hygiene may be incredibly difficult for patients with oral cancer or following head and neck surgery. Use of chlorhexidine gluconate (Corsodyl) and metronidazole (200 mg bd) reduces the smell associated with wound infection or tumour fungation even when there is no prospect of eliminating it entirely. Patients rarely develop the disulfiram reaction to metronidazole on this regimen which occurs with higher doses.

Pre-emptive analgesia Surgery is painful. Providing LA or systemic analgesia before surgery begins may reduce the overall requirements for analgesia.

Patient-controlled analgesia The optimal technique for severe postoperative pain. Small machine delivers a bolus (1–2 mg morphine) when patient presses a button. Can be repeated as needed.

Two variables: **1** bolus dose; **2** 'lock out time' time during which machine will not respond (allows time for opiod to achieve maximum effect and prevents overdosage).

Antibiotic prophylaxis

Prophylaxis is the prevention of an occurrence. Antibiotic prophylaxis is used in order to prevent the occurrence of bacterial infection, it is quite different from treating an established infection. There are two broad categories of patients requiring antibiotic prophylaxis, (1) those who have it in order to prevent a minor local bacteraemia causing a serious infection out of all proportion to the procedure, e.g. people at risk from SBE or the immunocompromised, and (2) those who receive it to prevent local septic complications of the procedure, e.g. wisdom-tooth removal.

Principles of prophylaxis The regimen used should be short, high dose, and appropriate to the potential infecting organisms. The aim is to prevent pathogens establishing themselves in surgically traumatized or otherwise at risk tissues, ∴ the antimicrobial must be in those tissues prior to damage or exposure to the pathogen. It must not, however, be present too long before this, as there is then a risk of pathogen resistance and damage to commensal organisms. The regimen should ∴ start immediately pre-op (i.e. <6 h) or perioperatively (e.g. with anaesthetic induction) and should be continued for 24–48 h maximum. When practicable select an antimicrobial as specific to the common pathogens for that procedure as possible, except in the immunocompromised, where broad-spectrum prophylaxis is appropriate.

Examples

• *SBE prophylaxis* (BSAC 1992)

Ordinary risk: Valvular disease with murmur, past rheumatic fever, shunts, coarctation.
LA–3 g amoxycillin PO 1 h pre-op
  –Penicillin allergic or course within last month—600 mg clindamycin PO 1 h pre-op.
GA–1 gm amoxycillin. IV/IM at induction—500 mg PO 6 h later, *or*
  –3 gm amoxycillin PO 4 h pre-op and 6 h post-op.

Penicillin allergic—as 'high risk'.

'*High risk*': treat in hospital LA or GA (e.g. past SBE, prosthetic valves).

| Amoxycillin 1 g | IV/IM with amoxycillin 500 mg PO |
| Gentamicin 120 mg | 6 h later. |

Penicillin allergic or course within last month.

| Teicoplanin 400 mg | IV, *or* |
| Gentamicin 120 mg | |

Clindamycin 300 mg IV with 150 mg PO 6 h later, *or*

| Vancomycin 1 g | IV separately. |
| Gentamicin 120 mg | |

Note Amoxycillin IM *hurts*—mix with 2.5 ml of 1% plain lignocaine. Vancomycin makes people feel terrible, they often faint, teicoplanin is slightly better. Pacemakers, coronary artery bypass grafts, and hip replacements *do not* need prophylaxis.

- *Immunocompromised* (e.g. severe leukopenia) 5 g azlocillin plus 120 mg gentamicin IV pre-op. Repeat qds for 24 h.
- *Dentoalveolar surgery* Simple extractions need not be covered but third-molar surgery often is. Regimens: metronidazole 400 mg PO (500 mg IV) pre-op followed by 400 mg tds for 24 h, amoxycillin 1 g PO/IV followed by 500 mg tds for 24 h.

Prophylactic anticoagulation To prevent deep vein thrombosis &/or pulmonary embolus in susceptible patients (e.g. women on the pill, prolonged surgery, iliac crest grafts) can be achieved using 5000 units heparin sc bd. TED antiembolism stockings will also ↓ dvt. Treatment of pulmonary embolus, p. 596.

Management of the diabetic patient undergoing surgery

▶ Many hospitals have a diabetic team who will advise on the management of these patients. Find out if this happens in your locality and make use of them.

Guidelines

- Know the type and severity of the diabetes you are dealing with.
- Inform the anaesthetist (and the diabetic team if available).
- Remember these patients are more likely to have occult heart disease, renal impairment, and ↓ resistance to infection, so get an ECG, U&Es, and be sensible with antibiotic prophylaxis.
- If nursing staff are experienced with blood glucose estimation using BM sticks or similar, 2–4 hrly BMs will suffice for monitoring control. Otherwise pre-, peri-, and post-operative blood glucose estimation is needed.
- If in doubt and you are on your own (although you never should be) use a GKI (see below).
- Always have diabetics first on the operating list.

Management non-insulin-dependent

If anything other than a minor, short procedure treat as insulin-dependent. If random blood glucose is >15 mmol/l treat as insulin-dependent.

Patients being treated under LA or LA and sedation should maintain their carbohydrate intake and any oral hypoglycaemic drugs as normal. Plan their surgery to fit their regular mealtimes. Have carbohydrate readily to hand if needed, and ensure adequate post-op analgesia, as pain or trismus can easily interfere with their usual intake. Remember antibiotic prophylaxis.

Patients for GA Halve the dose of oral hypoglycaemic 24 h prior to surgery and omit on day of surgery. Do BMs or blood glucose pre- and post-op (if >15 go to GKI). Halve normal dose of oral hypoglycaemic until able to take normal diet, then back to normal dose. Keep an eye on the K^+ concentration (U&Es pre- and post-op).

Management insulin-dependent

Examine the patient carefully. Get an ECG, FBC, U&Es, and random blood glucose. Look for heart and respiratory disease, leucocytosis indicating infection, anaemia, hypo or hyperkalaemia, and renal impairment. If these are acceptable, halve any long-acting insulin 24 h prior to surgery, otherwise normal insulin/carbohydrate intake up to and including evening meal the day before surgery. Starve following this, and place on an IV infusion of *glucose* (dextrose 5–10% 500 ml), *potassium* (10 mmol K^+ injected into bag of dextrose), and *insulin* (16 units Actrapid insulin in the same bag) infused via a controlled rate device over 5 h. This regimen will safely stabilize most diabetics over the necessary 24–48 h but allows for certain flexibility, in that the insulin can be increased or decreased in line with the BMs or blood glucose estimations and the K^+ can be raised or lowered according to the U&Es. Avoid, however, putting >20 mmol/l of K^+ into a 500 ml bag. An alternative is to infuse dextrose and potassium as above, but deliver the insulin via an infusion pump according to a sliding scale.

Sliding scale

BM	Insulin IU/h	
>22	8	
18–22	6	
14–18	4	
10–14	3	Target zone
6–10	2	
2–6	1	
<2	stop and give glucose/dextrose	

The scale details will vary according to the type of fingerprick glucose tests available in your locality, numbers are *not* absolute but the principle will always work. Choice of crystalloid is variable, ↑% dextrose ↑ carbohydrate and anabolism but tends to tissue veins quickly.

Management of patients requiring steroid supplementation

Principle These patients are unable to respond to the stress of surgery due to depletion or absence of their endogenous corticosteroid response.

Groups Patients with Addison disease, patients prescribed corticosteroids for modulation of the immune system, and steroid abusers from various 'sports'.

Addison disease is the rarest of these groups but requires the most aggressive supplementation as they are entirely reliant on prescribed steroids. Minor surgery can safely be covered with IM hydrocortisone sodium succinate 100 mg qds the day of surgery. Major surgery should be covered for 3 days using the same regimen.

Patients prescribed steroids Do you know why they are on these drugs? Consider the underlying disease and whether receiving it has any bearing on your treatment plan, e.g. a patient recieving steroids as part of a cytotoxic regimen will also be at risk from bleeding and infection. In an uncomplicated case before minor surgery, all that is needed is a single IM injection of hydrocortisone 100 mg 30 min prior to the procedure. Alternatives are an IV injection immediately pre-op followed by double the normal oral dose of steroid. Or relying on the patient to take a double dose on the day of the operation, normal dose plus 50% the day after, and normal plus 25% on the third day, thereafter returning to their usual dose. Those undergoing major surgery should have the IM regimen over 24 h or longer.

Steroid abusers are an unfortunately real and increasing problem. Although these drugs are taken for their androgenic effect, few 'sportspersons' have access to appropriate advice or drugs and many users will have a degree of suppressed steroid response. It is probably safest to cover procedures using the IM regimen as for those on prescribed steroids, although this could be considered overkill. Remember the risk of shared needles.

Common post-operative problems

General

Pain Use appropriate analgesia (p. 604).

Pyrexia A small physiological ↑ in temperature occurs post-operatively. Other causes: atelectasis, infection (wound, chest, urine), DVT, incompatible transfusion, allergic reactions.

Nausea and vomiting Antiemetics, e.g. IM prochlorperazine 12.5 mg or metoclopramide 10 mg. Ondansetrone for the recalcitrant.

Sore throat Common after intubation, needs reassurance and simple analgesia.

Muscle pain often follows suxamethonium use in anaesthetic induction. Again, reassurance and simple analgesics.

Hypotension Usually caused by autonomic suppression by a GA. Treat by placing 'head-down'. If necessary, speed up IV infusion for a short while.

Chest infection Chest X-ray, culture sputum, and start on ampicillin or cefuroxime until culture results available.

Confusion A symptom, ∴ look for the cause, e.g. infection, hypoxia, dehydration, and correct. Only consider sedation, e.g. haloperidol 5–10 mg, after action has been taken to deal with the cause and the patient constitutes a threat to themselves or others.

Rarer general complications

Urinary retention Comparatively rare, even after major maxillofacial surgery. Early mobilization and adequate analgesia helps; if not, temporary catheterization (p. 584) should be used.

Superficial vein thrombosis follows 'tissued' cannulae or irritant IV injections. Observe for infection, treat pain, and consider supportive strapping.

Deep vein thrombosis (DVT) Signs are painful, shiny, red, swollen calf, usually unilaterally but may be bilaterally. 'At risk' are immobile patients especially following pelvic surgery, women on the pill, the elderly, and the obese. Investigate with ascending venography. Ultrasound and isotope scanning occasionally used. Prevent by using heparin 5000 units sc bd pre-op and 5 days post-op &/or pressure stockings, and ensure early mobilization. Stop the pill or hormone replacement therapy prior to any major surgery. **Rx** consists of bed rest, leg strapping and elevation, analgesia and heparinization (give 10 000 units stat IV followed by 25 000 units in 50 ml normal saline by syringe driver infusion starting at 1000 units/h (2 ml/h) and adjust according to the KCT/APTT, keep betwen 1.5–2.5 times the control values). The major risk of DVT is the development of pulmonary embolus. This classically presents 10 days post-op when the patient has been straining at stool and may occur despite no apparent DVT. Symptoms are pleuritic chest pain, dyspnoea, cyanosis, haemopytisis, and a ↑ jugular venous pressure. Signs of shock are often present which range from very little in the young who can compensate, to cardiac arrest! Usually a clinical diagnosis, confirmed after heparinization, analgesia, and oxygen have been instituted by chest X-ray, blood gases, lung ventilation—perfusion scan &/or pulmonary angiography. ECG will sometimes show deep S waves in I, pathological Q waves in III, and inverted T waves in III (SI, QIII, TIII). Must be followed up by 3 months anticoagulation with warfarin so involve a haematologist.

Local complications following oral surgery

Local pain, swelling, infection, and trismus are the commonest post-operative problems following oral surgery. Antral complications may follow maxillary surgery. These are considered in the relevant chapters.

12 Therapeutics

Prescribing   600
Analgesics in general dental practice   602
Analgesics in hospital practice   604
Anti-inflammatory drugs   606
Antidepressants   608
Antiemetics   610
Anxyiolitics, hypnotics, sedatives, and
    tranquillizers   612
Antibiotics—1   614
Antibiotics—2   616
Antifungal and antiviral drugs   618
Antihistamines and decongestants   620
Miscellaneous   622
Alarm bells   624

Principal sources: BMA/RPSGB Sept 1993 BNF. Walton 1989 *Textbook of Dental Pharmacology and Therapeutics*, OUP. Consumers Association *Drugs and Therapeutics Bulletin*. HMSO *Prescribers Journal*.

Relevant chapters and pages: Medicine relevant to dentistry, Chapter 11; analgesia, anaesthesia, and sedation, Chapter 13; asepsis and antisepsis, p. 382; antiseptics and antibiotics in periodontal disease, p. 230; fluorides, p. 34; sugar-free medications, p. 128.

Generic—pharmaceutical name.
Proprietary—trade name.
Depending on patents a generic drug may have more than one proprietary name.

Prescribing

The following pages are a brief guide to the clinical use of some of the more commonly used and useful drugs in hospital and general dental practice. Doses are for healthy adults.

Prescribing in general dental practice Extremely useful information is available in the DPF, which is updated every 2 yrs. Use this as the first line of enquiry when unsure about any topic concerning drugs. Drugs listed in the DPF can be prescribed in the UK within the NHS on form FP14 (GP14 in Scotland, HS47 in Northern Ireland). Any other required drugs must be prescribed privately or via the patient's GMP. Some may be available more cheaply over the counter at pharmacies.

Prescribing in hospital The BNF is the definitive reference and should always be available for consultation. Use this to check dose alterations in children and the elderly, and for more detailed tables of drug interactions, C/I, and unwanted effects. Any drug in the hospital pharmacy may be prescribed by a hospital dentist for in-patients, patients being discharged, and out-patients. The only exception is controlled drugs for addicts which must be prescribed by specially licensed doctors, usually a psychiatrist.

In hospitals, there are three methods of prescribing: (1) a hospital drug kardex recording both prescriptions and dispensing kept on the ward for in-patients; (2) a take-home prescription form redeemable only at the hospital pharmacy for patients being discharged from the ward; (3) hospital out-patient prescriptions used in out-patient clinics and casualty redeemable at outside pharmacies.

Good prescribing Avoid abbreviations and write drug names legibly, using the generic whenever possible. Always describe the strength and quantity to be dispensed. When describing doses use the units micrograms, milligrams, or millilitres when possible. Do **not** abbreviate the term microgram or unit (when prescribing insulin) as these are easily misinterpreted.

Controlled drugs Each prescription must show in the prescriber's own handwriting in ink: the name and address of the patient, the form, strength, dose, and total quantity of the drug to be dispensed, in both words and figures. When written in general practice the prescription must also incorporate the phrase 'for dental treatment only'.

Prescribing in the elderly Doses should be substantially lower than for adults (often 50% lower).

Prescribing for children Children differ markedly from adults in their response to drugs, especially in the neonatal period when all doses should be calculated in relation to body weight (p. 129). Older children can usually be prescribed for in age ranges usually up to 1 yr, 1–6 yrs, and 6–12 yrs. All details of dosages should be checked in the BNF/DPF.

Prescribing in liver disease (p. 530) Try to avoid prescribing in patients with severe liver disease.

Prescribing in renal impairment (p. 532) Doses almost always need to be ↓ and some drugs are C/I completely.

Prescribing in pregnancy (p. 536) Avoid if possible.

Prescribing in terminal care, p. 588.

Drug information services Information on any aspect of drug therapy can be obtained free by calling the most appropriate number listed on p. xiv of the BNF.

Analgesics in general dental practice

▶ Consult the BNF for dosages in children. See also p. 600.

Most dental pain is inflammatory in origin and is hence most responsive to drugs with an anti-inflammatory component, e.g. aspirin and the NSAIDs.

Of the peripherally acting analgesics in the DPF, aspirin, paracetamol, and ibuprofen are available cheaper direct to the public from pharmacies.

Aspirin Used in mild to moderate pain, it is also a potent antipyretic, which should **not** be used in children <12 yrs (due to the rare but serious risk of Reye syndrome). Avoid in bleeding diathesis, G-I ulceration, and concurrent anticoagulant therapy. Ask about aspirin allergy, particularly in asthmatics. Often causes transient gut irritation (as do all NSAIDs). *Dose*: 600–900 mg 4 hrly PO.

Ibuprofen Also used for mild to moderate pain and has a moderate antipyretic action. Risks and side-effects are similar to that of aspirin but probably less irritant to the gut. *Dose*: 400–600 mg 4 hrly PO.

Paracetamol Similar in analgesic efficacy to aspirin but has no anti-inflammatory action and is a moderate antipyretic. Does not cause gastric irritation or interfere with bleeding times. Overdosage can lead to liver failure. *Dose*: 1000 mg 4 hrly PO.

Diflunisal is a prescription-only drug available on the DPF, which has similar properties and problems to aspirin. *Dose*: 250–500 mg bd with food PO.

Mefenamic acid is possibly slightly more effective than aspirin, but is more of a risk in asthmatics, has similar gastric effects and an ↑ incidence of diarrhoea. More nephrotoxic than other NSAIDs. *Dose*: 500 mg tds after food PO.

The addition of **codeine** to the minor analgesics, while never being proven to be of advantage, may have marginal benefits in some cases.

There are very few indications for the use of opioid analgesics in general dental practice. Although **dihydrocodeine** remains in the DPF it has been demonstrated to be hyperalgesic in certain types of dental pain.[1]

Pentazocine is prone to cause hallucinations and is a poor-quality analgesic. It is a mixed agonist/antagonist and is poorly reversed by naloxone.

Pethidine is even more likely, in tablet form (the only form in the DPF), than dihydrocodeine to produce side-effects.

---

1 R. A. Seymour 1982 *Lancet* **i** 1425–6.

Analgesics in hospital practice

In addition to those available in the DPF, some drugs available only within hospitals are of considerable value.

Diclofenac sodium is available in tablet, IM injection, suppository and in once-daily, slow-release form. It is a mid-potency NSAID and a useful alternative to high-dose lower potency NSAIDs or an opioid which has no anti-inflammatory effect. Dose for tablets: 50 mg tds after food; IM injection 75 mg bd no more than 2 days (is a painful injection); suppositories 100 mg PR od. Soluble tablets available.

Ketorolac trometamol 30 mg/ml injection, has advantage of ↓ volume of injection. However, ↑ unwanted effects have been noted.

Opioid analgesics

The opioids act centrally to alter the perception of pain, but have no anti-inflammatory properties. They are of value for severe pain of visceral origin, post-operatively (acting partly by sedation) and in terminal care. However, they all depress respiratory function and interfere with the pupillary response and are C/I in head injury. All opioids cause cough suppression, urinary retention, nausea, constipation by a reduction in gut motility and tolerance and dependence. The risk of addiction is, however, greatly overstated when these drugs are used for short-term post-operative analgesia and in the terminal care context. Fear of creating addicts should **never** cause you to withhold adequate analgesia.

Morphine in oral form (tablets, elixir or slow-release tablets MST) is the drug of choice in the management of terminal pain. Always prescribe a laxative (e.g. docusate 200 mg bd) and an antiemetic. Dose is dependent on previous analgesia, but often starts at 10 mg morphine 4 hrly or 30 mg MST bd.

Papaveretum A mixed opium alkaloid which is frequently prescribed, but does not appear to have any advantage over morphine. The presence of noscapine created a C/I in women of childbearing age. Noscapine-free equivalent (Omnopon 10+20) now available.

Buprenorphine is a mixed agonist/antagonist with similar problems to pethidine and pentazocine. It is unique in that it can be given sublingually. *Dose*: 200–400 micrograms 8 hrly.

Diamorphine (heroin) is the most powerful opioid analgesic. It causes less nausea and hypotension than morphine and is very soluble. This is a great advantage, allowing it to be infused in tiny amounts subcutaneously or IV via infusion pump. Its ability to mobilize venous reserves makes it the drug of choice in MI and heart failure. It is reversed by naloxone, dose 1–2 mg IV repeated up to 10 mg max (if that hasn't worked, you've got the diagnosis wrong).

▶ Chronic and post-operative pain requires *regular*, not as-required, analgesia, given in adequate amounts and within the therapeutic half-life of the drug.

Anti-inflammatory drugs

These are among the groups of drugs which may be either analgesics or co-analgesics (drugs which are not analgesic in themselves but may aid pain relief either directly or indirectly). The two major groups are the NSAIDs (p. 602) and the corticosteroids.

Steroids are used in various forms, topical, oral, intralesional, and parenteral, and all have uses in dentistry.

Topical steroids

Hydrocortisone lozenges 2.5 mg lozenges dissolved in the mouth qds.

Triamcinolone in carboxymethylcellulose paste 0.1% paste applied in a thin layer qds. Sticks only to dry mucosa and is rapidly rubbed off the palate and tip of tongue. Both these preparations are available in the DPF and are of use in the management of recurrent apthous ulcers, lichen planus, etc. They are low-potency steroids and are unlikely to have any of the systemic side-effects of steroids, such as exacerbation of diabetes, osteoporosis, psychosis, euphoria, peptic ulceration, immunosuppression, Cushing syndrome (p. 534), or adreno-cortical suppression.

Betamethasone phosphate tablets Prepared as a 0.5 mg soluble tablets (Betnesol) made into a 10 ml mouthwash rinsed qds, or a *betamethasone* inhaler designed for use in asthma, but can be used to spray on apthae (1 spray = 100 micrograms). Can be repeated to a maximum of 800 micrograms. Drops are also available.

Hydrocortisone 1% and oxytetracycline 3% Ointment or spray (hydrocortisone 50 mg oxytetracycline 150 mg per aerosol unit) qds, are useful treatments for apthae and related conditions seen in hospital.

Intralesional steroids

Methylprednisolone acetate 40 mg/ml injection up to 80 mg per month.

Triamincinolone acetonide 2–3 mg per week. These are of use in granulomatous cheilitis, intractable lichen planus, and keloid scars.

Intra-articular steroids

These can be used to induce a chemical arthroplasty in arthrosis of the TMJ (p. 498).

Hydrocortisone acetate 5–10 mg single injection.

Systemic steroids

Main indication is prophylaxis in those with actual or potential adrenocortical suppression. Occasionally used in erosive lichen planus, severe apthae, e.g. Behçet syndrome (p. 446) or arteritis (p. 540).

Hydrocortisone sodium succinate Used for prophylaxis, dose 100 mg IM 30 min pre-op.

Prednisolone 30 mg PO as enteric-coated tablets given with food in reducing dose. Regimen dependent on the condition treated.

Methylprednisolone Various regimens described for control of oedema, post major surgery.

Dexamethasone Various regimens described for control of oedema, post minor surgery.

Other immunosuppressants

Azathioprine and **Thalidomide** are sometimes used in specialist centres.

Antidepressants

This is another group of drugs which can be used as co-analgesics. In conditions such as atypical facial pain they may be used as the sole 'analgesic'. However, there are no antidepressants prescribable in the DPF.

In the past, there has been considerable debate about the potential interactions between the commonest antidepressants, the tricyclics and the monoamine oxidase inhibitors (MAOIs), and adrenalin contained in LA (which constitutes the most commonly professionally administered drug anywhere). To date, there is *no clinical evidence* of dangerous interactions between the adrenalin in LA preparations commonly used in dentistry and the tricyclics or the MAOIs.[1] In hospital practice, the two most commonly used antidepressants are amitryptilline (a sedative tricyclic antidepressant) and dothiepin (a related compound).

Amitriptyline This drug should be used with caution in patients with cardiac disease (as arrhythmias may follow the use of tricyclics) and be avoided in diabetics, epileptics, and pregnant or breast-feeding women. Amitriptyline can precipitate glaucoma, enhance the effect of alcohol, and cause drowsiness (which can impair driving). In common with other tricyclics it can cause sedation, blurred vision, xerostomia, constipation, nausea, and difficulty with micturition, although tolerance to these side-effects tends to develop as treatment progresses. There is often an interval of 2–4 weeks before these drugs reach a level which exhibits a clinically evident antidepressant effect. *Dose:* 50–75 mg either as a single dose nocte or in divided doses, maximum 150–200 mg daily. Children and elderly, half-dose.

Dothiepin has similar properties and unwanted effects to amitryptilline. It has, however, been demonstrated to be of value in the treatment of 'facial arthromyalgia' (a composite group of TMPDS and atypical facial pain patients).[2] *Dose*: initially 75 mg nocte, increasing to 150 mg daily if needed. Half-dose in elderly.

Nortriptyline is a less sedating tricyclic. Dose 10–30 mg nocte, can be ↑.

Tranylcypromine a MAOI which may be of value in treating facial pain unresponsive to tricyclics. Dose 10 mg tds before 16.00 hrs.[3]

Motival A combination drug which may be of value in atypical facial pain (fluphenazine 0.1 mg nortriptyline 10 mg) nocte.

MAOIs can precipitate a hypertensive crisis precipitated by dietary and drug interactions (sympathomimetics, opioids, especially pethidine, and foods containing tyramine, e.g. cheese, meat, or yeast extracts). LA is however, safe.

1 DPF 1988–90 In *BNF* **16** 1988. **2** C. Feinmann 1984 *Br Dent J* **156** 205. **3** M. Harris 1993 *BDJ* **174** 129.

Antiemetics

There are no antiemetics prescribable in the DPF; however, they form an essential part of in-patient hospital prescribing. The common indication is the control of post-operative nausea and vomiting, which may be due to the procedure, anaesthetic, post-operative analgesia, or by blood in the stomach.

Prochlorperazine A phenothiazine antiemetic which acts as a dopamine antagonist and blocks the chemoreceptor trigger zone. Avoid in small children as the drug's major side-effect, the production of extrapyramidal symptoms, is especially common in this group. *Dose*: 12.5 mg IM 6 hrly, 20 mg initially followed by 10 mg tds PO. Not licensed for IV use.

Metoclopramide has both peripheral and central modes of action. It increases gut motility, thus emptying the stomach. Acute dystonic reactions may occur especially in young women and children, a bizarre acute trismus is sometimes seen as one of the manifestations. *Dose*: 10 mg tds PO, 10 mg IM 6 hrly. High-dose intermittent and continuous IV regimens are used for anti-emesis in centres using cytotoxic chemotherapy.

Domperidone is less likely to cause central unwanted effects such as sedation and dystonia, as it does not cross the blood–brain barrier. Acts on the chemoreceptor trigger zone and is particularly useful for chemotherapy patients. *Dose*: 10 mg 4 hrly. May be used as part of a combination regimen, e.g. domperidone, prednisolone, and nabilone (a synthetic canabinoid).

Ondansetron is a newly licensed selective 5-HT$_3$ receptor antagonist which is very effective in prevention and treatment of postoperative nausea and vomiting. Dose 4 mg single IV dose or 8 mg PO.

Hyoscine, antihistamines, and major tranquillizers all have antiemetic properties but are rarely indicated. If unable to control emesis with one agent, use two; acting at separate sites *after* excluding intestinal obstruction, e.g. due to opioid constipation.

Cyclizine is often combined with opioid preparations as an anti-emetic—however, it can aggravate heart failure. Cheap. Dose 50 mg tds IM or PO.

Anxiolytics, sedatives, hypnotics, and tranquillizers

The short-term control of fear and anxiety associated with dental treatment is an entirely appropriate use of the benzodiazepines. It should not be confused with the long-term control of anxiety which is rife with problems of dependence and drug withdrawal. A benzodiazepine may also be a valuable adjunct in the management of TMPDS, where it acts as both a muscle relaxant and an anxiolytic (p. 478). IV and oral sedative techniques prior to surgery (p. 640).

Diazepam has a long half-life and is cumulative on repeated dosing. Like all benzodiazepines, can cause respiratory depression ∴ patients should be warned not to drive or operate machinery while on this drug. Dose for anxiety/TMPDS: 2 mg tds max 30 mg in divided daily doses. Paradoxical disinhibition may occur in children and its use in the <16s is not advised. Diazepam in lipid emulsion (Diazemuls) is the IV treatment of choice in status epilepticus (p. 566). A rectal preparation is popular for paediatric sedation in some countries (p. 638).

Midazolam is a water-soluble benzodiazepine of about double the potency of diazepam. Its main use is in IV sedation (p. 638).

Nitrazepam A long-acting hypnotic which tends to cause a hangover effect. *Dose*: 5–10 mg nocte.

Temazepam Shorter-acting hypnotic. *Dose*: 10–30 mg nocte. Main indication is pre-op or as pre-medication.

In hospital practice

The following may also be prescribed:

Chlordiazepoxide Sometimes used instead of diazepam in TMPDS. It has the same side-effect profile. *Dose*: 10 mg tds ↑ to maximum of 100 mg daily, in divided doses. It is also of value in the stabilization of alcohol-dependent in-patients.

Lorazepam Sometimes used as a pre-med by anaesthetists. *Dose*: 2 mg nocte, 2 mg 1 hr pre-op.

Chlormethiazole A hypnotic sometimes used to relieve severe insomnia in the elderly, dose 1–2 capsules nocte (each cap = 192 mg). Main indication is in the management of alcohol withdrawal. *Regimen*: day 1–3 caps qds; day 2–2 caps qds; day 3–1 cap qds. Acute withdrawal is sometimes managed by IV infusion; however, this is a dangerous technique, not to be undertaken lightly.

Zopiclone A new non-benzodiazepine hypnotic dose 7.5 mg nocte. Problems similar to benzodiazepines.

Haloperidol Very useful in the control of acute psychosis, in a dose of 10–30 mg IM. It is less painful and does the same job as chlorpromazine, but its main problem is extrapyramidal side-effects.

Flumazenil A specific benzodiazepine reversal agent (p. 638).
▶ Dosages should be ↓ in elderly. Avoid in children.

Trimeprazine an antihistamine, is a useful sedative for children.
Dose 2–3 mg/kg 1–2 h pre-op.

Antibiotics—1

Principles of antibiotic use

When prescribing, consider (a) the patient, (b) the likely organisms, (c) the best drug. Patients influence choice, in that they may be allergic to various drugs, may have hepatic or renal impairment, may be immunocompromised, may be unable to swallow, be pregnant or breast-feeding, or taking an oral contraceptive; consider also age and the severity of the infection. The infecting organism should ideally be isolated, cultured, and its sensitivity to antibiotics determined, but this is only feasible in hospital practice. In reality, most infections are treated blind ∴ it is essential to know the common infecting organisms in your field, and their sensitivities. The drugs' mode of action, absorption, unwanted effects, development of resistance, interactions, and techniques available for delivery, should all be considered. The best drug is the one which is safe in that patient, specific to the infecting organism, and can be given in a reliable convenient form. Remember prophylaxis (p. 590) differs from treatment with antibiotics, and antibiotics do not replace the drainage of pus in abscesses.

Benzylpenicillin is inactive orally and only used IM or IV. *Dose*: 600–1200 mg IV/IM qds. Drug of choice in streptococcal infections. Like all penicillins is bacteroicidal—interferring with cell-wall synthesis. Good tissue penetration except for CSF. Its most important unwanted effect is hypersensitivity, which is usually manifest as a rash, rarely as fatal anaphylaxis. Patients allergic to one penicillin will be allergic to all, 10% will be allergic to cephalosporins as well. A history of atopy (e.g. asthma) ↑ risk.

Phenoxymethyl penicillin Oral equivalent of above. *Dose*: 250–500 mg qds PO. Has a narrow spectrum, but is now largely superseded by

Amoxycillin, which has a broad spectrum similar to ampicillin, but it is better absorbed and achieves higher tissue concentrations. *Dose*: 250–500 mg tds PO. Both ampicillin and amoxycillin cause a maculopapular rash in patients with glandular fever or lymphatic leukaemia (this is **not** true penicillin allergy). Amoxycillin is drug of choice in prophylaxis against bacterial endocarditis (p. 390). May interfere with the action of oral contraceptives.

Tetracycline is one of a group of broad-spectrum antibiotics with a problem of ↑ bacterial resistance. It is likely to promote opportunistic infection with *Candida albicans*, particularly when used topically, as has been recommended for the treatment of apthae. Other problems are the deposition of tetracyclines in growing bone and teeth, causing staining and hypoplasia (∴ avoid in children <12 yrs and pregnancy) and erythema multiforme. It is also particularly likely to render the oral contraceptive ineffective. *Dose*: 250–500 mg qds PO. Absorption inhibited by chelation with milk, etc., ∴ should be taken well before food. May be of value in periodontal disease (p. 230).

Erythromycin has a similar spectrum to penicillin, but is bacteriostatic. Active against penicillinase-producing organisms. It was formerly an alternative to amoxycillin for endocarditis prophylaxis (superseded by clindamycin). Nausea is a major problem. *Dose*: 250–500 mg qds PO/IV.

Oral cephalosporins have little value in dental practice.

615

Antibiotics—2

Clindamycin should not be used in the management of dental infections, due to the risk of antibiotic induced colitis. It is useful in staphylococcal osteomyelitis in conjunction with metronidazole (which inhibits overgrowth with *Clostridium difficile*). Has replaced erythromycin for single-dose prophylaxis of bacterial endocarditis (p. 590).

Metronidazole is an anaerobicidal drug, and as such is effective in many acute dental and oral infections. Classical dose for AUG is 200 mg tds PO 3 days. For other anaerobic infections is more often used as 400 mg bd/tds (depending on severity) PO. Available in tablets, IV infusion or suppository. Main problem is severe nausea and vomiting if taken in conjunction with alcohol (disulfiram reaction).

Antibiotics of use in hospital practice

(in addition to the aforementioned)

Co-Amoxyclav is amoxicillin plus clauvulanic acid. The latter destroys beta-lactamase (penicillinase) and hence widens the range of amoxicillin to include the commonest cause of resistance in infections of the head and neck. *Dose*: 1–2 tabs tds PO or 600–1200 mg tds IV. Problems as for amoxicillin.

Cefuroxime A parenteral broad spectrum cephalosporin often used in combination with metronidazole for surgical prophylaxis in contaminated head and neck procedures. *Dose*: 750–1500 mg tds IV, (500 mg bd PO).

Doxycycline is a well-absorbed tetracycline. *Dose*: 200 mg day 1, then 100 mg od PO. Useful in sinus problems.

Gentamicin is a bacteroicidal aminoglycoside antibiotic, active mainly against gram −ve organisms. It is complementary to the penicillins and is available as a topical (ear use) or parenteral preparation. Major problem is dose-related ototoxicity and nephrotoxicity (monitor levels if used for >24 h. *Dose*: start at 80 mg bd/tds IV.

Co-trimoxazole has few indications in the head and neck which have not been replaced by trimethoprim alone (200 mg bd). It is used, however, in ear, sinus, and urinary infections, but is C/I in pregnancy and folate deficiency (as it is a folate antagonist).

Chloramphenicol is useful topically in bacterial conjunctivitis (0.5% eye drops, 1% eye ointment, apply 3 hrly). Systemic use is strictly limited due to toxicity. Ointment is an excellent wound dressing.

Vancomycin is a unique bacteriocidal antibiotic. Two main uses are orally in the treatment of antibiotic induced colitis (120 mg tds 10 days PO), and for prophylaxis of high-risk patients from bacterial endocarditis (p. 590). It is ototoxic, nephrotoxic, prone to cause phlebitis at infusion sites, and makes people feel generally unwell, ∴ not to be used lightly.

Teicoplanin similar to vancomycin. Lasts longer. Can be given IV or IM.

Antifungal and antiviral drugs

Antifungals

The main fungal pathogen in the mouth is *Candida albicans* (p. 438). All antifungal drugs available in the DPF are for topical use.

Amphotericin Best used as lozenges 10 mg dissolved slowly in the mouth qds 10–15 days. Double dose if needed.

Nystatin Available as pastilles or mixture. *Dose*: 100 000 units qds either by sucking a pastille qds or using 1 ml of the mixture and holding it in the mouth before swallowing.

Miconazole A useful drug, particularly in the management of angular cheilitis, as it is active against streptococci, staphylococci, and candida. Miconazole oral gel 25 mg/ml is of use in chronic mucocutaneous and chronic hyperplastic candidosis. *Dose*: place 5–10 ml in the mouth and hold near the lesions before swallowing, qds.

Hospital-based specialist treatment can involve systemic antifungals. **Ketoconazole** has been superseded, due to problems with sometimes fatal hepatotoxicity, by **Fluconazole** a triazole antifungal. This drug is available in both oral and IV formulations for severe mucosal candidosis in both normal and immunocompromised patients as a second line to topical preparations. Avoid in pregnancy. *Dose*: 50 mg od PO 7–14 days, 200–400 mg IV od. **Itraconazole** and **tioconazole** are other new antifungal drugs.

Antivirals

Most viral infections are treated symptomatically. Two drugs are available for the treatment of oro-facial viral diseases:

Acyclovir Active against herpes simplex and zoster, is relatively non-toxic and can be given systemically or topically. *Dose*: herpes labialis—acyclovir cream apply to site of prodromal or early lesion 4 hrly for 5 days, herpetic stomatitis—200 mg (400 mg in immunocompromised) PO 5 times daily for 5 days, herpes zoster—800 mg PO 5 times daily for 7 days.

Idoxuridine Available as idoxuridine 5% in dimethyl sulfoxide. Is indicated in herpes labialis. *Dose*: apply qds for 4 days (tastes terrible). Only effective if started during prodrome. Other antiviral drugs include: **Ganciclovir**, which is active against all herpes viruses, Epstein–Barr, and cytomegalovirus; however, it is very toxic and ∴ has no indications in dentistry. **Zidovudine**, which has been licensed for treating patients with serious manifestations of infection with HIV. Benefit has been shown in patients with established AIDS. There are no benefits in prophylaxis of AIDS in HIV +ve patients. **Inosine pranobex** is licensed for use in treating mucocutaneous herpes simplex in conjunction with podophyllin. **Interferon** is an interesting drug with no indications in dentistry.

Antihistamines and decongestants

Antihistamines

Rarely used in the usual range of dental practice. They are sometimes indicated in the management of allergy, especially hayfever, for pre-medication and sedation in children, occasionally as antiemetics, possibly in the management of overactive gag reflex and as part of the emergency treatment of angio-oedema and anaphylaxis (p. 562). Main differences between the antihistamines are the duration of action and the degree of accompanying sedation and antimuscarinic effects.

Chlorpheniramine A sedative antihistamine. *Dose*: 4 mg qds PO.

Promethazine Also a sedative antihistamine. *Dose*: 10 mg qds PO or 20–30 mg nocte when used as a hypnotic. Promethazine is on sale to the public as a hypnotic.

The sedative effects of these drugs potentiate alcohol and inhibit ability to drive or operate machinery safely. They should be used with caution in glaucoma, prostatic hypertrophy, and epilepsy.

Although these drugs are available in the DPF, probably the most useful antihistamines are not. These include:

Terfenadine, which has a ↓ incidence of sedative side-effects and antimuscarinic effects. *Dose*: 60 mg bd. Probably the drug of choice for the symptomatic relief of hay fever and allergic rashes. Avoid ketoconazole and erythromycin.

Trimeprazine A sedative antihistamine which may be of some value in the itching of uraemia, and is frequently used by anaesthetists as a pre-med in children. Adult dose is 10 mg bd/tds, paediatric pre-med 5–10 mg, dependent on age.

Decongestants

These are of value in the management of sinusitis and particularly of value in the closure of oro-antral fistulae.

Ephedrine nasal drops Produce vasoconstriction of mucosal blood vessels and ↓ the thickness of nasal mucosa, thus relieving obstruction. Avoid in patients taking MAOIs. Other problems are that prolonged use will lead to a rebound vasodilatation and a recurrence of nasal congestion, and long-term use results in tolerance and damage to the nasal cilia. *Dose*: ephedrine nasal drops 0.5–1%, 1–2 drops into the relevant nostril qds for 7–10 days. For symptomatic nasal decongestion and as an adjunct to the management of oro-antral fistulae (p. 422) inhalation of **menthol** and **eucalyptus** is valuable. *Dose*: 1 teaspoonful of menthol and eucalyptus inhalation BP is added to a pint of hot water, and the warm, moist air is inhaled with a towel over the head.

Xylometazoline Nasal drops 0.1% are an alternative to ephedrine, but are more likely to cause a rebound effect. Systemic decongestants are of dubious value and should not be prescribed as they contain sympathomimetics.

Miscellaneous

A number of drugs which do not fit into any specific category are of importance in managing oral and dental disease. These include:

Carbamazepine Primarily an antiepileptic drug which is of considerable value in the management of trigeminal and glosso-pharyngeal neuralgia. It is C/I in those sensitive to the drug, patients with atrioventricular conduction defects, porphyria, and should be used with extreme caution in patients on MAOIs, who are pregnant or have liver failure. It may interfere with the oral contraceptive. Common unwanted effects are G-I disturbances, dizziness and visual disturbances. Rarely, rashes may occur, as can leukopenia. Do a FBC 1 week after starting carbamezipine—blood dyscrasias usually occur in the first 3 months. *Dose*: 100–200 mg bd, can be increased gradually to 200 mg tds/qds. Maximum 1600 mg daily in divided doses. It is important to be sure of your diagnosis before starting patients on long-term carbamezipine (p. 544). A slow release preparation with ↓ unwanted effects is available.

Vitamins There is no indication for first-line treatment with vitamins in dental practice. Deficiency due to inadequate dietary intake in the UK is exceedingly rare. Although it can occur in the elderly and alcoholics, these people should be fully investigated and not treated empirically. Severe gingival swelling, stomatitis, glossitis, or pain should be fully investigated before using vitamin supplements. The only preparation available in the DPF is:

Vitamin B compound tablets strong, which is a combination of nicotinamide 20 mg, pyridoxine 2 mg, riboflavine 2 mg, thiamine 5 mg. *Dose*: 1–2 tablets tds.

Artificial saliva A valuable adjunct in the management of xerostomia, especially after radiotherapy and in Sjögren syndrome. It is a slightly viscous, inert fluid which may have a number of additives, such as antimicrobial preservatives, fluoride, flavouring, etc. Useful preparations are **Glandosane** and **Saliva-Orthana**, which are aerosol sprays used as required, usually 4–6 times per day. The latter contains fluoride.

Topical anaesthetics Two main uses:

1 For preparation of a site prior to injection, e.g. of LA. **Ligno-caine 5% ointment or spray** is the most useful.

2 To relieve pain from minor oral lesions (e.g. for 1° herpes or apthae). **Benzocaine lozenges** 10 mg dissolved in the mouth, can be useful, although **lignocaine viscous mouthwash** is probably more suitable. This contains 20 mg/ml lignocaine in a viscous base. *Dose*: 5 ml rinsed in the mouth.

Fluorides Fluoride supplementation is discussed on p. 36. It is important when using rinses, and particularly gels, that the fluid is not swallowed, since there is a risk of toxicity (p. 34).

Alarm bells

When prescribing any drug for patients already compromised by concomitant disease or drug therapy it is essential to exclude possible interactions. This can be achieved fairly quickly by consulting the comprehensive BNF. A more detailed account of such problems is given by Seymour.[1]

Interactions with the most commonly given drug (LA) are covered on p. 628.

Common drugs with relative contraindications in liver disease

Aspirin
All benzodiazepines
All opioids
All sedatives
All antihistamines
All NSAIDs
Erythromycin
Metronidazole (↓ dose)
Paracetamol
Tetracyclines
Terfenadine

Common drugs with relative contraindications in renal failure

Acyclovir (↓ dose)
All penicillins (↓ dose)
All opioids
Amphotericin
Cephalosporins (↓ dose)
Co-trimoxazole (↓ dose)
Benzodiazepines (↓ dose)
NSAIDs
Tetracyclines

Common drugs with relative contraindications in pregnancy

Aspirin
Benzodiazepines
Carbamezipine
All opioids
Co-trimoxazole
NSAIDs
Metronidazole
Tetracyclines

Common drugs with relative contraindications in breast-feeding

Antihistamines
Aspirin
Benzodiazepines
Carbamezipine
Co-trimoxazole
Metronidazole
Tetracyclines

624

These lists are not comprehensive and some of the drugs mentioned can be used in suitably modified dose or under specific circumstances. The aim of these tables is to make 'alarm bells' ring, and encourage you to both think and consult the BNF.

Adverse reactions

Almost any drug can produce these, and many are missed. Try to avoid polypharmacy, and never prescribe without being aware of a patient's full medical history. Always enquire about drugs, including self-prescribed medication.

Please report suspected adverse reactions, by phone to 'CSM Freefone' (dial 100) or in writing to:

1 R. A. Seymour 1988 *Adverse Drug Reactions in Dentistry*, OUP.

CSM, Freepost, London SW8 5BR
CSM West Midlands, Freepost, Birmingham B15 1BR
CSM Mersey, Freepost, Liverpool L3 3AB
CSM Northern, Freepost 1085, Newcastle NE1 1BR
CSM Wales, Freepost, Cardiff CF4 1ZZ

Report • **all** serious adverse reactions
• **any** reactions to 'black triangled' drugs in the BNF/
DPF.

13 Analgesia, anaesthesia, and sedation

Indications, contraindications, and
common sense   628
Local analgesia—tools of the trade   630
Local analgesia—techniques   632
Local analgesia—problems and hints   634
Sedation—relative analgesia   636
Sedation—benzodiazepines   638
Benzodiazepines—techniques   640
Anaesthesia—drugs and definitions   642
Anaesthesia and the patient on medication   644
Anaesthesia—hospital setting   646
Anaesthesia—practice setting   648

Principal sources: Walton 1989 *Textbook of Dental Pharmacology and Therapeutics*, OUP. *Department of Health Working Party on General Anaesthesia and Sedation in Dentistry* (Poswillo Report) 1990. Haglund 1984 *Local Anaesthesia in Dentistry*, Astra.

Relevant pages: LA for children, p. 84; emergencies in dental practice, p. 554.

Definitions

General anaesthesia (GA) is a state of unrousable unconsciousness to which analgesia and muscle relaxation is added to produce 'balanced anaesthesia'.

Analgesia is the absence of pain.

UK position in relation to general anaesthesia

Since the report of the committee chaired by Poswillo (1990) the general status and, inevitably, the medico-legal aspects of general anaesthesia and sedation in dentistry have altered considerably. The general principles, however, remain the same.

Indications, contraindications, and common sense

When dealing with LA, GA, and sedative techniques, indications and contraindications are often relative and the following should be thought of as guidelines rather than immutable laws.

LA is the technique of choice for simple procedures or when a GA is C/I. LA is contraindicated in:
- uncooperative patients (of any description);
- infection around the injection site;
- patients with a major bleeding diathesis;
- most major surgery;
- adverse reaction to LA is a contraindication, but in reality once allergy to preservatives in the solution is excluded, LA allergy probably does not exist.

Conscious sedation

This is an extension of LA technique using drugs and patient-management techniques. It is of benefit to anxious or mildly uncooperative patients and is a kind supplement to apicectomy or third-molar removal. C/I include:
- Cardiorespiratory, renal, liver or psychiatric pathology.
- An unescorted patient or one unable or unwilling to conform to the requirements of conscious sedation, p. 641.
- A demonstrated adverse reaction to sedative agents.
- Benzodiazipines should be avoided in pregnancy.

GA is indicated when LA or LA and sedation is ineffective or inappropriate (as above). C/I include:
- All those for conscious sedation.
- Presence of food or fluid in the stomach (most anaesthetists require at least 4 h between last feed and GA)

The anaesthetist usually prefers hospital admission for patients with the following:
- Cardiovascular or respiratory disease (especially MI <6 months ago).
- Uncorrected anaemia, sickle cell trait or disease.
- Severe liver or renal impairment.
- Uncontrolled thyrotoxicosis or hypothyroidism.
- Poorly controlled diabetes, adrenocortical suppression.
- Porphyria.
- Pregnancy.
- Neurological disorders, e.g. myopathy or multiple sclerosis.
- Cervical spine pathology such as rheumatoid arthritis or cervical spondylosis.
- Certain drugs, e.g. steroids, antihypertensives, MAOIs, anticoagulants, narcotic analgesics, antiepileptics (need to avoid using methohexitone), lithium, alcohol. Malignant hyperpyrexia, scoline apnoea, and other unwanted reactions to anaesthetic agents.
- Causes of upper airway obstruction such as angio-oedema, submandibular cellulitis, Ludwig's angina, and bleeding diathesis affecting the neck. (In fact, these are indications to secure the upper airway.)

Ask about previous GAs and any problems. Note drugs used as repeated administration of halothane may induce hepatitis. While all these conditions create problems with anaesthesia they may not preclude it absolutely within the hospital setting. They do, however, indicate the need for careful assessment and early prior consultation with the anaesthetist.

Local analgesia—tools of the trade

While any disposable needle and syringe system can be used to give LA, the vast number of LAs given in dental practice (>50 000/dentist/lifetime) has led to some very useful modifications.

LA cartridges Two sizes 1.8 and 2.2 ml, come pre-sterilized. Commonest solution used is lignocaine 2% with adrenalin 1 in 80 000.

Cartridge syringes Used with above, resterilizable. Used with ultrafine disposable needles. Major advantage is the ability to perform controlled aspiration during LA injection (although the consistency with which this is achieved has been questioned).

Lignocaine/adrenalin Most commonly used preparation, gives effective pulpal analgesia for 1.5 h and altered soft tissue sensation for up to 3 h. Extremely safe, maximum dose (adult) 500 mg (10×2.2 ml cartridges).

Prilocaine/octapressin Similar but slightly less duration and effect compared to lignocaine/adrenalin. May cause methaemoglobinaemia in excess. Maximum safe dose (adult) 600 mg (8×2.2 ml cartridges). In reality, there are few hard indications for the use of prilocaine over lignocaine.

Mepivacaine Short-acting LA advocated for restorative work but has not really caught on. Maximum safe dose 400 mg.

Bupivacaine Long-acting LA (6 h plain, 8 h with adrenalin), useful as a post-operative analgesic. Maximum safe dose 2 mg/kg. Only available in ampoules.

Topical analgesics Lignocaine is the only really useful topical analgesic amongst the above. It is available as a spray or a paste which is applied to mucosa several minutes prior to injecting. There is a high incidence of contact eczema in people frequently exposed to these preparations, so do not apply with bare fingers. Benzocaine in lozenge or paste form is used for mucosal analgesia. Amethocaine is a topical analgesic for use on mucous membranes. Cocaine 4% solution is used as a nasal mucosal analgesic and vasoconstrictor.

Emla cream, a eutectic mix of lignocaine and prilocaine, is an invaluable skin topical analgesic, used prior to venepuncture in children. Apply to puncture site and cover with 'opsite' or equivalent dressing for at least 30 min.

Handling equipment One cartridge and needle per patient. Discard cartridge if a precipitant is seen in the solution or if air-bubbles are present. Store in a cool dark place and use before expiry date. Warm cartridge to ↓ discomfort and load into the syringe immediately prior to use and before screwing the needle into place. Aspirate before injecting. The increasing risk associated with needlestick injuries has spawned a number of devices to aid resheathing the needle. It is simpler to hold the cover in a pair of artery forceps.

Local analgesia—techniques

The IDB and local infiltrations are the mainstay of LA technique; however, numerous others are available as alternatives, supplements and fallbacks.

Inferior dental block (IDB, inferior alveolar block) Technique of choice for mandibular molars, also effective for premolars, canines and incisors (the latter if supplemented by infiltration). Aim is to deposit solution around the inferior alveolar nerve as it enters the mandibular foramen underneath the lingula. The patient's mouth must be widely open. Palpate the landmarks of external and internal oblique ridges and note the line of the pteyergomandibular raphe. With the palpating thumb lying in the retromolar fossa, the needle should be inserted at the midpoint of the tip of the thumb slightly above the occlusal plane lateral to the pteyergomandibular raphe. The needle is inserted approximately 0.5 cm and if a *lingual nerve block* is required 0.5 ml of LA is injected at this point. The syringe is then moved horizontally about 40° across the dorsum of the tongue and advanced to make contact with the lingula. Once bony contact is made the needle is withdrawn slightly and the remainder of the LA injected. It should never be necessary to insert the needle up to the hub. Note that the mandibular foramen varies in position with age (children, see p. 84). In the edentulous, the foramen, and hence the point of needle insertion, is relatively higher than in the dentate.

Long buccal block The long buccal nerve is anaesthetized by injecting 0.5–1 ml of LA posterior and buccal to the last molar tooth.

Mental nerve block The mental nerve emerges from the mental foramen lying apical to and between the 1st and 2nd mandibular premolars. LA injected in this region will diffuse in through the mental foramen and provide limited analgesia of premolars and canine, and to a lesser degree incisors on that side. It will provide effective soft tissue analgesia. Place the lip on tension and insert the needle parallel to the long axis of the premolars angling towards bone, and deposit the LA. **Do not** attempt to inject into the mental foramen as this may traumatize the nerve. LA can be encouraged in by massage.

Sublingual nerve block An anterior extension of the lingual nerve can be blocked by placing the needle just submucosally lingual to the premolars, use 0.5 ml of LA.

Posterior superior alveolar block A rarely indicated technique. Needle is inserted distal to the maxillary 2nd molar and advanced inwards, backwards, and upwards close to bone for about 2 cm. LA is deposited high above the tuberosity after aspirating in order to avoid the ptyerygoid plexus.

Nasopalatine block Profound anaesthesia can be achieved by passing the needle through the incisive papilla and injecting a small amount of solution. This is extremely painful (hints on how to overcome pain on palatal injections below).

Infraorbital block Rarely indicated. Palpate the inferior margin of the orbit as the infraorbital foramen lies approximately 1 cm below the deepest point of the orbital margin. Hold the index finger at this point while the upper lip is lifted with the thumb. Inject in the depth of the buccal sulcus towards your finger, avoid your finger and deposit LA around the infraorbital nerve.

Infiltrations The aim is to deposit LA supraperiosteally in as close proximity as possible to the apex of the tooth to be anaesthetized. The LA will diffuse through periosteum and bone to bathe the nerves entering the apex. Reflect the lip or cheek to place mucosa on tension and insert the needle along the long axis of the tooth aiming towards bone. At approximate apex of tooth, withdraw slightly and deposit LA slowly. For palatal infiltrations, achieve buccal analgesia first and infiltrate interdental papillae; then penetrate palatal mucosa and deposit small amount of LA under force.

Intraligamentary analgesia Individual teeth can be rendered pain free by injecting LA along the periodontal membrane using small amounts of LA delivered via a specially designed system (high-pressure syringe and ultrafine needles). Has the advantage of rapid onset and specific analgesia to isolated teeth and is a useful adjunct to conventional LA and in some hands may replace it for minor procedures. Disadvantages include post-injection discomfort due to temporary extrusion and an apparent increased incidence of 'dry socket'.

Electronic dental anaesthesia (EDA) An aggressively marketed technique based on the principles of transcutaneous electrical nerve stimulation (TENS). Uses electrodes, buccally and lingually, which carry a minute electrical current to interfere with local nerve conduction and hence pain appreciation. May be of value in restorative procedures and others not requiring the profound analgesia or vasoconstrictive effects of LA.

633

▶ When developing LA technique there is no alternative to seeing, doing and doing again.

Local analgesia—problems and hints

Failure of anaesthesia

There is an enormous difference in individual response to a standard dose of LA, both in the speed of onset, duration of action, and the depth. Soft tissue analgesia is more easily obtained, needing a lower degree of penetration of solution into nerve bundles than does analgesia from pulpal stimulation. A numb lip does not ∴ indicate pulpal anaesthesia.

Causes of failure are:

- Poor technique and inadequate volume of LA.
- Injection into a muscle (will result in trismus which resolves spontaneously).
- Injection into an infected area (which should not be done anyway as this risks spreading the infection).
- Intravascular injection is clearly of no analgesic benefit. Small amounts of intravascular LA cause few problems. Toxicity p. 562.
- Dense compact bone can prevent a properly given infiltration from working. Counter by using intraligamentary or regional LA.
- Anastomosis from either aberrant or normal nerve fibres not transmitted with a blocked nerve bundle is an infrequent cause of failure.

Pain on injection

This is to a certain degree inevitable, but can be reduced by: patient relaxation; application of topical LA; stretching the mucosa; and slow, skilful, accurate injection, of slightly warmed solution in reasonable quantities. Causes of pain include:

- Touching the nerve when giving blocks, results in 'electric shock' sensation and followed by rapid analgesia (it is extremely rare for any permanent damage to occur).
- Injection of contaminated solutions (particularly by copper ions from a pre-loaded cartridge). Avoid by loading the cartridge immediately prior to use.
- Subperiosteal injections are painful and unnecessary ∴ avoid.

Other problems with administration

Lacerated artery May be followed by an area of ischaemia in the region supplied, but rarely any significant problem.

Lacerated vein Followed by a haematoma which resolves fairly quickly.

Facial palsy can be caused by incorrect distal placement of the needle tip, allowing LA to permeate the parotid gland. The palsy lasts for the duration of the LA.

Post-injection problems

Lip and cheek trauma Tell patient to avoid smoking, drinking hot liquids, biting lip or cheek. Assure them the sensation will pass in a few hours and that their face is not swollen (whether adult or child). If the advice goes unheeded and they return with traumatized mucosa, Rx: antiseptics/antibiotics and simple analgesia.

Needle tract infection Rare.

General points

Thick nerve trunks require more time for penetration of solution and more volume of LA. In nerve trunks autonomic functions are blocked first, then sensitivity to temperature, followed by pain, touch, pressure, and motor function. Concentration of analgesic rises rapidly around the nerve at first and provides soft tissue analgesia; however, this is reached substantially before the levels needed for pulpal analgesia which takes several minutes and which will wear off first (usually within an hour of a standard lignocaine/adrenalin LA). Disinfection of mucosa prior to LA is not required in reality; however, sterile disposable needles are absolutely mandatory due to risks of cross-infection. LA for children, see p. 84. Faints, p. 556.

Sedation—relative analgesia

Relative analgesia (RA) is the most commonly used and safest form of sedation in dentistry. It has two aspects: (a) the delivery of a mixture of nitrous oxide and oxygen, and (b) a semi-hypnotic patter from the sedationist. In dentistry, two different techniques of nitrous oxide sedation have been described: (1) inhalational sedation with a fixed concentration of nitrous oxide; and (2) RA in which nitrous oxide is titrated to the patient. In the authors' view RA is the most useful.

Nitrous oxide

Has both sedative and analgesic properties, the former is the most useful. The analgesic effect requires unnecessarily high levels of nitrous oxide. It is an inert gas which does not enter any of the body's metabolic pathways and is distributed as a poorly soluble solution. This allows very rapid distribution and peak saturation is reached within 5 mins and is similarly eliminated (90% elimination in 10 min).

Indications It is of particular value in anxious patients undergoing relatively atraumatic procedures and in children, for whom the benzodiazepines are less suitable.

Contraindications are few, but upper airway obstruction, e.g. a cold makes the procedure difficult and pre-existing B12 deficiency would C/I its use. Other C/Is, e.g. complex medical history, are relative and may limit nitrous oxide use in practice but not in hospital.

Nitrous oxide pollution is the major problem associated with RA. It is essential to have a scavenging system in place as nitrous oxide accummulation can lead to B12 deficiency and demyelination syndromes. There is a real potential hazard to pregnant staff working in confined conditions with this gas.

Aim

To produce a comfortable, relaxed, AWAKE, patient who is able to open their mouth on request with no loss of consciousness or the laryngeal reflex. During the procedure patients will experience general relaxation, a tingling sensation often in the fingers or toes, and describe feeling mildly drunk. There is often a sense of detachment and distortion of the sense of time. Rarely, patients may dream despite being awake, and these can be sexual fantasies, which is another reason for always having a second person in the surgery.

Technique

An RA machine will not deliver less than 30% oxygen. Start by delivering 100% oxygen via a nasal mask and set flow control to match their tidal volume (flow rates 6–8 l/min for adults, 4 l/min for child). Then give 10% nitrous oxide for 1 min, increase (if needed) to 20% for 1 min, increase to 30% (if needed) for 1 min, and so on. Most patients achieve adequate levels of sedation between 20 and 30%, some may require less, a few rather more. Remember that RA relies on the reassuring banter of the operator more than any of the other sedation techniques, and many view this as hypnosedation. Give LA and carry out treatment. To discontinue, turn flow to 100% oxygen and oxygenate patient for 2 min. Then remove mask and get the patient to sit in a recovery room for 10 min, by which time 90% of nitrous oxide will have been blown off and they will be safe to leave the surgery.

Sedation—benzodiazepines

General pointers, problems, and hints

Benzodiazepines are both sedative and hypnotic drugs which are useful for sedation of patients. Two techniques are generally used: oral and IV.

Elderly patients tend to be very sensitive to benzodiazepines and doses are best halved in the >60 age-group initially. Interestingly, children show not only resistance to these drugs but sometimes paradoxical stimulation, and benzodiazepine sedation is not recommended for the <16s.

Post-operative drowsiness is perhaps the biggest problem with these drugs, as patients may be influenced by, e.g. diazepam, for up to 24 h after administration (this time is reduced with midazolam). There is also a re-sedation effect caused by entero-hepatic recirculation.

Poswillo Committee Their report redefined 'sedation' as the 'single injection of a single drug'. While there are certain advantages to this definition, most practitioners would hold to the widely accepted definition, p. 636.

IV sedation is the most efficient and effective method of extending the use of LA in dental treatment; however, it requires substantial skill and confidence in venepuncture and administration of the drugs. Therefore an inexperienced operator, or a patient with needle phobia, form relative C/I. Other C/I include: inability to be accompanied, the need to be in a responsible position within 24 h of sedation (e.g. looking after young children on own, driving, etc.), patients with liver or renal impairment, glaucoma, psychoses, pregnancy, or a demonstrated allergy to the benzodiazepines.

Certain drugs interact, including cimetidine, disulfiram, anti-parkinsonian drugs, other sedatives, narcotic analgesics, anti-epileptic drugs, antihistamines, and antihypertensives. Patients on these drugs may be best treated (and then cautiously) in a hospital environment. Patients addicted to alcohol or other drugs may require a substantial dose modification; usually a big increase. Those being sedated need to give informed consent (preferably written). Always provide written post-operative instructions because the patients will not remember verbal ones. The operator sedationist is a perfectly sound and legal occupation, but note that a second appropriate person, i.e. one trained in cardiopulmonary resuscitation, must be present. At no time should the patient lose consciousness during IV sedation. The addition of any drug, especially opioids, converts a technique with a wide safety margin into one with a very narrow margin and confers *no* routine advantage. Multiple drug techniques should only be carried out by trained anaesthetists. There is no need to starve patients prior to being sedated, and oral medication should be continued. Rectal sedation is popular in some countries using a prepackaged solution of diazepam (stesolid) given PR dose 2–10 mg.

Flumazenil is a specific benzodiazepine reversal agent. Dose 200 micrograms IV over 15 sec followed by 100 micrograms at 60-sec intervals until reversal occurs. NB This drug has a shorter half-life than the drugs it will reverse, therefore multiple doses may be required. Patients **should never** be reversed then left unsupervised. This is an essential emergency drug for IV sedationists.

639

Benzodiazepines—techniques

Oral sedation

This is of use for managing moderately anxious patients. There are problems, however, with absorption time and the risk of sedation occurring too early or too late. There is again a risk of patients having sexual fantasies under the influence of these drugs. Two drugs are used:

Temazepam 30 mg, 1 h pre-treatment produces a degree of sedation similar to that seen with IV techniques.

Diazepam, either divided regimen, 5 mg night before, 5 mg morning of treatment, and 5 mg 1 h pre-treatment; *or* as 10–15 mg 1 h prior to treatment. Both drug and technique are highly variable in consistency and quality of sedation achieved.

IV techniques

A far greater control of the duration and depth of sedation can be consistently achieved with this approach. Gives excellent sedation with detachment for around 30 min with amnesia for the duration of treatment. The major disadvantage is a potential for respiratory depression, post-treatment drowsiness, and amnesia (a reversal agent is available but is not suitable for routine reversal of conventional sedation, p. 638). Skill with venepuncture is a prerequisite of the technique. Diazepam has been entirely superseded by diazepam in lipid emulsion (diazemuls) as diazepam is not water-soluble and is very irritant.

Midazolam A water-soluble benzodiazepine of roughly double the strength of diazepam. It has a much shorter half-life, with no significant metabolites creating a quicker and smoother recovery, and it is more amnesic than diazepam. It is supplied in both 2 ml and 5 ml ampoules both containing 10 mg midazolam (5 ml ampoule is preferable, as it is easier to control the increments, given at 1 mg/increment). Drug of choice.

Diazepam in lipid emulsion Diazepam is metabolized to desmethyldiazepam, which has a long half-life. It is presented in a 2 ml ampoule containing 10 mg diazepam, given in slow IV increments (usually 2.5 mg via a butterfly) until signs of sedation are observed. Suggested maximum dose 20 mg.

IV technique

Relax the patient and have them sitting in the chair. Place a tourniquet around the most convenient arm and ask them to dangle the arm at their side. Any useful veins on the dorsum of the hand will become quickly evident. While this is happening, get your equipment ready: sticky tape, spirit wipe, butterfly, and drug in syringe. Look at the hand; is there a reasonable straight segment of vein? (If not cut your losses and look elsewhere.) If there is, secure the hand with your non-dominant hand, palmar surface to palmar surface, with your thumb in a position to tense the skin overlying the selected vein. Stroke the back of the hand and tap the vein to engorge it. Prepare the skin with spirit wipe;

get your 'second appropriate person' to help remove the needle cover and inform the patient there will be a slight scratch. Introduce the butterfly tip at a shallow angle through skin, then reduce the angle further and move along the line of the vein until it is entered, as revealed by a flash back into the extension tubing. Then introduce the length of the needle along the vein carefully, so as to avoid cutting through, and secure with tape. Remove air by aspirating blood into the tubing and release tourniquet. Place the patient in the supine position and give a small bolus (2–3 mg diazepam, 1–2 mg midazolam). Wait for 1 min and then give further increments until adequate level of sedation is achieved. Leave the butterfly in to maintain venous access. Loss of laryngeal reflex constitutes oversedation and means you must stop and ensure the airway is patent, only proceeding if there is no respiratory depression and the airway can be protected.

Monitoring A 'second appropriate person' must be present. Pulse and BP should be monitored. Pulse oximetry is mandatory.

Post-sedation Allow time for recovery (30 min–1 h) in calm surroundings. The patient must be accompanied home and forbidden to drive or assume a responsible position for 12 h (midazolam) or 24 h (diazepam).

Anaesthesia—drugs and definitions

IV anaesthetic agents

Methohexitone (dose 1.5 mg/kg) Is effectively an ultra-short-acting barbiturate anaesthetic. Its half-life, however, is 1–5 h depending on the patient! It is an induction agent for out-patient dental anaesthesia, with or without supplemental inhalational agents. It is relatively non-irritant but can cause severe respiratory depression, dose-dependent cardiovascular depression, and sometimes laryngospasm. It is epileptogenic and can induce porphyria. Has no analgesic properties.

Thiopentone (dose 4–8 mg/kg) Again, although acting as an ultra-short acting barbiturate anaesthetic, its half-life is 6–12 h. It is a poor analgesic and relatively sparing to the laryngeal reflexes, requiring a greater depth of anaesthesia to prevent laryngospasm. It is highly irritant on injection but is cheap and very popular as an inpatient induction agent. It will abolish epileptic seizures in induction doses.

Propofol (dose 2.5 mg/kg) Is a true ultra short-acting anaesthetic as it is completely metabolized within minutes. It is very expensive but less irritant than the others.

Etomidate (dose 0.2 mg/kg) An IV induction agent often used for patients with compromised cardiovascular systems. Can be associated with involuntary movements, cough, and hiccup. Avoid in traumatized patients.

Ketamine (dose 1–2 mg/kg) Can be given IM (unique among anaesthetic agents for this). Tends to maintain the airway and causes little respiratory depression. Its use is limited by a high incidence of severe nightmare hallucinations in adults.

Inhalational anaesthetics

Nitrous oxide Excellent analgesic but weak anaesthetic, mainly used to supplement other inhalational anaesthetics or in RA.

Halothane Very useful and common anaesthetic. Weak analgesic. Causes hypotension and dysrythmias. The concomitant use of injected adrenalin should be avoided in non-ventilated patients breathing halothane as it ↑ risk of VF. Very rarely hepatotoxic after repeated anaesthetics in adults. Extremely rarely *de novo*.

Enflurane Weaker anaesthetic than halothane but less likely to produce dysrythmias or hepatitis. Avoid in epileptics.

Isoflurane Close relation of enflurane but more potent. Causes less cardiac problems than halothane and acts as a muscle relaxant. Very expensive.

Muscle relaxants

These are used to create laryngeal relaxation for intubation; this stops patients breathing, and they must then be ventilated until the agent wears off or is reversed.

Suxamethonium Short-acting depolarizing muscle relaxant; quick, good recovery but cannot be reversed. Main problems for the patient are muscle pains, which arise 24–48 h after administration. They can be very severe and are more likely in the ambulant. It can also cause severe bradycardia, especially on repeat dose in children. A particular but rare problem is suxamethonium sensitivity, p. 536.

Pancuronium, atracuronium, and vecuronium are all non-depolarizing muscle relaxants which are slower acting and longer lasting, but can be reversed using neostigmine.

Anaesthesia and the patient on medication

Certain drugs should ring alarm bells when patients require a GA, these include:

Monoamine oxidase inhibitors Should be stopped 2 weeks prior to GA.

Antiepileptic drugs Must be continued up to, during, and after GA; however, methohexitone should be avoided.

Antihypertensives Should be continued, but ensure the anaesthetist is aware they are being taken.

Bronchodilators Should be continued. Give inhaler with pre-medication and ensure nebulizer is available post-operatively.

Cardioactive drugs Should all be continued. Warn the anaesthetist, as these patients often benefit from a pre-operative anaesthetic assessment.

Cytotoxics Patients on these rarely need GA, but if they do they need FBC, U&Es, LFTs. Suxamethonium should be avoided.

Diabetic drugs For management, see p. 592.

Lithium Measure levels. Omit prior to major surgery.

Oral contraceptives Think about deep vein thrombosis prophylaxis although experience suggests the risk is remote.

Sedatives and tranquillizers If the patient takes a regular dose these can be maintained, but warn the anaesthetist.

Anaesthesia—hospital setting

With the exception of the out-patient GA service provided by dental hospitals, GA in a hospital setting is similar to that for any other surgical service. The provision of and the care of the patient immediately prior to, during, and after the anaesthetic is the domain of the trained anaesthetist. The input from the hospital junior is the same, whether in a dental or surgical specialty, and is mainly covered on p. 576. The fundamental problem for any dental, oral, or maxillofacial anaesthetic is that the surgeon and anaesthetist both need to have access to the same anatomical site. The means by which this is overcome in hospital practice is by endotracheal, particularly nasendotracheal intubation, muscle relaxation, and ventilation, none of which are practical propositions in a dental practice setting in the UK.

Endotracheal intubation secures the airway by placing a tube into the trachea either via the nose, mouth, or a tracheostomy. This tube has an inflatable cuff which prevents aspiration of debris and is connected to an anaesthetic machine to allow delivery of oxygen, nitrous oxide, and an inhalational anaesthetic. Most anaesthetists also use a throat pack to supplement the cuff **which must be removed at the end of the operation**. Intubation is a specialist skill and must be learned by practice; skilled practitioners can perform blind nasendotracheal intubation which is of enormous value in cases with trismus; most cases are, however, performed by direct vision of the vocal cords using a laryngoscope with the patients neck fully extended. The major risk of intubation is intubating the oesophagus **and not recognizing it**. Other complications include: traumatizing teeth; vocal cord granulomata; minor trauma to the adenoids; rarely, pressure necrosis of tracheal mucosa and laryngeal stenosis.

The laryngeal mask airway (LMA) The LMA has become an acceptable method of maintaining the airway. Structurally, it is a curved tube with a large cuff at its end. It is a device which is inserted orally without direct vision following the normal curve of the pharynx. Its onward progression is stopped by the upper end of the oesophagus and at this point the cuff is inflated forming a seal around the entrance to the larynx. Patients must be starved as it does not protect against aspiration. It occupies a substantial volume in the mouth. The necessity of movement of the LMA to allow surgical access may displace the cuff, possibly causing laryngeal obstruction. Expensive but autoclavable up to 40 times. It requires quite deep anaesthesia both to pass and maintain the LMA which may prolong anaesthesia inappropriately.

The LMA should only be used by those trained in intubation.

Muscle relaxation is essential for successful intubation in elective patients. These drugs are covered on p. 643. For most oral surgery, spontaneous breathing once intubated and anaesthetized is the norm, and a short-acting muscle relaxant, e.g. suxamethonium, is used.

Ventilation Requires a longer-acting muscle relaxant (non-depolarizing group). May be hand or mechanical. The latter is more precise and convenient for the anaesthetist. It involves a machine providing intermittent positive pressure ventilation with set tidal volume.

Monitoring There are increasingly sophisticated non-invasive techniques available: pulse oximetry to measure the percutaneous saturation of haemoglobin with oxygen (a normal person breathing air will have a saturation >95%); capnograph to measure end-tidal $PaCO_2$ (normal: 5.2 KPa or 40 mmHg); ECG and automatic blood pressure machines. It must be stressed, however, that these machines do not replace, only enhance, the ability of the anaesthetist to use clinical observation of pulse, colour, skin changes, and ventilatory pattern. The use of monitors requires both knowledge and skill—*do not* request a test or employ a machine you cannot use.

Anaesthesia—practice setting

In theory, there should be no difference between the anaesthetics given in a hospital and a practice setting. In reality, the two environments are totally different. There is an enormous demand for this type of service and there is no doubt that it cannot be met by a hospital-based service.

The ideal team to provide exodontia and simple dentistry under GA is a small skilled group, e.g. anaesthetist, dentist, and nurse or DSA, who have received appropriate postgraduate training and have suitable facilities.

Anaesthesia in GDP is given to ASA grade 1 or 2 patients (i.e. healthy), the majority of anaesthetics, from induction to recovery, are a few minutes. In hospital, patients are frequently medically compromised, the induction is performed in an anaesthetic room by anaesthetists who may be at any stage of their training and the surgery is often more demanding. In the hospital environment, the anaesthetics are, of necessity, longer than in GDP.

In a designated surgery environment both operator and anaesthetist are fully trained experienced practitioners. Patients are fit, generally more relaxed, and happier in a familiar environment often being dealt with by their own dentist. Total starvation and treatment time can usually be ↓

In practice, all drugs, gasses, and equipment are subject to strict financial limits and the onus is to use safe, well-tried techniques, rather than new potentially more expensive techniques available in the hospital service. Although newer drugs may have theoretical advantages over older ones, there have never been studies which can compare the two environments and demonstrate an advantage.

Dentistry under out-patient GA

This basically consists of extractions as more complex treatment would necessitate intubation. Rapid IV induction is performed by the anaesthetist. After induction, sufficient anaesthesia must be achieved to allow the mouth to be opened and a prop positioned, the oropharynx being protected by a suitable pack. The patient is now entirely dependent on their nasopharyngeal airway. The dentist now carries out the required work quickly and efficiently, working with the anaesthetist. Control any bleeding before handing back full control of the airway to the anaesthetist.

Inhalational induction may be preferred in small children. A nasal mask or a split nasal and mouth mask may be used.

Controversial points

1 **The upright position** In a practice setting many patients are anaesthetized and teeth removed in the sitting position. It has been felt that this is potentially dangerous as the patient may physiologically faint while anaesthetized with serious neurological sequelae. In reality this does not seem to occur and the sitting

position makes airway maintenance easier and aspiration of gastric contents much less likely.

2 **Pollution with anaesthetic gasses** In the hospital environment it is possible to scavenge the majority of gasses as patients are either intubated, have an LMA, or a relatively leak-proof face mask. This is not the case in a dental surgery as the patients are breathing through a nasal mask and, invariably, also through their mouths. The dental surgery, however, usually has the distinct advantage that the doors and windows can be opened!

3 **The hazards of inhaling anaesthetic gasses** are not well established. It is true that patients continuously breathing nitrous oxide for 24 h or so begin to have suppression of the bone marrow and that chronic recreational abuse can cause subacute combined degeneration of the spinal cord. The short-term and intermittent long-term effects of exposure to nitrous oxide, halothane, enflurane, and isoflurane are, however, simply not known. It remains common sense that no one should work in a polluted atmosphere. The current guideline is HC (76) 38.

Selected principle recommendations of the Poswillo Committee

Anaesthesia

1 The use of GA should be avoided whenever possible.
2 All anaesthetics should be administered by accredited anaesthetists.
3 The administration of GA, in dental surgeries and clinics equipped to the recommended standards of monitoring shall continue.
4 An ECG, pulse oximeter, and a non-invasive blood pressure device are essential.
5 A capnograph must be used when endotracheal anaesthesia is practised.
6 A defibrillator must be available.
7 Intravenous agents for anything other than ultra short procedures should be administered via an indwelling needle or cannula.
8 GA surgeries be subject to inspection and registration.
9 Appropriate training must be provided for those assisting the anaesthetist and dentist.
10 At no time should the recovering patient be left unattended.
11 Adequate recovery facilities must be available.
12 Good contemporaneous records of all treatments and procedures be kept.
13 Written consent be obtained on each occasion prior to the administration of an anaesthetic or sedation.
14 Patients should be provided with comprehensive pre- and post-treatment instructions and advice (GA and sedation).

649

Sedation

1 Sedation be used in preference to GA whenever possible.
2 For sedation by inhalation the minimum concentration of oxygen be fixed at 30%.
3 Flumazenil be reserved for emergency use.
4 Intravenous sedation be limited to the use of one drug with a single titrated dose to an end point remote from anaesthesia.
5 The use of IV sedation in children be approached with caution.
6 Dentists must be aware of the significance of pulse oximetry readings.
7 All patients treated with the aid of sedative techniques be accompanied by a responsible person.
8 Undergraduates should have experience of managing at least 5 cases of IV sedation, 10 cases of inhalation sedation, and be proficient in venepuncture.
9 Interested dentists should complete a recognized course in intravenous sedation within 2 years of qualification.

14 Dental materials

Properties of materials   654
Amalgam   656
Composites—constituents and properties   658
Composites—practical points   660
The acid-etch technique   662
Dentine bonding agents   664
Glass ionomers—properties and types   666
Glass ionomers—practical points   668
Cermets   668
Cements   670
Impression materials   672
Impression techniques   674
Casting alloys   676
Wrought alloys   678
Ceramics—dental porcelain   680
Ceramics—practical applications   682
Denture materials—1   684
Denture materials—2   686
Biocompatibility of dental materials   688

Relevant pages in other chapters: Materials used in endodontics, p. 316.

Principal sources: J. F. McCabe 1990 *Applied Dental
Materials*, Blackwell. B. G. N. Smith 1986 *The Clinical
Handling of Dental Materials*, Wright. E. C. Combe 1992 *Notes
on Dental Materials*, Churchill Livingstone.

Many materials texts list the requirements for an ideal cement,
filling material, impression material, etc., and then conclude that
it doesn't exist. We will try to resist this temptation.

Properties of dental materials

Definitions

Stress is the internal force set up in reaction to, and opposite to the applied force (force/area). Can be classified according to the direction of the force—tensile (stretching), compressive, or shear.

Strain is the change in size of a material that occurs in response to a force. It is calculated by dividing the change in length by the original length.

Yield strength (or elastic limit) is the stress beyond which a material is permanently deformed when a force is applied.

Elastic modulus is a measure of the rigidity of a material and is defined by the ratio of stress to strain (below elastic limit).

Stiffness Gives an indication of how easy it is to bend a particular piece of material without causing permanent deformation or fracture. It is dependent upon the elastic modulus, size, and shape of the specimen.

Toughness is the amount of energy absorbed up to the point of fracture. It is a function of the resilience of the material and its ability to undergo plastic deformation rather than fracture.

Resilience is the energy absorbed by a material undergoing elastic deformation up to its elastic limit.

Hardness is the resistance to penetration. A number of hardness scales are in use (e.g. Vickers, Rockwell). Between these scales hardness values are not interchangeable.

Creep is the slow plastic deformation that occurs with the application of a static or dynamic force over time.

Wear is the abrasion (± chemical) resistance of a substance.

Fatigue When cyclic forces are applied a crack may nucleate and grow by small increments each time the force is applied. In time the crack will grow to a length at which the force results in fracture through the remaining material.

Thermal conductivity is the ability of a material to transmit heat.

Thermal diffusivity is the rate with which temperature changes spread through a material.

Coefficient of thermal expansion is the fractional increase in length for each degree of temperature rise.

Wettability is the ability of one material to flow across the surface of another. Wettability is determined by the contact angle between the two materials and influenced by surface roughness and contamination. The contact angle is the angle between solid/liquid and liquid/air interfaces measured through the liquid.

Evaluation of a new material

Before it reaches the dental supply companies a new material should have undergone the following tests:

- *Standard specifications* (i.e. physical properties), e.g. compressive strength, hardness, etc. The actual values obtained are mainly of value in comparing the new material with those already in use and which are performing satisfactorily. Compliance with an international (ISO) standard indicates fitness for dental use.

- *Laboratory evaluation*. This should be relevant to the clinical situation, but this is easier said than done!
- *Clinical trials*. Usually conducted under optimal conditions. Many materials have been less successful under the conditions imposed by clinical practice, particularly those with demanding techniques.

Clinically, the important questions to ask the rep include:
- Shelf-life.
- Details of the chemical constituents.
- Handling characteristics, e.g. presentation, mixing, working time, setting time, and dimensional changes on setting.
- Performance in service.
- Cost.
- Does the material meet the relevant ISO standard?

Then decide whether this new material has any significant advantages over the material you are familiar with.

Amalgam

An amalgam is a mixture of mercury (Hg) with another metal. Dental amalgam is made by mixing together Hg with a powdered silver-tin alloy to produce a plastic mass that can be packed into a cavity before setting. Despite toxicity scares and the introduction of posterior composites, amalgam is still widely used.

Types of amalgam

Amalgam can be classified either by:

Particle shape Can be lathe-cut (irregular), spheroidal, or a mixture of the two. Spheroidal particles give a more fluid mix which is easy to condense, can be carved immediately, and takes 3 h to reach occlusal strength (compared to >6 h for lathe-cut amalgams). Spherical amalgams are preferable for large Class V and pinned restorations.

Particle composition The first (conventional) alloys introduced had a low copper content (5%). Research showed that the weakest (tin–mercury or gamma-2) phase of the set amalgam could be eliminated by ↑ the proportion of copper, so a variety of high copper (6–25%) amalgams have been introduced. These are more expensive, but are superior in terms of corrosion resistance, creep, strength, and durability of marginal integrity. There are two types of high copper alloy: (1) a single composition alloy of silver-tin-copper; or (2) a blended (dispersion) mix of silver–tin and copper–silver alloys. Of these, the former is the most resistant to tarnishing.

Handling characteristics

Mixing or trituration This is carried out mechanically, either:
1 Pre-encapsulated by manufacturer, with automatic vibrator.
2 Using an amalgamator which dispenses Hg and alloy in correct proportions and mixes them. However, the amalgamator requires re-filling by hand which ↑ likelihood of Hg spillage occurring. The duration of trituration varies from 5 to 20 sec.

Condensation should be carried out incrementally either by hand instruments (lathe-cut or spheroidal) or mechanically (lathe-cut only). Both are equally effective but the latter is quicker. Cavities should be overfilled so that the Hg-rich surface layer is removed by carving.

Carving With spherical alloys; this can be commenced immediately, but with lathe-cut a delay of a few minutes is advisable. Burnishing is now back in vogue.

Polishing Polished amalgams look good, but whether this step is necessary is still the subject of debate.[1] NB Maximum strength takes 24 h to develop.

---

1 I. W. M. Jeffrey 1989 *J Dent* **17** 55.

Marginal leakage

Whilst amalgam corrosion products will form a marginal seal in time, microleakage can be ↓ by the use of either a conventional cavity varnish (e.g. Copalite) or a bonding agent (e.g. Amalgambond or Panavia-Ex). The latter (an anaerobic resin adhesive) also bonds to set amalgam.[1] Alternatively, sealing over the completed restoration with a fissure sealant has been suggested, and this is a possible solution to the 'ditched', but caries-free amalgam.

Toxicity

'There is insufficient evidence to justify the claim that Hg from dental amalgam has an adverse effect on the vast majority of patients'.[2]

The greatest risk appears to be related to the inhalation of Hg vapour, ∴ attention should be paid to the following:
- Avoid spilling Hg (p. 716).
- Waste amalgam should be stored in a screw-top bottle under old X-ray fixing solution.
- When removing old amalgams, safety glasses, masks, and high-volume aspiration are a wise precaution.

Types of amalgam

Average composition (%)	Ag	Sn	Cu	Zn
Conventional	68	28	4	0–2
High copper	60	27	13	0

Types of amalgam currently available
Conventional—lathe-cut.
Conventional—spherical.
High copper dispersion—lathe + spherical.
High copper single—spheroidal.
High copper single—lathe-cut

657

1 D. C. Watts 1992 *J. Dent* **20** 245. **2** B. M. Eley 1993 *BDJ* **175** 161.

Composites—constituents and properties

The modern composite is a mixture of resin and particulate filler, the handling characteristics of which are determined largely by the size of the particles and method of cure.

Constituents

Resin Most composites are based on either Bis-GMA or urethane diacrylate, plus a diluent monomer, TEGDMA.

Filler Confers the following benefits on the composite:
- ↑ compressive strength, abrasion resistance, modulus of elasticity, and fracture toughness.
- ↓ thermal expansion and setting contraction.
- improved aesthetic qualities.

Composites can be subdivided according to particle size:

Macrofilled (or conventional) contains particles of radio-opaque barium or strontium glass 2.5 to 5 μm in size, to give 75–80% by weight of filler. This type has good mechanical properties but is hard to polish and soon roughens.

Microfilled contains colloidal silica particles 0.04 μm in size and 30–60% by weight. This type retains a good surface polish, but is unsuitable for load-bearing situations, has poor wear resistance, and ↑ contraction shrinkage.

Hybrid contain a mixture of conventional and microfine particles designed to optimize both mechanical and surface properties. Contains 75–85% by weight of filler.

Most composites are of the hybrid type.

Initiator/activator
1 Chemically cured—benzoyl peroxide (or sulfinic acid) initiator + tertiary amine activator. **2** Light cured—amine + ketone activated by blue light (460 to 470 nm).

Other constituents include pigments, stabilizers, and silane coupler to produce bond between particles and filler.

Important properties of composites

1. Polymerization shrinkage of 1–3%.
2. Thermal expansion is significantly greater than enamel or dentine and without acid-etch bond can result in marginal leakage.
3. Elastic modulus should be high to resist occlusal forces. Modulus of hybrid type is greater than other composites, amalgam, or dentine. However, composites are still brittle and # if used, in thin section.
4. Wear resistance is greatest in hybrids.
5. Radiopacity is particularly useful, especially for posterior composites.

Composites—practical points

Method of polymerization

Chemical (self-cure) No additional equipment required, but mixing of two components introduces porosity and the working time is limited.

Light activation Provides long working time, command set, and better colour stability, but requires a light source, has a limited depth of cure, and the temperature rise during setting can be as high as 40 °C. Three types of light source are currently available: fibre-optic (fibres liable to break), fluid filled and pistol grip.

Dual-cure Curing is initiated by a conventional light source, but continues chemically to help ensure polymerization throughout the restoration.

Practical tips:
- Visible light-cured composites can be cured by any blue-light source.
- Bulbs should be replaced every 6 to 12 months.
- Air attentuates light beam, ∴ position as close to tooth as possible.
- Cavities >2 mm depth should be cured incrementally.
- Precautions are necessary to protect eyes from glare, ∴ use safety specs, get the patient to close their eyes and the DSA to look away. Alternatively, a hand-held shield or one attached to the light tip can be used.
- Efficiency of light source can be tested by curing a block of composite. Practical depth of cure is half thickness of set material.
- The greater the intensity of the light source the greater the depth of cure.

Finishing Ideally, a Mylar strip, otherwise use microfine diamond finishing burs (under water spray) and polish with aluminium oxide coated discs (Shofu, Soflex). Shofu points or finishing pastes are useful for inaccessible concave surfaces.

Problems with composites

1 Difficult to obtain satisfactory contact points and occlusal stops.
2 Polymerization shrinkage.
3 Depth of cure of light-cured materials is limited. This is a particular problem in posterior teeth.

Indirect composite inlays may circumvent the last two problems p. 266).

Fissure sealants are composite resins containing little or no filler, which are either self- or light-cured. Clear or opaque types are available, the former having better flow characteristics (whether this is an advantage depends upon the position of the tooth). Success depends upon being able to achieve good moisture control for the acid-etch bond.

661

The acid-etch technique

Recent research would suggest that:

- Success depends upon adequate moisture control, as contact with saliva for as little as 0.5 sec will contaminate etch pattern.
- A prophylaxis prior to etching is not required unless abundant plaque deposits are present.
- 30–50% buffered phosphoric acid provides the best etch pattern.
- An etching time of 15–20 sec is adequate for both primary and permanent enamel.
- The etch pattern is easily damaged, ∴ using a probe to aid etchant penetration of pits and fissures, or applying etchant by rubbing vigorously with a pledget of cotton wool, is C/I.
- There is no difference in bond strength whether an etchant solution or gel is used. Gels take twice as long to rinse away but have the advantage of greater viscosity and colour contrast.
- Rinse for at least 15 sec.
- Remineralization of etched enamel occurs from the saliva, and after 24 h it is indistinguishable from untreated enamel.
- Etched enamel is porous and has a high surface energy. The etch pattern consists of three zones (from surface inwards):

etched zone (enamel removed)	10 micrometres
qualitative porous zone	20 micrometres
quantitative porous zone	20 micrometres

∴ composite resin tags may penetrate up to 50 micrometres into enamel to give micromechanical retention.

Dentine bonding agents

The advantages of bonding to dentine (e.g. preservation of tooth tissue) have fuelled considerable research effort. The problems that have had to be overcome include the high water and organic content of dentine; the presence of a 'smear layer' after dentine is cut; and the need for adequate strength, immediately following placement to withstand the polymerization contraction of composite. These difficulties have been approached in a number of ways, making the topic of dentine bonding confusing, a situation which has been exacerbated by the pace of new developments and by the claims of the manufacturers.

Indications

- Marginal seal where cavity margin in dentine or cementum, e.g. Class V, Class II box.
- Retention and seal of indirect porcelain and composite inlays.
- Dentine adhesives have also been used for repairing # teeth, cementing ceramic crowns and veneers, and as an endodontic sealer.

The smear layer consists of an amorphous layer of organic and inorganic debris, produced by cutting dentine. It ↓ sensitivity by occluding the dentine tubules and prevents loss of dentinal fluid. It purportedly has an ↑ inorganic content compared to dentine. The smear layer is partially or completely removed &/or modified during dentine bonding.

Types of dentine adhesive

1 Bonds via Ca^{2+} to the inorganic component of dentine (and enamel). Many are based on phosphate methacrylate. As the smear layer has an ↑ inorganic content it can be left *in situ*, ∴ technique is quicker. Will also bond to GI cement.
2 Bonds with amine or hydroxyl groups of the organic component of dentine. Many are based on isocyanate or aldehyde. With these adhesives the smear layer has to be removed with a cleanser (e.g. EDTA). Following this a hydrophilic primer which bonds to collagen (e.g. gluteraldehyde and HEMA) is placed. Finally, a sealer (of Bis-GMA) is used to provide a bond between the primer and the composite of the restoration.
3 Bonds with a disturbed and re-precipitated smear layer. First an acidic primer is used to partially dissolve the smear layer and expose the tubules. Then a mixture of bonding resins is applied to the dentine. The mechanism of bonding is not fully understood, but mechanical entanglement with the collagen fibrils is likely.

Practical points

- Follow manufacturer's instructions.
- Success with dentine bonding agents also depends upon cavity design. A V-shaped cavity shows less microleakage than a box-shaped design.

- For better results use a matched composite and adhesive system.
- When a dentine adhesive is used polymerization shrinkage of composite is more likely to result in cuspal deformation and ∴ post-operative pain. An incremental filling and curing technique will help to ↓ this problem.
- Pre-curing the adhesive bonding agent before placing the composite ↑ bond strength.[1]
- No technique produces zero microleakage.

For a concise review of adhesive systems the article by Watson and Bartlett is recommended.[2]

665

1 J. F. McCabe 1994 *BDJ* **176** 333.
2 T. F. Watson 1994 *BDJ* **176** 227.

Glass ionomers—properties and types

Glass ionomer has been officially renamed **Glass polyalkeno-ate**. However, we have continued with glass ionomer (GI) as abbreviating glass polyalkenoate could lead to confusion with gutta-percha.

Setting reaction

Alumino-silicate glass + polyalkenoic acid → calcium and
aluminium polyalkenoates
(polyacrylic = polyalkenoate = polycarboxylic)

The set material consists of unreacted spheres of glass surrounded by a silicaceous gel, embedded in metal poly-alkenoates. Calcium fluoride is present (20%) to give cario-static properties.

Presentation

1 Powder + liquid.
2 Powder (with anhydrous acid) + water.
3 Encapsulated.

Itaconic acid is added to ↑ rate of set, and tartaric acid to sharpen set. In some products maleic replaces polyalkenoic acid.

Properties

Adhesion To enamel and dentine by (1) ionic displacement of calcium and phosphate with polyacrylate ions, (2) absorption of polyalkenoic acid on to collagen. Some authors recommend pre-conditioning the dentine, e.g. with 10% polyalkenoic acid (GC Dentin Conditioner) for 30 sec. Whether this ↑ adhesion is controversial. GI cement also bonds to the oxide layer on stain-less steel and tin.

Cariostatic Due to fluoride release throughout the lifetime of the restoration. GI are also able to take-up fluoride when the IO concentration is raised.

Thermal expansion is similar to enamel and dentine.

Strength Brittle material. Tensile strength is only 40% of composite.

Radiolucent Except for Ketac-bond and the cermets.

Abrasion resistance is poor.

Biocompatibility This is being questioned following reports of pulpal inflammation when used as a luting cement.[1]

Applications

GI are unable to match the aesthetics and abrasion resistance of the composites; in addition, their brittleness limits their use to non-load-bearing situations. However, their adhesive and fluoride-releasing properties have resulted in a range of applica-tions and matching formulations.

1 C. G. Plant 1988 *BDJ* **165** 54.

Type I Luting cements for crowns, bridges, and orthodontic bands. Sub-line with calcium hydroxide over freshly cut dentine, and where this is not possible (e.g. vital PJC) GI are C/I.

Type II Restorative cements. There are two subtypes: (a) aesthetic, (b) reinforced. Can also be used as a fissure sealant, for the conservation of deciduous teeth (p. 86) and for repairing defective restorations.

Type III Fast-setting lining materials. Use of a sublining is a wise precaution. Defer placement of amalgam for at least 15 min and composite for 4 min. In load-bearing situations or where lining is exposed to the oral environment (e.g. in sandwich technique), use of a type II reinforced cement is preferable.

Type IV This category includes the light-cure and dual-cure GI (use of a light source optimizes the properties of the dual-cure materials although they will self-polymerize without). It has been suggested that the light-cured GI have higher bond strengths than self-cure GI.[1]

Newer materials which combine the adhesive and fluoride-releasing properties of GI with the abrasion resistance of composite, are being called, with a touch of originality, compomers. Some of these materials are used in conjunction with a primer of polyalkenoate acid and resin monomers which is light-cured prior to placement of the compomer.

667

---

1 R. van Noort (ed.) 1994 *J Dent* **22** 8.

Glass ionomers—practical points

Practical tips

- A dry field is essential.
- Hand-mixing should be carried out on a chilled glass slab, not paper, using a cobalt–chromium or agate spatula.
- Cement should be inserted before its sheen is lost.
- GI sticks to SS, ∴ use powder as a separator.
- A syringe system is useful for placing material in larger cavities, e.g. Ketac-fil (NB Nozzle can be bent), Centrix. Use a small sponge to tamp GI into place.
- Cellulose or soft metal strips provide the best finish. Burlew foil can be used to shape occlusal surface by burnishing or by the patient occluding on to it.
- Water balance during setting is critical. Absorption of water results in dissolution and dehydration leads to crazing, ∴ cement must be protected with waterproof varnish (NB Copalite is not waterproof). Alternatively, can use light-cured bonding resin, which acts as a lubricant for finishing and can then be cured.
- Although most manufacturers claim that trimming can be started 10–15 min after placement, it is better to defer for >24 h.
- GI restorations should be protected against dehydration, ∴ if isolation subsequently required, coat with bonding resin.

Cermets

These are similar to GI, except that the ion-leachable glass is fused with fine silver powder. Mixing with a polymeric acid gives a cement consisting of unreacted glass particles to which silver is fused, held together by a metal-salt matrix.

Properties

- Adhesion to enamel and dentine.
- Radio-opaque.
- ↑ wear resistance compared to GI, but equivalent strength.
- Cariostatic.

Applications Core build-ups; low-stress-bearing restorations; deciduous conservation.

Metal reinforced glass ionomer

An amalgam alloy powder is mixed with the glass powder, giving a grey restorative material with similar properties to the cermets.

Cements

Cements are used for a variety of purposes, including temporary dressings, cavity liners, and luting agents. With the exception of calcium hydroxide, the materials available are based on combinations of:

Powder—zinc oxide or fluorine-containing aluminosilicate glass (this releases fluoride and is stronger than zinc oxide).

Liquid—phosphoric acid (irritant) or eugenol (↑ solubility, obtundent) or polyalkenoic acid (adhesive).

Setting occurs by an acid–base reaction. The set cement comprises cores of unreacted powder in a matrix of reaction products.

Now thought that the ability of a cement to prevent microleakage is more important than its setting pH.

Based on zinc oxide eugenol (ZOE)

Zinc oxide eugenol Powder of pure zinc oxide is mixed (in a ratio of 3:1) with eugenol liquid to give zinc eugenolate and unreacted powder. Setting time is 24 h. This is the weakest cement, but the eugenol acts as an obtundent and analgesic, ∴ it is useful as a temporary dressing.

Acclerated ZOE, e.g. Sedanol. Addition of zinc acetate to the powder ↓ setting time to 5 min.

Resin-bonded ZOE, e.g. Kalzinol. Addition of 10% hydrogenated resin to the powder ↑ strength.

EBA, e.g. Staline, Opotow. Addition of ortho-ethoxybenzoic acid (62%) to the liquid ↑ strength.

Zinc phosphate, e.g. De Treys Zinc. The powder consists of zinc and magnesium oxides, and the liquid 50% aqueous phosphoric acid. The working time is ↑ by adding the powder in small increments. Popular because of its strength. Although its low setting pH theoretically C/I its use for vital teeth, in practice this does not seem to be a problem.

Zinc polycarboxylate, e.g. Poly-F, Durelon. The powder is a mixture of zinc and magnesium oxides and the liquid is 40% aqueous polyacrylic acid. Recently, anhydrous acid formulations have been introduced, which are mixed with water. The powder should be added quickly to the liquid. The temptation to remove excess cement should be resisted until it has reached a rubbery stage. Adheres to dentine, enamel, tin, and SS.

Calcium hydroxide Chemically curing types comprise two pastes which are mixed together in equal quantities. One paste contains the calcium hydroxide plus fillers in a non-reacting carrier and the other polysalicylate fluid. The set material consists of an amorphous calcium disalicylate complex plus calcium hydroxide and has a pH of 11. In addition to being bacteriostatic, calcium hydroxide can induce mineralization of adjacent pulp. Light-cured formulations which are resin-based are available. Have ↓ bacteriocidal properties, but ↑ strength.

Glass ionomer (p. 666). NB Remember to apply varnish to cement margins if using as a luting agent.

Strength Phosphate > EBA or polycarboxylate > resin-bonded ZOE > acclerated ZOE > calcium hydroxide.

Practical points
- Generally, the thicker the mix, the greater the strength, ∴ for cavity linings incorporate as much powder as possible.
- Heat ↓ setting time, ∴ a cooled slab is advisable.
- To stop cement sticking to the instruments during placement, dip in saliva (the patient's!) and then powder (except calcium hydroxide).
- When luting, apply cement to crown or inlay before tooth.

Choice of cement

Temporary restorations Choice depends upon how long the dressing needs to last and whether any therapeutic qualities are required. Pure ZOE is useful for a tooth with a reversibly inflamed pulp, but resin-bonded ZOE is stronger. Glass ionomer is preferable for semi-permanent dressings.

Luting cement Zinc phosphate, GI, and polycarboxylate are all popular as luting cements. EBA cement is C/I because of ↑ solubility. Composite-based luting systems are available which are often used in conjunction with a dentine bonding agent. Useful for cementing ceramic or porcelain inlays/onlays and ceramic veneers.

Lining cement Choice of lining depends upon the depth of the cavity and the material being used to restore it.

Amalgam: shallow—varnish; average—calcium hydroxide; deep—use a sub-lining of calcium hydroxide and any of the cements listed above.

Composite: Calcium hydroxide ('Life' is the most acid resistant) or GI. Eugenol content C/I ZOE.

Pulp capping Calcium hydroxide.

Sedative dressing ZOE ± calcium hydroxide.

Bacteriostatic dressing Calcium hydroxide plus ZOE or GI.

Impression materials

Classification

Non-elastic	Elastic	
Plaster	*Elastomers*	*Hydrocolloid*
Compound	Silicone	Reversible
ZOE paste	Polysulphide	Irreversible
Wax	Polyether	

Elastomers

These are indicated when accuracy is paramount, e.g. crown and bridge work.

Condensation-cured silicone, e.g. Xantopren, Optosil. This material is relatively cheap compared with other elastomers, but is prone to some shrinkage and should be cast immediately.

Addition cured silicone, e.g. Reprosil, President, Imprint. This type of silicone is very stable, which means that impressions can be posted or stored prior to casting. However, casting within 1 h is C/I. A perforated tray is advisable as the adhesives supplied are not very effective. Up to 5 viscosities are manufactured, allowing a range of impression techniques. NB Latex gloves can retard setting.[1]

Polysulphide, e.g. Permalastic. This material wets the preparation well, but is messy to handle. It is useful when a long working time is required. Should be used with a special tray and although stable, cast within 24 h.

Polyether, e.g. Impregum. A popular material because it uses a single mix and a stock tray (and smells of gin and tonic). The set material is stiff, and removal can be stressful in cases with deep undercuts or advanced periodontitis. Absorbs water, ∴ do not store with alginate impressions. Can cause allergic reactions.

Hydrocolloids

Reversible hydrocolloid is accurate, but liable to tear. Requires the purchase of a water bath.

Irreversible hydrocolloid (alginate) Setting reaction is a double decomposition reaction between sodium alginate and calcium sulphate. Popular because it is cheap and can be used with a stock tray. However, it is not sufficiently accurate for crown and bridge work. Impressions must be kept damp and cast within 24 h. Alginate can retard the setting of gypsum and affect the surface of the model.

Impression compound

This is available in either sheet form for recording preliminary impressions, or in stick form for modifying trays. The sheet material is softened in a water bath at 55 to 60 °C and used in a stock tray to record edentulous ridges. The viscosity of compound results in a well-extended impression, but limited detail.

1 W. W. L. Chee 1992 *J Prosthet Dent* **68** 728.

Impression waxes

(e.g. Korecta) These are produced in four grades. The very soft (orange) type is useful for the correction of small imperfections in ZOE impressions, or for recording -/P free-end saddles.

Zinc oxide pastes

These are dispensed 1:1 and mixed to give an even colour. They are used for recording edentulous ridges in a special tray or the patient's existing dentures, but are C/I for undercuts. Setting time is ↓ by warmth and humidity. To alter setting time: change brand.

Impression techniques

(for crown and bridge work)
▶ Time spent recording a good impression is an investment, as repeating lab work is costly.

Special trays Help the adaptation of impression material and reduce the amount required, i.e. ↑ accuracy and ↓ cost. Can be made in cold cure acrylic, or light-activated tray resin.

For crown and bridge work an accurate impression of the prepared teeth is required. Usually the palate does not need to be included, so design the special tray accordingly or use a lower stock tray. For the opposing arch, alginate in a stock tray will suffice.

Single mix technique (e.g. polyether) The same mix of medium viscosity material is used for both a stock tray and syringe. Although less accurate than other methods it is adequate for most tasks.

Double mix technique (e.g. polysulphide, addition-cured silicone) This is a single-stage technique necessitating the mixing of heavy and light bodied materials at the same time, and use of a special tray.
- Apply adhesive to tray.
- Mix light and heavy bodied viscosities simultaneously for 45 to 60 sec.
- Remove retraction cord and dry preparation, while DSA loads syringe with light-bodied material.
- Syringe light-bodied mix around prep. A gentle stream of air helps to direct material into crevice.
- Position tray containing heavy-bodied material.
- Support tray with light pressure until 2 min after apparent set.

Putty and wash technique (e.g. silicone) The putty and light-bodied viscosities can be used with a stock tray, either:

1 Single-stage, which is similar to the double mix technique above.
2 Two-stage, which is less liable to distortion than the single-stage technique. This involves taking an impression of the preparation with the putty, using polythene sheet as a spacer. This is then re-lined with the light-bodied material, which is also syringed around the preparation.

Cost, in decreasing order:
    addition-cured silicone using putty/reline
    polyether with special tray
    addition-cured silicone using double mix and special tray
    polysulphide with special tray
    condensation-cured silicone using putty/reline
    hydrocolloid

In general, a double-mix technique with a special tray is the best choice in terms of accuracy and probably also cost.[1]

1 C. M. Munson 1985 *Dental Update* **12** 163.

Automixing dispensers This double cartridge presentation is becoming popular. The 2 pastes are extruded and mixed in the nozzle when the trigger is pressed. Appears expensive, but ↓ waste.

Disinfection of impressions
Silicone and hydrocolloid: 10% v/v Domestos for 10 min.[1]
Polyether C/I. Alginate: 2% gluteraldehyde for 5 mins.

1 S. J. Wilson 1987 *Restorative Dentistry* **3** 86.

Casting alloys

An alloy is a mixture of two or more metallic elements. The chemistry of alloys is too complicated for this book (and its authors), so for a fuller understanding the reader is referred to one of the source texts.

The properties of an alloy depend upon:
- The thermal treatments applied to the alloy (including cooling).
- The mechanical manipulation of the alloy.
- The composition of the alloy.

NB The properties of an alloy may differ significantly from that of its constituents.

The main examples of alloys in dentistry include amalgam (p. 656), steel burs and instruments, metallic denture bases, inlays, crowns and bridges, and orthodontic wires.

Casting alloys

A warm moist mouth provides the ideal environment for corrosion. In order to overcome this problem, dental casting alloys comprise an essentially corrosion resistant metal (usually gold), with the addition of other constituents to enhance its properties. However, with the exception of titanium all have the potential to affect hypersensitive individuals.

Additions to gold alloys *Copper*—↓ density, ↓ melting point, ↑ strength and hardness, but ↓ corrosion resistance.
Silver—↑ hardness and strength, but ↑ tarnishing and ↑ porosity.
Platinum—↑ melting point, ↑ corrosion and ↑ tarnish resistance.
Palladium—similar properties to platinum, but less expensive.
Zinc or indium—scavenger, preventing oxidation of other metals during melting and casting.

Dental casting gold alloys Noble metal content must exceed 75% of which 65% must be gold. Four types are defined by proof stress and elongation values from type I (low strength for castings subject to low stress) to Type IV (extra-high strength).

Dental casting semi-precious alloys have noble metal content from 25% up to, but not including, 75%. The noble metals are gold, platinum, palladium, and iridium. Four types are defined by proof stress and elongation properties.

Silver palladium Palladium (>25%), silver, gold, indium, and zinc. Are cheaper than gold alloys and of equivalent hardness, but less ductile, more difficult to cast and prone to porosity.

Nickel chromium 75% nickel, 20% chromium.
Used in crowns and bridgework. In latter ↑ rigidity compared to gold alloys is an advantage. However, castings are less accurate than gold and nickel sensitivity can C/I its use.

Cobalt chromium 35–65%, cobalt 20–35% chromium.
Modulus of elasticity twice that of type IV gold alloys. A good polish is difficult to achieve, but durable. Used mainly for P/-.

Titanium Has good biocompatibility. Pure titanium casting is still being perfected.

Alloys for porcelain bonding

Requirements:
- Higher melting point than porcelain.
- Similar coefficient of thermal expansion to porcelain.
- Do not cause discoloration of the porcelain.
- High modulus of elasticity to avoid flexure and # of porcelain.

Indium is usually added to faciliate bonding to porcelain. Copper is C/I as it discolours the porcelain. A matched alloy and porcelain should be used.

High gold ↑ palladium or platinum content (to raise the melting point) compared to non-porcelain alloys.

Medium gold 50% gold, 30% palladium. Widely used.

Silver palladium Cheap, but care required to avoid casting defects.

Nickel chromium Very high melting point and modulus of elasticity, but casting more difficult. NB Some patients are sensitive to nickel.

Casting For gold (melting point <950 °C).
- Wax pattern and sprue are invested in gypsum bonded material.
- Wax burnt out by heating investment mould to 450 °C slowly.
- Alloy melted either by gas/air torch or electric induction heating and cast with centrifugal force.
- Casting allowed to cool to below red-heat.
- Quenched. Some alloys are used as cast and others heat-hardened.
- Cleaned with ultrasonics and acid immersion.

For nickel chromium and cobalt chromium alloys (melting point 1200–1500 °C) need silica or phosphate bonded investment and either oxy-acetylene torch or electric induction heating.

Casting faults A casting may be:
- dimensionally inaccurate;
- have a rough surface;
- be porous, contaminated, or incomplete.

Wrought alloys

Wrought alloys are hammered, rolled, drawn or bent into the desired shape when they are solid.

Stainless steel Steel is an alloy of iron and carbon. The addition of chromium (>12%) produces a passive surface oxide layer which gives SS its name. The SS used in dentistry is also known as austenitic steel (because the crystals are arranged in a face-centred cubic structure) or 18:8 steel (due to the chromium and nickel content). Available as:

1 Pre-formed sheets for denture bases. The SS is swaged on to the model by explosive or hydraulic pressure. This produces a thin (0.1 mm), light denture base, which is resistant to fracture.

2 Wires are produced by drawing the SS through dies of reducing diameter until the desired size is achieved. This work hardens the wire, but heat treatments are carried out to give soft, hard, or extra-hard forms. Manipulation of SS wire also work hardens the wire in the plane of bending, ∴ trying to correct a bend is more likely to result in fracture. Main applications are orthodontics and partial denture clasps. SS can be welded and soldered.

Soldering SS This requires the use of a flux (e.g. Easy-flo flux) to remove the passive oxide layer (this reforms after soldering). The reducing part of the flame should be used, but do not overheat as this can anneal and soften the components.

- Melt a small bead of low fusing silver solder (e.g. Easy-flo solder) on to the wire.
- Mix up the flux with water to a thick paste and apply to the item to be added.
- Heat up the solder so that wire underneath is a cherry red.
- Bring the fluxed wire into the molten solder and remove the flame at the same time.

This is not easy and requires practice, which explains the popularity of electrical soldering.

Cobalt chromium This has a similar composition to the cast form (p. 676) and is mainly used in wrought sections for surgical and dental implants.

Cobalt chromium nickel This alloy is used in orthodontics as an archwire material (Elgiloy). It has the advantage that it can be hardened by heat treatment after being formed. Also used for post fabrication in post and core crowns (Wiptam wire).

Titanium Pure titanium is used in the Bränemark implant system.

Titanium alloys Nickel and titanium alloy (Nitinol) is useful in orthodontics as it is flexible, has good springback and is capable of applying small forces over along period of time. However, it is not easy to bend without fracture. Titanium molybdenum alloy (TMA) is also used for archwires and has properties midway between SS and nitinol.

Gold Expense limits the application of wrought gold alloys to partial denture clasp fabrication.

Alloys for dentures

Cast cobalt chromium is the material of choice for partial denture connectors because of its high proof stress and modulus of elasticity—thin castings are strong, rigid and lightweight. Although wrought gold alloys are more suitable for clasps, the advantage of being able to cast connector and clasps in one, means that cobalt chromium is more commonly used (p. 338).

Ceramics—dental porcelain

Ceramics are simple compounds of both metallic and non-metallic oxides. Although many of the materials used in dentistry are ceramics, the term is commonly used to refer to porcelain and its derivatives. Dental porcelain actually more closely resembles a glass, and comprises feldspar, quartz (for strength and translucency), and kaolin (for strength and colour), plus pigments. Most dental porcelain is reinforced with alumina particles (40–50% by mass) to provide greater strength. Unfortunately, this increases the opacity, ∴ the proportion of alumina in enamel porcelains is reduced. In crown construction a platinum matrix is laid down on the die produced from the impression of the prepared tooth to act as a base. The porcelain powder is mixed with water to form a slurry, which is built-up in layers on to the foil until the desired shape is achieved. The porcelain is compacted by removing water from the preparation by blotting with absorbent paper or by flicking with a brush. This reduces firing shrinkage. The crown is then fired to reduce porosity, which ↑ strength and ↑ translucency. Glazing produces a glossy outer skin which resists cracking and plaque accumulation. It can be added as a separate layer or by firing at a higher temperature after the addition of surface glazes. The platinum foil lining is removed prior to cementation to give space for the cement lute.

Properties
- Firing shrinkage of 30 to 40%, ∴ crown must be overbuilt.
- Chemically inert provided the surface layer is intact.
- Low thermal conductivity.
- Good aesthetic properties.
- Brittle. The main cause of failure is crack propagation which almost invariably emanates from the unglazed inner surface. This can be ↓ by (1) fusion of the inner surface to metal, as in the platinum foil and metal bonded techniques; or (2) by the use of an aluminous porcelain core.
- High resistance to wear.
- Glazed surface resists plaque accumulation.

Ceramics—practical applications

Porcelain jacket crown In this type of crown a core of aluminous porcelain is laid down first on to which 'dentine' and then 'enamel' porcelains are built up. For strength a minimum porcelain thickness of 0.8 mm is required and a 90° butt joint at the margin.

Platinum bonded/McLean-Sced/twin foil crown In this technique 2 layers of platinum foil are laid down. The outermost one is tin-plated and then oxidized in order to bond to porcelain. This increases the strength of the crown by preventing crack propagation and in addition, the tin lining bonds to polycarboxylate and GI cements. Unfortunately, the strength gained still does not suffice for load-bearing teeth.

Porcelain fused to metal crown The porcelains used for bonding to a metal sub-structure have additional alkali oxides added to raise the coefficient of thermal expansion to almost match the alloys used. In addition, a porcelain which fuses below the melting point of the alloy is required. Bonding to the metal occurs by a combination of:
- mechanical retention;
- chemical bonding to the metal oxide layer on the surface of the alloy;

The ↑ strength of porcelain bonded to metal crowns is due to:
- the metal substructure supporting the porcelain;
- ↓ crack propagation by bonding of the inner surface of the porcelain to metal;
- by the outer surface of porcelain being under tension, thus ↓ crack propagation.

Glass ceramic or castable ceramics These are semi-crystalline glasses, e.g. tetrasilicic fluoramine glass (Dicor), produced by adding a nucleating agent on to which the glass crystals precipitate during cooling. This structure resists crack propagation, having a similar strength to core aluminous porcelain. An additional advantage of this material is that the lost wax process can be used, so compensation for firing shrinkage is not required. A wax pattern is built-up by the technician and invested and cast. Following cooling, the casting is then re-heated to allow crystal growth to occur. The castable ceramics have ↑ translucency compared to porcelain, but dentine colour has to be created with surface glazes, which requires time and skill. Glass ceramic crowns are more expensive than PJC, but are useful for young translucent teeth. Hot-pressed (IPS-Empress) and grass infiltrated (Inceram) ceramics do not match the strength of porcelain fused to metal.

Porcelain veneers consist of a thin shell of porcelain or castable ceramic approximately 0.5 to 0.8 mm thick (p. 280). Due to their thin section they are rather monochromatic in appearance.

Porcelain inlays Long-term evaluation, including the wear of opposing teeth, is still awaited.

Porcelain repairs can be carried out using composite and a silane coupling agent. A number of proprietary kits are available.

683

Denture materials—1

Acrylic resin

Acrylic is the most commonly used polymer for denture bases. Not only can it be re-lined, repaired, and added to comparatively easily, but it is aesthetic and lightweight. Acrylic is composed of a chain of methacrylate molecules linked together to give polymethylmethacrylate (PMMA).

Presentation This usually comprises a liquid and a powder which are mixed together. The liquid is stored in a dark bottle to ↑ shelf-life.

Powder	Liquid
PMMA beads (<100 μm)	methylmethacrylate monomer
initiator, e.g. benzoyl peroxide	cross-linking agent
pigments ± fibres	inhibitor, e.g. hydroquinone
	activator (self-cure only)

Manipulation The powder and liquid should be mixed together in a ratio of approximately 2.5:1 by weight. The mix passes through several distinct stages: sandy–string–dough–rubbery–hard (set).
The dough stage is the best for handling and packing.

In denture fabrication the wax pattern and teeth are invested in plaster. The wax is then boiled out and the plaster coated with sodium alginate as a separator. The resultant space is then filled to excess (to allow for contraction shrinkage of 7%) with acrylic dough under pressure. The acrylic is then polymerized.

Mode of activation

Self-cure acrylics show less setting contraction, but more water absorption than heat-cure. This may result in the final item being slightly over-sized thus ↓ retention. Self-cure acrylics are more porous, only 80% as strong, less resistant to abrasion, and contain a greater level of unreacted monomer compared to heat-cure. The main applications of self-cure acrylics are for denture repairs and re-lines, and orthodontic appliances, although for the latter the greater strength of heat-cure is preferable.

Heat-cure Conventionally polymerization requires heating in a hot-water bath for 7 h at 70 °C, then 3 h at 100 °C. The flask should be cooled slowly to minimize stresses within the acrylic. However, resins with different curing cycles (fast heat cure) are now available. Microwave energy can be used to cure acrylic resin, but (apart from lunchtime) has no advantage over a water bath.

Light-cure resins are supplied as mouldable sheets. Used for denture bases or special trays.

Properties
- The glass transition (or softening) temperature of self-cure acrylic is 90 °C and for heat-cure is 105 °C.
- Poor impact strength and low resistance to fatigue fracture.
- Abrasion resistance not very good, but usually adequate.

- Good thermal insulator. This is undesirable as it can lead to the patient swallowing foods which are too hot.
- Low specific gravity (i.e. not too heavy).
- Radiolucent. Attempts to ↑ radiopacity have not been very successful.
- Absorbs water, resulting in expansion. Drying out of acrylic should be avoided.
- Residual monomer (due to inadequate curing) weakens acrylic and can cause sensitivity reaction.
- Good aesthetics.

The strength of a denture depends upon:
1 Design, e.g. adequate thickness, avoidance of notches.
2 Strength of the acrylic, i.e. low monomer content, ↓ porosity, adequate curing. Can be ↑ by using high-impact resin.

Researchers are still evaluating methods of ↑ strength of denture resins. The addition of high-performance fibres (e.g. polyethylene fibres) appears promising.

Denture materials—2

Rebasing

Rebasing a denture base involves replacement of the fitting surface. Rebases can be either:

1 Hard: heat or self-cure.
2 Soft: permanent—heat or self-cure or light-cure.
        temporary (a) tissue conditioner—self-cure.
                (b) functional impression material—self-cure.

The properties of the self-cure materials are generally inferior, ∴ should only be used as a temporary measure.

Hard rebases Heat-cure PMMA is preferred, but requires the patient to do without the denture while it is being added. A self-cure material has obvious advantages, but even the higher acrylics, e.g. butylmethacrylate (Peripheral seal) should only be used as a temporary measure as they are weaker than PMMA and discolour.

Soft liners require a material with a glass transition temperature below or at that of the mouth so that it is soft and resilient. The majority of soft liners are either based on silicone or acrylic (polyethylmethacrylate or PMMA powder plus alkyl methacrylate monomer and a plasticizer), however, the recently introduced polyphosphazene fluoroelastomer liners look promising.[1]

Plasticized acrylic	Silicone polymers
Good bond to denture	Bond with denture base not reliable
Harden over time	Maintain resilience
More readily distorted	Absorb water—candida colonization
	More elastic

Heat, light or self-cure types are available. The self-cure materials have inferior properties, but all require replacement during the lifetime of the denture.[2] Cleaning, p. 360.

Tissue conditioners usually comprise powdered polyethylmethacrylate to which a plasticizing mix of esters and alcohol is added. No chemical reaction takes place, the liquids merely soften the powder to form a gel and leach out over time, resulting in hardening. To ensure maximum tissue recovery the lining should be a minimum thickness of 2 mm and replaced every few days. E.g. Coe-soft, Coe-comfort, Viscogel. Cleaning, p. 360.

Functional impression materials The tissue-conditioning materials are usually used for this purpose, an impression being cast of the fitting surface after a few days of wear. Pre-formed sheets of the co-polymer polyethylmethacrylate and polyethylacetate which are marketed as a soft liner are more suitable for functional impression materials (e.g. Ardee).

686

1 F. Kawano 1992 *J Prosthet Dent* **68** 368. 2 E. R. Dootz 1993 *J Prosthet Dent* **69** 114.

Biocompatibility of dental materials

Before a new material can be marketed it must successfully pass both laboratory and clinical trials to evaluate its biocompatibility. Yet, some adverse effects only become apparent after the material has been in clinical use. Unless used with care many materials may prove a hazard to the patient or the dentist and his staff.

Hazards to the patient
Systemic effects

Allergic reactions
- *Amalgam*. Although genuine cases of amalgam allergy exist, these are rarer than the tabloid press would suggest. For proven cases composite or GI should be used.
- *Nickel*, a constituent of some alloys can cause contact eczema. Sensitive patients often have a history of allergy to jewellery or watch casings. Alternative alloys are available.
- *Acrylic monomer* can cause an allergic reaction and should be considered in a patient complaining of a 'burning mouth'. The concentration of monomer is ↑ in poorly cured acrylic and is greater in self than heat cure. Extended curing, e.g. 24 h, may reduce the concentration of monomer to an acceptable level, if not a cobalt chrome or SS denture base will be required.
- *Epimine* in polyether impression material.

If an allergy is suspected, refer to a dermatologist.

Directly toxic
- Beryllium which is present in some nickel alloys is known to be a carcinogen. Provided the alloy is not ground any risks are confined to the production laboratory. Beryllium-free alloys are becoming ↑ available.
- Fluoride in excess can be toxic (p. 34).

Ingestion or inhalation of air-borne dust must be avoided.
Local effects

Eye damage Blue-light curing lamps can cause eye damage due to the glare. The simplest solution is to ask the patient to close their eyes.

Thermal injury can result (1) to the pulp, e.g. caused by exothermic setting reactions, or by filling materials which are good thermal conductors; (2) to the mucosa, e.g. caused by dentures which are thermal insulators, as the patient may swallow drink/food which is too hot, or by the setting reaction of self-cure denture re-line materials; (3) to the soft tissues, e.g. by hot instruments.

Chemical injury to the pulp can be caused by acidic cements or filling materials, or by inadequate dentine coverage when etchant is being used. Care is required to avoid chemical injury to the soft tissues.

Direct damage Can be caused by noxious chemicals (e.g. etchant, hydrogen peroxide) being allowed to come into direct contact with the tissues.

Hypersensitivity reactions Can occur in response to the materials that cause systemic allergy.

Hazards to staff

In the surgery
- Allergic reactions, e.g. topical anaesthetics, latex gloves, methyl methacrylate monomer.
- Eye damage from blue-light sources. Use eye protection or shielding.
- Alginate dust.
- Mercury vapour.
- Nitrous oxide.

In the laboratory
- Cyanide solution for electroplating.
- Vapours from low fusing metal dies.
- Siliceous particles in investment materials.
- Fluxes containing fluoride.
- Hydrofluoric acid used for etching porcelain veneers.
- Beryllium in some alloys.
- PMMA powders.
- Methyl methacrylate monomer.
- Casting machines.

15 Law and ethics

Legal processes	692
Complaints	694
Consent	696
Contracts	698
Negligence	700
The protection and defence organizations	702
Professional ethics and etiquette	702
The General Dental Council and the Dentists' Register	704
Wise precautions, or how to avoid litigation	706
Forensic dentistry	708

Relevant pages in other chapters: hiring and firing staff, p. 714.

Principal sources: J. Seear 1989 *Law and Ethics in Dentistry*, Wright. GDC 1993 *Professional Conduct and Fitness to Practice*. BDA 1993 *Ethical and Legal Obligations of Dental Practitioners*.

Definitions
Plaintiff (or in Scotland, *pursuer*)—the claimant in a civil action.
Defendant (or *defender*)—the person against whom a claim is made.
Legislation—the law laid down in Acts of Parliament.
Secondary legislation—the precise implementation of the general rules laid down in the Act (often published as Statutory Instruments).
Litigation—an action brought in a court of law.
Writ—a document setting out the details of a proposed action, which is served to the defendant.
Affidavit—a written statement on oath.
Serious professional misconduct—action or omission by a professional person which would be regarded as disgraceful or dishonourable by reputable colleagues. Examples of professional misconduct include fraud, indecent assault, and breaches of the Terms of Service.

▶ *On qualification*: (1) Register with the GDC (p. 704), (2) Join a defence organization (p. 702).

Legal processes

In England

Civil law in general, governs the rights and obligations between individuals and corporations, e.g. debt recovery, breaches of contract, negligence. It is also concerned with the functions of the State and of public authorities. In civil cases, the verdict is given on the balance of probabilities, i.e. each side argues their case and the decision is given to the version that is most likely to be true. The losing party is liable for the costs of both sides.

There are several tiers to this judicial system:
1 Small claims court—for claims <£500.
2 County court—for claims <£5000.
3 High court—for claims >£5000.

Appeals are referred to the Court of Appeal (civil division) and, occasionally, on to the House of Lords.

Criminal law Criminal prosecutions are undertaken when the 'law' has been broken, e.g. speeding, fraud, assault. In criminal law, the case must be proved beyond reasonable doubt. Depending upon the severity of the misdemeanour cases are heard:
1 Magistrates court. Appeals to Crown Court.
2 Crown Court (with a judge and jury). Appeals go to the Court of Appeal (criminal division).

Coroners court This straddles the two systems and meets to consider unnatural and unexpected deaths, e.g. a death in the dental chair. The process is investigative (as opposed to the plaintiff versus defendant stance taken in the other courts).

In Scotland

In parallel with many European countries, criminal prosecutions are brought by the Procurator Fiscal, who, in addition to prosecuting in the Sheriff and District courts, also fulfils a role similar to that of Coroner. The nomenclature of the courts in Scotland is quite different from that of England.[1] Another disparity between the two systems is that in criminal cases in Scotland, an additional verdict of 'not proven' is possible. In Scotland, there is no Coroner's Court, but in the event of an unnatural death in the dental surgery, a Fatal Accident Inquiry may be heard before a sheriff.

▶ Any unexpected death should be reported to the Coroner in England, Wales, and N. Ireland, or to the Procurator Fiscal in Scotland, either directly or through the police (the Registrar of Deaths will notify the authorities if this has not already been done).

Dentists are no longer automatically exempt from jury service. It is possible to apply for exemption (using the form sent with the summons).

---

1 B. Knight 1987 *Legal Aspects of Medical Practice*, Churchill Livingstone.

Complaints

The procedure adopted depends upon the complaint:

By a patient against GDP
(a) To Family Health Services Authority (NHS treatment only). This option has a time bar of 13 weeks from the alleged incident or within 6 months of completion of treatment, whichever is the sooner. This will involve either an informal investigation (for less serious cases) or a Dental Service Committee hearing. Even if a patient is successful no damages or compensation is awarded. Appeals can be made to the Regional Health Authority.
(b) To the GDC. A complaint of serious professional misconduct by a member of the public must be supported by one or more affidavits.
(c) Via a civil negligence action in the courts. This is the only route if financial recompense is sought for damage that occurred.

By a patient against Hospital/Community services or staff Most hospitals will have an officer appointed to deal with complaints by the public. If necessary he will refer a problem on to the Regional Medical Officer. Provided a complaint is not against an individual it can be referred to the Health Service Commissioner (Ombudsman).

By DPB against a GDP If this involves an alleged breach of the Terms of Service the case is referred to a Dental Service Committee. If involving an allegation of fraud, the Director of Public Prosecutions (or Proscurator Fiscal) may be informed.

By a GDP against the DPB An appeal should be made in writing (within 1 month) to the FHSA or Health Board, who will appoint two dentists to hear the case.

By a GDP to recover unpaid fees There are two approaches to this; either (1) issue a County Court Summons (the Citizens Advice Bureau have an explanatory leaflet), or (2) employ a debt-collection agency. The latter is considerably easier.

By GDP against GDP The Local Dental Committee will try to arbitrate. Advice can be sought from the defence organizations.

Consent

▶ Treatment without any consent = assault.

▶ Treatment with general consent, but without explanation of what is involved may = negligence.

Therefore, to be valid, consent should be 'informed', i.e. the patient should understand the treatment to be carried out (and any aftercare or precautions necessary) and be made aware of any alternative forms of treatment. An explanation of possible side-effects should be given, but these should be put in context. The story is told of an oral surgeon who managed to avoid having to carry out any orthognathic procedures by giving prospective candidates a detailed and graphic description of the full gamut of possible complications.

Consent can be given in writing, verbally, or be implied.

Written consent is preferable, especially when extensive treatment is planned and essential before a GA is given.

Verbal consent should be the minimum obtained for treatment. The benefits of having a third party present at all times are obvious.

Implied consent By attending an appointment and sitting in the dental chair a patient gives implied consent to a dental examination (but not treatment).

Some specific problems

• The legal age of consent is 16 years. A child <16 yrs may be competent to give consent to treatment on the basis of his/her understanding, but it is preferable to seek parental permission, except in an emergency.

• Where it is not possible to gain consent (e.g. a patient is unconscious), a dentist should act in the best interests of the patient; if necessary, seeking a second opinion. Treatment of an unconscious patient is valid under the legal doctrine of 'necessity', but is limited to emergency care.

• Special care is required when treating the mentally ill. The practitioner should assess the patient's ability to give informed consent. Legally, routine treatment of a person who is mentally handicapped is permissible, provided it is in their best interests, but if there is any doubt it is advisable to consult the patient's parent, guardian, or doctor.

Contracts

A contract is defined as an agreement between parties and can be either verbal or written. In law, both are equally binding, but as the parties may have differing recollections of what was said, the advantages of a written agreement are apparent.

Between dentist and patient Although in the past the majority of contracts were made verbally (as with consent), a written treatment plan and charge estimate which is signed by the patient puts the NHS contract between dentist and patient on a firmer legal footing. Once the time limits for Service Committee complaints are past an NHS form can be thrown away, unless it covers private treatment when it would be advisable to retain it for longer. For purely private patients it is wise to draw up a simple typewritten contract setting out the treatment to be carried out and the fee agreed.

Legally, the dentist is expected to exercise a reasonable degree of care, while the patient also has a duty to keep appointments and pay the agreed fee.

Between dentist and Family Health Services Authority or Health Board A dentist working within the General Dental Services is in contract with this body for every patient accepted for treatment under the NHS Terms and Conditions of Service. Breaches of this contract are investigated by a Service Committee and, if applicable, the clinical freedom of the dentist may be restricted or a fine levied.

Between principals and associates, or between partners Because of their importance these should be written with the help of a solicitor. The BDA (address p. 762) has sample agreements for partnerships, assistantships, associateships, and for expense-sharing arrangements, and in cases of dispute can arrange arbitration. A barring-out clause is a common inclusion designed to prevent unfair competition by a rival practice being set up in the near vicinity. However, these need to be fair (e.g. 1 mile radius in a busy suburban area for 1 yr) to be enforceable. Consideration should be given to emergency arrangements; what is to happen to a dentist's continuing care and capitation patients when he/she leaves, and the new NHS payments (maternity pay, sick pay, and course allowances). Arrangements for payment of fees for any remedial or replacement work also need to be agreed and included.

Between dentist and staff This should include a job description, pay (national minimum rates of pay are set for DSAs, technicians, and hygienists), holiday entitlements and arrangements, bonuses for additional qualifications and loyalty, sickness allowances, pensions, disciplinary rules, and termination procedures with details of notice required by each party. Advice sheets are available from BDA.

Negligence

Professional negligence is defined as a failure to exercise reasonable care, in one's professional capacity.

Often these cases hinge on what 'reasonable care' constitutes; however, if a dentist can show that his actions were in line with those of a large number of his colleagues, he is unlikely to be held negligent (Bolam test).

For a negligence case to win the onus is on the plaintiff's or pursuer's lawyer to prove the following:
- The dentist owed a duty of care to the patient.
- That there was a breach of that duty.
- Damage occurred to the patient as a result (no damage, no case).

Where no demonstrable loss has occurred a patient has no recourse in the law and can only voice his complaint to the FHSA or Health Board. The ploy of attempting to shift the onus of proof on to the defendant by pleading *res ipsa loquitur* (the facts speak for themselves), is rarely successful.

If the case is proved, then financial compensation will be awarded by the court, taking into account the damage that occurred and the steps necessary to put it right.

Criminal negligence For criminal proceedings to be started the negligent action must be very serious and have some accentuating factor, e.g. the dentist was drunk or drugged or disregarded well-known safety principles.

Contributory negligence When the actions of a patient have been partially (or completely) to blame for the damage that occurred, then contributory negligence can be pleaded; e.g. failure to follow post-operative instructions.

Vicarious liability An employer can be held responsible for any negligence by an employee which occurs during his employ. This means that a dentist is responsible for the actions or omissions of his/her staff; e.g. DSA, receptionist, technician. However, as every individual is responsible for their own acts, a charge of negligence could be brought against both employee and employer. Vicarious liability and Crown Indemnity are discussed on p. 702. A GDP is also responsible for the safety of the practice premises (p. 716).

Time bar This is usually 3 years; however, this does not begin to run until the patient knows or ought to have known of the damage that is said to have been done. For children the time bar does not start until 18 years of age. However, patient may petition the Court to set aside the time bar if they can demonstrate reasonable grounds for this to be done.

The steps involved in a civil action for damages

- Usually the first intimation of trouble is a solicitor's letter (known as a letter before action). This should be dealt with only after consultation with a defence society. Notification of a Legal Aid application should also be passed on.

- If the patient's solicitor proceeds, then a writ or summons will be served. As there is a time bar (usually 14–21 days) within which notification of a defence to the claim must be given to the court it is important to respond quickly via a defence organization.
- The defence society will then make their own investigation into the claim, if necessary seeking the advice of an 'expert'. Should the case appear not to be defensible, then the society will try to settle the matter out of court. Where they judge that no negligence has occurred the case will be allowed to proceed to a full hearing.
- Prior to the trial, counsel for the defence will need to establish the full facts of the case. This can be time-consuming.
- At the end of the trial the verdict will be given on the balance of probability and the level of compensation set as appropriate.

There but for the grace of God go I? It must be remembered that doctors and dentists are human beings and not infallible. The law does realize that unforeseen accidents can happen. The defence organizations advise: 'In circumstances where complications and errors arise it is proper that objective, factual information, with appropriate clinical reassurance is provided. Adequate explanations assist in reducing fear and uncertainty which may give rise to complaints and claims'.[1]

The importance of clear, concise and contemporaneous notes cannot be over-emphasized. Record cards, X-rays, and study models (where appropriate) are invaluable.

701

---

1 Medical Protection Society, *Annual Reports 1986* and *1987*.

The protection and defence organizations

These non-profit making organizations serve to provide the medical and dental professions with advice, indemnity, and defence in legal proceedings involving professional matters (in addition they also contribute significantly to the social life of medical and dental students). In the UK there are three:

Medical and Dental Defence Union of Scotland.
Medical Defence Union.
Dental Protection (part of Medical Protection Society).

Their addresses are given p. 763.

All dentists in the GDS should join one of these organizations upon qualifying. Because of the rising cost of medical litigation, in 1989 the Health Authorities took over liability for the acts or omissions of doctors and dentists working in their hospitals and the Community Service (Crown Indemnity) and this has subsequently passed directly to those hospitals which are Trusts. However, dentists in the Hospital and Community Services are advised to maintain some personal cover as Crown Indemnity does not cover 'good Samaritan' deeds, representation at Coroner's inquests, Fatal Accident Inquiries, and private practice. In addition, health authority or Trust lawyers may try to settle without going to court, thus preventing defendants from proving their innocence of the charge. The defence societies have ∴ introduced several levels of cover.

Professional ethics and etiquette[1]

▶ Do as you would be done by.

Although not imposed by legislation, ethical behaviour is necessary, to maintain the standing of the profession in the eyes of the public. The above precept should apply in all aspects of life, but the following are more specific examples of how it pertains to dentistry:

- The needs of the patient should be of overriding concern.
- Professional confidentiality should be observed except where to withhold information could be construed as acting against the public interest.
- Do not criticize colleagues.
- Do not solicit patients from other dentists.
- When consulted for emergency treatment by a colleagues' patient: treat as necessary to alleviate pain and refer the patient back with an explanation of what was done.
- Refer to suitably qualified colleagues, cases which require specialist advice and treatment.
- Avoid false certification and misleading statements.

1 BDA 1982 *Ethical and Legal Obligations of Dental Practitioners*, BDA, London.

- Although the restrictions on advertising have now been lifted, all adverts must still be legal, decent, truthful, and have regard for professional propriety.
- It is the duty of a dentist leaving a practice to make the necessary arrangements for the continuing care of his patients.

703

The General Dental Council and the Dentists' Register

▶ The practice of dentistry is limited by law to registered dental and medical practitioners and enrolled ancillary workers.

▶ A dentist shall be liable to have his name erased from the Dentists' Register if, either before or after he is registered, he has been convicted of a criminal offence or has been guilty of serious professional misconduct (see p. 690). Section 27, Dentists Act 1984.

The GDC is a statutory body set up by the Dentists Act to regulate the practice of Dentistry. It is composed of 44 members, which include 34 elected dentists, 6 members of the GMC, and 4 lay persons.

Functions of the GDC

To maintain the Dentists' Register This is a list of those who have attained the appropriate qualification and paid their annual retention fee. Additional qualifications can be included in the entry beside a dentist's name. A name may be erased if a person is found guilty of certain offences or if the retention fee is not paid.

To supervise the standards of Dental Education This involves regular visits to the UK dental schools with the power to withdraw recognition if standards are not met.

Discipline Procedure:

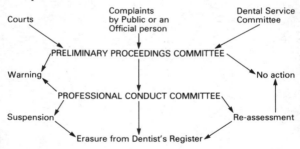

704

Appeals against erasure or suspension can be made by the dentist to the Privy Council. A dentist whose name has been erased may apply to the Professional Conduct Committee for reinstatement >10 months after erasure.

▶ It is unwise for a dentist to enter a plea of guilty in criminal matters, for a charge for which he/she has a defence, as this will be taken as an admission of guilt by the GDC.

Wise precautions, or how to avoid litigation

▶ When in doubt contact your defence organization.

● Good records are invaluable. They should be concise, factual, and objective, and contain details of any treatment and advice given (especially that which the patient is unwilling to accept). All appointments made and failed should be noted. Keep your opinions about difficult patients to yourself; committing them in writing on their dental records is unwise. The Access to Health Records Act 1991 allows patients access to manually held records made since that date. Patients can apply for a copy of their records (a fee: < £10 and the cost of copying can be charged) and to have incorrect data removed.

● Take adequate radiographs whenever clinically necessary.

● Records (and X-rays) should be kept for at least 11 years, except for children where they should be retained until age 25, or for 11 years, whichever is the longer. There is no legal requirement to keep models, but they may be required by the DPB.

● Protect the patient's eyes and airway.

● Check the medical history at every recall.

● Keep up-to-date. Continuing with outmoded techniques could render a dentist liable to a change of professional misconduct. Similarly, be wary of using controversial or unproven procedures or materials.

● When treating a patient it is wise to have a third person present at all times to act as a chaperon. This is particularly important if using sedation, as reports of patients having erotic fantasies under the influence of nitrous oxide, diazepam, or midazolam can no longer be regarded as amusing anecdotes.[1]

● For any expensive or complicated procedures obtain written consent.

● Refer a patient, when it is in their best interests.

● Take adequate precautions against cross-infection for patients, staff, and yourself (p. 734).

● Where a GA is being administered this must be by a (suitably trained) medical or dental practitioner other than the dentist treating the patient and with adequate facilities for resuscitation readily available.

● Read the 'Notice for the Guidance of Dentists', issued by the GDC.

● Comply with the Ionising Radiation Guidelines (p. 736).

▶ If you are sued, make copies of all relevant documents and records. The originals should then be kept in a safe place. One copy should be forwarded to your defence society. It is also wise to compile a diary of any relevant events, noting down all that you remember. Never alter records under the threat of litigation. Significant alterations, made at any time, should be dated and initialled.

---

1 M. Fields 1990 *BDJ* **169** 4.

Advice lines
Dental Protection: 0171-323 6555
Medical Defence Union: London 0171-486 6181
Manchester 0161-428 1234
Medical and Dental Defence Union of Scotland 0141-221 5858

Forensic dentistry

Theoretically, forensic dentistry encompasses all aspects of dentistry and the law, but here we will confine ourselves to the application of dental science to criminal investigations.[1]

Identification As the dental tissues survive the effects of fire, water, and time well, dental identification is helpful where other means of distinguishing a person have been lost, e.g. following incineration or burial. Not only can the teeth be used to indicate the approximate age of a victim, but the condition of the dentition, including any filled or missing teeth, can be compared with dental records to aid identification. Occasionally, the dental chart of an unknown person is publicized in the dental press with an appeal from the police for help in identification. In this situation it must be remembered that dental treatment may have been carried out subsequently and that additional restorations or missing teeth compared with the GDP's records of a suspected victim do not necessarily exclude identification. Information can also be obtained from studying the skull and jaw bones, e.g. age, sex, racial origin, and the time elapsed since death. Comparative radiographs may be as individual as fingerprints.

From a forensic point of view the value of accurate up-to-date records, and identification marks in dentures, are self-evident.

Occasionally, a practitioner may be requested by the police to reveal the dental records of a live, but missing person. The advice of a defence society should be sought, as legally, a patient's confidential records should not be revealed without a Court Order.

Bite-marks in inanimate objects left at the site of a crime have on occasion contributed to the conviction of the perpetrator. If the 'evidence' is a perishable foodstuff, then a permanent record can be made either by casting in stone or rubber base material, following photography.

Bite marks in human skin can occur in assault or child abuse cases. Considerable aggression is required to penetrate the skin. It is important to first establish any identifying features of the teeth that caused the bite and then compare them with the dentition of any suspects. Good photographs, including a linear scale, are essential, and if the victim is alive, need to be repeated 24 hrly as the clarity of the bite may improve with time. In addition, any saliva associated with the mark can be analysed to see if it indicates the blood group of the assailant.

708

1 D. K. Whittaker 1989 *A Colour Atlas of Forensic Dentistry*, Wolfe.

16 Practice management

Management skills   712
Hiring and firing staff   714
Health and safety   716
Financial management   718
Running late   720
Marketing   722
Practice leaflets   724
Computers and dental practice   726
Independent and private practice   728
Vocational training   730
Audit and peer review   732
Prevention of cross-infection   734
X-rays—the statutory regulations   736
X-rays—practical tips and helpful hints   738

Relevant pages in other chapters: Contracts, p. 698; Biocompatibility of dental materials, p. 688.

Principal sources: K. J. Lewis 1989 *Practice Management for Dentists*, Wright. D. W. Crosthwaite 1982 *A Handbook of Dental Practice Management*, Churchill Livingstone. J. O. Forrest 1984 *A Guide to Successful Dental Practice*, Wright. The BDA have an excellent range of leaflets covering many of the topics in this chapter.

Management skills

Management technique courses are now big business. Although it would be easy to dismiss them as a way of making a quick profit (for the person giving the course), in fact, underneath the jargon there is a good deal of sound business sense.

What are the benefits of good management? ↑ efficiency, ↓ stress and ↑ job satisfaction—for the whole dental team. A happy practice environment is not only more pleasant to work in, but the *bonhomie* will also be transmitted to patients.

Keys to successful management

Good communication While the benefit of good communication with patients is acknowledged, the importance of discussion with other members of the dental team receives less emphasis. A few jokes about the receptionist's latest boyfriend (strange though he may be) is not communication. For teamwork to be successful the opportunity for all staff to discuss problems and ideas for improvements needs to be created. This usually means setting aside some time within the working week for regular staff meetings. It is helpful if these meetings have some sort of structure—which does not mean that the dentist dominates proceedings, rather that all staff are encouraged to submit ideas for an agenda. On occasion, some guidance may be necessary to prevent these meetings developing into fractional warfare, but with care the opportunity to discuss problems will ensure the smooth running of the practice. One approach is to prepare a handbook of procedures within the practice, for the benefit of new staff.

Delegate those tasks that do not require your training and expertise. In addition to ↓ stress and freeing you for the more demanding work this also ↑ job satisfaction for ancillary staff, provided they are given the training and the time to cope with their new responsibilities; e.g. getting the hygienist to do the fieldwork involved in deciding which new ultrasonic scaler to buy.

Teamwork The importance of building a mutually supportive team can readily be appreciated by trying to work in an environment where everyone has been forced to protect their corner. Successful leadership involves encouraging staff to develop their potential both as an individual and as a valued member to the team, and encouraging discussion as to what the goals are to be and how to achieve them. Motivation to work as a team can be fostered by monetary incentives linked to the performance of the practice.

Staff training This should not only involve the newly appointed; by training existing staff to extend their skills more tasks can be delegated. Patient management and communication skills of both reception and nursing staff can also be developed. A manual of procedures and routines applicable to the practice can be useful and staff should be encouraged to contribute and help update this. In-house training days with speakers either from within the practice, or invited, are appreciated.

Adequate training is mandatory for staff involved in sedation, GAs, and taking radiographs.

Pay Motivation can often be enhanced by financial incentives.[1] Therefore by structuring payment to comprise (1) a fixed hourly rate; (2) an individual bonus which is related to attendance, sickness record, and productivity paid as a percentage of the hourly rate; and (3) a group bonus which is a fixed proportion of the profits of the practice; all staff have an inducement to reduce overheads and improve efficiency in the practice.

713

1 C. M. Scola 1990 *BDJ* **169** 334.

Hiring and firing staff

Hiring

1 Define what tasks the practice team would like the new member of staff to perform (but do not limit possible applications only to Superman). Decide on the criteria for an ideal candidate, as this will aid selection later.

2 Draw up a job description. Consider including details of practice, the role of the new member of staff in the team, required skills, in-job training to be provided, hours of work, pay, and other benefits.

3 Advertise post in local press and hospitals. Remember to include a realistic closing date for applications.

4 Short-list candidates.

5 Interview, preferably with two or three people on the panel. This should be structured so that candidates are asked the same questions, to aid comparison. Notes should be made, because after several interviews the candidates may begin to merge! Hopefully, a suitable person will be found and they should be offered the job in writing, subject to references, as soon as possible. If no one is acceptable, go back to **1** and re-assess requirements.

6 Draw up a provisional contract (usually 6–8 weeks is long enough) including how assessment is to be carried out at the end of the trial period. Both employee and employer should retain a signed copy.

7 Orientate and train the new member of staff, giving plenty of time for feedback in both directions.

8 Before end of trial period, re-assess, and if progress satisfactory draw up a formal contract (p. 698).

The BDA have an excellent advice sheet on recruitment.

Firing

The BDA have advice sheets on dismissal and redundancy and the Advisory, Conciliation and Arbitration Service (ACAS) will also give guidance. Both from a practical and emotional point of view, dismissal of staff is not easy and if taken to tribunal, can be expensive. The Trade Union and Labour Relations acts allow dismissal only for unsatisfactory conduct, ineptitude, and contravention of the law.

If problems are encountered with an employee, to prevent a claim of unfair dismissal the procedure to follow is:

- Tell employee formally that their conduct is unsatisfactory, giving a timetable for improvement, advice, and help.
- Written warning that if no improvement, dismissal will follow.
- Written notice as per contract.

The notice required depends upon length of service, >26 weeks = 1 week, >2 yrs = 2 weeks, >5 yrs = 4 weeks. For gross misconduct, instant dismissal is acceptable.

Redundancy is dismissal for reasons other than the personal behaviour of the employee. Redundancy payments are required for staff who have been employed for >2 yrs' continuous employment and >21 h/week. The amount paid depends upon length of service and age. Staff in this category are also entitled to paid time off to look for a new job or undergo training.

Health and safety[1]

Health and Safety At Work Act 1974

This Act states 'It shall be the duty of every employer to ensure, so far as it is reasonably practicable, the health, safety, and welfare at work of all his employees'.

In addition to requiring that all equipment and systems of work are safe and that information, training, and supervision are provided, the Act demands a written statement of policy with regard to health and safety for premises employing >4 members of staff. In addition, to the working environment, the Act requires practice premises to be maintained in a safe condition (see RIDDOR below).

Practice inspections are usually carried out by appointment, although this is not legally required. Employees are also expected to take reasonable care for their own and other people's safety—refusal to comply is grounds for dismissal.

COSHH

The Control of Substances Hazardous to Health regulations were introduced in 1989. Employers are required to assess all substances (e.g. vapours, microorganisms), in the workplace, which are potentially hazardous, and take steps to prevent or reduce any risks to the health of employees. The following procedure is recommended:
1 Identify hazardous substances.
2 Record frequency and quantity of use.
3 Assess the risk.
4 Monitor the medical condition of staff, where necessary.
5 Prevent or control risk.
6 Staff training.
7 Record the assessment.

Mercury spillage Use of Hg should be confined to impervious surfaces where any spillage will be limited, ideally a lipped tray lined with foil. Staff should wear gloves when handling Hg-containing substances. Lead from radiographic packets can be used for absorbing small spillages along with a paste of equal parts calcium hydroxide, flowers of sulphur and water, or a proprietary spillage kit can be purchased. Any waste Hg or amalgam should be stored in a sealed, labelled container under radiographic fixer or potassium permanganate.

X-rays (see p. 736).

Cross-infection control (see p. 734).

Hepatitis B—Employer's responsibilities Dentists are responsible for ensuring that any of their staff who carry out procedures that could bring them into contact with blood (e.g. Assistants, Hygienists, DSAs) should be immunized against Hepatitis B. Any health care worker who becomes *e*-antigen positive is obliged to cease practising.

---

1 *BDA Advice Sheet*, Health and Safety Law for Dental Practice.

RIDDOR

In accordance with the Reporting of Injuries, Diseases and Dangerous Occurrences Regulations, any serious accidents or injuries to staff or patients should be immediately reported to the Health and Safety Executive, and if made verbally followed up in writing on form F2508 (from HMSO bookshops). Less serious incidents which result in absence from work >3 days, should also be notified within 7 days. A written record should be kept of all accidents in the practice.

Disposal of waste

Under the Environmental Protection Act, dentists are responsible for segregating waste, storing it safely, and arranging for its disposal. The Health and Safety Executive recommend that clinical waste for incineration should be stored in yellow containers and disposed of by a registered collector (e.g. FHSA or Rentokil). Non-clinical waste should be stored in black containers.

Employers liability

A certificate of insurance must be displayed on the premises.

FHSAs are now empowered to carry out practice inspections—the above issues are likely to be high on the agenda of any such inspection.

Financial management

Find a good accountant and a friendly bank manager, preferably on the recommendation of another practitioner.

It is advisable to develop a structured system for dealing with fees and estimates, tailored to the individual practice, which is understood and adhered to by all staff.

Delegation of many of the aspects of calculating and collecting fees from patients to trained and motivated staff, should make the practice more cost-effective. E.g., the DSA can complete the patient's treatment plan sheet and the NHS estimate forms; the receptionist can then calculate the patient's contribution, collect it, issue a receipt, and record the transaction. However, failure to monitor the situation adequately can, at best, result in a false sense of security.

Book keeping is time-consuming, but necessary. Many book keeping tasks can be performed by computer either with an integrated practice management system or standalone software. Suggested minimum:
- Fees due and fees received.
- Bank deposits.
- Patient lists.
- List of all FP17s sent to the DPB and their individual value. This can then be checked against schedules when they are returned. As it is a major headache if these forms are lost in the post, therefore the DPB recommend sending a few at a time, by registered or recorded delivery, or via computer line.
- Income/expenditure. This should include all monies received and all bills paid (e.g. lab fees, wages) and be compiled monthly by the principal, in order to develop a feel for the financial situation. Suitable ledgers (e.g. Admor) can be purchased from stationers. It is wise to seek the advice of your accountant as to the methodology, since accurate accounts will make his job easier (and therefore cheaper).
- Petty cash transactions should be recorded, together with relevant receipts. The money (<£50) is best stored in a separate locked box. For larger sums a practice cheque can be cashed.
- Wages.
- Staff absences and sickness record.

Banking It is helpful to bank all monies at the end of each day, as the bank statement then indicates the daily takings. To encourage settlement of fees, it is wise to accept payment in any form, i.e. cash, cheque, or credit card. Although credit cards incur a 3% commission, patients with cash-flow problems may be happy to accept this form of payment as a face-saver; then their lack of funds becomes the credit companies' problem. It is good policy to negotiate overdraft facilities in advance to cover those occasions where cash-flow problems are experienced.

Budgeting An annual forecast and budget should be prepared jointly with the accountant.

Bad debts These can often be prevented by having a definite House Policy which is widely advertised to patients and adhered to; e.g. payment in part at the beginning of treatment and the balance on completion; or payment in full, up front. At the examination appointment patients should be given a written estimate and reminded when payment is due. If a patient forgets, at the last visit they should be asked to sign a form confirming that the treatment has been satisfactorily completed and that they agree to pay (£x) within 7 days. If payment is still not forthcoming, reminders (with an ↑ chill factor) should be sent out at 7, 14, and 28 days. If there is still no joy, send in the debt collectors, but beware of a counter claim of negligence.

Tax This is really where a good accountant comes in. By providing him with information on income and expenditure on a monthly basis, he will be able to provide advice on what to do before the end of the financial year to minimize the tax man's percentage.

Insurance is essential for property, contents, equipment, indemnity, staff, loss of income, and personal insurance.

Consumer Credit Act 1974 requires those extending credit to the public by allowing them to pay for goods or a service in instalments, to obtain a licence. Provided payment is in <4 instalments it is possible to gain exemption. For further details, contact the Office of Fair Trading.

VAT registration Dentists are now able to register for VAT, so that VAT can be reclaimed on purchases made by the practice. VAT must be recorded in the income/expenditure book (or computer file). Claims are made quarterly.

Running late

Running late happens occasionally to everyone, usually when you were hoping to finish early to rush off to do something else. A common reaction is to cut corners, a strategy which can misfire and as a result waste even more time. Another response to being overstressed is to try too hard to hurry, not pausing to plan time effectively. There are no immediate simple solutions, but if another member of staff is free, e.g. hygienist or DSA, you may be able to delegate some simple tasks.

If running late is becoming a habit, it is imperative to stop and re-assess your working practices, either before your patients get fed up waiting or your BP rises too far. The first step is to re-evaluate the times booked for each common procedure; are you being realistic? If the answer is yes, then the problem lies with the receptionist's perceptions of your superhuman abilities, and some gentle re-education is necessary. After erring on the generous side for the length of time required for each treatment, go through the appointment book and block out some buffer zones in each session for emergencies, catching-up with paperwork or, if for no other reason, a breather. The frequency and length of these will depend upon the severity of the original problem.

Some additional hints to reduce everyday stress:
- Have an appointment book, divided into 5 min blocks to provide flexibility.
- A working day comprising a longer morning and a shorter afternoon is more productive.
- If you are so busy that longer procedures have to be booked well in advance, designate some specific sessions for them each month (thus reducing the temptation to squeeze them in).
- For busy sessions, try and utilize two surgeries.
- Schedule less stressful work (e.g. check-ups) for the end of sessions.
- For last-minute cancellations, have a list of patients who are willing to come in at short notice.
- Define the working day and try not to extend beyond this.
- Coffee-breaks and lunch-time should not always be used to catch up on other work; be kind to yourself and have a rest occasionally.
- It is very common to run late after holidays. It is helpful if you ask the receptionist to pretend that you are having at least an extra couple of days off and book these last of all. Then you will be able to cope with fitting in urgent patients on your return and not feel stressed.

Emergencies can be defined as the patient who is willing to attend at any time. The patient who tries to stipulate when they are seen is not an emergency. If buffer zones are built into your timetable, fitting in the true emergencies should not be a problem. For emergencies out of hours there should be a number to contact for advice and, if appropriate, treatment.

Marketing

With the introduction of the recent NHS reforms, the concept that the general public are consumers of health care has resulted in the busy practitioner having to add marketing to his skills. Whether NHS or independent, the success or otherwise of a practice is going to depend on its ability to attract and keep patients.

Advertising Relaxation of the GDC restrictions on advertising has opened up a range of opportunities that you need to grasp before the opposition does. Advertising does not need to be brash; after all, it is merely a means of letting the public know about the existence of a practice and the services that are available. Apart from the 'yellow pages', this information can be disseminated via:

1 Leaflets (p. 724). These can be distributed to existing patients and possible sources of new recruits, e.g. nurseries, doctors' waiting rooms.
2 Open days. These allow apprehensive patients to find out more about painless dentistry without the need to have an examination or treatment.
3 Adverts in local publications: but follow the GDC guidelines.

First appearances count When the prospective patient arrives at the surgery the external and internal decor, together with the welcome they receive, will play a role in determining whether or not an appointment is actually made. It is relatively cost-effective to decorate non-dental areas, therefore ensure that the exterior of the premises looks well cared-for, with a professional-looking plate. The reception and waiting areas should appear cosy and inviting. Choose light, warm colours and comfortable seating to give a relaxing ambience. Plants, provided someone remembers to water them, also help. A small area for children to play in with toys and some books, or if possible a crèche, are good practice-builders. A range of interesting magazines (look at the range available at your opticians', rather than the doctors', waiting room) or practice leaflets on different aspects of dental care/health should be available.

Staff The receptionist must be friendly and helpful (even on Monday mornings). It is worth spending some time with the receptionist, deciding on stock responses to some of the more common problems that arise (e.g. dealing with the angry patient) and to ↑ their knowledge of the techniques available so that patients' queries can be answered correctly. It is also helpful if the receptionist can ask new patients how they heard about the practice, so as to better target future marketing strategies.

The presentation and attitude of all the staff is of vital importance. An attractive and functional uniform in the practice colour or bearing the practice logo is bound to impress.

Emergencies Although it is tempting to exclude non-registered patients from receiving out-of-hours emergency treatment, it is a good practice-builder to see any patient in pain, as some of them will become regular attenders.

Remember, the most important marketing aid is without doubt the personal touch.

Practice leaflets

What information to include?

Compulsory information (in NHS practice):
- Practice name and address and telephone number.
- Names of dentists, with qualifications.
- Hours, and appointment system.
- Whether any dentist provides only orthodontic treatment.
- Whether suitable for patients who cannot climb stairs.
- Special facilities, e.g. availability of hygienist, access for wheelchair, domiciliary visits.

It is not necessary to include full details of emergency arrangements, since these leaflets are intended for potential as well as actual patients.

The *optional* information contained in the leaflet depends upon whether or not it is to form part of a series of booklets:
- Practice philosophy.
- Map showing location of practice.
- Emergency arrangements and contact telephone number.
- Further information on the special interests of the dentists, both dental and non-dental.
- Illustrations of the practice.
- Details of charges—for broken appointments also, if applicable.
- Methods of payment accepted.
- Any foreign languages spoken (holiday Spanish probably doesn't count!).
- Whether both NHS and private treatment available.
- Special facilities and treatments available, e.g. sedation, crèche.

How to set about producing a leaflet Broadly speaking there are two approaches: either get professional help (e.g. designer, photographer, printer) or DIY, using a desk-top printing package on the practice computer. However, the two are not mutually exclusive and all practices should, at least, consider taking advice from a designer. Before seeking help it is important to have some idea of what you want. For example:
- Only one leaflet, or the first of a series?
- What is your potential market (young families with small children, older professionals and their families)?
- Black and white, or two or more colours?
- Glossy booklet or a folded sheet of A4?
- How much money do you want to spend?
- How many copies do you need (to avoid being lead astray by the bulk discounts)?
- How is the leaflet to be distributed?

It is wise to shop around and examine the work of several professionals before choosing.

Design and layout The aim of the leaflet is to create the impression of a caring practice, and to inform. The key to success is simplicity. One helpful tip is to have a practice logo &/or house style (or colours) which are replicated in other items of practice stationery. This idea of a corporate image is not new, but has worked well in the business world. A designer will be able to suggest styles and layouts best suited to your projected market, as well as help with the wording of the text. Photographs and illustrations will increase the cost, but also the impact, as will using more than one colour in the printing of the leaflet.

Sponsorship is worth considering, to give full flight to your creative aspirations. Try approaching toothbrushing manufacturers, dental supply companies, or local firms that you deal regularly with.

Points to watch

- Don't say that a dentist has specialist expertise; rather, that their practice is mainly devoted to a particular type of treatment.
- Don't advertise other services or goods.
- Don't include names of practice staff other than dentists.
- Be legal, decent, honest, and truthful.

Computers and dental practice

These wonders of modern technology can, with the right software and some skill, function as a secretary, accountant, appointment book, filing cabinet, calculator, publisher, etc., etc., BUT when they go wrong the repercussions can be quite spectacular. It is not possible in a book of this size to cover the subject adequately, the intention is rather to point the reader in the right direction.

▶ Befriend a computer buff, particularly one who is readily available by phone.
▶ Garbage in—garbage out.

Definitions
Hardware—computer equipment, i.e. display screen, central processing unit (CPU), keyboard, and printer.
Software—computer programs.
Hard disc—computer's memory, usually located within CPU itself.
Floppy disc—additional, removable source of memory and data.
Disc drive—slot in CPU into which floppy discs are entered and read.
Byte—one letter or space in text. Gauges size of memory.
RAM—Random Access Memory, is the part of computer which stores programs. A progam will not run if its requirements exceed RAM.
User-friendly—means that a program is easy to run (often a misnomer).
Back-up—additional copies of important information.

Functions of a computer are determined by the software that is used. A wordprocessing package converts the computer to a super-typewriter, making letters and reports easy to produce. A database package will compile and manage lists of patients, thereby facilitating recalls and continuing care payments. A spreadsheet package can be used to produce accounts. An integrated program allows several functions to operate simultaneously without having to change from one package to another.

Where to go for advice
1 Dental computer specialist—generally worth the extra expense.
2 Business computer specialist.
3 Local retailer.
4 Teach yourself; but be warned, computers can be addictive.

Choosing a system It is important to decide what you want the computer to do and how much you are prepared to spend. If an (expensive) multifunction practice management system is being considered, then the choice of hardware is less important and is often dictated by the software package. However, if a more cautious approach is planned (e.g. starting with wordprocessing only) then the choice of computer and operating system is more important. In this case, an IBM or IBM-compatible computer with a hard-disc drive is probably the best choice, as the memory can be expanded and there is a wide choice of programs available. An inkjet printer is usually adequate, and cheaper than the laser type. An agreement providing on-site maintenance is a necessity in case of breakdown, and training and back-up are essential. It is worth asking other practices with computers which system they wish they'd bought instead!

Computerization and the DPB A national scheme with a direct computer link between practices and the DPB came into operation in April 1991. The network is accessed by a modem which connects the computer to the telephone lines, thus allowing electronic FP17s to be sent to the Board, and messages to be sent and received. Only systems certified by the DPB can be used, so check with them before installing any software, and take advantage of their starter packs. For security, each dentist is issued with a personal identity number (PIN) which has to be used with every claim. Advantages include ↓ clerical work, ↓ errors, 'faster' prior approval and better cash flow.

NB The protection societies still advise keeping a written copy of all important information.

NB If personal information of any kind is kept on computer, you must register with the Data Protection Registrar, and comply with the regulations of the Data Protection Act 1984.[1] This is not a reason for avoiding computers; as from November 1991 patients are allowed access to manual health records as well.

---

1 The BDA produce a useful fact sheet.

Private and independent practice

Independent is the term preferred by many for private practice, as it sounds less avaricious. An increasing proportion of dentists are already turning to other methods of remuneration than the NHS, indeed the pace of change in this sphere is so fast that it is difficult to provide information that will necessarily be relevant in the future. Therefore this page is limited to discussing general principles.

Researching the market

In order to develop the potential of a practice it is necessary first to fully evaluate its present position. An appreciation of the existing patient base can be gained simply by going through the manual or computer records and looking at the geographic spread and socio-economic groups. If most patients are part of a family group, then future developments need to provide advantages for parents and children, for example, if changing to independent practice, then a family-based capitation scheme might be more applicable.

Also, the potential for attracting new patients to the practice and the competition from other practices in the area for those and existing patients, should be assessed.

It is also important to research and take into consideration the staffs' views about any changes to be made to the practice and to inform them of the advantages these changes will bring.

Business planning

If the financial basis of a practice is to be changed a business plan will need to be drawn up and discussed with the practice's accountants and bankers. Two fashionable business planning tools are:

1 The planning cycle:

2 A SWOT analysis is a consideration of the Strengths, Weaknesses, Opportunities, and Threats, to and of a business.

Fee setting

This difficult exercise should be carried out in conjunction with the advice of the practice accountant. The following will need to be calculated:
1 Practice overheads.
2 Profit desired (realistically!).
3 Inflation.
4 Number of sessions worked per week.

From this the target hourly rate can be calculated. However, local market conditions need to be considered, which may necessitate a little rounding down of the profits desired, or a reduction in overheads if that is possible. Once a realistic hourly rate has been calculated, *either* fees can then be determined for each procedure by the average amount of time taken, to give a set price list, *or* patients charged according to the time taken. Laboratory charges and hygienist fees, as indicated, should be additional.

Types of independent practice

There are basically 4 different approaches:
1 A combination of independent and NHS practice. For example, exempt patients are seen on the NHS, whereas other adults are seen privately. In practice, this two-tier system can cause difficulties and friction.
2 Low-cost independent practice.
3 Traditional 'private' practice.
4 Insurance-based schemes.

Private dental schemes

This is one of the real growth areas in modern dentistry. At present there are 4 types:
1 Capitation based (e.g. Denplan, Smile). Patients pay a monthly fee to the company (which is passed on to the GDP minus an administration fee), usually determined by their dental status and the practice's overheads. The dentist then provides treatment as indicated, but certain items, e.g. orthodontics and laboratory fees, may be excluded.
2 Insurance schemes where treatment costs are reimbursed.
3 Corporate dental schemes.
4 Savings schemes. At present these are restricted to Orthodontics.

Prescribing for private and independent patients

When a private patient requires medicines as part of their treatment, a private prescription should be provided. Dentists can also supply or sell medicines to any patient, but are bound by pricing and other dispensing rules. Alternatively, common drugs can be given as required and their cost included in the overall fee.

Complaints

If patients are paying more for their treatment they will naturally have raised expectations of the service provided. In a large proportion of cases, a potential problem can be dealt with by better communication. Therefore, if patients are provided with the means to voice any complaints within the practice setting an opportunity is given to resolve the situation early.

Vocational training

Vocational training, or VT as it is known, is designed to gently introduce the newly qualified graduate to general dental practice.

Trainees work as salaried assistants and are ∴ not under any pressure to achieve a high volume turnover. Each region organizes a day-release scheme (30 days per year) which covers the clinical and administrative aspects of NHS practice, as well as patient and financial management. But each trainee will also receive on-the-job training and supervision from their trainer. Trainees do not have their own NHS number, so all patient fees go to the trainer. Salary is set at $\frac{1}{2}$ TANI, and the contract includes 4 weeks of holiday a year. On satisfactory completion of VT, a certificate is awarded.

Trainers To be accepted, a practice needs to satisfy the criteria set by the VT Committee and the trainer needs to be present not less than 3 days a week. Trainers are themselves trained in 14 sessions designed to improve their teaching and assessment skills as regular tutorials are a formal requirement. An allowance of 15% of TANI is paid and the trainees' salary is reimbursed in full. However, the FHSA deduct a proportion of the fees earned by the trainee at source. Trainers are vicariously liable for their trainees.

Procedure Suitable trainers with >4 yrs' experience are selected after a visit to the practice to check particularly health and safety protocols and an interview. Details of each regional scheme are advertised to final-year students or are available from the Regional Adviser in General Practice. Potential trainees are responsible for contacting practices on the approved list and arranging visits and interviews.

Contract A standard contract, which both parties are required to sign is available from the BDA and course organizers. The contract runs for 12 months, at the end of which each party is free to make their own arrangements. It includes a binding-out clause which prevents the trainee subsequently accepting as a patient someone he has treated at the practice, should they move to another practice.

Community Vocational Training This is also compulsory, and extends over 2 yrs. Each region has a nominated trainer. There is often some tie-in with the GDS VT course.

Audit and peer review

This has become a buzz word for the nineties—which is unfortunate, because clinical audit is simply an extension of good clinical practice.

What it is Medical audit (medical = dental) is the systematic critical analysis of the quality of clinical care. The analysis covers procedures used for diagnosis and treatment, the use of resources, and the outcome for the patient. Clinical audit is an extended term sometimes used to embrace the activities of all health-care personnel; this implies that medical (dental) audit is confined to the assessment by peer review of medical (dental) care.

What it is not Medical audit is not the same as its commercial equivalent, which is the assessment of an institution's financial status against best practice. It is not a management tool for dictating clinical policy to clinicians. It is not judgemental, and should not be used for disciplinary measures.

Aims and objectives These are, primarily, improvement of the quality of care provided. This is achieved by identifying less-than-adequate care and raising it to the standard of the agreed best.

General principles The basis is frank and open discussion without fear of criticism. It is essential to identify agreed standards against which the quality of care provided can be compared. These standards should not be fixed and immutable, so as to allow evolutionary change. It is imperative that the results of such discussion should lead to changes in practice when indicated. Confidentiality is an absolute prerequisite. Other crucial elements are genuine motivation and interest amongst the participating staff, honesty, and adequate data. Appropriate information technology, while of undoubted benefit is not absolutely essential. An 'audit cycle' is illustrated opposite.

Frequency Audit should be conducted regularly and not less than once a month, weekly being regarded as ideal.

Peer review The object of peer review is to improve the quality of care provided in dental practice by encouraging communication. It is ∴ audit for general dental practice. Possible topics include, cross-infection control, appointment systems, clinical methodology.

British Standards 5750 BS5750 can be gained by any manufacturer or service sector organization that can demonstrate quality management systems. A number of dental practices have successfully achieved this standard.

Audit cycle

Meeting to agree a 'gold standard' of care for a
specific condition.

$\downarrow$

Prospective analysis of care for that condition.

$\downarrow$

Meeting to compare actual standard with 'gold standard'.

Identify areas where deficiencies exist and make
recommendations.

Agree to implement recommendations for improvement.

$\downarrow$

Prospective analysis of care following implementation of
new recommendations.

$\downarrow$

Meeting to compare actual standard with 'gold standard'.

'Gold standard' not met—re-enter cycle.

'Gold standard' met—is 'gold standard' good enough?

If yes—recognize, and periodic review only.

If no—agree new standard and re-enter cycle.

Prevention of cross-infection

Cross-infection is the transmission of infectious agents between patients and staff within the clinical environment. Potential risks include not only hepatitis and HIV, but also other viruses (e.g. herpes) and bacteria (e.g. *Streptococcus pyogenes*). Transmission can occur by innoculation or inhalation. The BDA recommend:

1 A standard cross-infection control policy for all patients. This is advisable because it may not be possible to distinguish the healthy carrier. Also an effective policy will reduce the risks to the dentist and his staff, as well as being a practice builder.
2 Carriers of blood-borne viruses (hepatitis, HIV) can be treated in general dental practice, provided routine procedures are implemented rigorously.
3 Patients with manifestations of immunosuppression should be referred for specialist hospital care.

Routine cross-infection procedures for all patients

Immunization This is necessary against hepatitis B for all clinical staff, with a booster after 3 years (NB only 95% seroconvert). Also ? tuberculosis, rubella, tetanus.

Medical and social history Tact and discretion is required if truthful answers are to be obtained to sensitive questions.

Gloves[1] Should be worn routinely by dentist and DSA. The BDA recommend a new pair of gloves for every patient. Hands should be washed before gloving in a disinfectant solution and any cuts covered with a waterproof dressing. Hand cream should be used at the end of a session to prevent drying and cracking of the skin. Those with an allergy to latex can try vinyl instead, or wear undergloves of silk or nylon.

Surgery design and equipment[2] New and existing surgeries should include separate areas for dentist and DSA, within which are designated 'clean' zones. Layout and equipment must be planned to minimize the number of surfaces touched, e.g. taps or lights that can be turned on with infrared light switches or foot controls. Dental chairs and units are now designed with easy to clean surfaces.

Cleaning and sterilization of instruments Disposable instruments and cleaning materials should be used wherever possible. Manufacturers are exhibiting considerable ingenuity in this area (e.g. single-use diamond burs, disposable handpieces). Reusable items (including handpieces) must be cleaned and sterilized after use (p. 360). Chemical solutions only disinfect and should be restricted to those articles which cannot be sterilized by conventional methods. The use of plastic sheeting to cover handpiece and 3-in-1 tubing and equipment handles is recommended

734

1 M. A. Baumann 1992 *Int Dent J* **42** 170. 2 L. S. Worthington 1988 *BDJ* **165** 226.

by the BDA and is increasingly accepted by dentists, however, its use does increase the time taken between patients.

Treatment of work surfaces During use, instruments should only be placed on a sterilizable tray or disposable covering. Care is required to avoid contamination of areas which are difficult to disinfect.
- Apparently clean surfaces—70% alcohol.
- Contaminated non-metallic surfaces—10% chloros for 3 min.
- Contaminated metallic surfaces—2% gluteraldehyde for 3 min.
- Spillage of blood—hypochlorite granules (left to gel).
- At end of session—as for contaminated surfaces.

Aerosols These should be minimized by high volume suction. Gloves and masks should be worn. Aspirators and tubing should be flushed through daily with recommended disinfecting agent.

Disposal of sharps Care is required to prevent needlestick injuries, and preferably a re-sheathing device should be used. Sharps must be placed in a rigid box. Disposal can be arranged via FHSA. Needlestick injuries, p. 383.

Laboratory items All impressions and appliances should be rinsed before going to the lab. Alginates which contain a disinfectant are available. Impressions of known carriers should be disinfected, therefore rubber base or impression compound materials are preferable (see p. 675).

X-rays—the statutory regulations

The Ionising Radiation Regulations 1985

The Ionising Radiation (Protection of persons undergoing medical examination or treatment) Regulations 1988

With the introduction of the above on to the statute books, dentists now have to comply with the following:

1 All practitioners must have a certificate to show training in 'The Core of Knowledge', and are also responsible for ensuring that any staff who take X-rays also have adequate training.

2 Need to complete an inventory of X-ray equipment (including age, manufacturer, model, and serial number) and notify Health and Safety Executive.

3 Ensure that all equipment is regularly serviced and a radiation safety assessment carried out at least every 3 yrs (this can be done by post by the National Radiological Protection Board).

4 Appoint a Radiation Protection Supervisor to monitor day-to-day running of X-ray facilities. Can be either the dentist or a suitably trained person.

5 Ensure staff are not exposed to an X-ray dose >7.5 microsieverts per hour, to avoid the need to appoint a Radiation Protection Adviser. Whether this dose is exceeded can be determined by a radiation survey.

6 Have a set of local rules which must include name of Radiation Protection Supervisor, details of controlled areas and a contingency plan in the event of an equipment malfunction.

7 Having a designated controlled area, though in dental practice this is required only when an exposure is being made. For <70 kV machines the controlled area is a radius of 1 metre and for >70 kV 1.5 metres, except in the direction of the beam, where it extends until the beam is attenuated.

8 Keep radiation dose As Low As Reasonably Practicable (ALARP). This has several aspects:
- Be able to justify taking every radiograph.
- Avoid repetition by recording all X-rays taken in the patient's records, sending X-rays with patient for referrals.
- Use minimum skin to focus distance <70 kV = 10 cm, >70 kV = 20 cm.
- Good technique to ↑ diagnostic yield and ↓ need for repeat X-rays.
- Analyse reasons for poor films and rectify faults.

X-rays—practical tips and helpful hints

Practical tips

- When an exposure is made the raised dot on the packet should face the direction of the X-rays. When the processed film is viewed from the side with the raised dot, the patient's right is shown on the left of the film.
- In order of radiopacity: air, soft tissues, cartilage, immature bone, tooth-coloured fillings, mature bone, dentine and cementum, enamel, metallic restorations.
- View films in subdued lighting against an illuminated background.
- The oblique lateral mandible is a useful view for the angle of the mandible and the molar region that can be carried out with a dental X-ray set. Two views (left and right) can be taken on a single film if a lead sheet is placed over half the cassette. *Technique*: get patient to extend neck and protrude mandible (to ↓ superimposition of cervical spine), position cassette parallel to lower border of mandible and 2 cm below it, and centre tube 2 cm below and behind the contralateral angle of the mandible.
- For soft-tissue views (e.g. after trauma) a very short exposure (often below the lowest setting on the X-ray set) is required. A slower occlusal film may be more practical.
- The incisive foramen will have a parallax shift in relation to an incisor apex.

Processing

Manual This should be carried out either in a dark-room or a daylight processing tank. It is important to always keep the developer and fixer baths in the same order.

Automatic There are several dental types available which can process intra-oral and some extra-oral films.

Method

1 Standard regimen:
- Remove packaging and put film carefully in holder.
- Place in developer. Length of time varies according to the manufacturer and the temperature, but is usually between 3–4 min.
- Rinse.
- Place in fix for twice as long as it takes for film to clear.
- Wash. The recommended time is 10 min, but 2–3 min is adequate.

2 If a film is urgently required:
- Cut developing time to 30 sec.
- Rinse.
- Fix until clear.
- Wash.

Once film has been viewed it should be replaced in the developer for the standard time, and the remainder of the standard regimen followed.

NB Avoid contamination of the developer with fix and change the chemicals at least once a month.

Automatic film This is so called because developer and fix are combined in one solution (called a monobath) which is contained within an extension of the X-ray packet. Following exposure, the monobath is released into the film compartment to develop the X-ray. Although this method obviates the need for processing equipment, the results are inferior.

Film faults

Film dark	• fogged film—out of date film, poorly stored film
	• over-exposure
	• overdevelopment
Film pale	• under-exposed
	• underdeveloped—impatience, exhausted chemicals
Poor contrast	• overdevelopment
	• developer contaminated with fix
	• inadequate fixation &/or washing
Poor definition	• patient movement

The most common faults are over-exposure and underdevelopment.

17 Syndromes of the head and neck

Preface
Definitions
Albright syndrome
Apert syndrome
Behçet syndrome
Binder syndrome
Chediak–Higashi syndrome
Cri-du-chat syndrome
Cleidocranial dysostosis
Crouzon syndrome
Down syndrome
Eagle syndrome
Ehlers–Danlos syndrome
Frey syndrome
Gardener syndrome
Goldenhar syndrome
Gorlin–Goltz syndrome
Graves' disease
Heerfordt syndrome
Hemifacial microsomia
'Histiocytosis–X'
Horner syndrome
Hurler syndrome
Klippel–Feil anomalad
Larsen syndrome
Lesch–Nyhan syndrome
MAGIC syndrome
Marfan syndrome
Melkersson–Rosenthal syndrome
Multiple endocrine neoplasia
Papillon–Lefevre syndrome
Patterson-Brown–Kelly syndrome (Plummer–Vinson syndrome)
Peutz–Jeghers syndrome
Progeria
Ramsay Hunt syndrome
Reiter syndrome
Romberg syndrome
Sicca syndrome
Sjögren syndrome
Stevens Johnson syndrome
Sturge–Weber anomalad
Treacher–Collins syndrome
Trotter syndrome
von Recklinghausen neurofibromatosis

Principal sources: R. J. Gorlin 1990 *Syndromes of the Head and Neck*, OUP.

Preface

The aim of this section is neither to bemuse the reader nor to demonstrate esoteric knowledge, although both may appear to occur. The real importance behind the learning of names associated with conditions which may be of relevance to, or be picked up by, clinical examination of the head and neck, is that, once learned, some difficult diagnostic problems can be quickly solved and appropriate treatment instituted. We have retained eponyms where relevant, although this is not in keeping with current fashion as they are, we feel, easier to learn and certainly more fun. Examiners have a tendency to remember their favourite eponymous syndrome and it helps to at least agree on the name. We have, however, avoided the use of the possessive when using eponyms because invariably others were involved in describing or elucidating the condition and the syndrome does not belong to the individual(s) associated with it. The following list is in no way comprehensive but takes you through conditions met by the authors either in their clinical practice or in examinations and could therefore be considered worth knowing about.

Definitions

Malformation is a primary structural defect resulting from a localized error of morphogenesis.

Anomalad is a malformation *and* its subsequently derived structural changes.

Syndrome is a recognized pattern of malformation, presumed to have the same aetiology but not interpreted as the result of a single localized error in morphogenesis.

Association is a recognized pattern of malformation not considered to be a syndrome or an anomalad, *at the present time*.

Syndromes

Albright syndrome (McCune–Albright syndrome) Consists of polyostotic fibrous dysplasia (multiple bones affected), patchy skin pigmentation (referred to as *café-au-lait* spots) and an endocrine abnormality (usually precocious puberty in girls). Facial asymmetry affects up to 25% of cases.

Apert syndrome is a rare developmental deformity consisting of a craniosynostosis (premature fusion of cranial sutures) and syndactyly (fusion of fingers or toes). Severe mid-face retrusion leads to exopthalmos of varying severity. Early surgical intervention may be indicated for ↑ intracranial pressure or to prevent blindness from subluxation of the globe of the eye.

Behçet syndrome (pronounced behh-chet, after Turkish doctor Hulusi Behçet) is classically, oral ulceration, genital ulceration, and uveitis. Clinical diagnosis can be made on finding any two of these. It is, in fact, a multisystem disease of immunological origin, although no hard diagnostic tests are yet available. It tends to affect young adults, especially males and there is an association with HLA-B5. It undergoes spontaneous remission, although a variety of drugs including thalidomide are in use to treat it. See also p. 440.

Binder syndrome Maxillonasal dysplasia, severe midfacial retrusion and absent or hypoplastic frontal sinuses are the main features. There is no associated intellectual defect.

Chediak–Higashi syndrome is a combination of defective neutrophil function, abnormal skin pigmentation and ↑ susceptibility to infection (leading to severe gingivitis, periodontitis and aphthae in young children). It is a genetic disease.

Cri-du-chat is a chromosomal abnormality caused by deletion of part of the short arm of chromosome No. 5, resulting in microcephaly, hypertelorism, a round face with a broad nasal bridge and malformed ears. Associated laryngeal hypoplasia causes a characteristic shrill cry. There is associated severe mental retardation.

Cleidocranial dysostosis (cleidocranial dysplasia) is an autosomal dominant inherited condition consisting of hypoplasia or aplasia of the clavicles, delayed ossification of the cranial fontanelles, with a large, short skull. Associated features are shortness of stature, frontal and parietal bossing, failure to pneumatize the air sinuses, a high arched palate ± clefting, midface hypoplasia, and failure of tooth eruption with multiple supernumerary teeth. Many of the teeth present have inherent abnormalities such as dilaceration of roots or crown gemination. Hypoplasia of 2° cementum may occur. The condition mainly, though NOT exclusively, affects membraneous bone.

Crouzon syndrome is the commonest of the craniosynostoses. It is an autosomal dominant condition consisting of premature fusion of cranial sutures, mid-face hypoplasia and, due to this, shallow orbits with proptosis of the globe of the eye. Radiographically the appearance of a 'beaten copper skull' is characteristic. The enlarging brain is entrapped by the prematurely fused sutures, and raised intracranial pressure can lead to cerebral damage and resulting intellectual deficiency. This and the risk of blindness justify early craniofacial surgery to correct the deformity.

Down syndrome (trisomy 21) is the commonest of all malformation syndromes, affecting up to 1 in 600 births. Risk ↑ with maternal age. Downs' children account for ⅓ of severely mentally handicapped children. Facial appearance is characteristic with brachycephaly, mid-face retrusion, small nose with flattened nasal bridge and upward sloping palpebral fissures (mongoloid slant). There is relative macroglossia and delayed eruption of teeth. Major relevant associations are heart defects, atlanto-axial subluxation, anaemia, and an ↑ risk of leukaemia. Most Downs' children and adults are extremely friendly and co-operative.

Eagle syndrome is dysphagia and pain on chewing and turning the head associated with an elongated styloid process.

Ehlers–Danlos syndrome Comprises a group of disorders characterized by hyperflexibility of joints, ↑ bleeding and bruising, and hyperextensible skin. There appears to be an underlying molecular abnormality of collagen in this inherited disorder. Bleeding is commoner in type IV, early onset periodontal disease in type VIII. Pulp stones may be seen in all types.

Frey syndrome is a condition in which gustatory sweating and flushing of skin occurs. It follows trauma to skin overlying a salivary gland and is thought to be due to post-traumatic crossover of sympathetic and parasympathetic innervation to the gland and skin respectively. Its frequency following superficial parotidectomy ranges from 0–100% depending on which surgeon you are talking to, but is almost certainly present in all cases to some degree if looked for carefully enough.

Gardener syndrome This comprises multiple osteomas (particularly of the jaws and facial bones), multiple polyps of the large intestine, epidermoid cysts, and fibromas of the skin. It shows autosomal dominant inheritance. The discovery on clinical or X-ray examination of facial osteomas mandates examination of the lower gastrointestinal tract, as these polyps have a tendency to rapid malignant change. This is a highly 'worthwhile' syndrome.

Goldenhar syndrome is a variant of hemifacial microsomia and consists of microtia (small ears), macrostomia, agenesis of the mandibular ramus and condyle, vertebral abnormalities (e.g. hemivertebrae), and epibulbar dermoids. Also, cardiac, renal or skeletal abnormalities can occur. Up to 10% of patients may have mental handicap—in other words, 90% DON'T.

Gorlin–Goltz syndrome (multiple basal cell naevi syndrome) see naevus, p. 546. Consists of multiple basal cell carcinomas (epitheliomas), multiple jaw cysts (odontogenic keratocysts), vertebral and rib anomalies (usually bifid ribs), and calcification of the falx cerebri. Frontal bossing, mandibular prognathism, hypertelorism, hydrocephalus, eye and endocrine abnormalities have also been noted.

Graves' disease Autoantibodies to thyroid stimulating hormone (TSH) causes hyperthyroidism with opthalmopathy, women 30–50 yrs commonly affected, exophthalmos can be helped by craniofacial approach to orbital decompression.

Heerfordt syndrome (uveoparotid fever) is sarcoidosis with associated lacrimal and salivary (especially parotid) swelling, uveitis and fever. Sometimes there are associated neuropathies, e.g. facial palsy.

Hemifacial microsomia Prevalence: 1 in 5000 births. Bilateral in 20% of cases. Congenital defect characterized by lack of hard and soft tissue on affected side(s) usually in the region of the ramus and external ear (i.e. 1st and 2nd branchial arches). Wide spectrum of ear and cranial deformities found.

Histiocytosis—X Really 3 broad groups of diseases with the histological feature of tissue infiltration by tumour-like aggregates of macrophages (histiocytes) and eosinophils.
1 Solitary eosinophilic granuloma, mainly affects males <20. Mandible a common site. Responds to local treatment.

2 Hand–Schüller–Christian disease, multifocal eosinophilic granuloma causing skull lesions, exopthalmos and diabetes insipious. Affects younger group. May respond to cytotoxic chemotherapy.

3 Letterer–Siwe disease, rapidly progressive disseminated histiocytosis. Associated pancytopenia and multisystem disease can be fatal.

Horner syndrome Consists of a constricted pupil (miosis), drooping eyelid (ptosis), unilateral loss of sweating (anhydrosis) on the face, and occasionally sunken eye (enopthalmos). It is caused by interruption of sympathetic nerve fibres at the cervical ganglion secondary to, e.g. bronchogenic carcinoma, invading the ganglion or neck trauma. Scores high on the 'worthwhile' rating.

Hurler syndrome is a mucopolysaccharidosis causing growth failure and mental retardation. A large head, frontal bossing, hypertelorism, and coarse features give it its classical appearance. Multiple skeletal abnormalities (dysostosis multiplex), corneal clouding, and serum and urinary acid mucopolysaccharide abnormalities also occur.

Klippel–Feil anomalad is the association of cervical vertebrae fusion, short neck, and low-lying posterior hairline. A number of neurological anomalies have been noted and unilateral renal agenesis is frequent. Cardiac anomalies sometimes occur.

Larsen syndrome is a mainly autosomal recessive condition, with a predilection for females, consisting of: cleft palate, flattened facies, multiple congenital dislocations, and deformities of the feet. Sufferers are usually of short stature.

Lesch–Nyhan syndrome is a defect of purine metabolism causing mental retardation, spastic cerebral palsy, choreoathetosis, and aggressive self multilating behaviour (particularly involving the lips).

MAGIC syndrome stands for Mouth And Genital ulcers and Interstitial Chondritis and is a variant of the Behçet syndrome group.

Marfan syndrome is an autosomal dominant condition characterized by tall, thin stature and arachnodactyly (long, thin spider-like hands), dislocation of the lens, dissecting aneurysms of the thoracic aorta, aortic regurgitation, floppy mitral valve, and high arched palate. Joint laxity is also common. This condition is highly prevalent among top-class basketball and volleyball players, for obvious reasons.

Melkerson–Rosenthal syndrome Consists of facial paralysis, facial oedema, and fissured tongue. It is probably a variant of the group of conditions now known as oro-facial granulomatosis.

Multiple endocrine neoplasia are a recently re-classified group of conditions affecting the endocrine glands. MEN 3 (sometimes called 2b) is of particular relevance as it consists of multiple mucosal neuromas which have a characteristic histopathology, pheochromocytoma, medullary thyroid carcinoma, and a thin wasted appearance. Calcitonin levels are elevated if medullary thyroid carcinoma is present. Index of suspicion should be high in tall, thin, wasted-looking children and young adults presenting with lumps in the mouth. Biopsy is mandatory and if histopathology is suggestive the thyroid must be adequately investigated. This is another 'worthwhile' syndrome.

Papillon–Lefevre syndrome is palmoplantar hyperkeratosis and juvenile periodontitis which affects both 1° and 2° dentition. Normal dental development occurs until the appearance of the hyperkeratosis of the palms and soles, then simultaneously an aggressive gingivitis and periodontitis begin. The mechanism is not well understood.

Patterson-Brown–Kelly syndrome (Plummer–Vinson syndrome) is the occurrence of dysphagia, microcytic hypochromic anaemia, koilonychia (spoon-shaped nails), and angular cheilitis. The dysphagia is due to a post cricoid web, usually a membrane on the anterior oesophageal wall, which is pre-malignant. The koilonychia and angular cheilitis are secondary to the anaemia but may be presenting symptoms. The main affected group are middle-aged women, and correction of the anaemia may both relieve symptoms and prevent malignant progression of the web.

Peutz–Jeghers syndrome is an autosomal dominant condition of melanotic pigmentation of skin (especially peri-oral skin) and mucosa and intestinal polyposis. These polyps, unlike those of the Gardener syndrome, have no particular propensity to malignant change, being hamartomatous, and are found in the small intestine. They may, however, cause intussusception or other forms of gut obstruction. Ovarian tumours are sometimes associated with the condition (10% woman with Peutz–Jeghers).

Progeria is probably a collagen abnormality. It causes dwarfism and premature ageing. Characteristic facial appearance occurs due to a disproportionately small face with mandibular retrognathia and a beak-like nose, creating an unforgettable appearance; death occurs in the mid teens.

Ramsay Hunt syndrome is a lower motor neuron facial palsy with vesicles on the same side in the pharynx, external auditory canal, and on the face. It is thought to be due to herpes zoster of the geniculate ganglion.

Reiter syndrome Consists of arthritis, urethritis, and conjunctivitis. There are frequently oral lesions which resemble benign migratory glossitis in appearance, but affect other parts of the mouth. The condition is probably an unwanted effect of an immune response to a low-grade pathogen; however, some still believe it to be a sexually transmitted disease, although there is no hard evidence for this.

Romberg syndrome (hemifacial atrophy) Consists of progressive atrophy of the soft tissues of half the face, associated with contralateral Jacksonian epilepsy and trigeminal neuralgia.

Rarely, half the body may be affected. It starts in the first decade and lasts about 3 yrs before it becomes quiescent.

Sicca syndrome (primary Sjögren syndrome) is xerostomia and keratoconjunctivitis sicca, i.e. dry mouth and dry eyes. There is an ↑ risk of developing parotid lymphoma with this condition. Interestingly, although Sicca syndrome has certain serological abnormalities in common with systemic connective tissue disorders such as rheumatoid arthritis, it does not have any of the symptomatology (unlike Sjögren syndrome).

Sjögren syndrome (secondary Sjögren syndrome) in addition to dry eyes and dry mouth has both serology and symptomatology of an autoimmune condition, usually rheumatoid arthritis, but sometimes systemic lupus erythematosus, systemic sclerosis, or primary biliary cirrhosis. Actual swelling of the salivary glands is relatively uncommon and late onset swelling of the parotids may herald the presence of a lymphoma.

Stevens Johnson syndrome is a severe version of erythema multiforme, a muco-cutaneous condition probably autoimmune in nature and precipitated particularly by drugs. Classical signs are the target lesions, concentric red rings which especially affect the hands and feet. Stevens Johnson syndrome is said to be present when the condition is particularly severe, and is associated with fever and multiple mucosal involvement. Viral infections, e.g. Herpes simplex, are the second commonest cause.

Sturge–Weber anomalad is due to a hamartomatous angioma affecting the upper part of the face which extends intracranially. There may be associated convulsions, hemiplegia (on the contralateral side of the body), or intellectual impairment. The risks of surgery are obvious.

Treacher–Collins syndrome (mandibulofacial dysostosis) Basically involves defects in structures derived from the first branchial arch. It is inherited as an autosomal dominant trait with variable expressivity, and consists of downward sloping (antimongoloid slant) palpebral fissures, hypoplastic malar complexes, mandibular retrognathia with a high gonial angle, deformed pinnas, hypoplastic air sinuses, colobomas in the outer third of the eye, middle and inner ear hypoplasia (and hence deafness). 30% have cleft palates and 25% have an unusual tongue-like projection of hair pointing towards the cheek. Most have *completely normal intellectual function* which may be missed because they are deaf and are 'funny-looking kids' (how would you like to be called a funny-looking dentist/doctor?) These people may well miss out on fulfilling their potential because of society's (and some professionals') attitude to deformity. This syndrome is a prime indication for corrective craniofacial surgery.

Trotter syndrome is unilateral deafness, pain in the mandibular division of the trigeminal nerve, ipsilateral immobility of the palate, and trismus, due to invasion of the lateral wall of the nasopharynx by malignant tumour. Pterygopalatine fossa syndrome is a similar condition where the first and second divisions of the trigeminal are affected.

von Recklinghausen neurofibromatosis/syndrome Multiple neurofibromas with skin pigmentation, skeletal abnormalities, central nervous system involvement, and a predisposition to malignancy are the basics of this syndrome. It undergoes autosomal dominant transmission and has a large and varied number of manifestations. Lesions of the face can be particularly disfiguring.

18 Useful information and addresses

Tooth notation   752
Some qualifications in Medicine and
  Dentistry   754
Bur numbering systems   756
File sizes   756
Gas cylinder colour coding   758
Wire gauges   758
Passing exams   760
Useful addresses   762

Tooth notation

FDI

Permanent teeth

$$R \quad \frac{18\ 17\ 16\ 15\ 14\ 13\ 12\ 11 \quad | \quad 21\ 22\ 23\ 24\ 25\ 26\ 27\ 28}{48\ 47\ 46\ 45\ 44\ 43\ 42\ 41 \quad | \quad 31\ 32\ 33\ 34\ 35\ 36\ 37\ 38} \quad L$$

Deciduous teeth

$$R \quad \frac{55\ 54\ 53\ 52\ 51 \quad | \quad 61\ 62\ 63\ 64\ 65}{85\ 84\ 83\ 82\ 81 \quad | \quad 71\ 72\ 73\ 74\ 75} \quad L$$

Zsigmondy–Palmer, Chevron, or Set Square system

Permanent teeth

$$R \quad \frac{8\ 7\ 6\ 5\ 4\ 3\ 2\ 1 \quad | \quad 1\ 2\ 3\ 4\ 5\ 6\ 7\ 8}{8\ 7\ 6\ 5\ 4\ 3\ 2\ 1 \quad | \quad 1\ 2\ 3\ 4\ 5\ 6\ 7\ 8} \quad L$$

Deciduous teeth

$$R \quad \frac{e\ d\ c\ b\ a \quad | \quad a\ b\ c\ d\ e}{e\ d\ c\ b\ a \quad | \quad a\ b\ c\ d\ e} \quad L$$

European

Permanent teeth

$$R \quad \frac{8+\ 7+\ 6+\ 5+\ 4+\ 3+\ 2+\ 1+ \quad | \quad +1\ +2\ +3\ +4\ +5\ +6\ +7\ +8}{8-\ 7-\ 6-\ 5-\ 4-\ 3-\ 2-\ 1- \quad | \quad -1\ -2\ -3\ -4\ -5\ -6\ -7\ -8} \quad L$$

Deciduous teeth

$$R \quad \frac{05+\ 04+\ 03+\ 02+\ 01+ \quad | \quad +01\ +02\ +03\ +04\ +05}{05-\ 04-\ 03-\ 02-\ 01- \quad | \quad -01\ -02\ -03\ -04\ -05} \quad L$$

American

Permanent

$$R \quad \frac{1\ 2\ 3\ 4\ 5\ 6\ 7\ 8 \quad | \quad 9\ 10\ 11\ 12\ 13\ 14\ 15\ 16}{32\ 31\ 30\ 29\ 28\ 27\ 26\ 25 \quad | \quad 24\ 23\ 22\ 21\ 20\ 19\ 18\ 17} \quad L$$

Deciduous

$$R \quad \frac{A\ B\ C\ D\ E \quad | \quad F\ G\ H\ I\ J}{T\ S\ R\ Q\ P \quad | \quad O\ N\ M\ L\ K} \quad L$$

Some qualifications in Medicine and Dentistry

BA	Bachelor of Arts
BCh	Bachelor of Surgery
BChD	Bachelor of Dental Surgery
BDS	Bachelor of Dental Surgery
BS	Bachelor of Surgery
ChM	Master of Surgery
DCH	Diploma in Child Health
DDOrth	Diploma in Dental Orthopaedics
DDPH	Diploma in Dental Public Health
DDR	Diploma in Dental Radiology
DDS/DDSc	Doctor of Dental Surgery/Science
DGDP	Diploma in General Dental Practice
DOrth	Diploma in Orthodontics
DPhil	Doctor of Philosophy
DRD	Diploma in Restorative Dentistry
DSc	Doctor of Science
FDS	Fellowship in Dental Surgery
FFD	Fellow in Faculty of Dental Surgery
FRCGP	Fellow of Royal College of General Practitioners
FRCP	Fellow of Royal College of Physicians
FRCS	Fellow of Royal College of Surgeons
LDS	Licentiate in Dental Surgery
MB	Bachelor of Medicine
MCCD	Membership in Clinical Community Dentistry
MCDH	Master of Community Dental Health
MD	Doctor of Medicine
MDOrth	Membership in Dental Orthopaedics
MDS	Master of Dental Surgery
MDentSci/MDSc	Master of Dental Science
MGDS	Membership in General Dental Surgery
MOrth	Membership in Orthodontics
MPhil	Master of Philosophy
MRCGP	Member of Royal College of General Practitioners
MRCS	Member of Royal College of Surgeons
MSc	Master of Science
PhD	Doctor of Philosophy

Bur numbering systems

Maximum diameter of bur head	ISO	UK		
		round	inverted cone	flat fissure
0.6	006	$\frac{1}{2}$	$\frac{1}{2}$	
0.8	008	1	1	$\frac{1}{2}$
0.9	009			1
1.0	010	2	2	2
1.2	012	3	3	3
1.4	014	4	4	4
1.6	016	5	5	5
1.8	018	6	6	6
2.1	021	7	7	7
2.3	023	8	8	8
2.5	025	9	9	9
2.7	027	10	10	10
2.9	029	11		11
3.1	031	12		12

File sizes (for endodontic therapy)

Size	Tip diameter	ISO colour
08	0.08	grey
10	0.10	purple
15	0.15	white
20	0.20	yellow
25	0.25	red
30	0.30	blue
35	0.35	green
40	0.40	black
45	0.45	white
50	0.50	yellow
55	0.55	red
60	0.60	blue
70	0.70	green
80	0.80	black
90	0.90	white
100	1.00	yellow
110	1.10	red
120	1.20	blue
130	1.30	green
140	1.40	black

Gas cylinder colour coding

Name of gas		Colour of cylinder body	Colour of valve end (where different from body)
Oxygen		Black	White
Nitrous oxide		Blue	
Cyclopropane		Orange	
Carbon dioxide		Grey	
Ethylene		Violet	
Nitrogen		Grey	Black
Oxygen + Carbon dioxide	mix	Black	White and grey
Oxygen + Nitrous oxide	mix	Blue	Blue and black
Air (medical)		Grey	White and black

Wire gauges

Size on wire gauge	Diameter	
	Imperial (inches)	Metric (mm)
14	0.080	2.0
19	0.040	1.0
20	0.036	0.9
21	0.032	0.8
22	0.028	0.7
23	0.024	0.6
24	0.022	0.55
25	0.020	0.5
26	0.018	0.45
27	0.0164	0.4
28	0.0148	0.37
29	0.0136	0.35
30	0.0124	0.3

18 Useful information and addresses

Gas cylinder colour coding

Passing exams

Preparation First, there is unfortunately no substitute for the sheer hard work of revising, and to have a chance of being adequate this should start well before the exam. Beware of those who try to impress by saying they have passed every exam, first time, having only started revising the night before. These people are not super-intelligent (in fact, their behaviour suggests the reverse) they have probably only been lucky . . . to date!

Equally well the human brain is not a computer and revising for hours, day in day out, could result in 'burn-out'. Each person needs to find their own method, but a break of 5–10 min every hour and a complete hour off after 3 hours' work, has succeeded for others.

It is important to be sure that you are revising the correct material in sufficient depth. One way of ensuring this is to go over past questions. If some are tackled under exam conditions, this also gives practice in exam technique.

Immediate pre-exam period If your nerve is strong enough it is probably wise to stop working 2–3 days before the exams start and have a rest, doing something that you enjoy. Most people baulk at this suggestion, but see the logic of a marathon runner resting before an important race. Studying the night before is akin to an élite runner going out the night before the olympic marathon and running 26.2 miles just to make sure he can cover the distance.

Written exams It is important to know beforehand how many questions you have to answer and ∴ how long can be spent on each. Do not be tempted to overrun. Spend a few minutes deciding which questions to answer and in what order. It is better to tackle your best question second, thus allowing the butterflies to settle. Read each question carefully, and periodically stop to check that you are answering the question set, not the one you'd like to answer. A rough plan helps to produce a more logical essay and provides an aide memoire in case of panic halfway through. Write neatly.

Multiple choice exams are beloved of examiners, but feared by candidates. As they involve negative marking, a careful approach is required for success.
1 Do the necessary revision, it is impossible to waffle hopefully round the subject in a multiple choice exam.
2 Go through the paper, only answering those questions of which you are completely sure.
3 Count up the number you have answered and see if it is over the pass mark. If yes, stop there.
4 If you have not answered enough, go back through the paper, only answering those questions which you are nearly sure of. DO NOT look at the questions you have previously answered as usually your first instinct is right.
5 Stop there, as the next step is to guess—which is usually counter-productive.

Clinical exams Turn up neatly dressed, armed with a clean white coat. Although some patients are regularly wheeled out for exams, many will be almost as nervous about what is going to happen as you are. Many patients can give helpful hints towards their diagnosis and management if asked the right questions. Some candidates spend the whole of their time with the patient, scribbling furiously, and do not leave themselves any time to think. Present the patient to the examiners in a logical fashion, i.e. starting with their name and age, not the radiographs.

GOOD LUCK!

Useful addresses

Birmingham Dental School, St. Chad's Queensway, Birmingham B4 6NN. Tel: 0121-236 8611.

Bristol Dental School, Lower Maudlin Street, Bristol BS1 2LY. Tel: 0117 9230050.

British Council, 65 Davies Street, London W1Y 2AA. Tel: 0171-499 8011.

British Dental Association, 64 Wimpole Street, London W1M 8AL. Tel: 0171-935 0875.

British Dental Health Foundation, Eastlands Court, St. Peter's Road, Rugby, Warwickshire CV21 3QP. Tel: 01788-546 365.

British Library, Boston Spa, Wetherby, N. Yorkshire LS23 7BQ. Tel: 01937 843434.

British Medical Association, BMA House, Tavistock Square, London WC1H 9JP. Tel: 0171-387 4499.

British Postgraduate Medical Federation, 33 Millman Street, London WC1N 3EJ. Tel: 0171-831 6222.

Cardiff Dental School, Heath Park, Cardiff CF4 4XY. Tel: 01222 755944.

Cork Dental School, John Redmond Street, Wilton, Cork, Eire. Tel: 010-353-21 545100.

Council for Postgraduate Medical Education, 8 Marylebone Road, London NW1 5HH. Tel: 0171-323 1289.

Dental Practice Board, Eastbourne, East Sussex BN20 8AD. Tel: 01323 417000.

Department of Health, Chief Dental Officer, Richmond House, 79 Whitehall, London SW1A 2NS. Tel: 0171-210 5247.

Dublin Dental School, Dublin University, Trinity College, Dublin 2. Tel: 0001-772941.

Dundee Dental School, Park Place, Dundee DD1 4HN. Tel: 01382 60111.

Edinburgh Dental School, Chambers Street, Edinburgh EH1 1JA. Tel: 0131-225 9511.

European Dental Society, 10 Pike End, Eastcote, Pinner HA5 2EX. Tel: 0181-868 0837.

Faculty of General Dental Practoners (UK), The Royal College of Surgeons of England, 35–43 Lincolns' Inn Fields, London WC2A 3PN. Tel: 0171-831 6999.

General Dental Council, 37 Wimpole Street, London W1M 8DQ. Tel: 0171-486 2171.

General Dental Practitioners' Association, GDPA House, High Street, Thorpe-le-Soken, Clacton-on-Sea, Essex CO16 0DY. Tel: 01255 861829.

Glasgow Dental School, 378 Sauchiehall Street, Glasgow G2 3JX. Tel: 0141-332 7020.

Guy's Hospital Dental School, Guy's Tower, London Bridge, London SE1 9RT. Tel: 0171-955 5000.

Health Education Authority, Hamilton House, Marbledon Place, London WC1H 9TX. Tel: 0171-383 3833.

HMSO, 51 Nine Elm's Lane, London SW8 SDR. Tel: 0171-873 0011.

London Hospital Medical College Dental School, University of London Dental School, Turner Street, London E1 2AD. Tel: 0171-377 7000.

Institute of Dental Surgery, Eastman Dental Hospital, 256 Gray's Inn Road, London WC1X 8LD. Tel: 0171-915 1000.

King's Healthcare Dental Hospital, Caldecot Road, Camberwell, London SE5 9RW. Tel: 0171-346 3262.

Leeds Dental Institute, Dental and Medical Building, Clarendon Way, Leeds LS2 9LU. Tel: 0113 2440111.

Liverpool Dental School, Pembroke Place, PO Box 147, Liverpool L69 3BX. Tel: 0151-706 2000.

London School of Hygiene and Tropical Medicine, Keppel Street, London WC1E 7HT. Tel: 0171-636 8636.

Manchester Dental School, University Dental Hospital of Manchester, Bridgeford Street, Manchester M15 6FH. Tel: 0161-276 4396.

Medical and Dental Defence Union of Scotland, 144 West George Street, Glasgow G2 2HW. Tel: 0141-221 5858.

Medical Defence Union, 3 Devonshire Place, London W1N 2EA. Tel: 0171-486 6181.

Medical Protection Society, 50 Hallam Street, London W1N 6DE. Tel: 0171-637 0541.

National Radiological Protection Board, Chilton, nr Didcot, Oxon OX11 0RQ. Tel: 01235 831600.

Newcastle Dental School, Framlington Place, Newcastle-upon-Tyne NE2 4BW. Tel: 0191-222 6000.

Queen's University of Belfast, School of Dentistry, Grosvenor Road, Belfast BT12 6BP. Tel: 01232 240503.

Royal College of Physicians and Surgeons of Glasgow, 234–242 St. Vincent Street, Glasgow G2 5RJ. Tel: 0141-221 6072.

Royal College of Surgeons of England, 35–43 Lincolns' Inn Fields, London WC2A 3PN. Tel: 0171-405 3474.

Royal College of Surgeons of Edinburgh, Nicolson Street, Edinburgh EH8 9DW. Tel: 0131-556 6206.

Royal College of Surgeons in Ireland, 123 St. Stephen's Green, Dublin 2, Eire. Tel: 0001-780 200.

Royal Society of Medicine, 1 Wimpole Street, London W1M 8AE. Tel: 0171-408 2119.

Scottish Dental Practice Board, Trinity Park House, South Trinity Road, Edinburgh EH5 3SG. Tel: 0131-552 6255.

Sheffield Dental School, Charles Clifford Dental Hospital, Wellesley Road, Sheffield S10 2SZ. Tel: 0114 2670444.

The Fluoridation Society, 63 Wimpole Street, London W1M 8AL. Tel: 0171-486 7007.

World Health Organization, Avenue Appia, CH-1211, Geneva 27, Switzerland. Tel: 022-91 3453.

Index

abdominal examination 10
abducens nerve (VI) 542
abrasion, tooth 304
abscess
  acute periapical, *see* peri-
    apical abscess, acute
  formation during pulp
    therapy 104
  periodontal 220–1, 256,
    408
  phoenix 328
absorbents 258
abutment teeth
  bridges 292, 293–4
  caries 371
  overdentures 370, 371
  periodontally involved 296
  pier 296
  taper and parallelism 294
  tilted 296–7
acanthosis nigricans 549
accessory nerve (XI) 542
Access to Health Records Act
  (1991) 706
achondroplasia 538
acid-etch composite tip
  technique 112
acid-etch technique 662
acid pumice abrasion
  technique 78
acinic cell tumour, salivary
  glands 506
acne 548
acquired immune deficiency
  syndrome, *see* AIDS
acromegaly 468, 534
acrylic (resin) 684–5
  allergic reactions 688
  complete dentures 354,
    358
  heat-cure 684
  laminate veneers 288
  light-cure 684
  partial dentures 336–7
  plasticized 686
  properties 684–5
  self-cure 684
actinic cheilitis 453
actinic keratosis 546
Actinobacillus
    actinomycetemc-
    omitans 202, 208,
    222, 223
Actinomyces 38
Actinomyces israelii 202
actinomycosis 409
acyclovir 128, 436, 618
Adams crib 182, 183
Addison disease 468, 534,
  594

addresses, useful 762–3
adenoameloblastoma 420
adenocarcinoma, salivary
  glands 506
adenoid cystic carcinoma,
  salivary glands 456,
  506
adenolymphoma, salivary
  glands 456, 506
adenomas, salivary glands
  456, 506
ADH, inappropriate
  secretion 534
adrenalin 554, 562, 608
adrenocortical hypofunction,
  primary (Addison
  disease) 468, 534,
  594
advanced trauma life support
  (ATLS) 484–5
adverse drug reactions, *see*
  drug reactions
advertising 703, 722
aerosols 382, 735
affective disorders 550
affidavit 690
age
  changes 374–5
  restoration of primary teeth
    and 82
AIDS 552–3
  dental care 473
  oral features 439, 472–3
  treatment 473, 618
  see also HIV infection
airways
  chronic obstructive disease
    (COAD) 526
  inhaled foreign bodies 572
  management, resuscitation
    484, 486, 560
  protection 259
Albright syndrome 538, 742
alginate 672
allergic contact eczema 548
allergic reactions
  dental materials 688, 689
  drugs 459, 562
alloys
  casting 676–7
  for dentures 336–7, 354,
    678, 679
  porcelain bonding 677
  semi-precious 676
  soft, crowns 306
  wrought 678–9
  see also gold (alloy)
alveolar bone
  fractures 116, 388
  permeability in children 80

alveolar ridge
  augmentation 426–7, 500
  combination mucosal flap
    and epithelial inlay
    500
  epithelial inlay 500
  flabby (fibrous)
    impression taking 351
    surgical repair 426
  grossly resorbed lower 363
alveolotomy, interseptal 426
alveolus, closure of cleft
    anterior 502
amalgam 656–7
  allergic reactions 688
  bonding agents 657
  cavity preparation 261
  class I restorations 88, 262
  class II restorations 90–1,
    264
  class V restorations 268
  handling characteristics
    656
  lining cements 671
  marginal leakage 657
  spillage 716
  toxicity 657
  types 656, 657
ameloblastic fibroma 421
ameloblastoma 420
amelogenesis imperfecta 74–
    5
American tooth notation
    system 752
amethocaine 630
amitriptyline 608
amoxycillin 128, 614
  dento-oral infections 221,
    223, 408, 409
  prophylactic use 590, 591
amphotericin 128, 364, 618
ampicillin and cloxacillin
    (Ampiclox) 128
amyloidosis 520
anaemia 466, 518
anaesthesia
  general, see general
    anaesthesia
  local, see local anaesthesia
  topical 622, 630
anaesthetic agents
  inhalational 642, 649
  intravenous 642
analgesia 588–9
  acute pain 588
  definition 626
  maxillofacial trauma 486,
    588
  oral surgery 380
  patient-controlled 589

  post-operative 577, 588
  pre-emptive 589
  relative (RA) 64–5, 636–7
  terminal disease 588
  see also local anaesthesia
analgesics 588
  general dental practice 602
  hospital practice 604–5
  sugar-free 128
  topical 622, 630
anaphylactic shock 562
anchorage 180–1
  loss 181
  re-inforcing 180–1
ancillary staff, see staff
Andrew's straight wire
    appliance 190
'angina bullosa
    haemorrhagica' 466
angina pectoris 461, 524,
    558
angio-oedema
  allergic 453, 562
  hereditary 453
angular cheilitis 438, 466
ankyloglossia 452
ankylosing spondylitis 541
ankylosis
  primary teeth 111
  re-implanted tooth 121
  temporomandibular joint
    498
anodontia 72
anorexia nervosa 551
antacids, sugar-free 128
anterior nasal spine 141
anterior open bite (AOB)
    136, 170
Ante's law 293–4
antibacterial agents,
    toothpastes 37
antibiotic-induced colitis
    528–9, 616
antibiotics 614–17
  commensal oral flora and
    458
  hospital practice 616–17
  maxillary sinusitis 422
  periodontal disease 228,
    230–1
  principles of use 614
  prophylaxis 577, 590–1
  sugar-free 128–9
antibiotic/steroid paste
    (Ledermix) 103, 271,
    316
anticalculus agents 37
anticoagulants 522–3, 590
anticonvulsant drugs 128,
    644

anticurvature filing 320
antidepressants 480, 608
antidiuretic hormone (ADH),
    inappropriate
    secretion 534
antidotes, fluoride 35
antiemetics 577, 588, 596,
    610
antiepileptic drugs 128, 644
antifungal drugs 364, 618
antihistamines 128, 620
antihypertensive drugs 644
anti-inflammatory drugs
    606–7
antimetabolites, periodontal
    disease 228
antisepsis 382
antiseptics
    periodontal disease 230–1
    plaque control 38
    root canal therapy 316
antiviral drugs 618–19
anxiety
    neurosis 551
    oral sedation 640
    preventing uptake of dental
      care 50
anxiolytics 612–13
anxious child 64–5
Apert syndrome 742
apex locators 318
aphthous ulcers
    major 440
    minor 440
    recurrent 440–1, 466, 476
apical abscess, see periapical
    abscess, acute
apically repositioned flap
    236–8
apical transportation
    (zipping) 319
apicectomy 402–3, 404
A point 141
appliances, orthodontic, see
    orthodontic appliances
appointments
    running late 720
    timing, elderly patients 376
archwires, fixed appliances
    188
Ardee liner 359
argon laser 430
arterial puncture 578
artery, lacerated, local
    anaesthesia causing
    634
arthrography 23
arthroscopy, temporo-
    mandibular joint 480–
    1

articulators 277, 353
artificial saliva 622
artificial teeth
    mould and shade 352
    position on dentures 352,
      354–5
asepsis 382
aspirator 384
aspirin 128, 558, 588, 602
assistantships/associateships
    698
asthma 526
    acute 570
atracurium 643
attrition, tooth 304
audit 732–3
    clinical 732
    cycle 733
    medical (dental) 732
auroscopy 12
autoimmune disease 552
Automixin dispenser 675
auxiliary personnel, see staff
AVPU test 484
avulsion 120–1
    delayed presentation 121
    management 120
    primary teeth 110
    prognosis 120
    sequelae 121
azathioprine 607
azlocillin 590

bacteria
    methods of control 38–9
    role in caries 38
bacterial infections
    children 126
    mouth 434–5
bacteriology 17
bacteriostatic dressings 671
Bacteroides 408
Bacteroides gingivalis
    (melaninogenicus), see
    Porphyromonas
    gingivalis
Bacteroides intermedius, see
    Prevotella intermedia
balanced force technique
    320, 324
Ballard's conversion 143
bands, fixed appliances 188
banking of takings 718
basal cell carcinoma
    cryosurgery 412
    skin 546
basal cell naevi syndrome,
    multiple 744
baseplate, removable
    appliances 182

Basic Periodontal
  Examination (BPE)
  210
beechwood creosote 101,
  105
Begg appliance 190
behaviour management,
  children 64
behaviour problems 65
behaviour shaping 64
Behçet syndrome 440–1, 742
Bell's palsy 461
benzocaine lozenges 622,
  630
benzodiazepines 554, 612,
  638–41
  drug interactions 638
  intravenous 566, 638,
    640–1
  oral sedation 640
  reversal 639
benzylpenicillin 614
beryllium toxicity 688
betamethasone 606
betel nut/quid chewing 448,
  450
bilirubin, urinary 16
bimaxillary proclination 136
Binder syndrome 743
biochemistry, blood 16–17
biocompatibility, dental
  materials 688–9
Bionator appliance 194
biopsy 410
bisecting angle technique,
  X-rays 20
bite-marks 708
bitewing radiographs 20, 32
bladder catheterization 584
bleaching 310–11
  non-vital 310–11
  vital 310
bleeding disorders 522
bleeding/haemorrhage
  intraoral, leukaemia 466
  management 484, 486,
    580
  mid-face fractures 492
  post-operative 390–1
  retrobulbar 492
bleeding index 210
blisters
  intraepithelial 442
  preceding oral ulcers 476
  subepithelial 442
blood
  autologous, for transfusion
    582
  urinary 16
  whole, for transfusion 582

blood count, full 16, 576
blood cultures 17
blood pressure (BP) 16, 488
blood tests 16–17
blood transfusion 582–3
bone
  basal fractures 388
  non-tumour lumps 416
  sequestrae 426
  smoothing irregularities
    426
  surgical removal 380
bone cysts
  aneurysmal 419
  solitary 419
bone disease 538
bone grafts 514
  alveolar augmentation 500
  cleft palate 199
bone marrow depression,
  drug-induced 458
bone morphogenic protein
  238
book keeping 718
Borrelia intermedius 220
Borrelia vincenti (refringens)
  202, 220
Bowen's disease 546
B point 141
bracing, partial dentures 336,
  343
brain tumours 544
branchial cyst 513
branchial fistula 513
breast-feeding
  caries 98
  drugs contraindicated 624
bridges 292–303, 334
  definitions 292
  design 296–7
  failures 300
  practical stages 298–9
  removing old 300
  resin-bonded 302–3
  temporary 298, 307
  treatment planning 296
  trial cementation 299
  types 292–3
British Dental Association
  (BDA) 698, 734, 762
British National Formulary
  (BNF) 600, 624
British Standard 5750
  (BS5750) 732
brittle bone disease 538
broaches, root canal 314
'broken instrument'
  technique 394–5
bronchodilators 644
bronchogenic carcinoma 526

brown tumours of hyper-
        parathyroidism 414,
        416
bruxism 304, 478
buccal advancement flap
        423, 424
budgeting 718
bulimia nervosa 551
bulla 442
bupivacaine 630
buprenorphine 604
burning mouth 362, 468
burning (sore) tongue 452,
        466
burns, chemical, oral mucosa
        458
burs
    Gates–Glidden 314
    numbering systems 756
    oral surgery 384
    removal of third molars
        400
    restorations of primary
        teeth 86
business planning 728

calcification
    of root canals 123
    times in children 69
calcium hydroxide 271
    cements 670, 671
    endodontic use 316–17
    non-setting paste 316
calculus 206
    removal 228
    subgingival 206
    supragingival 206
Caldwell–Luc antrostomy
        422
cancrum oris 220
Candida albicans 202, 438
candidal endocrinopathy 439
candidiasis (candidosis) 438–
        9, 466
    acute atrophic 438
    acute pseudomembranous
        (thrush) 126, 438
    AIDS patients 472
    children 126
    chronic atrophic (denture
        stomatitis) 126, 364–
        5
    chronic hyperplastic 439
    chronic mucocutaneous
        439
    dentures and 364–5
    elderly 375
    granulomatous skin 439
canines
    bridge replacement 297

buccally displaced
    maxillary 158
    orthodontic extraction 150
palatally displaced
    maxillary 160–1
    removal of unerupted 396
    transposition 161
Capnocytophaga 222
carbamazepine 128, 460,
        622
carbamide peroxide 310
carbon dioxide laser 430
carboxymethylcellulose pads
        258
carcinoma-in-situ, skin 546
cardioactive drugs 644
cardiopulmonary resuscita-
        tion (CPR) 560–1
cardiorespiratory arrest 560–
        1
cardiovascular disease 524–5
cardiovascular system
    age changes 374
    examination 10
caries 28–33
    abutment teeth 371
    approximal 32
    arrested 28
    dentine 28
    diagnosis 32–3
    electronic detectors 32
    enamel 28
    immunization against 39
    nursing 98
    nursing bottle 98
    pit and fissure 32
    plaque and 38, 204
    prevention 30, 34, 36–41,
        44
    progression 28
    radiation 98
    rampant 98
    root 30, 269, 372
    saliva and 28–30
    secondary (recurrent) 87,
        272
    smooth surface 32
    stabilization 253
    statistics 56, 57
    sugar and 42
    susceptible sites 28
    see also restorative
        dentistry
carotid aneurysm 512
carotid body tumour 512
cartridges, local anaesthetic
        630
carving, amalgam 656
castable ceramics 682
casting alloys 676–7

catheterization, urethral 584
cavity design
　class II restorations 90,
　　262
　primary teeth 86–7, 90
　principles 260–1
　root canal access 318, 319
cavity varnishes 657
cefuroxime 616
cementomas 421
cementopathia 223
cements 670–1
　lining 667, 671
　luting 667, 671
　restorative 667
　temporary 307, 671
　see also specific cements
cementum, structural distur-
　bances 75
central nervous system (CNS)
　examination 10
　infections 544
　tumours 544
centric occlusion (intercuspal
　position; ICP) 274,
　277, 352
centric relation (retruded
　contact position;
　RCP) 274, 277, 352
centric stops 274
cephalometrics 21, 140–3
　analysis and interpretation
　　142
　commonly used points
　　141
　pitfalls 140
　tracing 140
cephalosporins 615
ceramics 680–3
　applications 682
　castable 682
　see also porcelain
cerebral palsy 52
cerebrospinal fluid (CSF)
　leaks 486
cerebrovascular accidents
　544
cermets 668
　class II restorations 92,
　　264
cervical lymphadenopathy
　512
cervical rib 513
cervical spine immobilization
　484, 486
cervico-facial lymphadeno-
　pathy 474–5
chancre 434
Chediak–Higashi syndrome
　743

cheek trauma
　denture wearers 363
　local anaesthesia causing
　　634–5
cheilitis
　actinic 453
　angular 438, 466
　exfoliative 453
　granulomatous 453, 464
chemical burns, oral mucosa
　458
chemical injury, dental
　materials 688
chemotherapy, cytotoxic 511,
　520, 644
cherubism 416, 538
chest, examination 10
chest infection 526, 596
chest pain, acute 558
Chevron tooth notation
　system 752
chewing gum, remineralizing
　potential 30
chicken-pox 126, 436
child abuse 108
children 58–129
　aims of treatment 60
　anxious 64–5
　behaviour problems 65
　common ailments affecting
　　mouth 126–7
　communication with 87
　consent by 696
　dental health statistics 57
　dental trauma 106–23
　drug prescribing 128–9,
　　600
　first visit 60
　fluoride supplementation
　　37
　keeping records 706
　local anaesthesia 84–5
　management of 60
　maxillofacial fractures
　　496–7
　neck lumps 512
　periodontitis 222–3
　sedation 64–5, 638
　toothache 66–7
　treatment planning 62
　see also primary teeth
chisels 384, 400
chloramphenicol 458, 616
chlordiazepoxide 612
chlorhexidine 38, 220, 224,
　230, 588
chlormethiazole 612
chlorpheniramine 554, 562,
　582, 620
choking 572

choline salicylate 129
Christmas disease (haemo-
philia B) 522
chronic obstructive airways
disease (COAD) 526
Chvosteck's sign 534
cimetidine 128
cirrhosis, liver 464
Clark's rule 129
clasps
partial dentures 338, 339,
342
removable orthodontic
appliances 182
cleaning
dentures 360
instruments and equipment
382, 734–5
laboratory items 735
root canals 318
work surfaces 735
cleft lip/palate (CL(P)) 198–
9
management 198–9
problems 198
surgery 502, 504
cleidocranial dysostosis
(dysplasia) 538, 743
clerking, in-patients 576
clicking
dentures 362
temporomandibular joint
478
clindamycin 590, 591, 616
clinical exams 761
coagulation defects 522
co-amoxiclav 616
cobalt chromium
cast 676, 679
clasps 338
wrought 678
cobalt chromium nickel
678
cocaine 630
codeine 128, 588, 602
coeliac disease 464, 528
cold sore (secondary herpes)
126, 436, 618
colitis, antibiotic-induced
528–9, 616
collapse
corticosteroid users 564
sudden, management 574
see also unconscious
patients
colonic cancer 528
'commando' procedure 510
Committee on Safety of
Medicines (CSM)
624–5

communication
child patients 87
elderly patients 376
with staff 712
Community Periodontal
Index of Treatment
Needs (CPITN) 210
Community Service 50
complaints against 694
community vocational
training 730
complaints 694, 729
compomers 269, 667
composite(s) 658–62
acid-etch technique 662
bonding fixed appliances
188
class I restorations 262–3
class II restorations 92,
266
class III restorations 268
class IV restorations 96,
268
class V restorations 269
constituents 658
facing, with glass ionomer
(sandwich technique)
269
inlays 266–7
lining cements 671
polymerization (curing)
methods 660
problems 660
properties 658
veneers 288
compressed air 258
compulsive behaviour 551
computed tomography (CT)
22, 488
computers 726–7
concrescence 75
concussion, tooth 110, 116
condensation
amalgam 656
gutta-percha 326
condylar fractures 490, 496
condylar hyperplasia 498
confidentiality 702
conformative approach 276
confusion
post-operative 596
toxic 550
connective tissue diseases
462–3, 540
connective tissue tumours,
mouth 420, 451
connectors
bridges 292
partial dentures 336, 340,
343

Conn syndrome 534
consciousness
  Glasgow coma scale 488–9
  sudden loss of, manage-
    ment 574
consent 696
  general anaesthesia 576
  implied 696
  verbal 696
  written 696, 706
Consumer Credit Act (1974)
  719
contracts 698
  staff 698, 714
  vocational training 730
Control of Substances
  Hazardous to Health
  (COSHH) regulations
  716
convulsions (fits)
  diagnosis and management
    566
  febrile 544
  types 544
Coroner's court 692
corticosteroids, see steroids
cost
  impressions 674
  preventing uptake of dental
    care 50
co-trimoxazole 129, 616
cotton-wool rolls 258
courts, law 692
coxsackie virus infections
  436–7
cracked tooth syndrome 255
Craddock classification,
  partial dentures 336
cranial (temporal) arteritis
  460, 540
cranial nerves 542
craniofacial anomalies,
  surgery 502–3
creatinine, plasma 17
creep, dental materials 654
cri-du-chat syndrome 743
Crohn's disease 464, 528
crossbites 136, 176–9
  anterior 176
  buccal 136
    bilateral 178
  lingual 136
    bilateral 178
  posterior 176–8
  unilateral 176–8
cross-infection
  prevention 382–3, 706,
    734–5
  routine procedures 734–5
Crouzon syndrome 743

crown form, primary teeth
  80
crown fractures
  permanent teeth 112–13
  primary teeth 110
Crown Indemnity 702
crowns
  full veneer gold 290
  glass ceramic/castable
    ceramic 682
  platinum bonded/McLean-
    Sced/twin foil 282,
    682
  polycarbonate 306
  porcelain fused to metal
    (PFM) 282, 284, 290–
    1, 682
  porcelain jacket (PJC)
    280–1, 282, 284, 682
  post and core, see post and
    core crowns
  posterior 290–1
  root-filled posterior teeth
    291
  soft metal alloy 306
  stainless steel 94–5
  telescopic 297
  temporary 280, 306–7
  three-quarter gold 291
  vital anterior teeth 280–2
    common problems 282
    fitting 281
    preliminary treatment
      280
    removing old 282
cryosurgery 412
CSF leaks 486
curettes 384
Cushing disease 534
Cushing syndrome 468, 534
cusps
  functional 274
  non-supporting 275
  supporting 274
cyanosis 10
cyclizine 610
cystic fibrosis 526
cystic hygroma 513
cysts
  bone 419
  branchial 513
  calcifying odontogenic 419
  dentigerous 418
  dermoid 414, 512
  eruption 68
  fissural 419
  inflammatory dental 418
  jaws 418–19
  sebaceous 512
cytology 17

cytotoxic chemotherapy 511,
520, 644

dam, rubber 86, 258–9
Data Protection Act (1984)
727
deaf, manual alphabet 53
deaths, reporting unexpected
692
debridement 380
Debris Index 210
debts, bad 694, 719
deciduous teeth, *see* primary
teeth
decongestants 620
deep circumflex iliac flap 515
defence organizations 701,
702
defendant (defender) 690
deflective contacts 275
dehydration 10
delirium 550
deltopectoral flap 514
dementia 550
dens in dente 76
dental arches, development
146
dental care 50–1
AIDS patients 473
barriers to uptake 50–1
domiciliary 376–7
elderly 376–7
handicapped 52–3
liver disease 530–1
provision and receipt 50–1
recall intervals 33
dental hard tissues, *see* hard
tissues
dental health
education 48–9
indices 57
statistics 56–7
dental history 6
dental materials 652–89
biocompatibility 688–9
definitions 654
evaluation of new 654–5
hazards to patient 688–9
hazards to staff 689
properties 654–5
root canal therapy 316–17
dental panoramic tomograph
(DPT) 21
Dental Practice Board (DPB)
762
complaints by 694
computer links 727
Dental Practitioner's Formulary (DPF) 600

Dental Protection (Medical
Protection Society)
702, 707, 763
dental surgery assistant
(DSA) 330, 734
dental team 54–5
dental therapists 54
dentigerous cysts 418
dentine
bonding agents 664–5
caries 28
fractures 112
hereditary opalescent 75
hypersensitivity 255
pins 308
smear layer 664
structural disturbances 75
dentinogenesis imperfecta 75
Dentists' Register 704
dentition, normal development 146
dento-alveolar compensation
136
dento-alveolar surgery 394–
407
aids to endodontics 402–4
antibiotic prophylaxis 590
helping the orthodontist
406–7
removal of roots 394–5
removal of third molars
398–401
removal of unerupted teeth
396–7
dento-facial infections 408–9
dentures
alloys for 678, 679
base materials 336–7, 354,
678, 684–5
candida and 364–5
complete 346–59
advice to patients 356
after healing 334
common problems 348–
9, 354–5
copying 366–7
fitting 356–7
immediate 334, 346–7,
367–8
impressions 350–1
maintenance 358–9
occlusal recording 352–
3
patient assessment 348
principles 348–9
treatment planning 335
trial insertion 354–5
copying 366–8, 377
elderly 372, 377
indications 334

dentures (*cont.*):
  materials 684–6
  overdentures 370–1
  partial, *see* partial dentures
  preprosthetic surgery 426–7, 500
  problems and complaints 362–3
  rebasing 345, 358
  recurrent fracture 363
  relining 358, 686
  soft liners 359, 686
  strength 685
  tissue conditioners 358–9, 686
  treatment planning 334–5
denture stomatitis (denture sore mouth) 364–5
depression 550
dermatitis 548
  herpetiformis 444, 549
dermatology 548–9
dermatomyositis 463, 540
dermoid cysts 414, 512
desensitization, anxious child 64
de-sensitizing agents 37
devitalization pulpotomy, two-visit 102–3
dexamethasone 607
dextrose, intravenous 554, 574, 580, 592–3
diabetes insipidus 534
diabetes mellitus 552
  hypoglycaemia 568
  oral features 468
  surgery 592–3
diagnostic mounting 276, 277, 298
diagnostic waxing 298
diamorphine 558, 588, 604
diastema, median 156
diazepam 612, 640
  intravenous 566, 612, 640, 641
  rectal 638
diclofenac 588, 604
diet, maxillofacial trauma 497
dietary advice 46–7
dietary analysis 46
differential diagnosis 24–5
diflunisal 602
digital imaging 22–3
digit (finger) sucking 146, 162
dihydrocodeine 602
dilaceration 76–7
diodontic implants 404
discoloured teeth 78

bleaching 310–11
  extrinsic staining 78
  intrinsic staining 78
  post-traumatic 110
disinfection 382
  impressions 675, 735
  laboratory items 735
  work surfaces 735
dislocation, temporo-mandibular joint 498
diverticular disease 528
domiciliary care 376–7
domperidone 610
Doppler ultrasound (US) 23
dothiepin 608
double teeth 76
Down syndrome 743
doxycycline 409, 616
dressings
  bacteriostatic 671
  sedative 671
  temporary 306
drooling, surgery for 508
drug eruptions, fixed 458
drug history 376
drug interactions 624
  benzodiazepines 638
  general anaesthesia 644
drug reactions
  emergency management 562
  oral 458–9
  reporting 624–5
drugs
  controlled 600
  emergency 554
  generic 598
  information services 601
  proprietary 598
  sugar-based 98
  sugar-free preparations 128–9
drug therapy 598–625
  alarm bells 624
  children 128–9, 600
  elderly 600, 638
  general dental practice 600
  hospital 600
  local periodontal 228
  prescribing guidelines 600–1
  private/independent practice 729
  TMPDS 480, 608, 610
dry socket 408–9
dysaesthesia, oral, drug-induced 458
dyskeratosis congenita 462
dysphagia 528

Eagle syndrome 743
ears
  cleft lip and palate 502
  examination 12
Easlick's devitalizing paste
    100
EBA cements 670, 671
eczema 548
edentulous patients 56
  maxillofacial trauma 496
  preprosthetic surgery 426–
    7, 500
  see also dentures
edgewise fixed appliance 190
Ehlers–Danlos syndrome
    462, 743–4
Eikenella corrodens 214,
    222
elastic limit 654
elastic modulus 654
elastics, removable
    appliances 184
elastomers 672
elbow formation, root canal
    preparation 319
elderly 372–7
  age-related changes 374–5
  clinical techniques of value
    377
  delivery of care 376–7
  drug prescribing 600, 638
  general management
    problems 376
  periodontal problems 372
  prosthetic problems 372
  restorative problems 372
electric pulp tester 18
electronic caries detectors 32
electronic dental anaesthesia
    (EDA) 633
electrosurgery 259
elevators 384
  periosteal 384
  root removal 394
emergencies
  dental 702, 720, 722–3
  medical 554–74
  useful kit 554
Emla cream 630
employees, see staff
employer's liability 717
enamel
  abnormal structure 74–5
  caries 28
  chronological hypoplasia
    74
  fractures 112
  hypomineralization (hypo-
    calcification) 74
  hypoplasia 74

opacities 78
  primary teeth 80
encephalitis, herpetic 544
endocarditis, subacute
    bacterial (SBE) 524,
    590–1
endocrine disease 468, 534–
    5
endocrine-related problems
    536
endocrine system, age
    changes 375
endocrine tumours 536
endocrinopathy, candidal
    439
endodontic implants 404
endodontics, see root canal
    therapy
endoscopic sinus surgery,
    functional (FESS) 422
endotracheal intubation 646
enflurane 642
En masse appliance 154, 155
enteral feeding 586
eosinophilic granuloma 744
ephedrine nasal drops 620
epidermolysis bullosa 444
epiglottitis, acute 526
epilepsy 544, 566
epimine allergy 688
epithelial inlay 500
  and combination mucosal
    flap 500
epithelial tumours
  mouth 420
  salivary glands 506
epithelioma (basal cell
    carcinoma) 412, 546
epulis
  congenital 414
  giant cell (granuloma) 414,
    416
  pregnancy 414
equipment
  cleaning and sterilizing
    382, 734–5
  local anaesthesia 630
erosion, tooth 304
erosions, mucous membranes
    442
eruption (of teeth)
  abnormalities 68–70
  cyst 68
  delayed 68
  ectopic 70
  failure 68
  normal sequence 69
  times 69
  see also unerupted teeth
erythema marginatum 548

erythema migrans 452
erythema multiforme 444–5,
    476, 548
erythema nodosum 548
erythroleukoplakia 448
erythromycin 129, 615
erythroplakia 448
ethics, professional 702–3
ethyl chloride, vitality testing
    18
etiquette, professional 702–3
etomidate 642
European tooth notation
    system 752
examination
  head and neck 12–13
  medical 10
  mouth 14
  occlusal 14, 276
  orthodontic 138–9
exams, passing 760–1
exarticulation, see avulsion
exfoliation of teeth
  abnormalities of 68–70
  premature 70
exfoliative cheilitis 453
exfoliative stomatitis 458
exostoses 416
exposure, unerupted tooth
    406
external fixation, mid-face
    fractures 496
extraction(s) 386–9
  alveolar bone preservation
      426
  balancing 137
  compensating 137
  complications 388–9
  denture provision after
      334, 346
  forceps 384
  maxillary antrum and 422
  orthodontic 150–2
  out-patient general
      anaesthesia 648
  periodontal disease 249
  primary teeth 386
  serial 148–9
  vs. restoration 82–3
  retained roots 394–5, 426
  technique 386
  third molars 386,
      398–401
  unerupted teeth 396–7
  wrong tooth 388
extra-oral traction, ortho-
    dontics 180–1
extrusion
  orthodontic 114
  traumatic 110, 116

eyes
  damage 688, 689
  examination 12

face
  appearance 12
  skin lesions 12
facial arthromyalgia 478–81,
    608
  see also temporo-
    mandibular pain
    dysfunction syndrome
facial fractures 490–7
facial nerve (VII) 470, 542
facial pain 460–1, 588
  atypical 461, 608
  intractable, cryosurgery
      412
facial palsy, local anaesthesia
    causing 634
facial paralysis, neurological
    causes 470–1
factor VIII deficiency 522
factor IX deficiency 522
fainting 556
Family Health Services
    Authority (FHSA)
    694, 698, 717
Fatal Accident Inquiry 692
fatigue, dental materials 654
FDI tooth notation system
    752
febrile convulsions 544
fees
  setting 728–9
  unpaid 694, 719
Fermit 307
fibroepithelial polyp 414
fibroma 420
  ameloblastic 421
  ossifying 420
fibrous bands, excision 426
fibrous dysplasia 416, 538
fibula flap, free 515
files
  root canal 314
  sizes 756
filling materials, root canals
    316
financial management 718–
    19
finger spreaders 315
finger (digit) sucking 146,
    162
fissural cysts 419
fissure sealants 40–1, 657,
    660
  application technique 41
  efficacy 40
  types 40–1

fistulae
  branchial 513
  oro-antral 422–4, 620
  salivary duct/gland 455
fits, *see* convulsions
fixed appliances (FA) 188–91
  anchorage 180
  components 188
  principles 188
  types 190
flaps
  apically repositioned 236–8
  apicectomy 402, 403
  buccal advancement 423, 424
  free 514–15
  head and neck reconstruction 514–15
  modified Widman 234–5
  mucoperiosteal 380
  palatal rotation 424, 425
  periodontal disease 238
Flex-o-file 314
flossing 224
flucloxacillin 409
fluconazole 438, 618
fluids, intravenous 580
flumazenil 554, 613, 639
fluoridation
  milk 36
  salt 36
  water 36, 37
fluoride 34–7, 622
  antidotes 35
  drops and tablets 36, 37
  mechanisms of action in reducing decay 34
  professionally applied 36
  recommended daily supplementation 37
  rinsing solutions 36
  safety and toxicity 34–5, 688
  statistics 56
  systemic therapy 36
  toothpastes 36–7
  topical 36–7
fluorosis 34–5
folate deficiency 466, 518
folic acid 129
forceps
  dissecting 384
  extraction 384
  root 394
forehead flap 514
foreign bodies
  inhaled 572
  maxillary antrum 422

forensic dentistry 708
formocresol 100, 102, 103
four-handed dentistry 330
Frankel appliance 194
Frankfort plane 136, 141
freeway space (FWS) 274, 352, 353
frenectomy 406
Frey syndrome 744
frictional keratosis 446
fulcrum axis 336
functional appliances 192–5
  practical tips 194–5
  rationale and mode of action 192–3
  types 194
functional endoscopic sinus surgery (FESS) 422
functional impressions 351, 686
fungal infections, oral 126, 438–9
furcation involvement, periodontal disease 246, 248–9
furcation plasty 248
fusion, tooth 76
Fusobacterium 202
Fusobacterium fusiformis 220

gancyclovir 618
Gardener syndrome 744
gas cylinder colour coding 758
gastric carcinoma 528
gastrinomas 536
gastrointestinal disease 464, 528–9
gastrointestinal system, examination 10
Gates–Glidden burs 314
gemination 76
general anaesthesia (GA) 642–50, 706
  anxious child 65
  definition 626
  diabetic patients 592
  drug interactions 644
  drugs and definitions 642–3
  hospital admission for 628–9
  hospital setting 646–7
  indications and contra-indications 628–9
  medical examination 10
  out-patient 648
  Poswillo Committee recommendations 649

general anaesthesia (*cont.*):
    practice setting 648–50
    removal of roots 394–5
    removal of third molars
        398
    UK position 626
General Dental Council
    (GDC) 694, 704, 762
general dental practitioners
    (GDPs)
    complaints procedures 694
    contracts 698
    vocational training 730
General Dental Service
    (GDS) 50
genioplasty 504
genitourinary system,
    examination 10
gentamicin 590, 591, 616
geographic tongue 452
gerodontology 372–7
    see also elderly
giant cell granuloma, peri-
    pheral 414, 416
gigantism 534
gingival hyperplasia 458
gingival recession 246
    mucogingival surgery 240
gingival retraction 259
gingivectomy 114, 236
gingivitis
    acute streptococcal 221
    acute ulcerative (necrotiz-
        ing; AUG) 126, 220,
        472
    chronic 212–13
    chronic atrophic
        (desquamative) 213
    chronic hyperplastic 212
    chronic oedematous 212
    chronic ulcerative 220
    HIV 472
    microbiology 208
    statistics 57
gingivostomatitis, primary
    herpetic 126, 436
Glandosane 622
glandular fever 126, 437
Glasgow coma scale (GCS)
    488–9
glass ceramic crowns 682
glass ionomers (GI) 666–8,
    671
    cavity preparation 261
    class II restorations 92,
        264
    class III restorations 96,
        268
    class V restorations 96,
        268–9

    with composite facing 269
    light-cured 269
    metal reinforced 668
    practical tips 668
    primary teeth 86, 92, 96
    properties and types 666–
        7
glass polyalkenoate, *see* glass
    ionomers
glaucoma 461
glossitis 452
    atrophic 466
    benign migratory 452
    median rhomboid 439
glossodynia (sore tongue)
    452, 466
glossopharyngeal nerve (IX)
    542
glossopharyngeal neuralgia
    460, 622
gloves 382, 734
glucagonomas 536
glucose
    blood 17
    urinary 16
    see also dextrose
glyceryl trinitrate 558
goitre 534
gold (alloy) 676
    casting 676
    casting technique 677
    clasps 338
    full veneer crowns 290
    porcelain bonding 677
    restorations, cavity
        preparation 261
    three-quarter crown 291
    veneer pinledge
        restorations 308
    wrought 679
Goldenhar syndrome 744
gonion 141
gonorrhoea 435
Goretex membranes 238
Gorlin–Goltz syndrome
    744
gout 498, 541
grafts
    bone, *see* bone grafts
    head and neck reconstruc-
        tion 514–15
    mucogingival 240–1
    skin 446, 514
granular cell myoblastoma
    420
granuloma(ta) 415
    annulare 549
    peripheral giant cell 414,
        416
    pyogenic 414

granulomatous cheilitis 453, 464

granulomatous skin candidosis 439

Graves' syndrome 744

group function 274

'guardsman's' fracture 490

guided tissue regeneration (GTR) 238, 242, 429

guide planes, partial dentures 336

gumma 434

gum-stripper 336

Gunning splints 496–7

gutta-percha (GP) 316, 326–7

  coated carriers 327

  lateral condensation 326

  thermomechanical compaction 326

  thermoplasticized injectable 326–7

  vertical condensation 326

habits, orthodontics and 146, 170

haemangioma

  oral 412, 415

  salivary glands 456, 506

haematocrit 16

haematological disease 466, 520–3

haematological malignancy 520, 552

haematoma, nasal septal 494

haemoglobin 16

haemophilia A 522

haemophilia B 522

haemorrhage, see bleeding/ haemorrhage

hairy leukoplakia 472

haloperidol 612

halothane 642

hand, foot and mouth disease 126, 437

handicapped patients 52–3

handpiece, miniature 86

Hand–Schüller–Christian disease 745

hardness, dental materials 654

hard tissues

  age changes 374

  non-tumour lumps 416

Harvold appliance 194

head

  examination 12–13

  reconstructive surgery 514–15

headache 543

cluster (periodic migrainous neuralgia) 460, 543

  tension 543

head and neck syndromes 740–8

head-gear, orthodontic appliances 180–1

head injuries 486, 488–9

health

  dental, see dental health

  sugar and 42

health and safety 716–17

Health and Safety at Work Act (1974) 716

Health Board, contract with 698

health education, dental 48–9

heart disease, ischaemic 524

heart failure 524

heat, vitality testing 18

Hedstroem file 314, 323

Heerfordt syndrome 744

Heimlich manoeuvre 572

helium-neon laser 430

hemifacial atrophy 746–7

hemifacial microsomia 744

hemifacial spasm 471

hemisection

  root 404

  tooth 249

heparin 522, 590, 596

hepatic disease, see liver disease

hepatitis B 530

  employer's responsibilities 716

  immunization 382, 734

  prevention of cross-infection 382–3

herpangina 126, 437

herpes simplex 436, 618

  encephalitis 544

  primary herpetic gingivo-stomatitis 126, 436

  secondary (herpes labialis; cold sore) 126, 436, 618

herpes zoster 375, 436, 476, 618

herpetiform oral ulcers 440

hinge axis 274

histiocytosis-X 744–5

histoplasmosis 439

history 2–8

  child with toothache 66

  dental 6

  drug 376

  medical 8, 376

  presenting complaint 4

HIV infection 472, 552–3, 618
  acute ulcerative gingivitis 220, 472
  gingivitis 472
  periodontitis 472
  prevention of cross-infection 382–3
  see also AIDS
Horner syndrome 745
hospital
  admission
    for anaesthesia 628–9
    maxillofacial trauma 486–7
  complaints against 694
  drug prescribing 600, 604–5
  see also in-patients, dental
Hospital Service 50
human immunodeficiency virus infection, *see* HIV infection
Hurler syndrome 745
Hutchinson incisors 434
hydrocolloid impressions 672
hydrocortisone
  intra-articular injection 606
  lozenges 606
  medical emergencies 554, 562, 564, 570, 574
  and oxytetracycline ointment or spray 606
  sodium succinate 594, 607
hydrogen peroxide 310, 311
hydroxyapatite, subperiosteal injection 426–7
hygienists 54–5
hyoid bone, prominent 513
hyperaldosteronism, primary 534
hypercementosis 75
hyperdontia (supernumerary teeth) 72–3, 396–7
hyperparathyroidism 468, 534
  brown tumours 414, 416
hyperplastic lesions
  cryosurgery 412
  irritation (denture) 414, 426
hyperpyrexia, malignant 536
hypersensitivity, dentine 255
hypersensitivity reactions, dental materials 689
hypertension 524
hyperthyroidism 468, 534
hypnosis, anxious child 65

hypnotics 612–13
hypochlorite solutions 316, 360
hypodontia 72, 156
hypoglossal nerve (XII) 542
hypoglycaemia 568
hypoparathyroidism 468, 534
hypopharynx, examination 12
hypophosphatasia 223
hypopituitarism 534
hypotension, post-operative 596
hypothyroidism 468, 534
hypovolaemic shock 524

ibuprofen 588, 602
identification, dental 708
idoxuridine 618
IgA deficiency, selective 552
imaging techniques, advanced 22–3
immune system, age changes 374
immunization
  caries 39
  hepatitis B 384, 734
  staff 734
immunocompromised patients 552–3, 590
immunodeficiencies
  acquired 552–3
  congenital 552
immunology 17
immunosuppressive drugs 458, 552, 607
impetigo 126
implants 428–9
  endodontic (diodontic) 404
  insertion techniques 428–9
  missing incisors 125
  peri-implantitis 238–9
  salvage/bone augmentation 429
impression compound 672
impressions 672–5
  bridges 298–9
  complete dentures 346, 350–1
  costs 674
  crowns 281
  disinfection 675, 735
  double mix technique 674
  functional 351, 686
  materials 672–3
  muco-compressive vs. muco-static 350
  partial dentures 344

putty and wash technique 674
reciprocal denture technique 350–1
single mix technique 674
impression trays, special 350, 674
impression waxes 673
incisions, oral surgery 380
incisors
congenital absence 124
Hutchinson 434
missing 124–5
orthodontic extraction 150
independent practice 728–9
Index of Orthodontic Treatment Need (IOTN) 134–5
indices
dental health 57
periodontal 210–11
toothwear 304
infections
central nervous system 544
cervical lymphadenopathy 512
control of 382–3
dento-facial 408–9
immunosuppressive 552
local anaesthesia causing 635
oral mucosal 434–9
presenting as neck lumps 513
skin 548
urinary tract 532
see also cross-infection; specific infections
infectious mononucleosis (glandular fever) 126, 437
inferior dental block (IDB; inferior alveolar block) 84, 632
inferior dental canal, extruded paste 404
infestations, skin 548
infiltration anaesthesia 84, 633
inflammatory dental cysts 418
infraorbital block 633
inlays
composite 266–7
porcelain 266–7, 682
temporary 307
inosine pranobex 618
in-patients, dental 576–97
investigations 576–7

post-operative care 577
pre-operation 576
instructions to patients
complete dentures 356
removable appliances 187
instruments
cleaning and sterilizing 382, 734–5
endodontic 314–15
fractured in root canal 328–9
oral surgery 384
restorations of primary teeth 86
transfer, four-handed dentistry 330
insulin, intravenous infusion 592–3
insulinomas 536
insurance 719
intercuspal position (ICP) 274, 277, 352
interdental cleaning 224
interferences 275
interferon 618
intermaxillary fixation (IMF) 496
post-operative care 497, 577
inter-maxillary traction 180, 188
internal fixation
mid-face fractures 496
open reduction and fixation (ORIF), facial fractures 496, 497
interproximal box 90
intracranial pressure, raised 543
intraligamentary analgesia 84–5, 633
intra-maxillary traction 180
intravenous fluids 580
intravenous infusions 578
intravenous sampling 578
intrusion, tooth 110, 116
intubation, endotracheal 646
investigations
advanced imaging techniques 22–3
general 16–17
in-patients 576–7
specific 18–19
see also X-rays
Ionising Radiation Regulations 736
iron-deficiency anaemia 466, 518
iron edetate 129
irrigants, root canal 316

irritable bowel syndrome 528
ischaemic heart disease 524
isoflurane 642
isolation
  primary teeth 86
  restorative dentistry 258–9
  root canal therapy 318
isthmus 90
itraconazole 618

jaundice 10, 530
jaws
  cysts 418–19
  osteosarcoma 451
jet injection 85
joint disease 540–1
jury service 692
juvenile periodontitis 222–3

Kaposi's sarcoma 472, 546
Kennedy classification, partial
    dentures 336, 337
keratocysts 418–19
keratosis
  actinic 546
  frictional 446
  smoker's 446
Kesling's set-up 125
ketamine 642
ketoconazole 618
ketones, urinary 16
ketorolac trometamol 604
K-files 314, 319
K-flex file 314
Klippel–Feil anomalad 745
Koplik spots 437

laboratory items, infection
    control 735
Lactobacillus 38, 202
large bowel disorders 528–9
Larsen syndrome 745
laryngeal mask airway
    (LMA) 646
laryngocele 512
larynx, examination 12
lateral oblique view 21
latissimus dorsi myo-
    cutaneous flap 514
law 690–708
  civil 692
  criminal 692
  processes 692
lead aprons 21
leaflets, practice 722, 724–5
Ledermix 103, 271, 316
leeway space 136
Le Fort classification, mid-
    face fractures 492,
    493

Le Fort surgical procedures
    504–5
legal processes 692
legislation 690
  secondary 690
lentigo maligna 546
lentigo simplex 546
Lesch–Nyhan syndrome 745
Letterer–Siwe disease 745
leukaemia 466, 520
leukoplakia 448, 449
  candidal 439
  cryosurgery 412
  hairy 472
  panoral 446
  speckled 448
  syphilitic 446
liability
  employer's 717
  vicarious 700
lichenoid eruptions, drug-
    induced 458, 462
lichen planus 375, 444, 462
  erosive 448, 476
light activation, composites
    660
lignocaine
  topical preparations 622,
    630
  viscous mouthwash 622
lignocaine/adrenalin 630
limb movements, head injury
    488
lingual appliances 190
lingual nerve block 632
lining cements 667, 671
lipoma 420, 512
lips 452, 453
  cancer 450
  cleft, *see* cleft lip/palate
  competent 136
  dry sore 453
  incompetent 136
  persistent median fissure
    453
  repair in cleft lip 198, 502
  trauma due to local
    anaesthesia 634–5
listening 2
lithium 644
litigation 690, 694
  avoiding 706–7
  negligence actions 694,
    700–1
liver disease 464, 530–1
  drug therapy 601, 624
local anaesthesia (LA) 630–5
  apicectomy 402
  cartridges 630
  children 84–5

diabetic patients 592
equipment 630
failure 634
indications and contra-
    indications 628
investigative 19
pain on injection 634
periodontal surgery 233
problems 634–5
removal of third molars
    398
techniques 632–3
toxic reactions 562
local anaesthetics 630
long buccal block 632
looseness, dentures 362
lorazepam 612
lower facial height (LFH)
    136
increased 194
lung cancer 526
lupus erythematosus
chronic discoid 463
systemic (SLE) 462–3
luting cements 667, 671
luxation, teeth 110, 116
lymphadenitis, staphylo-
    coccal 409
lymphadenopathy
cervical 512
cervico-facial 474–5
persistent generalized 472
lymphangioma
oral 415
salivary glands 456, 506
lymphoepithelial lesion 456
lymphoma 472, 520

McCune–Albright (Albright)
    syndrome 538, 742
macrodontia 76
macroglossia 452
MAGIC syndrome 745
magnetic resonance imaging
    (MRI) 22
malar fractures 494, 496,
    497
malar osteotomy 505
malocclusion
class I 136
    management 164
class II 136
    management 164–5,
        168, 192, 194
class III 136
    management 168, 172
extraction of primary teeth
    and 83
prevalence 132
management, practice 710–39

financial 718–19
skills 712–13
mandible
posteroanterior (PA) view
    21
surgical procedures 504
mandibular deviation 137
mandibular displacement
    137
mandibular fractures 490–1
complications 497
diagnosis and assessment
    490–1
treatment 491, 496, 497
mandibular plane 141
mandibulofacial dysostosis
    747
marble bone disease 538
Marfan syndrome 745
marketing 722–3
Masseran kit 286, 329
masseter muscle flap 514
materials, dental, see dental
    materials
maxilla, surgical procedures
    504–5
maxillary antrum 422–5
displaced teeth/roots 422
pathology 422
removal of extruded paste
    404
sinus lift 500
maxillary expansion, rapid
    178
maxillary fractures 492–3,
    496, 497
maxillary plane 141
maxillary sinusitis 422
maxillofacial injuries 485,
    486–97
children 496–7
complications 497
open reduction and internal
    fixation (ORIF) 496
pain control 486, 588
primary management 486–
    7
treatment 496–7
maxillofacial surgery 482–
    515
clefts and craniofacial
    anomalies 502–3
facial fractures 496–7
flaps and grafts 514–15
oral cancer 510–11
preprosthetic 500
salivary glands 508
temporomandibular joint
    498–9
mean cell volume 16

measles 126, 437
median diastema 156
Medical and Dental Defence Union of Scotland 702, 707, 763
Medical Defence Union 702, 707, 763
medical examination 10
medical handicap 52
medical history 8, 376
Medical Protection Society (Dental Protection) 702, 707, 763
medications, *see* drugs
medicine
  in dentistry 516–97
  oral 432–81
Medium Opening Activator 194, 195
mefenamic acid 602
megadontia 76
melanoma, malignant 546
Melkersson–Rosenthal syndrome 452, 745
meningitis
  bacterial 544
  viral 544
menopause 468, 536
menstrual cycle 468
mental handicap 52, 696
mental illness, consent to treatment 696
mental nerve
  block 632
  repositioning 500
menthol and eucalyptus inhalation 620
menton 141
mepivacaine 630
mercury spillage 716
metabolic bone disorders 538
metal alloys, *see* alloys
metastases, skin 546
methohexitone 642
methylprednisolone 606, 607
metoclopramide 577, 588, 596, 610
metronidazole 616
  dento-oral infections 220, 221, 230, 408
  prophylactic use 590
  terminal cancer 588
miconazole 129, 364, 618
microbiology 17
  dento-facial infections 408
  oral 202
  periodontal disease 208, 214, 220, 222, 223
microdontia 76

Miculicz disease 456
Miculicz syndrome 456
midazolam 612, 640, 641
mid-face fractures 492–3, 496
migraine 543
migrainous neuralgia, periodic 460, 543
milk, fluoridation 36
Miller's classification, tooth mobility 244
misconduct, serious professional 690, 704
mobility, tooth 18, 244
modelling, behaviour 64
models
  diagnostic mounting 276, 277, 298
  orthodontic 139
  orthognathic surgery 196
modified Widman flap 234–5
moisture control, restorative dentistry 258–9
molars
  first
    impacted upper 146
    orthodontic extraction 152
  mulberry (Moon) 434
  primary, *see* primary molars
  second, orthodontic extraction 150
  third, *see* third molars
monitoring
  general anaesthesia 647
  intravenous sedation 641
monoamine oxidase inhibitors (MAOIs) 608, 644
monoclonal gammopathies 520
morphine 588, 604
Motival 608
mouth
  benign tumours 420–1
  burning 362, 468
  common ailments in children 126–7
  examination 14
  microbiology 202
  non-tumour hard-tissue lumps 416
  non-tumour soft-tissue lumps 414–15
  see also oral medicine; oral mucosa; oral surgery
mouthwashes
  bacterial control 38–9
  fluoride 36
mucoceles 412, 414

mucoepidermoid tumour,
    salivary glands 506
mucogingival surgery 240–1
    free grafts 240–1
    pedicle grafts 241
    vestibular extension
        procedures 240
mucoperiosteal flaps 380
multiple choice exams 760
multiple endocrine neoplasia
    (MEN) syndromes
    536, 745–6
multiple sclerosis 461, 545
mumps 126, 454
murmurs, heart 524
muscle(s)
    abnormal movements 471
    age changes 375
    attachments, surgical
        displacement 426
    diseases 540
    post-operative pain 596
muscle relaxants 642–3,
    646–7
muscle tumours 513
muscular dystrophy 471, 540
musculoskeletal system,
    examination 10
myasthenia gravis 470–1,
    545
mycobacteria, atypical 409,
    434
Mycobacterium tuberculosis
    434
mycosis fungoides 546
myeloma, multiple 466, 520
myeloproliferative disorders
    520
myoblastoma, granular cell
    420
myocardial infarction
    facial pain 461
    management 558
myocutaneous flaps 514, 515
myotonic disorders 540
myxoedema, pretibial 549
myxoma 421

naevi 546
    dysplastic 546
    white spongy 446
naloxone 604
nasal decongestants 620
nasal deformity
    cleft lip and palate 502
    rhinoplasty 505
nasal floor, extruded paste
    404
nasal fractures 494, 496, 497
nasal septal haematoma 494

nasendoscopy, cleft palate
    502
nasion 141
nasoethmoid fractures 494,
    496
nasogastric intubation 586
nasolabial flap 514
nasopalatine block 632
natal teeth 68
nausea, post-operative 596
nebulizers 570
neck
    examination 12–13
    lumps 512–13
    reconstructive surgery
        514–15
necrobiosis lipoidica 549
needle holders 384
needles, suture 392
needlestick injuries 383
negligence 700–1
    civil action for 694, 700–1
    contributory 700
    criminal 700
    time bar 700
neodymium-yttrium
    aluminium garnet (Nd-
    Yag) laser 430
nephrotic syndrome 532
nervous system, age changes
    374
neurofibroma 420, 506
neurofibromatosis, von
    Recklinghausen 748
neurolemmoma 420
neurological disorders 470–
    1, 542–5
neuroses 551
neutral zone 348
neutropenia, cyclical 466
nickel allergy 688
nickel and titanium alloy
    (Nitinol) 678
nickel chromium alloy 676,
    677
Nikolsky's sign 442
nitrazepam 612
nitrogen, liquid 412
nitrous oxide
    anaesthesia 642
    pollution 636, 649
    sedation 64–5, 636–7
noma 220
non-accidental injury (NAI)
    108
non-steroidal anti-
    inflammatory drugs
    (NSAIDs) 228, 588,
    602, 604
nortriptyline 608

notation, tooth 752
nursing bottle caries 98
nursing caries 98
nutrition
  elderly 375
  enteral 586
  parenteral 586
  post-operative 577
nystatin 364, 438, 618

obsessional neurosis 551
occipitomental view 21
occlusal analysis 276
occlusal key 90
occlusal plane, position 352
occlusal records 277–8
  bridges 299
  complete dentures 346,
    352–3
  partial dentures 344
occlusal table, primary teeth
    80
occlusal vertical dimension
    (OVD) 275
  edentulous patients 352,
    354
occlusal views 20
occlusion 274–8
  balanced 274
  canine guided 274
  centric (intercuspal
    position; ICP) 274,
    277, 352
  complete dentures 356
  definitions 274–5
  elderly patients 377
  examination 14, 276
  functional 274
  ideal 136, 274
  normal 136
  periodontal therapy and
    244
  restorative dentistry and
    276–7
  TMPDS and 275, 480
oculomotor nerve (III) 542
odontogenic cysts, calcifying
    419
odontogenic tumours 420–1
  calcifying epithelial 421
odontomes 421
oedema (swelling), post-
    operative 381, 400–1
oesophageal disorders 528
olfactory nerve (I) 542
oligodontia 72
oliguria, post-operative 580
ondansetron 610
open bite, anterior (AOB)
    136, 170

open days, practice 722
open reduction and internal
    fixation (ORIF), facial
    fractures 496, 497
ophthalmoscopy 12
opioid analgesics 588, 602,
    604
optic nerve (II) 542
oral cancer 446, 450–1, 476
  biopsy 410
  children 127
  histopathology 450–1
  management 510–11
  premalignant lesions 448–
    9
  reconstructive surgery
    514–15
  staging 450
  survival 510–11
  terminal care 588
  verrucous carcinoma 451
oral contraceptives 644
oral hygiene (OH) 224
  advanced cancer 588
  elderly 376
  handicapped patients 53
  instruction (OHI) 53, 224
Oral Hygiene Index 210
oral medicine 432–81
oral mucosa
  disease in elderly 375
  drug-induced lesions 458–
    9
  examination 14
  infections 434–9
  premalignant lesions 448–
    9
  vesiculo-bullous lesions
    442–5
  white patches 446
oral surgery 378–430
  asepsis and antisepsis 382–
    3
  cryosurgery 412
  implants 428–9
  instruments 384
  lasers 430
  local complications 597
  minor preprosthetic 426–7
  post-operative bleeding
    390–1
  principles 380–1
  suturing 392–3
  see also dento-alveolar
    surgery; extraction(s)
oral ulceration
  approach to 476
  children 126–7
  herpetiform 440
  recurrent 440–1, 466, 476

orbitale 141
orbital floor (blow-out)
    fractures 494
orbital injuries 492
orbital malformations 502
organic brain syndromes 550
oro-antral fistula 422–4, 620
orofacial granulomatosis 453
oropharynx, examination 12
orthodontic appliances
    anchorage 180–1
    fixed (FA) 180, 188–91
    functional 192–5
    removable (URA) 180,
        182–7
orthodontic auxiliaries 54
orthodontics 130–99
    anterior open bite 170
    assessment 138–9
    buccally displaced canines
        158
    cleft lip/palate 198–9
    crossbites 176–9
    definitions 136–7
    developing dentition 146–
        9
    distal movement of upper
        buccal segment 154,
        155
    extractions 150–2
    increased overbite 166–8
    increased overjet 162–5
    index of treatment need
        (IOTN) 134–5
    indications 132
    orthognathic surgery and
        196–7
        post-surgical 197
        pre-surgical 196–7
    palatally displaced canines
        160–1
    reverse overjet 172–5
    spacing 156
    surgical procedures aiding
        406–7
    timing 132–3
    treatment planning 144–5
    treatment providers 132
orthognathic surgery 199,
    504–5
    diagnosis and treatment
        planning 196
    orthodontics and 196–7
orthopantomograph (OPG)
    radiograph 21
osseointegration 428
osseous surgery, periodontal
    disease 238
ossifying fibroma 420
ostectomy 114

mandibular body 504
periodontal disease 238
osteoarthrosis 540
    temporomandibular joint
        498–9
osteogenesis imperfecta 538
osteoma 420
osteomalacia 538
osteopetrosis 538
osteoplasty, periodontal
    disease 238
osteoporosis 538
osteosarcoma, jaws 451
osteotomy
    intra-oral vertical sub-
        sigmoid, of mandible
        504
    inverted L- and C-shaped,
        of mandible 504
    malar 505
    sagittal split, of mandible
        504
    subapical mandibular 504
otorrhoea, CSF 486
out-patient dental anaes-
    thesia 648
overbite 136
    complete 136
    incomplete 136
    increased 166–8
        aetiology 166
        approaches to reducing
            166
        management 168
        stability of reduction
            166
overdentures 370–1
overjet 136
    increased 162–5
        aetiology 162
        management 164–5
        stability of reduction
            162
    reverse 172–5
        aetiology 172
        assessment 172
        management 174–5
        treatment planning 172
oximetry, pulse 641, 647
Oxyguard 303

packed cells, for transfusion
    582
packed cell volume 16
packs, periodontal 233
paedodontics 58–129
Paget's disease of bone 416,
    538
pain
    acute chest 558

pain (*cont.*):
  control, *see* analgesia
  dental (toothache) 254–7
    children 66–7
    non-dental causes 257
  denture wearers 362
  extractions causing 388
  facial, *see* facial pain
  oral ulcers 476
  periapical/periradicular 255
  post-operative 400, 596
  as presenting complaint 4
  primary teeth 82
  pulpal 254–5
  root canal instrumentation causing 328
palatal finger springs 184
palatal rotation flap 424, 425
palate
  cleft, *see* cleft lip/palate
  closure of cleft 198, 502
Panavia Ex 302, 303
pancreatic cancer 529
pancreatitis, acute 529
pancuronium 643
panoramic radiographs 21
papaveretum 588, 604
papillomas, squamous cell 415, 420
Papillon–Lefevre syndrome 746
paracetamol 128, 588, 602
paralleling technique, X-rays 20
paranoid states 550
Parapost system 285
parenteral feeding 586
parents
  accompanying child 62
  motivation/co-operation 82
Parkinson's disease 471, 545
parotid duct calculi 508
parotidectomy 508
parotid gland tumours 456, 506
partial dentures 336–45
  acrylic vs. metal 336–7
  class IV 343
  classification 336, 337
  clinical stages 344–5
  components 338–40
  copying 367–8
  'creeping' 346
  definitions 336
  design 342–3
  dysjunct 336
  indications 334
  principles 336–7

problems and complaints 362–3
  rebasing 345
  surveying 342
  swinglock 336
  treatment planning 334–5
  two-part 336
partnerships 698
paste, removal of extruded 404
patient-controlled analgesia 589
Patterson–Brown–Kelly syndrome 746
pay, staff 713
pectoralis major myo-cutaneous flap 514
peer review 732
pellicle, acquired 204
pemphigoid 444
  benign mucous membrane 375, 444
pemphigus 375, 442
  benign familial chronic 442
penicillins 562, 614
pentazocine 602
peptic ulceration 528
percussion, tooth 18, 66
perforations, root 319, 329, 404
periapical abscess, acute 256, 328, 408
  pulpotomy techniques 104
  vs. periodontal abscess 221
periapical/periradicular pain 255
periapical views 20
pericision 406
pericoronitis 408
peri-implantitis 238–9
periodontal abscess 220–1, 256, 408
periodontal disease
  acute 220–1
  aetiology 208–9
  assessment 14, 216
  bridges and 296
  calculus and 206
  diagnostic tests 218
  elderly 372
  epidemiology 210–11
  furcation involvement 246, 248–9
  host factors 209
  immunopathology 208–9
  indices 210–11
  microbiology 208, 214, 220, 222, 223
  monitoring 218
  plaque and 204, 208

prevention 224–5
pulpal effects 246
statistics 57
tooth mobility 244
see also gingivitis; periodontitis
periodontal packs 233
periodontal pockets 216
assessment 216
drug delivery into 228
false 216
pulpal involvement 246
true 216
periodontal therapy
apically repositioned flap
236–8
bony infill in osseous
defects 242
gingivectomy 236
guided tissue regeneration
238
minimally invasive 232–5
modified Widman flap
234–5
mucogingival surgery 240–
1
non-surgical 228–31
occlusion and 244
osseous surgery 238
phases 226
practical tips 233
principles 226
reattachment/new attach-
ment 242
splinting 244
surgical 232–9
periodontitis
aetiology 208
children 222–3
chronic 214–15
diagnosis 214
HIV 472
juvenile (JP) 222
periapical, pulpal necrosis
with 255–6
prepubertal 222–3
rapidly progressive 215
statistics 57
periodontium, age changes
374
periodontology 200–49
perio-endo lesions 246–7
persistent generalized
lymphadenopathy 472
personality disorders 551
personnel, *see* staff
Perthes disease 541
petechiae, intraoral 466
pethidine 602
Peutz–Jeghers syndrome 746

phaeochromocytoma 535
pharyngeal diverticulum 512
pharyngoplasty, cleft palate
502
phenobarbitone 128
phenoxymethyl penicillin
(penicillin V) 221,
614
phobia 551
phoenix abscess 328
photographs 196, 708
physical handicap 52
pier abutments 296
pigmentation, oral 459
Pindborg tumour 421
pinledge preparations, gold
veneer 308
pinned restorations 308
pituitary tumours 535
plaintiff 690
planning cycle 728
plaque 38–9, 204
attachment 204
caries and 38, 204
chemical removal 38–9
development 204
index 211
periodontal disease and
204, 208
physical removal 38, 228
platelet disorders 522
platelets 16
platinum bonded/McLean-
Sced/twin foil crown
282, 682
pleomorphic adenoma,
salivary glands 456,
506
Plummer–Vinson syndrome
746
pockets, periodontal, *see* periodontal pockets
police, revealing records to
708
polishing
amalgam 656
composites 660
teeth 228
pollution
anaesthetic gases 649
nitrous oxide 636, 649
polyarteritis nodosa 463
polycarbonate crowns 306
polyether impression
material 672, 688
polymethylmethacrylate
(PMMA) 684, 686
polymyalgia rheumatica 540
polymyositis 540
polyposis coli, familial 528

polyps, fibroepithelial 414
polysulphide impression material 672
polyuria, post-operative 580
pontics 292, 294
porcelain 680–3
  alloys for bonding 677
  inlays 266–7, 682
  properties 680
  repairs 683
  slips 289
  veneers 288, 289, 682
porcelain fused to metal (PFM) crowns 282, 284, 290–1, 682
porcelain jacket crowns (PJC) 280–1, 282, 284, 682
porion 141
Porphyromonas gingivalis (formerly *Bacteroides gingivalis/ melaninogenicus*) 202, 208, 214, 220
positioning
  dental surgery assistant 330
  dentist 330
  elderly patients 376
  for extractions 386
  general anaesthesia 648–9
  patients 330
post and core crowns
  anterior 284–6
  failure 286
  multirooted teeth 291
  parallel sided/tapered 284
  practical tips 286
  prefabricated/custom-made systems 284, 285
  preliminary preparation 284
  removing old 286
  temporary 306–7
  threaded/smooth/serrated posts 284–5
post-dam 354
posterior nasal spine 141
posterior superior alveolar block 84, 632
posteroanterior (PA) view, mandible 21
post-nasal space 12
post-operative bleeding 390–1
post-operative care
  common problems 596–7
  intermaxillary fixation (IMF) 497, 577

  maxillofacial trauma 497
  pain control 577, 588
  in-patients 577
  removal of third molars 400–1
post-operative oedema (swelling) 381, 400–1
Poswillo Committee 626, 638, 649–50
potassium
  intravenous 592–3
  plasma 16
practice management 710–39
pre-centric record 277–8
precision attachments 297, 370–1
prednisolone 607
pre-emptive analgesia 589
pregnancy 468, 536
  drug therapy 601, 624
  epulis 414
premalignant lesions 448–9
premises
  practice 722
  surgery design 734
premolars
  orthodontic extraction 150
  removal of unerupted 396
pre-operative care, in-patients 576
preprosthetic surgery 426–7, 500
prepubertal periodontitis 222–3
prescribing, *see* drug therapy
presenting complaint 4
preventive dentistry 26–57
  approaches 30
  caries vaccines 39
  dental health education 48–9
  dietary analysis/advice 46–7
  fissure sealants 40–1
  fluoride therapy 36–7
  periodontal disease 224–5
  plaque removal 38–9
  recall intervals 33
  reducing sugar intake 44
preventive resin restoration (PRR) 262
Prevotella denticola 202, 208
Prevotella intermedia (formerly *Bacteroides intermedius*) 202, 208, 214, 222, 223
prilocaine/octapressin 630
primary anterior teeth
  class III, IV and V restorations 96

pulp therapy 104
primary molars
  anatomy 80, 81
  class I restorations 88–9, 92
  class II restorations 90–2
    amalgam 90–1
    composite 92
    glass ionomer/cermet 92
  pulp therapy 100–4
    materials 100–1
    non-vital pulps 100, 104, 105
    principles 100
    vital pulps 100, 102–3
  stainless steel crowns 94–5
  submerged 68–70, 146, 396
primary (deciduous) teeth
  anatomy 80–1
  class III, IV and V restorations 96
  extraction 386
    serial 148–9
    vs. restoration 82–3
  injuries 110–11
    management 110
    sequelae 110–11
  notation 752
  premature loss 148
  restoration 86–96
    cavity design 86–7
    instruments 86
    materials 86
    reasons for failure 87
    useful tips 87
  retained 146
principals, contracts with associates 698
private dental schemes 729
private practice 728–9
probes
  blunt 32
  periodontal 210, 216
  transillumination 32
probing attachment level 216
prochlorperazine 577, 596, 610
proclination, bimaxillary 136
Procurator Fiscal 692
professional ethics and etiquette 702–3
professional misconduct, serious 690, 704
progeria 746
promethazine 620
propofol 642
prosthetics 332–71
  elderly 372

treatment planning 334–5
  see also dentures
protection (defence) organizations 701, 702
protein, urinary 16
Pro-temp 306
psoriasis 548
psoriatic arthritis 498, 540–1
psychiatry 550–1
psychopathy 551
psychosis 550
pterygopalatine fossa syndrome 748
pulmonary embolus 590, 596–7
pulmonary system, age changes 375
pulp
  age changes 372, 374
  death, post-traumatic 111, 122
  fractures involving 112–13
  materials for maintenance of vitality 271
  periodontal disease and 246
  primary teeth 80
  sequelae of trauma 122–3
  stones 328
  vitality testing, see vitality testing
pulpal necrosis, with periapical periodontitis 255–6
pulpal pain 254–5
pulp capping 271, 671
  crown fractures 112
  indirect 270, 271
pulpectomy
  definition 100
  root canal therapy 318
pulpitis
  irreversible 254–5
  reversible (hyperaemia) 254
pulpotomy 271
  crown fractures 112–13
  definition 100
  non-vital primary molars 100, 104, 105
  vital primary molars 100, 102–3
pulp therapy
  deciduous molars 100–4
  primary anterior teeth 104
pulse 16, 488
pulse oximetry 641, 647
pupils
  head injuries 488
  light responses 12

purpura 466
pursuer 690
pus, swabs 17
putty and wash impression
technique 674
pyoderma gangrenosum 549
pyogenic granuloma 414
pyostomatitis vegetans 464
pyrexia, post-operative 596

quad helix appliance 178
qualifications, medical and
dental 754
questionnaire, medical 8
questions, history-taking 2, 4

radial forearm flap 515
radiation
caries 98
risks 21
Radiation Protection Super-
visor/Adviser 736
radiographs, see X-rays
radiotherapy, oral cancer
510
Ramsay Hunt syndrome 746
ranitidine 128
ranula 415
Raynaud's phenomenon 463
reamers, root canal 314, 319
rebasing, dentures 345, 358
recall intervals 33
receptionists 722
records, dental 706, 708
rectus abdominus flap, free
515
redundancy 714–15
re-inforcement, child
behaviour 64
Reiter syndrome 437, 541,
746
relative analgesia (RA) 64–5,
636–7
relining, dentures 358, 686
removable appliances (upper;
URA) 182–7
active components 182,
184–5
anchorage 180, 182
baseplate 182
design 182
fitting 186
instructions to patient 187
monitoring progress 186–7
retention 182
renal disorders 532–3
renal failure 446
acute (ARF) 532
chronic 532–3, 552
drug therapy 601, 624

renal transplantation 533
reorganized approach 276
Reporting of Injuries,
Diseases and
Dangerous Occur-
rences Regulations
(RIDDOR) 717
repositioning (transplantation), tooth 125, 161,
406–7
resilience, dental materials
654
resins
bonding amalgam restora-
tions 657
in composites 658
see also acrylic (resin);
composite(s)
resistance, bridges 292
resorption
external 123
internal 122
post-traumatic 122–3
re-implanted avulsed tooth
121
unerupted canines causing
161
respiratory disease 526–7
respiratory rate 16, 488
respiratory system
age changes 375
examination 10
respiratory tract infections
lower 526, 596
upper 526
restorations
class I 261, 262–3
primary molars 88–9, 92
class II 261, 264–7
primary molars 90–2
class III 261, 268
primary teeth 96
class IV 261, 268
primary teeth 96
class V 261, 268–9
primary teeth 96
deep carious lesions 270–1
failure 272
gold veneer pinledge 308
handicapped patients 53
nomenclature 261
pinned 308
preventive resin (PRR)
262
primary teeth 86–96
secondary caries 272
survival 272
temporary 306–7, 671
restorative dentistry 250–
330

bridges 292–303
crowns 280–6, 290–1
elderly 372
isolation and moisture
control 258–9
occlusion and 276–7
principles of cavity
preparation 260–1
treatment planning 252–3
veneers 288–9
see also caries; root canal
therapy
rest position 274
rests, partial dentures 338
resuscitation
cardiopulmonary (CPR)
560–1
major trauma 484–5
retainers
bridges 292, 293
partial dentures 336
retching
dentures causing 363
impressions causing 351
retention
bridges 292, 293
complete dentures 348
indirect 336, 342
overdentures 370–1
partial dentures 342
retraction, mucoperiosteal
flaps 380
retractors 384
retrobulbar haemorrhage
492
retrobulbar neuritis 461
retruded arc of closure 274
retruded contact position
(RCP) 274, 277, 352
reverse Townes position 21
rheumatoid arthritis 462,
498, 540
juvenile 540
rhinoplasty 505
rhinorrhoea, CSF 486
rib, cervical 513
rickets 538
ridge, alveolar, *see* alveolar
ridge
Robert's retractor 184
rodent ulcer (basal cell
carcinoma) 412, 546
Romberg syndrome 746–7
root(s)
hemisection 404
perforations 319, 329, 404
planing 232, 246, 248
primary teeth 80
removal of retained 394–5,
426

resection 248–9
root canals
anatomy 312–13
post-traumatic calcification
123
primary teeth 81
sclerosed 328
root canal therapy (RCT)
312–29
assessment 313
canal obturation 326–7
canal preparation 318–24
access 318, 319
aims 318
common errors 319
isolation 318
modified stepback
technique with orifice
enlargement 322–4
preparation of tooth 318
pulp extirpation 318
simple stepback
technique 322
techniques 319–24
follow-up 327
fractured instruments 328–
9
indications 312
instruments 314–15
materials 316–17
problems and their
management 328–9
removing old fillings 329
surgical aids 402–4
teeth with immature apices
122
root caries 30, 372
treatment 269
root fillers, spiral 314
root fractures
permanent teeth 114
primary teeth 110
vs. periodontal disease 246
rubber dam 86, 258–9

saddles
bridges 292
free-end 336
lower bilateral 343
multiple bounded 343
partial dentures 336, 338,
342
salbutamol 129, 570
saline, intravenous 580
saliva
artificial 622
caries and 28–30
surgery for drooling 508
Saliva-Orthana 622

salivary duct
 calculi 508
 fistulae 455
salivary gland(s)
 diseases 454–6
 enlarged 512
 fistulae 455
 minor 454, 508
  tumours 506
 surgery 508
 tumours 456, 506
salt, fluoridation 36
sandwich technique, class V
  restorations 269
sarcoidosis 453, 526
sarcomas
 Kaposi's 472, 546
 oral 451
scaling 228
 deep subgingival 232
 and root planing 232, 248
scalpels 384
scarlet fever 434
schizophrenia 550
schwannoma 420
scissor bite 178
scissors 384
screws, removable appliances
  185
sealers, root canal 316
sebaceous cysts 512
seborrhoeic eczema 548
'second appropriate person'
  641
sedation 636–41
 benzodiazepines 638–41
 chaperons during 706
 children 64–5, 638
 conscious 628
 diabetic patients 592
 intravenous 638, 640–1,
  650
 nitrous oxide 64–5, 636–7
 oral 640
 Poswillo Committee recommendations 650
 rectal 638
sedative dressings 671
sedatives 612–13, 644
seizures, see convulsions
Selenomonas 220
sella 141
sensory alterations, neurological causes 470
sensory handicap 52
set square tooth notation
  system 752
sex hormones 468
sexual fantasies 636, 640,
  706

shaping, root canals 318
sharps disposal 735
shingles (herpes zoster) 375,
  436, 476, 618
shock
 anaphylactic 562
 hypovolaemic 524
sialadenitis 454
 acute bacterial 454
 chronic bacterial 454
 recurrent 508
 viral 454
sialography 23, 508
sialorrhoea 454
sialosis 456
sicca syndrome 456, 747
sickle cell anaemia 518
sign language 53
silicone
 addition-cured 672
 condensation-cured 672
 denture soft linings 686
silver palladium alloy 676,
  677
sinusitis, maxillary 422
sinus lift 500
Sjögren syndrome
 primary 456, 747
 secondary 456, 747
skin
 bite marks 708
 grafts 446, 514
 infections 548
 infestations 548
 neoplasms 546
skin disease 548–9
 facial lesions 12
 neck lesions 512
 oral features 462–3
small bowel disorders 528
smear layer, dentine 664
smoker's keratosis 446
social class, uptake of dental
  care and 50
sociopathy 551
sodium, plasma 16
sodium perborate 311
soft linings, dentures 359,
  686
soft tissues
 age changes 374
 intraoral
  examination 14
  non-tumour lumps 414–
   15
 swellings in children 127
soldering, stainless steel 678
sore throat
 post-operative 596
 streptococcal 126, 526

sore (burning) tongue 452, 466
Southend clasp 182
space closure, missing incisors 124, 125
space-maintenance/opening, missing incisors 125
spacing of teeth 156
speech problems
  cleft palate 502
  denture wearers 362
spiral root fillers 314
spirochaetes 202, 208
splinting 116, 118
  avulsed tooth 120
  periodontology 244
  TMPDS 480
sponsorship 725
spring dividers 352
springs
  buccally approaching 184
  palatal finger 184
  removable appliances 184
  strap 184
  whip 191
sputum 17
squamous cell carcinoma (SCC)
  oral 450, 476
  skin 546
squamous cell papilloma 415, 420
stability
  complete dentures 348
  orthognathic surgery 505
staff 54–5
  complaints against 694
  contracts 698, 714
  dismissal/redundancy 714–15
  hazards to 689
  hiring 714
  meetings 712
  pay 713
  presentation/attitude 722
  training 712–13
  vicarious liability for 700
staining
  extrinsic 78
  intrinsic 78
stainless steel (SS) 678
  crowns 94–5
  soldering 678
staphylococcal lymphadenitis 409
statistics 56–7
status epilepticus 566
stepback technique 320
  modified, with orifice enlargement 322–4

simple 322
stepdown technique 320
Stephan curve 43
sterilization 382, 734–5
'sternomastoid tumour' 513
steroid abusers 594
steroids 606–7
  collapse of patients on 564
  intra-articular 499, 606
  intralesional 606
  surgery in patients on 594
  systemic 606–7
  topical 606
Stevens Johnson syndrome 444, 747
stiffness, dental materials 654
Still's disease 540
stomach disorders 528
stomatitis
  denture 364–5
  exfoliative 458
  recurrent aphthous 440–1, 466, 476
stops, apical 315, 322, 324
strain, physical 654
strap springs 184
streptococcal gingivitis, acute 221
streptococcal sore throat 126, 526
Streptococcus 408
Streptococcus milleri group 202
Streptococcus mutans 38, 202, 204
Streptococcus salivarius 38, 202
Streptococcus sanguis 202
stress
  physical, dental materials 654
  surgical, patients on steroids 564, 594
  TMPDS 478
stress-breaker 336
stroke 544
Sturge–Weber syndrome 747
subacute bacterial endo-carditis (SBE) 524, 590–1
sublingual gland tumours 506
sublingual nerve block 632
subluxation, teeth 110, 116
submandibular duct calculi 508
submandibular gland tumours 506
submentovertex view 21
submucous fibrosis 448–9
sucrose 42

suction
  high volume 258
  low volume 258
sugar 42–4
  alternatives to 44
  denture stomatitis and 364
  -free medications 128–9
  health and 42
  reducing intake 44
  role in caries 42
  statistics 56
sulcus deepening 427, 500
supernumerary teeth 72–3,
    396–7
support
  bridges 292, 293–4
  partial dentures 336, 342
surgery
  complications 596–7
  diabetic patients 592–3
  liver disease 530
  maxillofacial 482–515
  orthognathic 196–7, 199,
    504–5
  post-operative care, see
    post-operative care
  pre-operative care 576–7
  preprosthetic 426–7, 500
  steroid-treated patients
    594
  temporomandibular joint
    480–1, 498–9
survey line, partial dentures
    336
sutures 384, 392–3
  knot tying 392, 393
  materials 392
  removal 392–3
suturing techniques 233,
    392–3
suxamethonium 643
  sensitivity 536
sweeteners, artificial 44
swelling (oedema), post-
    operative 381, 400–1
SWOT analysis 728
symptoms, presenting 4
syncope, vaso-vagal 556
syphilis 434
syphilitic leukoplakia 446
syringes, cartridge 630
systemic lupus erythematosus
    (SLE) 462–3
systemic sclerosis 463

tantrum, temper 65
tartar, see calculus
taurodontism 77
tax 719

team, dental 54–5, 712
teething 68
temazepam 612, 640
temperature, body 16
temper tantrum 65
temporal (cranial) arteritis
    460, 540
temporalis flap 514
temporary bridges 298, 307
temporary cements 307, 671
temporary crowns 280, 306–
    7
temporary dressings 306
temporary restorations 306–
    7, 671
temporomandibular joint
    (TMJ) 498–9
  ankylosis 498
  arthritides 498
  arthrography 23
  condylar hyperplasia 498
  dislocation 498
  examination 13
  osteoarthrosis 498–9
  surgery 480–1, 498–9
  trauma 498
  tumours 498
temporomandibular pain
    dysfunction syndrome
    (TMPDS)
  aetiology 478
  clinical features 478
  drug therapy 480, 608,
    610
  management 478–81
  occlusal factors 275, 480
  surgery 480–1, 499
terfenadine 128, 620
terminal care 588
terminal hinge axis 274
tetanus toxoid 486
tetcoplanin 591, 617
tetracycline 222, 223, 230,
    614
thalidomide 440–1, 607
therapeutics 598–625
  see also drug therapy
thermal conductivity 654
thermal diffusivity 654
thermal expansion 654
thermal injury 688
thiopentone 642
third molars
  extraction 386, 398–401
    anaesthesia 398
    assessment 398
    post-operative care
      400–1
    technique 400, 401
  orthodontic extraction 150

thrombocytopenic purpura,
  idiopathic (ITP) 466
thrombophlebitis migricans
  549
thrush 126, 438
thyroid gland
  enlarged 512, 534
  lingual 534
tioconazole 618
tissue conditioners 358–9,
  686
tissue growth factor β-2
  238
tissue transfer by micro-
  vascular re-
  anastomosis, free
  514–15
titanium 676, 678
  alloys 678
titanium molybdenum alloy
  (TMA) 678
TMPDS, *see* temporo-
  mandibular pain
  dysfunction syndrome
tobacco chewing 448, 450
tongue 452
  depapillation 452
  fissured 452
  flaps 514
  geographic 452
  hairy 452
  sore (burning) 452, 466
  thrusts 170
  tie 452
tonsils, examination 12
toothache, *see* pain, dental
toothbrushing 224
tooth form, abnormal 76–7,
  421
tooth numbers, abnormal
  72–3
toothpastes 36–7, 224
tooth structure, abnormal
  74–5
tooth surface loss, *see* toothwear
toothwear 304–5, 372
  diagnosis 304
  indices 304
  management 304–5
topical analgesia 622, 630
tori (torus) 416, 426
toughness, dental materials
  654
toxicity
  amalgam 657
  beryllium 688
  fluoride 34–5, 688
tracheostomy 577

training
  staff 712–13
  vocational (VT) 730
tranquillizers 612–13, 644
transfer coping technique
  277
transplantation, tooth 125,
  161, 406–7
transposition, canines 161
tranylcypromine 608
trauma
  advanced trauma life
    support 484–5
  deaths 484
  dental 106–23
    classification of injuries
      107
    complicating extractions
      388
    missing incisors 124
    permanent teeth 112–23
    prevention 107
    primary teeth 110–11
    principles of treatment
      106–7
    pulpal sequelae 122–3
  extractions causing 388
  head 486, 488–9
  maxillofacial 485, 486–97
  non-accidental (child
    abuse) 108
  temporomandibular joint
    498
Treacher–Collins syndrome
  747
treatment planning 24–5
  bridges 296
  children 62
  handicapped patients 52–3
  orthodontics 144–5
  orthodontics and ortho-
    gnathic surgery 196
  prosthetics 334–5
  restorative dentistry 252–3
Treponema 214, 220
Treponema pallidum 434
triamcinolone
  acetonide 606
  in carboxymethylcellulose
    paste 606
trigeminal nerve (V) 470,
  542
trigeminal neuralgia 460, 622
trimeprazine 613, 620
Trim (poly-*n*-butyl methacrylate) 306
trismus 400, 498
trisomy 21 (Down syndrome)
  743
trituration, amalgam 656

trochlear nerve (IV) 542
Trotter syndrome 747–8
T-springs 184
tuberculosis (TB) 434, 526
tumours
  benign oral 420–1
  central nervous system
    (CNS) 544
  malignant oral, *see* oral
    cancer
  presenting as neck lumps
    512
  salivary gland 456, 506
  temporomandibular joint
    498
tuneable dye laser 430
tunnel preparation 248
Turner tooth 77
tyramine sulphate 478

ulcerative colitis 464, 528
ulcers 442
  herpetiform 440
  see also aphthous ulcers;
    oral ulceration
ultrasonic scalers 228
ultrasound (US) 23
unconscious patients
  Glasgow coma scale 488–9
  hypoglycaemia 568
  legal aspects of treating
    696
  see also collapse
unerupted teeth
  exposure 406
  extraction 396–7
units, bridges 292
upper buccal segments, distal
    movement 154, 155
upper removable appliances,
    see removable appli-
    ances
urea, plasma 17
urethral catheterization 584
urinalysis (urine tests) 16,
    532
urinary retention, post-
    operative 596
urinary tract infections 532
urobilinogen, urinary 16
uveoparotid fever 744

vaccines, caries prevention
    39
vagus nerve (X) 542
valproate, sodium 128
vancomycin 591, 616–17
varicella (chicken-pox) 126,
    436

vascular malformations, *see*
    haemangioma
vaso-vagal syncope 556
VAT registration 719
vecuronium 643
vein, lacerated, complicating
    local anaesthesia 634
veneers 288–9
  acrylic laminate 288
  composite resin 288
  partial preparations 297
  porcelain 288, 289, 682
  temporary 307
venepuncture 578
venous thrombosis
  deep (DVT) 590, 596–7
  superficial 596
ventilation
  cardiopulmonary arrest
    560
  general anaesthesia 647
  trauma patients 484
ventricular fibrillation (VF)
    560, 561
verrucous carcinoma, oral
    451
vesicle 442
vesiculo-bullous lesions 442–
    5
  intraepithelial 442
  subepithelial 444–5
vestibulocochlear nerve
    (VIII) 542
vicarious liability 700
viral infections
  children 126
  mouth 436–7
virology 17
vitality testing 18
  children 66
  post-traumatic 122
vitamin B compound tablets
    622
vitamin B12 deficiency 466,
    518
vitamin D deficiency 538
vitamins 129, 622
vitiligo 548
vocational training (VT) 730
Vomer flap 502
vomiting, post-operative 596
von Recklinghausen syn-
    drome 748
von Willebrand disease 522

warfarin 522–3
warts, viral 126, 412, 415
Wassmund procedure 504
waste disposal 382–3, 717
water, fluoridation 36, 37

waxing, diagnostic 298
wax squash bite 277
wear resistance, dental materials 654
wettability, dental materials 654
wheelchairs, access for 376
whip spring appliances 190, 191
white cell count 16
white patches, oral mucosa 446
Wickham's striae 462
Widman flap, modified 234–5
Willis gauge 352
Wiptam technique 284
wire
  gauges 758
  stainless steel 678
wisdom teeth, *see* third molars
Wolinella recta 214
working length 318, 322, 324
work surfaces, cleaning and disinfection 735
World Health Organization periodontal probe 210
writ 690
wrought alloys 678–9

xerostomia 454, 622
X-rays (radiographs) 18–19, 20–1, 736–9
  bitewing 20, 32
  cephalometry 21, 140–3
  children 66
  digital imaging 22–3
  extra-oral views 20–1
  film faults 739
  intra-oral views 20

maxillofacial trauma 490–1, 493
  occlusals 20
  orthodontics 139
  orthognathic surgery 196
  panoramic 21
  periapical 20
  periodontal disease 218
  practical tips 738
  processing 738–9
  retaining 706
  root canal therapy 313, 322
  safety 21
  statutory regulations 736
  trauma 484
xylometazoline 620

yield strength 654
Young's rule 129

zidovudine 473, 618
Zimmer frames, access for 376
zinc oxide eugenol (ZOE) 271, 306, 670, 671
  accelerated 670, 671
  resin-bonded 670, 671
  temporary cement 307
zinc oxide pastes 673
zinc phosphate cement 670, 671
zinc polycarbonate cement 670, 671
zipping, apical 319
zopiclone 612
Zsigmondy–Palmer tooth notation system 752
Z-spring 184
zygoma (malar) fractures 494, 496, 497